AF787162

Sustainable Nanomaterials for Energy Applications

Online at: https://doi.org/10.1088/978-0-7503-3531-7

Sustainable Nanomaterials for Energy Applications

Edited by
Ana Cremades and David Maestre
Department of Materials Physics, Faculty of Physics, Complutense University of Madrid, Madrid, Spain

IOP Publishing, Bristol, UK

© IOP Publishing Ltd 2023

All rights reserved. No part of this publication may be reproduced, stored in a retrieval system or transmitted in any form or by any means, electronic, mechanical, photocopying, recording or otherwise, without the prior permission of the publisher, or as expressly permitted by law or under terms agreed with the appropriate rights organization. Multiple copying is permitted in accordance with the terms of licences issued by the Copyright Licensing Agency, the Copyright Clearance Centre and other reproduction rights organizations.

Permission to make use of IOP Publishing content other than as set out above may be sought at permissions@ioppublishing.org.

Ana Cremades and David Maestre have asserted their right to be identified as the editors of this work in accordance with sections 77 and 78 of the Copyright, Designs and Patents Act 1988.

ISBN 978-0-7503-3531-7 (ebook)
ISBN 978-0-7503-3529-4 (print)
ISBN 978-0-7503-3532-4 (myPrint)
ISBN 978-0-7503-3530-0 (mobi)

DOI 10.1088/978-0-7503-3531-7

Version: 20230901

IOP ebooks

British Library Cataloguing-in-Publication Data: A catalogue record for this book is available from the British Library.

Published by IOP Publishing, wholly owned by The Institute of Physics, London

IOP Publishing, No.2 The Distillery, Glassfields, Avon Street, Bristol, BS2 0GR, UK

US Office: IOP Publishing, Inc., 190 North Independence Mall West, Suite 601, Philadelphia, PA 19106, USA

Contents

Preface

Today, we face many global challenges that have become even more pronounced during the global COVID-19 pandemic. As a positive aspect among the more numerous catastrophic effects of this health and economic crisis, public awareness of environmental, climate change, energy, and health-related issues and the role of science as an irreplaceable solution provider has risen significantly. In this window of opportunity, resolute action by policy makers and governments, industrial leaders, and the research and innovation community is demanded at different levels. This book is a modest contribution written in an attempt to build a stronger and wider research and education community to address the development of sustainable solutions based on nanomaterials for the energy sector.

Recent progress in functional materials has been driven by the global need for more sustainable technological development and the production of clean and affordable energy while meeting the demanding standards of high-performance devices. Today, the focus is placed on concepts such as portable, low-power, and smart materials, which are going to be fully integrated into the 'Internet of Things.' Nanomaterials are the best candidates to meet these challenges, and a special category among them, the sustainable oxide nanomaterial family, is especially suitable, as most of these oxides are low cost, nontoxic, biocompatible, and abundant, while exhibiting multifunctionality and improved properties as compared to their bulk counterparts, as well as having proven applicability in the energy sector. However, other emerging nanomaterials or nanocomposites including graphene, polymers, and hybrid biomaterials are also gaining a place in energy sector applications and will also be included in this book.

The development of nanomaterials is a hot topic, with meaningful applications in energy generation, transformation, and storage, which are now in great demand by society. Their impact on the sustainability of our planet will increase over time, and the basic aspects, such as those covered by this book, will be introduced in advanced courses at different levels of education (grade, master, and doctorate) and even in some optional courses at high-school level. This book presents both research results from the cutting edge as well as the fundamentals, thereby creating a well-balanced reference text. The reader will benefit from a complete perspective not only on the materials and applications involved but also on other considerations related to sustainability, which are sometimes missed in pure scientific books. Our approach will enable the reader to develop more informed research, in line with the current policies developing in the EU, USA, Japan, and other strategic areas involved in technical developments. The materials and applications are quite diverse, but they have methodologies that may be shared or extended to other materials and devices.

For those who want to use this book as a textbook for advanced courses on nanomaterials, materials growth and design, functional materials, electronic materials, energy materials, materials properties, etc. this book is a source of examples of oxide nanomaterial synthesis, properties, and applications as well as other emerging alternatives, and we provide end-of chapter summaries, questions, and/or student work assessments.

The book is organized into 11 chapters that cover multiple fundamentals as well as more applied issues, such as those related to selected synthesis techniques for nanomaterials with different dimensionalities, the properties of the most interesting nanomaterials for technological applications, and their development in applications for the generation, storage, and transformation of energy. The latter are widely understood to be a result of the interaction of external stimuli such as light, electrons, or temperature with matter at the nanoscale. The materials of choice are a broad selection of metal oxides, metal halides, organic dyes, single and binary semiconductors, graphene, Xenes and MXenes, rare-earth phosphors, and perovskites, depending on the intended application. The applications include solar cells, batteries and supercapacitors, energy harvesting and thermoelectric materials, electro- and photochromic materials for smart applications, luminescence, electronic devices, and photocatalysis.

In order of appearance in this book, our co-authors are Antonio Vázquez-López, Gezira de Ávila, Javier García-Alonso, María Taeño, Francisco Domínguez-Adame, Bianchi Méndez, Aleksandra B Djurišić, Annie Ng, Gaukhar Nigmetova, Fangzhou Liu, Zhilin Ren, Yi Zhang, Gao Liu, Neslihan Yuca, Ömer Suat Taşkın, Sebahat Topal, Ozan Karakaya, Sema Topal, Kasper Wenderich, Marco Altomare, Mario Amado, David Sánchez, José Montero, Sara Ferrara, Rubén D Costa, David Rogers, Ferechteh Teherani, Eric Sandana, and Philippe Bove. They are materials science researchers in different specialities such as physics, chemistry, and engineering; whether their approaches are applied or theoretical, they are highly recognized in their work field, or early career researchers with huge potential. They are professionally active in different regional areas (Europe, North and South America, and Asia) and provide relevant insights in their corresponding chapters.

The editors truly thank their coauthors for accepting the invitation to participate in this book even though they were dealing with a difficult time due to the global pandemic. We value all the efforts and the excellent results obtained and presented to you, our reader, in this book. We hope that this book fully satisfies your need for scientific education or any other expectations. We are also grateful for the IOP's invitation to write this book and for its helpful editorial staff, especially Caroline Mitchell, who followed this adventure from the very beginning to its completion.

We hope you enjoy this book!

David Maestre and Ana Cremades

Acknowledgements

We would like to express our sincerest gratitude to all the contributors in this book for their commitment, efforts and patience during the completion and revisions of the chapters, even more so considering the difficult working conditions due to the COVID-19 pandemic.

We are also very grateful for the IOP's invitation and the helpful support from the editorial staff during the preparation and revisions of this book, in particular from Caroline Mitchell.

Finally, we would like to thank all the researchers, pioneers in their respective fields, whose figures, tables and research are cited in this book in the search of providing a global view of the development of sustainable nanomaterials for the energy sector.

Editor biographies

Ana Cremades

Ana Cremades is a full professor at the Material Physics Department of the Complutense University of Madrid, Spain. She belongs to the research group 'Physics of electronic nanomaterials' (https://www.ucm.es/fine). She currently holds the post of Director General for Research and Technology Innovation of the Regional Government of Madrid. Ana Cremades has great experience in the characterization of the materials and nanostructures of diamond, GaN and related alloys, porous silicon, transparent conducting oxides (such as In_2O_3, SnO_2, TiO_2, and NiO), and composites with polymers and graphene derivatives with applications in the energy sector, such as photovoltaics and batteries. She has supervised nine doctoral theses (two of which are ongoing); she is the coauthor of more than 150 publications and also the coinventor of five patents, two of which were finalists at the 2015 Emerging Technologies Competition of the Royal Society of Chemistry (RSC, London). She has been the principal investigator of national and international projects on functional materials and a co-organizer of international conferences and workshops in related subjects. ORCID ID: orcid.org/0000-0002-0473-5093.

David Maestre

David Maestre is full professor at the Materials Physics Department of the Complutense University of Madrid, Spain. His main research work has been carried out as a member of the research group 'Physics of electronic nanomaterials' (https://www.ucm.es/fine) in the fields of the fabrication and study of the micro- and nanostructures of diverse semiconducting oxides (SnO_2, TiO_2, In_2O_3, NiO, etc.) with applications in optoelectronics and energy-related devices. David Maestre is the coauthor of more than 80 scientific contributions and two book chapters, and he is the coinventor of five patents. He has been a co-supervisor of four PhD theses on nanomaterials for energy-related applications and he has also been awarded the Fonda-Fasella award (2010) and the FEMS Lecturer award (2016). (ORCID ID: 0000-0001-9898-2548).

List of contributors

Marco Altomare
PhotoCatalytic Synthesis (PCS) Group, MESA+ Institute for Nanotechnology, University of Twente, Enschede, The Netherlands

Mario Amado
Group of Nanotechnology, USAL-NANOLAB, Universidad de Salamanca, E-37008 Salamanca, Spain

José Montero Amenedo
Department of Materials Science and Engineering, The Ångström Laboratory, Uppsala University, Sweden

Philippe Bove
Nanovation, 8 route de Chevreuse, 78117 Chateaufort, France

Rubén D Costa
Technical University of Munich, Schulgasse, Straubing, Germany

Gezira De Ávila Montiel
Chemical Engineering Department, Faculty of Engineering, University of Cartagena, Cartagena, Colombia

Aleksandra B Djurišić
Department of Physics, the University of Hong Kong, Pokfulam Road, Hong Kong

Francisco Domínguez-Adame
Department of Materials Physics, Faculty of Physics, Complutense University of Madrid, Madrid, Spain

Sara Ferrara
Technical University of Munich, Schulgasse, Straubing, Germany

Javier García-Alonso
Department of Materials Physics, Faculty of Physics, Complutense University of Madrid, Madrid, Spain

María Taeño González
Department of Materials Physics, Faculty of Physics, Complutense University of Madrid, Madrid, Spain

Ozan Karakaya
Istanbul Technical University, Istanbul, Turkey

Fangzhou Liu
Department of Physics, the University of Hong Kong, Pokfulam Road, Hong Kong

Gao Liu
Lawrence Berkeley National Laboratory (LBNL), Berkeley, CA, USA

Bianchi Méndez
Department of Materials Physics, Faculty of Physics, Complutense University of Madrid, Madrid, Spain

Annie Ng
Department of Electrical and Computer Engineering, Nazarbayev University, Nur-Sultan, Kazakhstan

Gaukhar Nigmetova
Department of Electrical and Computer Engineering, Nazarbayev University, Nur-Sultan, Kazakhstan

Zhilin Ren
Department of Physics, the University of Hong Kong, Pokfulam Road, Hong Kong

David Rogers
Nanovation, 8 route de Chevreuse, 78117 Chateaufort, France

David Sánchez
Institute for Cross-Disciplinary Physics and Complex Systems IFISC (UIB-CSIC), Palma de Mallorca, Spain

Eric Sandana
Nanovation, 8 route de Chevreuse, 78117 Chateaufort, France

Ömer Suat Taşkın
Istanbul University, Istanbul, Turkey

Ferechteh Teherani
Nanovation, 8 route de Chevreuse, 78117 Chateaufort, France

Sebahat Topal
Istanbul Technical University, Istanbul, Turkey

Sema Topal
Istanbul Technical University, Istanbul, Turkey

Antonio Vázquez-López
Department of Materials Physics, Physics Faculty, Complutense University of Madrid, Madrid, Spain

Kasper Wenderich
PhotoCatalytic Synthesis (PCS) Group, MESA+ Institute for Nanotechnology, University of Twente, Enschede, The Netherlands

Yi Zhang
Department of Physics, the University of Hong Kong, Pokfulam Road, Hong Kong

Neslihan Yuca
Istanbul Technical University, Istanbul, Turkey

IOP Publishing

Sustainable Nanomaterials for Energy Applications

Ana Cremades and David Maestre

Chapter 1

Why push sustainable nanomaterials for energy applications?

Ana Cremades, David Maestre and Antonio Vázquez-López

1.1 Introduction

In this chapter, we introduce the basic concepts of sustainability and the need for sustainable nanomaterials to meet the requirements of the energy sector, taking into account the avoidance of critical elements and highlighting the current challenges and recent progress achieved in this area. The issues related to sustainability in materials development are complex and quite challenging, as they can impact over-warming and climate change, resource scarcity, the environment, and the circular economy. A crucial point is related to how we produce, transmit, utilize, and store energy and our capacity to match the increasing demands for energy due to new and more power-consuming devices, such as the electric automobile, among others. Although it involves an apparent contradiction, a certain degree of access to elements such as Li, In, Ga, Pb, and Ge, which are included on the critical elements list, is necessary to achieve the goals of global sustainability. Taking into account their relative risk, the possibilities for substitution, their advantages for low-power devices, and the small quantity of material employed in nanodevices, these critical elements will be also considered in this book due to the potential technological impact of presenting alternatives that eliminate or reduce their contents in the materials of interest, whenever possible.

1.2 Sustainability: a global challenge

From the study of natural systems and how they evolve, to reaching a balance from the ecological point of view, to the application of analog concepts in technology, to seeking a low impact on the environment and raw resources, sustainability is a broad concept related to different disciplines, such as environmental sciences, technological development, and social sciences. Despite the contributions of early environmental and social movements, the institutionalization of the term *sustainable development*

 © IOP Publishing Ltd 2023

started in 1987 with the United Nations World Commission on Environment and Development [1] and its report 'Our Common Future,' also known as the 'Brundtland Report,' which defined the term *sustainable development* as 'development that meets the needs of the present without compromising the ability of future generations to meet their own needs.'

The point is to reach a compromise between meeting our most energy- or resource-hungry needs and the protection of our natural world. Historically, the concept of sustainable development was related to environmental issues when referring to the *sustainable* part, and to economic aspects when related to the *development* one. During the 1980s, a social leg gradually emerged in the academic literature, and three pillars [2, 3] were identified and graphically represented by the popular three-circle diagrams or their close conceptual variants, as shown in figure 1.1.

As a result, in 2005, during the World Summit on Social Development [4] celebrated under the auspices of the United Nations, these three pillars were consolidated and included in the agendas of many governments.

In September 2015, 17 Sustainable Development Goals (SDGs) were issued by United Nations, expanding the previous restricted visions within a 2030 Agenda for Sustainable Development; see figure 1.2. In December of the same year, the Paris Agreement on Climate Change was also adopted, creating a landmark year for global multilateral agreements. Although sustainability science has been considered a substantive field in the last 20 years [5], the concept of sustainability is very wide and open to interpretation. However, it is clear that the significant growth of developing countries and the further global industrialization due to affordable energy in recent decades have had a measurable impact on climate change due to greenhouse gases, which has produced alarming global warming [6]. Our ability to produce energy is the driving force of our current progress, as the production, transmission, storage, and distribution of clean energy in the pursuit of energy efficiency and a low impact on the environment will be the focus of a sustainable future.

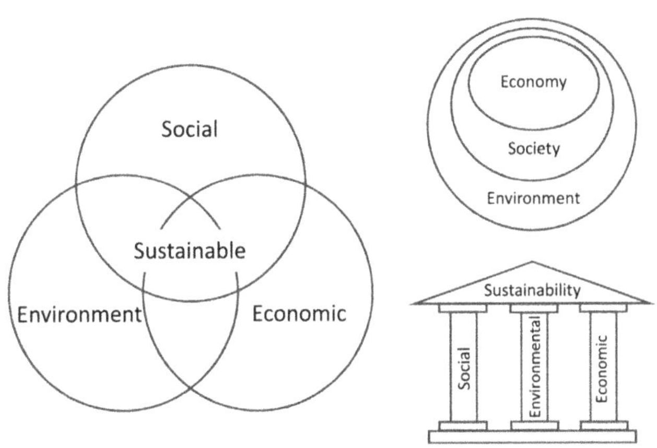

Figure 1.1. Diagrams representing the three pillars of sustainability. Reproduced from reference [3]. CC BY 4.0.

Figure 1.2. Sustainable Development Goals. Reproduced with permission from [7]. © United Nations. The content of this publication has not been approved by the United Nations and does not reflect the views of the United Nations or its officials or Member States.

1.3 The role of materials development in sustainability

All the technologies that allow environmental damage and climate change to be controlled are based on the use of specific materials which should be abundant and recyclable; they should ideally have a long life cycle with minimal impact and be fabricated by cheap, scalable, and clean processes.

Materials-based solutions can be offered that improve materials design and devices in different fields, such as solar energy for clean energy generation, batteries for the storage of energy in electric vehicles and the grid, photocatalysis for the conversion of light into chemical energy, water splitting for hydrogen generation, CO_2 reduction, and others as developed in the following chapters. Several methodologies can be used to assess the sustainability of materials: for example, the most common tool is life cycle assessment (LCA), [8] although it presents certain limitations, as it exclusively evaluates the environmental impact of a product through its life cycle, sometimes more by prediction than by the use of real data [9]. The LCA includes five different stages, namely the extraction of the raw materials, manufacture, distribution, use, and recycling and final disposal.

An alternative perspective that can be used to measure the sustainability of a technology is through the use of the IPAT equation ($I=P\times A\times T$) [10], which relates the impact of population P, affluence A (or consumption per person), and technology T (the impact per unit of consumption) on the environment, focusing more on the interaction of the combined issues than on the isolated aspects. Thus, to lower the impact, three different strategies can be used, namely, to lower the number of consumers, to lower the consumption, or to produce in a green way, including the macroeconomics and the socioeconomics of the technology in the sustainability value of the materials.

1.3.1 Critical raw materials

The supply chains of the most relevant technologies include critical raw materials (CRMs) [11] that are key enablers, so that it is very important to provide some clues about the evolution of future developments. The executive report on 'Critical Raw Materials for Strategic Technologies and Sectors in the European Union (EU): A Foresight Study' [12] published by the EU Commission in 2020 studies nine technologies in three strategic sectors, such as renewable energy, e-mobility, and defense and aerospace, including trends and conclusions that can be extended globally. The energy sector is one of the most globally relevant today; in particular, technologies related to e-mobility and the production and storage of renewable energy will be applied to provide solutions to present and future challenges related to climate change and sustainability in order to achieve a climate-neutral economy. The COVID-19 pandemic and its effects on supply and demand as well as on the implementation of a climate-neutral economy are going to be present for some time, although they are difficult to foresee at the moment.

Focusing on the raw materials, their availability, reasonable cost, and responsible outsourcing or other geopolitical aspects are the main issues to be considered for a viable technology. A materials supply bottleneck does not always arise from a high-growth-rate sector, as many issues other than price increases can be triggered by high demand. Higher prices could interestingly drive exploration and mining as well as the alternative substitution and recycling strategies, making investments attractive over a long period. For example, lithium, graphite, cobalt, and nickel are the essential raw materials for the batteries and fuel cells used in e-mobility and they are also used to store the energy produced by photovoltaic (PV) and eolian means, while some rare-earth elements, such as neodymium, dysprosium, and praseodymium, are considered to be essential raw materials for the motors of electric cars and also for wind turbines. They can be also in demand in other sectors such as defense, aerospace, and digitalization-related hardware. By 2050, the expected EU demand, compared to present consumption, is estimated to grow by up to almost 60 times for Li, 15 times for graphite, Co, and Dy, almost four times for Ni and Nd, and around three times for Pr.

In the case of Ni, the performance of batteries involves the use of high-purity Ni, which is a technological challenge; it also compromises the geological resources, and this goal is expected to encounter trade barriers. The rare-earth element (REE) market is dominated by China, and this is also the case for natural graphite. In addition, REEs, in particular Dy, are scarce materials. Although the demand for Li is higher, its prospects are less compromised in the short term, although the availability of Li could be reduced beyond 2025. In the case of Co, the geopolitical situation in the Democratic Republic of the Congo, which is the main global provider, could compromise the market. Taking these considerations into account, the supply risk is summarized for the different materials involved in the technologies of interest in figure 1.3.

A more detailed analysis of the technologies treated in this book follows below, paying attention to the bottlenecks that arise due to the current supply risks of raw

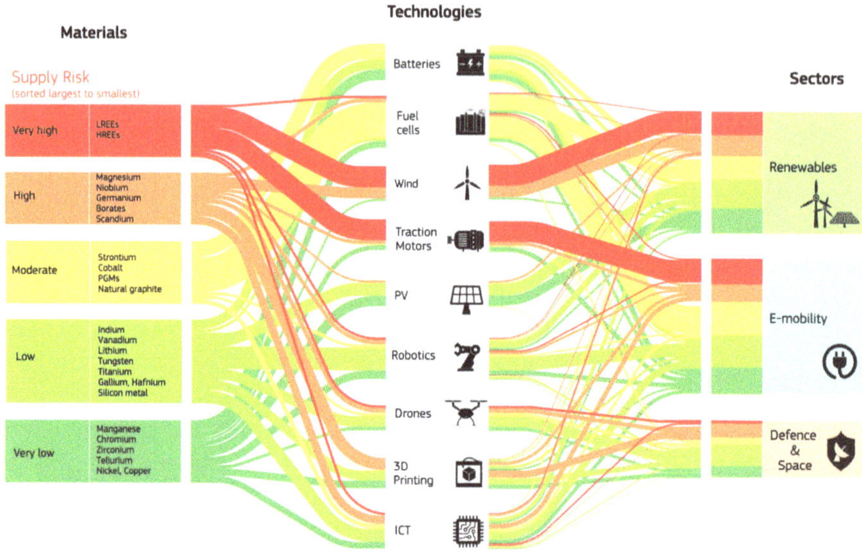

Figure 1.3. Semiquantitative representation of raw materials and their supply risks for the nine technologies and three sectors. Reproduced from [13]. © European Union, 2020.

materials, processed materials, components, and assemblies and including in some cases a demand scenario.

1.3.2 Batteries

Li-ion batteries are leading the energy storage sector, as they provide improved power and better performance compared to other contemporary technologies. They enable higher specific energies to be obtained in reduced volumes, thereby reducing weight and charging times. Most of the research in this area is devoted to improving the capacity and cyclability of batteries to increase durability and improve safety; very recently, the self-healing concept has also been included in the global equation to obtain improved performance. As a strategy, new anode and cathode materials are being designed, including the key anode elements lithium, silicon, titanium, and niobium and the cathode elements niobium, nickel, and aluminum, thereby reducing the cathodic Co content. A second alternative is to reduce the volume by enabling closer packing using less electrolyte and thinner separators and current collectors. The main raw materials used in Li-ion batteries are represented in figure 1.4.

As mentioned above, Li, graphite, and Co are the critical materials with a higher supply risk involved in the fabrication of batteries. Today, Si, Ti, and Nb are increasingly used in new electrode designs and are also classified as critical raw materials.

The global bottlenecks present in the raw materials and Li-ion cell production stages are mainly due to the predominant role of China, as illustrated in figure 1.5. China provides a huge amount of Co (46% of the refined cobalt), Li (45% of the hard rock mineral refining facilities), and natural graphite (producing the majority of the

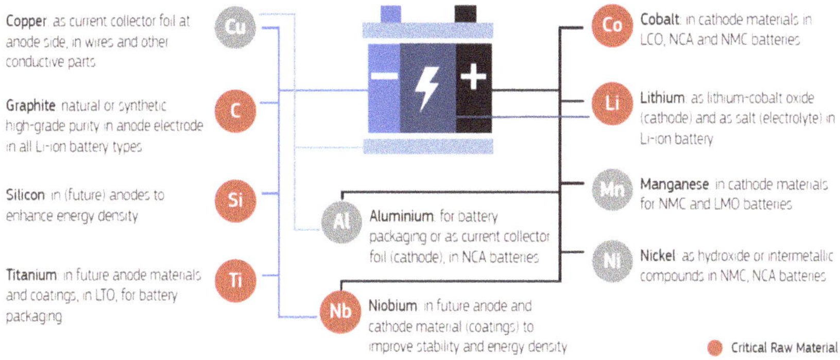

Figure 1.4. Raw materials used in Li-ion batteries. Reproduced from [13]. © European Union, 2020.

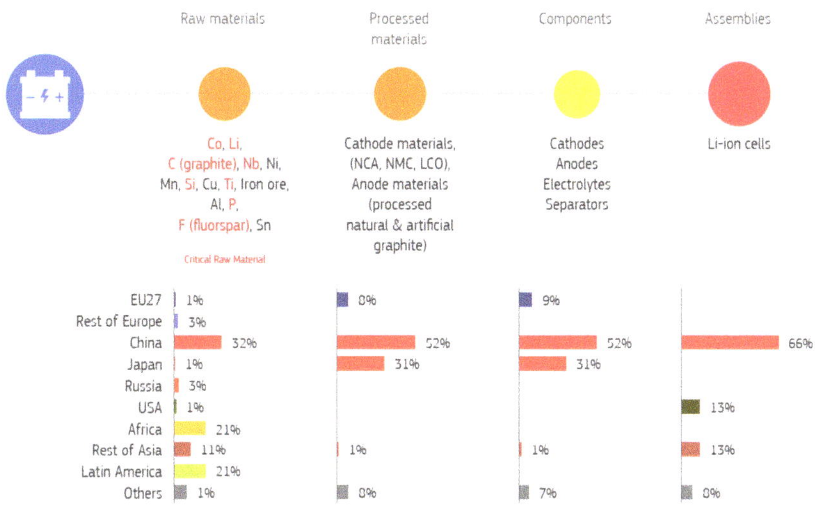

Figure 1.5. An overview of the supply risks, bottlenecks, and key players along the supply chain of Li-ion batteries. Reproduced from [13]. © European Union, 2020.

world's spherical graphite). China is also a major supplier of anode materials as well as processed cathode materials and accounts for 66% of global cell production. Asia (mainly represented by China, Japan, and South Korea) delivers 86% of the processed materials and components for Li-ion batteries worldwide. However, the global production capacity is increasing in other economic regions, where it is also driven by Asian investments. As an example, the EU intends to increase its capacity from 3 GWh in 2018 to 40 GWh by the end of 2023.

Cobalt is mainly produced in the Democratic Republic of the Congo, and there is a current research trend to reduce the content of Co in cathode materials in an attempt to avoid environmental impacts as well as other uncertainties in the provider chain. Chile (32%) and Argentina (20%) supply refined lithium from brine operations, which are essential to maintain a low price for lithium. However, several analysts [14] consider that an increment in the price would be beneficial, as it would support new investments in long-term production capacity. The global supply of nickel is inadequate for the fabrication of batteries, as a purity of more than 99.8% is required, and its production depends on the production of nickel sulfate. Major investments should be made in refining capacity in order to supply the battery market in the near future.

From a demand point of view, batteries are increasingly needed in the e-mobility sector as the energy storage systems associated with renewable energy, defense, and aerospace applications, as well as in the digitalization-related sector. The demand for Li-ion batteries is expected to increase yearly by more than 30% over the next decade to cover civilian and defense applications, so the cell production capacity should be expanded considerably worldwide together with the development of new policies for the substitution, recycling, and reuse of battery materials such as cobalt, lithium, manganese, and nickel.

1.3.3 Photovoltaics

PV technologies create electricity from solar energy and are regarded as one of the main future renewable resources. The increasing global interest in these technologies drives the exploitation of materials for this purpose, some of which are CRMs. The main raw materials used in PVs are represented in figure 1.6, where it can be observed that some of them, such as silicon metal, indium, gallium, germanium, and boron, are in high demand.

Today, among the most employed materials, silicon is the main protagonist. Despite the advances achieved in the photovoltaic field, crystalline silicon solar panels are still the dominant technology, accounting for as much as 95% of the global PV capacity, as silicon is employed in different varieties such as crystalline silicon (c-Si) and amorphous silicon (a-Si), while other relevant technologies which

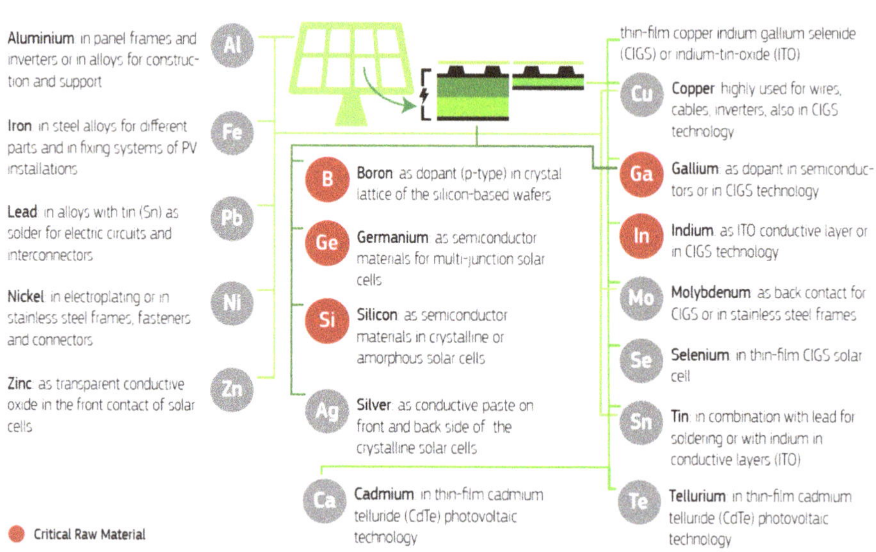

Figure 1.6. Raw materials used in solar PV technologies. Reproduced from [13]. © European Union, 2020.

also use CRMs are copper indium gallium (di)selenide (CIGS)–disulfide- and cadmium telluride (CdTe)-based PV cells [12]. In particular, in 2015, 69.5% of the total global production of PV systems was based on polycrystalline silicon technology, 23.9% on monocrystalline silicon, and 6.6% on CdTe.

However, many other materials at a lower supply risk are used in the assembly of PV cells. Among these, some materials can be highlighted, such as Ag, which is commonly used as an electrode in these technologies, in addition to many other structural and electric components such as steel, plastic, and copper.

Today, a diversified set of innovative technologies is emerging into a market still dominated by silicon panels; thus, the raw materials needed for the third and fourth generations of PV devices are an increasing list and therefore a variety of suppliers have entered the supply chain. However, the relevance of China is still huge, as it accounts for half of the market; it is especially monopolistic in the component stage and therefore creates a high global supply risk, as can be observed in figure 1.7. Currently, the EU has a silicon-based PV assembly capacity of 1%, partially due to the former dumping practices of China, which created a deep crisis in the silicon-related PV European market whose industries were not able to compete with the prices offered by Chinese PV producers. These practices also affected other regions, such as the United States of America (USA) and other parts of Asia, although with less impact.

1.3.4 Fuel cell and photoelectrochemical materials

Fuel cells (FCs) constitute an energy conversion technology which is based on electrochemical devices capable of converting fuel, such as hydrogen, into electricity without the need for combustion. This takes place because the hydrogen reaction with oxygen produces water and releases electrons which create electricity in an

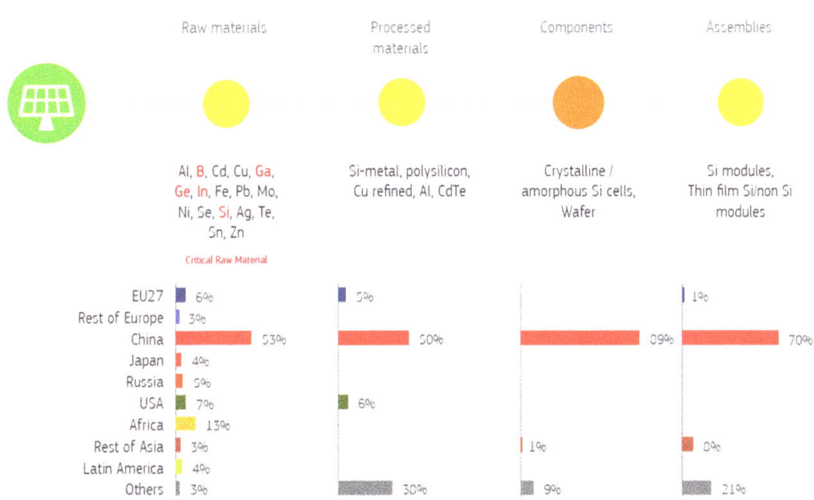

Figure 1.7. An overview of the supply risks, bottlenecks, and key players along the supply chain of PV technologies. Reproduced from [13]. © European Union, 2020.

external circuit. Among the different FCs, the most notable technology is the polymer electrolyte membrane fuel cell (PEMFC), while others such as solid oxide FCs (SOFCs) or alkaline FCs (AFCs) are less used. PEMFCs are characterized by a high power density; they work at lower temperatures than other FCs. FCs, in combination with hydrogen fuel supplies, are the basis of a potential environmentally friendly energy solution. However, they require large amounts of platinum-group metals (PGMs) as catalysts for the fuel-to-power conversion, and the cost of the platinum is as much as half of the cost of the fuel cell. The high price of this element remains one of the main challenges for the development of FC technologies. The main producers of Pt are South Africa, Russia, and Zimbabwe, among which, the former is by far the largest producer. The main supply demand comes from the USA and Canada (48%), while Japan and South Korea (51%) dominate FC production, and the EU produces less than 1% of commercial FCs.

Figure 1.8 shows the raw materials related to FCs; most of them, as mentioned above, are classified as CRMs. Apart from Pt, Sr or Ti is used in SOFCs, while Co is used in PEMFCs. The key players in FC technology are quite dispersed among different geographical areas, as one can observe from the data shown in figure 1.9. Africa has a notable role (48%) as the main provider of the raw materials, followed by China (17%), whereas the EU and the USA lead in processing the materials, and the USA and the rest of Asia (excluding China) fabricate the components and final assemblies.

1.3.5 Advanced electronics materials

Advanced electronic materials encompass a broad group of optoelectronic and digital materials of growing interest in today's society. Included in this group are computers, smart phones, drones, and all the physical objects which possess sensors and software and exchange data throughout the Internet, i.e. the so-called Internet of Things (IoT), e.g. robots, autonomous vehicles, and sensors. The immense group of these materials, which is fundamental to today's society, includes different components such as hard disk drives (HDDs), magnetic RAM, flash memories, and flat and touch screens that make use of CRMs, which are summarized in figure 1.10. Smart phones, tablets, and other computer devices have boosted the use of touch screens, which has increased the consumption of indium-based thin films, such as the transparent electrodes formed of indium tin oxide (ITO) films. All of the

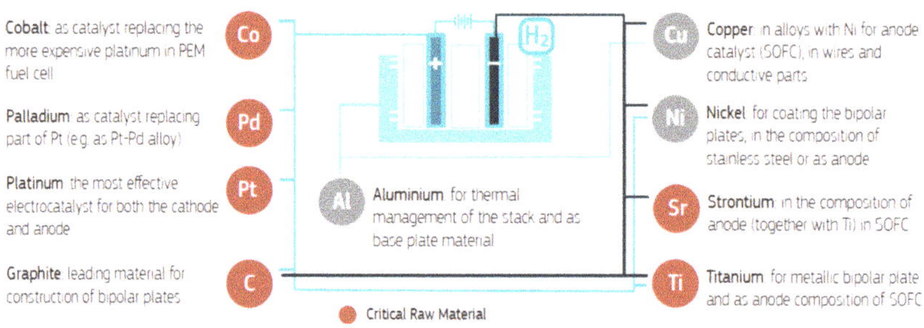

Figure 1.8. The raw materials used in fuel cells. Reproduced from [13]. © European Union, 2020.

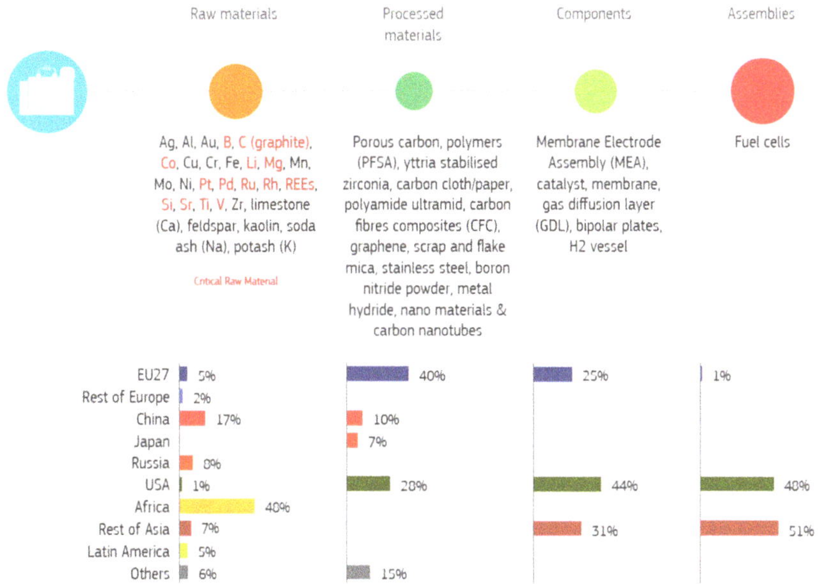

Figure 1.9. An overview of the supply risks, bottlenecks, and key players along the supply chain of fuel cells and hydrogen technologies. Reproduced from [13]. © European Union, 2020.

aforementioned devices use physical objects as memories, such as HDDs based on magnetic CRMs such as Co or other materials such as borates, while the most novel solid-state disks (SSDs) employ Si. Similarly, rare earths such as Nd or Dy are used in NdFeB permanent magnets, which are one of the main components of HDDs due to their resistance to high temperatures.

This group of advanced materials includes a large number of devices, which means that most of the elements of the periodic table are present in one application or another, as depicted in figure 1.10. Particularly noteworthy is the use of metals such as copper, gallium, gold, indium, platinum, and germanium, as well as groups of elements such as the PGMs and REEs. China (41%) and African countries (30%) are the dominant suppliers, creating a strong dependency in Europe, the USA, and other Southeast Asian countries, as they are the main bottleneck in the supply chain of these materials (see figure 1.11).

Although [12] only assessed the raw materials bottleneck, this does not mean that there are not others. In fact, at the time of writing, the common use of chips not only in computers but also in the automobile industry, home and personal appliances, and a number of other high-performance products (see table 1.1) has caused a global shortage of semiconductor-based chips [15], creating a bottleneck in many industries and limiting the production of new automobiles, smart phones, and computer components by producers all over the globe (such as Apple, LG, BMW, Ford, etc). The largest chip manufacturer is located in Taiwan, and other relevant suppliers are in South Korea; however, the USA is the second-biggest producer. European production accounts for less than 10% of the global production, and its capacity

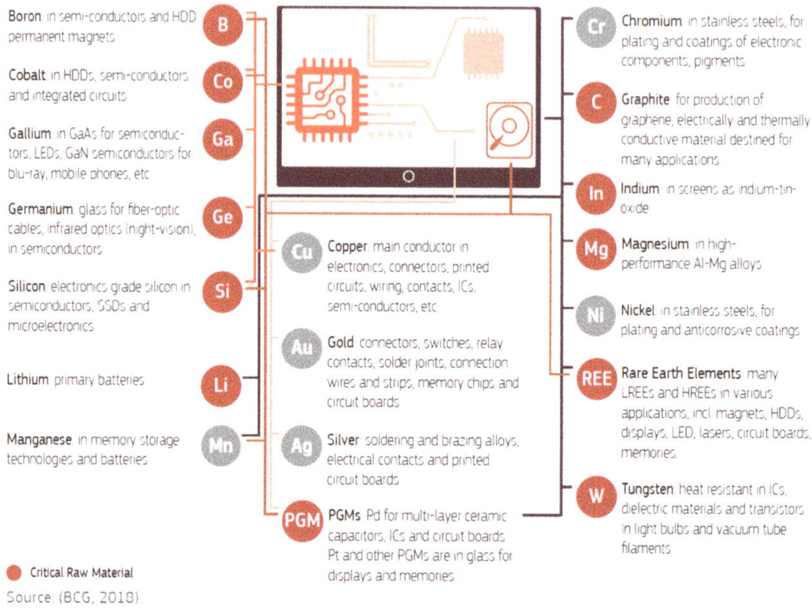

Figure 1.10. The raw materials used in digital technologies. Reproduced from [13]. © European Union, 2020.

Figure 1.11. An overview of the key players along the supply chain of raw materials only for production of advanced electronic materials. Reproduced from [13]. © European Union, 2020.

is slowly increasing, as Europe is trying to boost that figure to 20% in the next few years in order to gain technical sovereignty and become more self-reliant for critical technologies. This bottleneck is basically the result of an unbalanced demand-to-supply ratio partially induced by the reduction of production during the pandemic; it is not correlated with a lack of resources and will require an expansion of the

Table 1.1. Chip demand by sector. Data retrieved from reference [15].

Chip demand by revenue	U.S. $ (billions)
Communications infrastructure	36.3
Industrial	41.6
Automotive	39.5
Consumer	60.1
Wireless	126.7
Computing	160.2

production capacity in other geographical areas. Although the use of chips by the auto industry is not the highest among the technology sectors, it is increasing rapidly, and as this sector is key for many economies, the effects of this bottleneck on the automotive industry have recently been analyzed [16].

1.4 The circular economy

The circular economy consists of an environmental, economic, and social model in the pursuit of sustainability; it was proposed in 1990 by the economists Pearce and Turner [17]. The circular economy model is in line with the sustainable development that was formulated by the United Nations in 1987, enabling the creation of an economic prosperity that 'meets the needs of the present without compromising the ability of future generations to meet their own needs.' The first implementation worldwide was achieved in Germany in 1996 with the enactment of a law called the 'Closed Substance Cycle and Waste Management Act.' The main idea is to use certain strategies such as reduce, reuse, and recycle (also known as the 3R rule) to create a closed-loop model, maximizing the use of different materials and minimizing the creation of waste. An illustrative concept diagram of the circular economy can be observed in figure 1.12.

Recently, the circular economy approach has widely been introduced in different governmental and supra-regional agendas together with the design of roadmaps for the implementation of a clean and climate-neutral economy through investments in environmentally friendly technologies, the reduction of contamination, and biodiversity restoration. As an example, the European Green Deal is a roadmap of reference [19]. The main idea behind these plans is that 'the products of today should be the raw materials of the future' [20]. This ranges from production to consumption, through repair and remanufacture, by controlling waste generation, and through recycling. Circular use offers multiple benefits. As an advantage, this model secures the supply of CRMs due to the use of recycling, while reducing the contamination due to gas emissions and water use, since secondary CRMs typically need much less treatment than the extraction of primary CRMs; in addition, they create less impact on the environment and the biosphere, as no extraction is needed, and they also produce less waste per tonne of reused material. A key concept is to

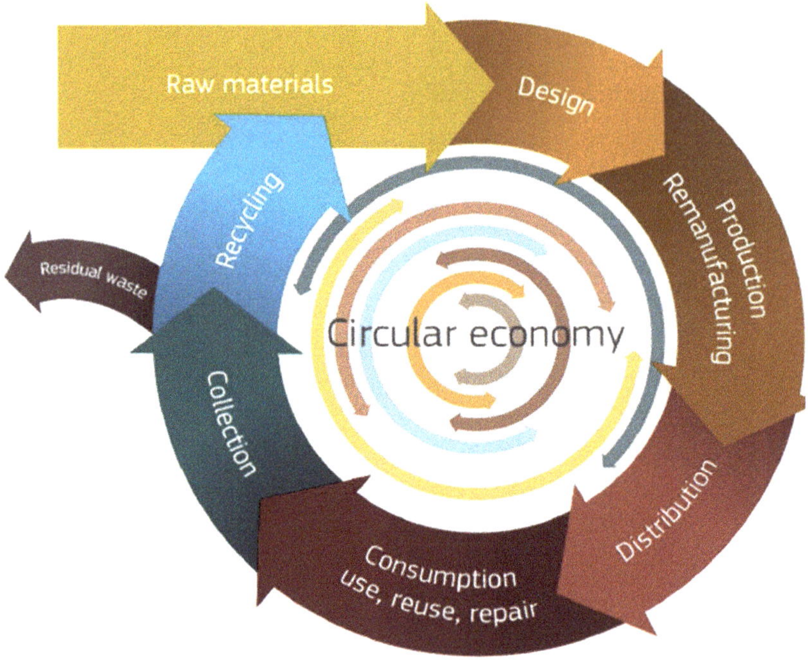

Figure 1.12. Diagram illustrating the circular economy. Reproduced from [18].

include all the measures required to obtain a sustainable product or technology in the design of new products and materials instead of mitigating the impact afterwards, as was common with conventional non-sustainable industries. In fact, the new materials should be sustainable and circular, and at the same time safe, functional, cost-effective, compliant with regulations and accepted by consumers, as indicated in the scheme shown in figure 1.13.

In 2015, by accepting an action plan and a set of measures for the circular economy, the EU agreed to 'the transition to a more circular economy, where the value of products, materials and resources is maintained in the economy for as long as possible, and the generation of waste minimized, is an essential contribution to the EU's efforts to develop a sustainable, low carbon, resource efficient and competitive economy' [11]. However, not only global policy makers and regulators are involved in this transition. Regulations and efficient laws might facilitate the implementation of the circular economy in many sectors; however, the key players are the economic agents such as industries and consumers, who will ensure the path toward innovation in this field.

One approach toward circular use consists of the recycling of CRMs; however, this is generally low, which is mainly attributed to the lack of recycling technologies at a competitive cost, the current CRM assets that are supply blocked in long-life uses, and the growing demand for CRMs, which recycling cannot meet. Nonetheless, a few materials such as vanadium, cobalt, antimony, and tungsten have high recycling input rates, as observed in figure 1.14.

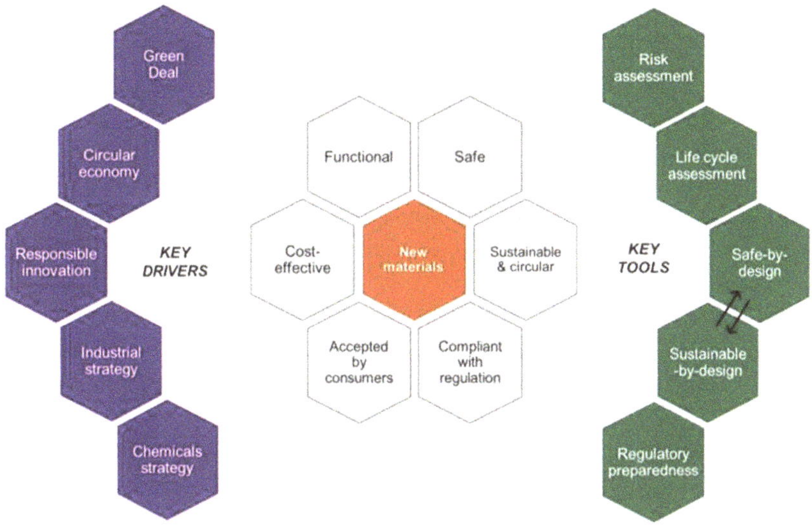

Figure 1.13. The European Union's policy initiatives (in purple) for the development of new materials and the key tools to ensure their safety and sustainability. Reproduced from [21]. CC BY 4.0.

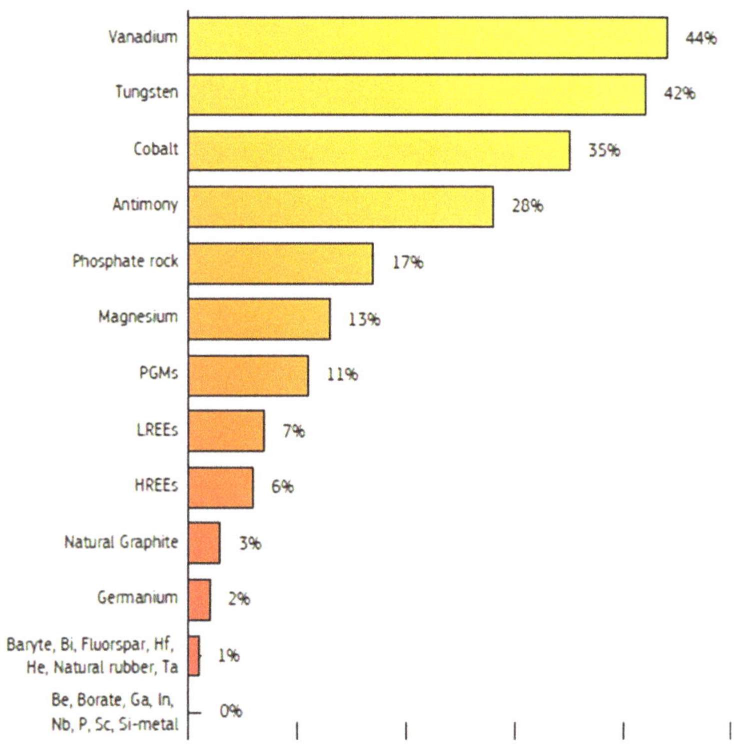

Figure 1.14. The contributions due to the recycling of different CRMs: end-of-life recycling input rates (EOL-RIRs). Reproduced from [18].

A group of general policy measures has been accepted for circular use. In the last period of Horizon 2020 (2018–20), more than €250 million were dedicated to actions on raw materials, including more than €100 million under a Circular Economy Focus Area. The new Horizon Europa initiative speeds this transformation, supporting large-scale systemic solutions and regional demonstration projects such as circular cities [22] and delivering an action plan in 2020 [23].

Lately, other regions, such as North America (including Canada, the United States, and Mexico) have included the circular economy as a priority, and discussion papers have been recently published such as the 'Discussion Paper & Event Summary Circular North America: Accelerating the Transition to a Thriving and Resilient Low-carbon Economy' [23], in which the circular economy is also highlighted as a trillion-dollar opportunity. However, some conditions such as low population density, which contributes to inhabitant dispersity, increase the cost of supporting the infrastructures and delivery models of the circular economy. The intensive consumption culture also does not create the necessary consumer support. The economy is only considered to be circular in less than 10% of the world, although performance indicators are lacking for North America. Added to the fact that the USA generates some of the highest amounts of waste per capita in the world (2.21 kg per day) [24], the COVID-19 pandemic has also created a global waste crisis due to the disposal of plastic bags, takeout containers, and personal protective equipment such as masks, gloves, and face shields, which should be resolved too.

On the other hand, Asia has already moved toward the circular economy for different reasons. In Japan, for example, the interest is driven by the lack of land for waste disposal as well as the reduced availability of raw materials in the country and the desire to prevent further environmental deterioration [25, 26]. As early as 2002, the Basic Law for Establishing a Recycling-Based Society came into force in Japan [27]. In China, which has had increasing economic growth and high productivity in recent decades, the circular economy is seen as a way of increasing resources and feedstocks as well as reducing pollution and other environmental problems and has been implemented at nationwide level [28, 29].

1.5 The advantages of nanomaterials

Nanomaterials can be designed to be sustainable due to their wide range of origins and applications. Some nanomaterials can be prepared from recycled or renewable ingredients, reducing some bottlenecks, or via processes with low environmental impact by the synthesis referred to as 'green synthesis,' which involves green chemistry principles in the design, manufacture, use, and end of life of nanomaterials and nano-enabled devices [30]. Moreover, some nanomaterials are considered to be nontoxic and their products are considered to be innocuous to human or environmental health [31]; however, some questions still remain open regarding the environmental and toxicological behaviors of nanomaterials, which have driven the regulatory initiatives of different governments or other organizations [32] and intensive research during the last 15 years [20, 21, 33]. Novel, stimuli-responsive ('smart') nanomaterials are used in food, food packaging, cosmetics, and agriculture

[21], offering advantages in terms of design and sustainability. Progress towards sustainable energy is also an important factor, aiming for green paths to obtain sustainable energy [34], in which nanomaterials offer advantages.

1.5.1 Nanomaterials in energy

Energy-wise, nanomaterials are advantageous, as they offer higher surface-to-volume ratios and improved transport properties and confinement effects due to their nanoscale dimensions; therefore, they are more suitable for energy sector applications. Solar cells, thermoelectrics, hydrogen storage, catalysts, thermoelectrics, and Li-ion batteries are only a few of the applications in which these materials have found application [17, 35, 36]. The vast family they belong to makes these materials well suited for these applications. As shown in this book, nanomaterials comprise carbon-based materials, inorganic nanomaterials (metal based and metal oxides), and organic materials. Their wide variability and the ability to engineer nanomaterials allow materials to be produced for each application with the desired electronic, chemical, physical, optical, and magnetic properties, which makes these materials suitable for energy applications.

1.6 Conclusions

This chapter has provided a general overview of the sustainability concept and its relationship with the role played by the materials used in several energy sector technologies. As a matter of fact, a range of characteristics should be considered for the materials that enable key technologies before we can consider that a technology is sustainable; to mention just a few of them: they should be abundant and recyclable, they should ideally have a large life cycle with a minimal impact, and they should be fabricated by cheap, scalable, and clean processes.

Raw material are crucial assets for the development of emerging energy technologies which could contribute to a clean and climate-neutral industry. The supply chain may contain several bottlenecks associated with different raw material risks, based on which, the materials are classified into a list of critical materials. Most of the materials needed for energy sector applications, such as batteries, photovoltaics, fuel cells, and photoelectrochemical technologies as well as those needed for the automotive, defense, and other sectors are included in this list, meaning that their abundancy, monopolistic outsourcing, and other bottlenecks in their provider chains, as well as geopolitical issues in some cases, could influence the cost-effective development of a viable industry that is sustainable from social, economic, and environmental perspectives.

Today, the circular economy model is a priority in different government agendas in order to fulfill the Sustainable Development Goals. The reduction, recycling, and reuse of materials will help to implement the circular economy principles, together with other options such as the redesign of materials and devices while avoiding the use of the higher-risk raw materials when possible. Sometimes, a compromise must be reached when the use of certain materials, although classified as CRMs, is necessary to attain the sustainability goals, giving rise to high-performance devices with a lower use of energy and resources and less impact on the environment. This is where technologies mainly based on nanomaterial implementations could offer some advantages.

Nanomaterials could offer a reduction in the intensive use of materials that would otherwise be used as bulk materials in conventional technologies. Reduced amounts of materials are used in nanodevices, which nonetheless offer new functionalities and high performance. In addition, advantage can be taken of the novel properties that appear at the nanoscale, which can lead to more efficient devices. Not only improved synthesis methods but also an in-depth understanding of the properties of the targeted nanomaterials should be achieved to face the modern challenges in sustainability and energy applications.

As a conclusion, intensive research into nanomaterials for energy sector applications could really make a key contribution toward the sustainability of the planet, enabling the design of a more inclusive system of production with higher efficiency. As a researcher in materials science, one should be aware of these strategies for the design of new materials, processes, and devices within the rules of the circular economy model, taking responsibility for the wellbeing of our planet and future generations.

1.7 Summary

- *The concept and evolution of the sustainability principles.* The institutionalization of the term *sustainable development* started in 1987, with the United Nations World Commission on Environment and Development and its report 'Our Common Future,' also known as the 'Brundtland Report.' In 2005, during the World Summit on Social Development celebrated under the auspicious of the United Nations, three pillars (social, economic, and environmental) were consolidated and included in the agendas of many governments. In September 2015, 17 Sustainable Development Goals were issued by the United Nations, expanding the previous restricted visions within a 2030 Agenda for Sustainable Development.
- *The role of materials in sustainability.* All the technologies which allow the control of environmental damage and climate change are based on the use of specific materials which should be abundant and recyclable; they should ideally have a large life cycle with a minimal impact and be fabricated by cheap, scalable, and clean processes. To lower the impact, according to the PAT approximation, three different strategies can be used, namely, lower the number of consumers, lower the consumption, or produce in a green way, including the macroeconomics and the socioeconomics of the technology in the sustainability value of the materials.
- *Raw materials.* Focusing on the raw materials: their availability, reasonable cost, and responsible outsourcing or other geopolitical aspects are the main issues to be considered for a viable technology. A materials supply bottleneck does not always arise from a high-growth-rate sector, as many other factors in addition to price increases can be triggered by high demand.

- *Raw materials for batteries.* Li, graphite, and Co are the critical materials with a higher supply risk involved in the fabrication of batteries. Today, Si, Ti, and Nb are increasingly used in new electrode designs and are also classified as critical raw materials. The main bottlenecks are due to the increasing role of Asia in materials extraction and production (mostly represented by China, Japan, and South Korea), which delivers 86% of the processed materials and components for Li-ion batteries worldwide. The demand for Li-ion batteries is expected to increase yearly by more than 30% over the next decade to meet the demands of civilian and defense applications, so that the cell production capacity should be significantly built up worldwide together with new policies to substitute, recycle, and reuse battery materials such as cobalt, lithium, manganese, and nickel.
- *Raw materials for PVs.* The main raw materials that are part of photovoltaics are silicon metal, indium, gallium, germanium, and boron, which are in high demand. Today, a diversified set of innovative technologies is emerging into a market still governed by silicon panels. Thus, the raw materials needed for the third and fourth generations of photovoltaic devices are an increasing list and therefore a variety of suppliers have entered the supply chain. However, the relevance of China is still huge, as it accounts for half of the market, it is especially monopolistic in the component stage, and it therefore introduces a high global supply risk.
- *Raw materials for FCs and photoelectrochemical materials.* These require a large amount of PGMs as catalysts for the fuel-to-power conversion; the cost of the platinum can be as high as half the cost of the fuel cell. The high price of this element remains one of the main challenges for the development of FC technologies. The main producers of Pt are South Africa, Russia, and Zimbabwe, among which, the former is by far the largest producer. The main supply demand comes from USA and Canada (48%), whereas Japan and South Korea (51%) dominate FC production, and the EU provides less than 1% of commercial FCs.
- *Raw materials for advanced electronic materials.* This group of advanced materials includes a large number of devices, which means that most of the elements of the periodic table are present in one application or another. The use of metals such as copper, gallium, gold, indium, platinum, or germanium is noteworthy, as is the use of groups of elements such as PGMs and REEs. China (41%) and African countries (30%) are the dominant suppliers, creating a strong dependency in Europe, the USA, and other Southeast Asian countries; this is the main bottleneck in the chain supply of these materials. At the time of writing, various sectors are suffering from a global shortage of chips due to an unbalanced demand-to-supply ratio, partially induced by a reduction in production during the COVID-19 pandemic, which will require an expansion of production capacity to other geographical areas.
- *Nanomaterials offer advantages.* Nanomaterials can be designed to be sustainable due to their wide range of origins and applications and the reduced amount of material included in the final devices. New technological designs based on the use of nanomaterials as building blocks allow for a

better choice of materials and improved performance. New materials properties associated with the reduction in scale could also lead to the introduction of new functionalities and new combinations of materials and dimensionalities, raising the utility of specific technologies.

- *The circular economy.* The main idea is to use strategies such as reuse, remanufacturing, and recycling to create a closed-loop model, maximizing the use of different materials and minimizing the creation of waste in the production and use of different devices and technologies. Some advantages are that the circular economy model secures the supply of CRMs due to recycling, while reducing contamination (gas emissions and water use), as secondary CRMs typically need much less treatment than that of primary CRM extraction. Some drawbacks are the low amount of recycled CRMs, which is mainly attributed to the lack of recycling technologies at a competitive cost, the currently available CRM assets, and the growing demand for CRMs, which recycling alone cannot meet. Every region worldwide has implemented plans for the transformation of its economy into a circular economy model, driven by different necessities and backgrounds.

1.8 Questions and work assessments for the student

1. What does sustainability mean, and how has this concept evolved in recent decades?
2. List some relationships between some of the Sustainable Development Goals and the uses of materials involved in energy sector applications.
3. Give some examples of the LCA of energy-related technologies found in the research literature. For example: *A Review of Battery Life-Cycle Analysis: State of Knowledge and Critical Needs* by J L Sullivan and L Gaines, Center for Transportation Research, Energy Systems Division, Argonne National Laboratory, 2010 (https://publications.anl.gov/anlpubs/2010/11/68455.pdf) Comment on the main results.
4. Define the concept of CRMs.
5. Which are the main CRMs involved in the different energy-related technologies?
6. What are the roles of the EU, the USA, and Asian regions in the different energy-related technologies? Complete a table indicating whether each region is the raw material provider, the producer of components or systems, the main market, etc. What are the roles of developing countries or emergent areas?
7. Give examples of the major bottlenecks in the different energy-related technologies mentioned in this chapter; comment on their origins and possible solutions. Add others; for example, the current shortage of integrated circuits.
8. Why could nanomaterials offer advantages as compared to bulk materials in the sustainable development of energy-related technologies? Give three examples for different energy-related technologies.

9. Find information about the 3R rule and its extension to the 7R or even 10R rules involved in the circular economy concept of materials. Compile a list of the 3R-, 7R- and 10R-related concepts (reduce, reuse, recycling, etc).
10. Search in the literature for the application of the 3R rule to the cases of Li, REEs, Pb, and Pt in energy sector applications. Draw a schematic of each kind of material application with the most relevant facts.

References and further reading

[1] World Commission on Environment and Development 1987 *Our Common Future* (Oxford: Oxford University Press)
[2] Barbier E B 1987 The concept of sustainable economic development *Environ. Conserv.* **14** 101
[3] Purvis B, Mao Y and Robinson D 2019 Three pillars of sustainability: in search of conceptual origins *Sustain. Sci.* **14** 681–95
[4] *The 2005 World Summit, High-Level Plenary Meeting of the 60th Session of the UN General Assembly* 14–16 September 2005 (New York: UN Headquarters) https://un.org/en/events/pastevents/worldsummit_2005.shtml
[5] Kajikawa Y, Tacoa F and Yamaguchi K 2014 Sustainability science: the changing landscape of sustainability research *Sustain. Sci.* **9** 431–8
[6] Edwards P N 2010 *A Vast Machine: Computer Models, Climate Data, and the Politics of Global Warming* (Cambridge, MA: MIT Press)
[7] 2022 UN 17 goals https://www.un.org/sustainabledevelopment
[8] van der Giesen C, Kleijn R, Kramer G J and Guinée J 2013 Towards application of life cycle sustainability analysis *Rev. Métall. – Int. J. Metall.* **110** 29–36
[9] Ayres R U 1995 Life cycle analysis: a critique *Resour. Conserv. Recycl.* **14** 199–223
[10] Ehrlich P R and Holdren J P 1972 Critique: one dimensional ecology *Bull. At. Sci.* **28** 16–27
[11] Mathieux F *et al* 2018 *Report on Critical Raw Materials and the Circular Economy* (Brussels: Office of the European Union)
[12] Bobba S, Carrara S, Huisman J, Mathieux F and Pavel C 2020 *Critical Raw Materials for Strategic Technologies and Sectors in the EU. A Foresight Study* (Brussels: Office of the European Union)
[13] 2020 European Commission, Critical materials for strategic technologies and sectors in the EU - a foresight study 2020 https://rmis.jrc.ec.europa.eu/uploads/CRMs_for_Strategic_Technologies_and_Sectors_in_the_EU_2020.pdf
[14] Roskill R 2018 *Lithium-Ion Batteries Market Development & Raw Materials* 2nd edn https://www.cetem.gov.br/antigo/images/eventos/2018/iii-litio-brasil/apresentacoes/lithium-batteries-xevs.pdf
[15] Moore S K 2021 How and when the chip shortage will end *IEEE Spectr.* https://spectrum.ieee.org/chip-shortage
[16] The Global Semiconductor Chip Shortage Crisis and Its Effects on the Automotive Industry ID: 5381931 Report July 2021 Region: Global (Santa Clara, CA: Frost & Sullivan) 66 pages. https://researchandmarkets.com/reports/5381931/the-global-semiconductor-chip-shortage-crisis-and?utm_source=BW&utm_medium=PressRelease&utm_code=kxh92l&utm_campaign=1568309+-+Global+Automotive+Chip+Ecosystem+Market+Report+2021+with+Focus+on+Semiconductor+Chip+Shortage+Crisis+and+Its+Effects+on+the+Industry&utm_exec=chdo54prd
[20] Chu S, Cui Y and Liu N 2016 The path towards sustainable energy *Nat. Mater* **16** 16–22

[18] 2018 Report on Critical Raw Materials and the Circular Economy, 2018 https://op.europa.eu/en/publication-detail/-/publication/d1be1b43-e18f-11e8-b690-01aa75ed71a1

[19] Turner R K 1990 *Economics of Natural Resources and the Environment* (London: Harvester Wheatsheaf)

[20] Colvin V 2003 The potential environmental impact of engineering nanomaterials *Nat. Biotechnol* **21** 1166–70

[21] Gottardo S, Mech A, Drbohlavová J, Małyska A, Bøwadt S, Sintes J R and Rauschera H 2021 Towards safe and sustainable innovation in nanotechnology: state-of-play for smart nanomaterials *NanoImpact* **21** 100297

[22] European Commission 2019 *Communication from the Commission to the European Parliament, the European Council, the Council, the European Economic and Social Committee and the Committee of the Regions* The European Green Deal. Brussels, 11.12.2019, COM(2019) 640 final

[23] Circular cities and regions initiative. European Comission, Research and Innovation 2022 https://ec.europa.eu/info/research-and-innovation/research-area/environment/circular-economy/circular-cities-and-regions-initiative_en

[24] https://ec.europa.eu/environment/topics/circular-economy/first-circular-economy-action-plan_en

[25] Discussion Paper & Event Summary Circular North America: Accelerating the Transition to a Thriving and Resilient Low-Carbon Economy Date: May 2021 Cat. No.: En4-413/2021E-PDF EC8017 https://canada.ca/content/dam/eccc/documents/pdf/circular-economy/north-america-paper/WCEF-Circular-North-America_Report_2021_EN.pdf

[26] World Bank. What a Waste (2.0) Report (2018). See page 20: https://openknowledge.worldbank.org/handle/10986/30317

[27] Morioka T, Tsunemi K, Yamamoto Y, Yabar H and Yoshida N 2005 Eco-efficiency of advanced loop closing systems for vehicles and household appliances in Hyogo Eco-town *J. Ind. Ecol* **9** 205e221

[28] METI 2004 *Handbook on Resource Recycling Legislation and 3R Initiatives* (Tokyo, Japan: Ministry of Economy, Trade and Industry)

[29] Yong R 2007 The circular economy in China *J. Mater. Cycles Waste Manag.* **9** 121–9

[30] Dahl J A, Maddux B L S and Hutchison J E 2007 Toward greener nanosynthesis *Chem. Rev.* **107** 2228–69

[31] Lynn L 2013 Bergeson sustainable nanomaterials: emerging governance systems *ACS Sustain. Chem. Eng* **1** 724–30

[32] International Risk Governance Council (IRGC) Nanotechnology Risk Governance Recommendations for a Global, Coordinated Approach to the Governance of Potential Risks (2007)

[33] Elsaesser A and Vyvyan Howard C 2012 Toxicology of nanoparticles *Adv. Drug Deliv. Rev.* **64** 129–37

[34] Zhang Q, Uchaker E, Candelariaz S L and Cao G 2013 Nanomaterials for energy conversion and storage *Chem. Soc. Rev.* **42** 3127–71

[36] Baig N, Kammakakam I and Falath W 2021 Nanomaterials: a review of synthesis methods, properties, recent progress, and challenges *Mater. Adv.* **2** 1821–71

[33] Wang H, Liang X, Wang J and Xue S J D 2020 Multifunctional inorganic nanomaterials for energy applications *Nanoscale* **12** 14–42

[37] Su B, Heshmati A, Geng Y and Yu X 2013 A review of the circular economy in China: moving from rhetoric to implementation *J. Clean. Prod.* **42** 215–27

IOP Publishing

Sustainable Nanomaterials for Energy Applications

Ana Cremades and David Maestre

Chapter 2

Selected synthetic methods for nanomaterials with different dimensionalities

David Maestre, Antonio Vázquez-López, Gezira De Ávila Montiel, Javier García-Alonso and María Taeño González

2.1 Introduction

In response to the worldwide energy demand and the necessary escalation of ecological concerns in modern society, the fabrication of nanomaterials increasingly requires the use of cost-effective, scalable, and environment-friendly synthesis routes that can achieve controlled dimensions, morphologies, and compositions and tailored properties. In recent decades, tremendous advances have been achieved in the synthesis of nanomaterials through a variety of well-established routes, including top-down and bottom-up approaches; however, recent trends envisage sustainable and energy-saving methodologies involving scalability and higher-efficiency processes. In that sense, great efforts are focused on the development of improved methods and alternative synthesis methodologies based on sustainability, low energy consumption, and the avoidance of toxic and critical elements as far as possible.

This chapter is devoted to some of the physical and chemical methods usually employed for the synthesis of nanomaterials with diverse morphologies and dimensions; it pays special attention to the processes developed in recent years that include sustainable aspects, nontoxic or green precursors, low reaction temperatures, or ambient processes as well as the use of recycled or renewable materials. Although more specific details of the corresponding synthesis methods are provided in later chapters, this chapter presents a general overview of some of the most relevant and commonly employed methods, taking into account that not all of them can be included here due to the vast number of methodologies.

Among the chemical methods, this chapter pays special attention to those processes with low environmental impacts, referred to as 'green synthesis'; interest in these is increasing at a staggering pace. In addition, some other chemical routes that avoid noxious elements and reagents are reviewed. Physical methods are also

doi:10.1088/978-0-7503-3531-7ch2

© IOP Publishing Ltd 2023

considered that provide eco-friendly paths for the fabrication of nanomaterials while avoiding the use of toxic precursors and high energy consumption. Finally, the use of sustainable synthesis methods to fabricate organic and inorganic materials as well as composites in the form of thin films is reviewed in this chapter.

2.2 Physical methods

Recent trends in the synthesis of nanomaterials based on physical methods have tried to evolve toward lower-energy-consumption routes involving lower temperatures and faster processes as well as environment-friendly methodologies that avoid toxic materials and subproducts. In the search for improved and alternative methodologies, different approaches have been pursued. On the one hand, classical physical methods such as pulsed laser deposition, molecular beam epitaxy, or sputtering, to name a few, have been used to support lower-energy methodologies and optimized processes, while also avoiding the use of critical materials. On the other hand, alternative methodologies have also been used in order to achieve faster and lower-energy synthesis processes that combine scalability and sustainability. In this section, some representative examples of physical methods commonly employed for the synthesis of nanomaterials for energy-related applications are reviewed, with especial emphasis on those processes involving low energy consumption, easy scalability, and sustainability.

2.2.1 High-energy ball milling

High-energy ball milling is one of the simplest techniques for the synthesis of nanoparticles and other nanomaterials from a variety of metals, alloys, oxides, carbides, sulfides, and other compounds; it is based on a mechanical milling process. Using this method, a large number of nanoparticles can be obtained by a single-step process and without the need for any solvent or wash, making it suitable for the large-scale and sustainable production of nanomaterials. In this synthesis method, the material or materials of interest are added to a container in the form of powder or powders, along with a selected number of milling balls. The container can be also filled with an inert or reactive gas if necessary. It is then rotated around one or more axes, depending on the type of mill, for a certain amount of time, during which the material suffers collisions by being trapped between the different balls and between the balls and the walls of the container. Models of this process have been extensively studied, but as it is a dynamic process with a large number of variables, these models become extremely complex and hard to calculate [1]. The container and the hardened balls can be of different materials (SiO_2, Al_2O_3, SiN_4, ZrO_2, WC, etc.) and are generally harder than the material being milled, while the container is usually rotated at hundreds of rpm during an optimized time. During the milling, a rise in temperature (100–1000 °C) is likely to occur and the material can undergo different physicochemical process. The simplest process is mechanical milling, during which, numerous collisions inside the container generate dislocations in the particles that accumulate and result in a decrease in the size of the milled material. After a determined amount of time, this process is balanced by the tendency of particles to nucleate, and a minimum particle size is attained. Another process that can occur is the mechanical alloying of two compounds due to the highly energetic collisions that happen inside the container. Another possible process is the generation of chemical reactions between the different compounds used and/or with the gas filling the container, i.e. oxidation and size reduction of metals can happen simultaneously.

As an example of mechanical milling in the fabrication of nanomaterials for energy-related applications, some articles have reported the generation of nanoparticles for electrodes in Li-ion batteries. Del Prado *et al* [2] used this method to mill powders of TiO_2 and SnO_2 for 60 min at 400 rpm in order to create nanoparticles that, in conjunction with a graphene oxide nanocomposite, formed a lithium-ion battery (LIB) anode. Ulissi *et al* also used this method in the field of Li-ion batteries, in this case to generate a nanocomposite of C–FeS_2–S capable of substituting for the Ni–Co–transition-metal cathodes usually used [3]. For this purpose, they milled activated carbon, elemental sulfur, and FeS_2 at 360 rpm for 13 h with 15 min pauses every 45 min until a nanocomposite was formed for the all-solid-state cathode. Kahimbi *et al* followed this method to generate a mechanochemical reaction between Ni powder and graphene oxide powder by milling them at 1200 rpm for 1 h to oxidize the Ni and reduce the graphene oxide, thereby generating a NiO–reduced graphene oxide (rGO) nanocomposite with promising characteristics for use in the fabrication of supercapacitors [4].

2.2.2 Vapor-phase methods

The main component of these methods is the evaporation of a selected material via different routes such as sputtering, laser ablation, and arc discharge, to name a few. If the evaporation of the material source is paired with a controlled atmosphere through some form of vacuum and a gas injection system, a vapor phase of the material can be created and controlled. This vapor phase is unstable and eventually condenses due to a supersaturation process, allowing the nucleation of the vapor molecules to form nanoparticles and diverse nanostructures. Moreover, this process can be altered by multiple factors, depending on the method and the system used, in the search for the variable control of synthesized nanomaterials. As an example, the use of catalysts is a very common strategy, as these materials can act as nucleation sites to accelerate the growth or alter it in order to obtain different nanostructures. The list of materials that can be used with these methods is almost endless, hence the possibilities for the synthesis of nanostructures with variable morphologies, dimensions, and compositions are extremely varied. For zero-dimensional structures, nanoproducts such as nanoparticles can be synthesized by different processes such as homogeneous nucleation in the gas phase, most of the syntheses in inert gas condensation systems, or by a process of heterogeneous nucleation and agglomeration over a substrate based on diffusion and a Volmer–Weber process [5, 6]. An example of one of these types of growth is the generation of nanoparticles through a chemical vapor deposition process such as the one used by Huang *et al*, in which MoO_2 nanoparticles were obtained over a substrate of graphene and used in Li-ion batteries [7]. In the case of two-dimensional structures, thin films are formed by following the same nucleation process in a substrate but through a Frank–van der Merwe process which favors laminar growth [6], or some other processes. Using the same technique, MoO_2 thin films can be obtained instead of nanoparticles [8]. For one-dimensional nanostructures, the growth usually involves a combination of diverse physical mechanisms underlying the growth process [9]. These methods allow for the condensation of material to occur within a single direction, creating long and thin structures of different morphologies. Furthermore, complex structures and 3D architectures can be also produced by this method.

2.2.2.1 Thermal evaporation methods

One of the most commonly reported methodologies leading to the growth of nanostructures of many compounds with controlled dimensions and morphologies is based on thermal evaporation methods. These methods can lead to the growth of nanowires, rods, belts, tubes, and a large variety of morphologies of single or complex oxides as well as doped materials. Two of the main synthesis methods involving this growth mechanism are the vapor–liquid–solid (VLS) and vapor–solid (VS) methods. In the former, metal catalysts are used as nucleation sites for the anisotropic growth of elongated nanostructures, while in the latter, no catalyst is required, leading to the growth of nanostructures that follow different driving forces such as self-catalytic growth, growth based on surface energy, and defect-induced growth. By controlling some of the parameters involved in the evaporation–deposition methods, such as the selected precursors, atmosphere, temperature, or duration of the treatment, tailored morphologies, dimensions, and compositions can be achieved. This method has been used for the synthesis of different structures of an enormous variety of compounds such as ZnO, SnO_2, In_2O_3, Ga_2O_3 [10], and NiO [11, 12], among others, allowing for the creation of very complex nanostructures such as p–n heterostructures or highly conductive nanowires that can be used in applications such as solar cells.

Another approach in this vapor-phase methodology is inert gas condensation; this is based on the vaporization of material by heating it in a crucible or by any other method of vaporization (e.g. sputtering or an ion beam) under high vacuum conditions and deposition on a cold substrate under an inert gas flow. This method, commonly employed for the synthesis of nanoparticles [13], requires a high vacuum, has a high energy consumption, and involves high costs, but a poor yield of nanoparticles is obtained; hence, in most cases, its use is avoided in favor of other techniques. However, this method has recently found a new use in the recycling of materials used in different electronic systems thanks to the different temperatures of vaporization of their components; among others, it has been used in the extraction of Zn and Pb from Zn–Mn batteries and printed circuit boards, respectively [14, 15]. The main advantage of this approach is that it not only extracts these materials but also generates nanoparticles for new applications at the same time.

However, most of the abovementioned methods are time- and energy-consuming, as they usually require long treatments (from tens of minutes to tens of hours) and high temperatures (in most cases over 1000 K, depending on the selected precursors). In order to overcome these problems, alternative fast and scalable synthesis methods based on thermal oxidation, such as the resistive heating method, have recently been considered.

2.2.2.2 Resistive heating

The thermal evaporation and oxidation of metals is a well-known method for the fabrication of mainly oxide nanomaterials via a low-cost and easy method. The direct thermal oxidation of a metallic wire or plate in air or another oxidizing atmosphere is a rapid and simple method as compared to other evaporation processes; it also leads to the growth of nanowires and plates, among other nanostructures, in diverse oxides, such as CuO, ZnO, MoO_3, and TiO_2, among others [16–18]. The application of an external electric field during thermal oxidation and Joule heating can improve the growth processes, increasing the efficiency and also allowing greater control of the synthesis as reported by different authors. In most cases, the growth is based on the Joule heating of a metallic wire, in which cationic diffusion, thermal stress gradients, electromigration, and other VS and VLS mechanisms could be involved. The underlying mechanisms of this versatile and simple method are yet not well understood, and still require further study. Despite the fact that high temperatures are also involved in this synthesis method, the duration of the treatment is usually in the range of tens of seconds or a few minutes, which leads to very fast growths and large energy savings. Moreover, this method is highly scalable and there is still room for improvement in the optimization of this facile and rapid route.

2.2.3 Other physical methods

2.2.3.1 Flame synthesis

Flame synthesis is commonly used to grow nanoparticles of different metals and oxides in the industrial field. Furthermore, this simple one-step process is relevant in the synthesis of materials such as carbon black and fumed silica, to name a few. Flame synthesis is very attractive for the sustainable production of large quantities of nanoparticles and thin films due to its high yield and simplicity. The method consists of a precursor unit that is used to evaporate the source material, a flame maintained by a selected gas, usually a combination of oxygen, methanol, or ethylene, and a system of pumps and filters to collect the particles. In the case of thin films, a substrate cooled by water is used for the deposition of the particles that form the film. The precursor can be dispersed by a dispersion gas, usually oxygen, but can also be directly present in the gas injected for the feeding of the burner. Different variables affect the process, the main one being the temperature of the flame; however, the residence time of the particles and the amount of precursor used can also alter the synthesis, due to an increase in the collisions between particles. The process can be facilitated by the use of additives that can minimize the aggregation tendency of the particles or by the use of electric fields that generate repulsive forces between them. A variation of this technique has recently gained strength: flame spray pyrolysis. In this variation, the source material is dissolved in a combustible solution that is sprayed by an air or oxygen gas disperser to be ignited by the burner [19].

Apart from its use in the industrial synthesis of large quantities of different ceramic nanoparticles, flame synthesis has also found a place in other fields such as the production of hydrogen to be used as a clean fuel. Liu *et al* applied this technique to the synthesis of Mn_3O_4 and Mn_2O_3 through the use of a solution of $Mn(acac)_3$, acetonitrile, and 2-ethylhexanoic acid, an oxygen flow, and a premix of methane and oxygen for the fuel. The nanoparticles were used as catalyst for water splitting and resulted in an activity increase of 350% in the process with respect to a commercially available catalyst [20]. This method was used by the same group with other non-noble metals such as Fe, which they used to grow thin films of iron oxide over a substrate of fluorine doped tin oxide (FTO) [21]. This enabled the production of large quantities of hematite photoanode for water splitting by a simple one-step method.

2.2.3.2 Arc discharge

Arc discharge, also known as arc deposition, is a technique that can be used for the synthesis of different compounds and metal oxides, but is mostly applied for the synthesis of different carbon nanostructures. This method allows the synthesis of relatively large quantities of nanomaterial with good crystallinity. The system used in this method usually comprises a vacuum chamber that can be filled with a selected gas (which can be inert or reactive, depending on the product desired) and two close electrodes. At least one of the electrodes acts as a source of material, while a DC current of usually 50–100 A and 10–15 V is applied. This applied current and the small gap between the electrodes generate an arc discharge that vaporizes the material and starts the synthesis. The nanomaterial can be deposited on the walls of the chamber or on the electrodes. In addition, different catalysts can be applied to the process to accelerate or modify the process [22]. Not only can gases fill the chamber but also solutions or different liquids such as water can be used to perform doping or alter the synthesized nanostructures.

This process has been applied in numerous fields, including in energy storage technology, as a nanomaterial for anodes in Li-ion batteries. As an example, Sun *et al* used a Sn anode and a graphite anode in an arc discharge system to create SnO_2 particles wrapped in graphite that were able to adjust to the volume changes of SnO_2 during charge and discharge and allowed for high charge mobility [23]. Another example is the work by Ashkaran *et al* on the synthesis of ZnO nanoparticles decorated with graphene nanosheets for photocatalytic applications; this group used two Zn electrodes and rGO dissolved in a solution for the synthesis. The nanoparticles obtained presented good charge separation capacity due to the trapping of the electrons in the graphene [24].

2.2.3.3 Pulsed laser deposition

Despite the fact that pulsed laser deposition (PLD), also known as laser ablation, is a synthetic technique capable of generating a variety of structures such as nanowires and nanoparticles, this method is commonly employed for the growth of thin films of metals, oxides, and complex compounds with high crystallinity and controlled composition. The PLD setup comprises a laser beam that vaporizes the source material; the lasers used are usually excimer lasers that operate in the UV region, but visible and infrared lasers are also used. The laser is focused on the source material in a vacuum chamber with, at a minimum, high vacuum conditions, although it can be filled with a small quantity of reactant gas. The evaporated material is deposited on a substrate opposite the source. The incident laser generates multiple physical processes that can be classified as primary and secondary processes. The primary processes are implicated in the direct vaporization of the source by the laser, while the secondary processes are responsible for the evaporation of particles due to the backscattering of previously evaporated particles. These processes can be modified by varying the laser used, the controlled atmosphere, the distance between the source and the substrate, and the temperature of the chamber and the substrate, among other parameters.

Although PLD has traditionally been employed as a research technique, much work has recently been done to bring it in line with industrial standards. To give some examples, it has been applied for the easy and relatively low-cost fabrication of thin films for multiple purposes: solar energy, high-temperature superconductivity, multiferroics, etc. Khamlich *et al* used this technique to deposit thin films of Cr_2O_3 in a few layers of graphene, obtaining high crystallinity and good contact between the two layers that made them useful as anodes for Li-ion batteries due to their good electrical conductivity [25]. PLD has also been useful in the creation of thin films for different parts of perovskite solar cells. Qiu *et al* used it to synthesize a NiO layer on some perovskite film to act as a hole transport layer due to the high crystallinity and good contact produced by this technique [26]. In the same field, Wang *et al* used it to create the perovskite layer of their solar cells by depositing $CsPbBr_3$ over a substrate of TiO_2 while managing to maintain the stoichiometry of the compound [27].

2.2.3.4 Chemical vapor deposition

Chemical vapor deposition (CVD) is a hybrid technique, i.e. both physical and chemical, that is mainly used to synthesize thin films and two-dimensional materials. In recent decades, this method has been refined to the point at which the large-scale production of thin films, especially of carbon nanostructures, can be achieved. In this method, a reactant gaseous material is introduced into a vacuum chamber, usually in the form of a tube furnace, where it encounters a substrate at a high temperature (300 °C–1000 °C). At that high temperature, the gaseous material separates into different products that diffuse through surface of the substrate, leading to the formation of thin films by means of diverse chemical reactions and nucleation processes. The gas carrying the material to be grown can undergo multiple chemical reactions, not only on the surface of the substrate but also before deposition. These reactions can be promoted by the heating of not only the substrate but the chamber as a whole. The most important part of the system is probably the substrate used. The substrate is not only relevant for film deposition but it also acts as a catalyst for the reactions that occur in it. Its composition (for example, different transition metals) can yield different graphene depositions; its microstructure, lattice structure, and even orientation play a role in the synthesis. One of the main drawbacks of this method is its high energy consumption.

To solve some of the limitations of this technique, different variations of the traditional CVD method have appeared, such as metal–organic CVD (MOCVD) and plasma-enhanced CVD (PECVD). MOCVD solves the problem of evaporating some materials from a solid source by using a metal–organic compound as the reactant gas. However, this method introduces the problems of the high prices and toxicities of metal–organics. PECVD addresses the high temperatures needed and their high energetic and economic costs through the use of a plasma generator that excites the precursor gas, favoring chemical reactions in it and avoiding the need for a high substrate temperature in order for synthesis to occur.

As mentioned above, the use of this technique is mainly focused on the synthesis of thin films of different materials; in particular, it has also been found useful for the synthesis of carbon nanostructures such as graphene or few-layer graphite. Tian *et al* created N-doped graphene combined with single-wall carbon nanotube films, using FeMoMgAl flakes as a substrate and ethylene and ammonia as reactant gases, to be used for the oxygen reduction and evolution reactions [28]. Tang *et al* created hierarchical porous graphene for use as the cathode in lithium–sulfur batteries, using CaO as a substrate and a temperature of 950 °C. The porous material obtained allowed for short diffusion pathways; it had a low internal resistance and a robust framework that resisted volume changes [29]. Qi *et al* used a similar process, using Ni foam instead of CaO to create hierarchical graphene at 1000 °C. When combined with paraffin wax, this material was capable of storing thermal energy due to its high thermal energy density and its thermal and chemical stability due to the highly stable structure produced by CVD [30].

2.2.3.5 Molecular beam epitaxy

For decades, molecular beam epitaxy (MBE) has been the preferred technique for the synthesis of highly crystalline and highly pure thin films, mainly of group III–V compounds and other semiconductors such as Si and Ge. The rate of growth achieved by this technique allows for the generation of epitaxial thin films with deposition control down to a single layer. This controlled growth makes this method commonplace in diverse research fields, but it still is not widely applied in industrial-scale synthesis. This system generally comprises a vacuum chamber equipped with different vacuum pumps, individual effusion cells as material sources, and a heated substrate. The cells of materials are separated from the chamber containing the substrate by shutters that can be closed at any moment. Thanks to the ultrahigh vacuum (UHV) and the cooling of the chamber walls, the flow of particles from the sources to the substrate has a sufficiently long free path that the particles do not undergo any reactions before arriving at the substrate. Once the atoms and molecules arrive at the substrate, they can undergo diverse processes, such as physi- or chemisorption, diffusion, reevaporation, agglomeration, etc. The flow of atoms can be changed, leading to a large degree of control over the stoichiometry of the synthesized compounds. Due to the slow growth rate that allows for such high control, the chamber must be kept under a UHV regime, as even a slow flow of foreign atoms results in the extensive contamination of each monolayer grown. The effusion cells traditionally used are Knudsen cells, as these cells contain the material and resist the extremely high temperatures necessary to vaporize it (usually in the range of 800 °C–1200 °C or even higher). The source of material can also be in a gaseous form, which prevents the use of high temperatures but can generate residues; they can also be vaporized by other processes, such as heating with an electron beam. The walls of the chamber must be kept at very low temperatures to avoid the vaporization of the particles deposited on them. It is important to control the composition and orientation of the substrate, due to its interaction with the particles deposited on it. Moreover, the substrate must be heated to a sufficient temperature to allow the diffusion of the deposited atoms or molecules and it is usually rotated throughout the process to prevent irregularities in the growth due to the inhomogeneity of the particle flow.

The requirements of this technique, namely vacuum conditions and high temperatures, make it economically and environmentally costly. To give some examples of the use of MBE, it has been used in studies of the synthesis of hematite thin films used as photoanodes in the splitting of water [31], in the analysis of the relevance of structural changes in silicene and silicon mono- and bilayers during the adsorption of lithium, and due to its ability to prevent the formation of lithium clusters in Li-ion battery anodes [32].

2.3 Chemical methods

2.3.1 Green chemistry

For decades, there has been a deterioration in the environment that has created the need to investigate alternatives focused on environmental sustainability. One of these techniques is 'green chemistry,' a method that links process and product design while focusing on reducing waste and decreasing the formation of toxic substances, as well as improving the use of natural resources, reducing energy consumption, and being environmentally friendly [33, 34]. Today, green technologies or sustainable routes are preferred to physical and chemical methods in the synthesis of nanoparticles because of their excellent advantages: they are fast and easily scaled processes, they are biocompatible and sustainable, they use nontoxic substances, improve the stability of nanoparticles as well as other nanostructures, and require low energy consumption. The physical and chemical synthesis methods are usually high-cost processes, since they consume a lot of energy and form toxic by-products which are released into the environment [35]. Due to their nature, size, and surface area, nanomaterials present excellent and tailored properties in different application areas: optical, thermal, chemical, biological, and electrical, to name a few [36]. The green route commonly uses a bottom-up approach, in which atoms or molecules are introduced or adjusted into specific arrangements by chemical forces until the nanoscale product is obtained. Among the common chemical routes, the green method has been preferred in recent years, since it has advanced eco-friendly and cheap methods for the elaboration of nanoparticles and other nanostructures [37]. Today, there is an increase in the demand for environmentally friendly and sustainable chemical processes in industry; therefore, green chemistry has received the support of the Environmental Protection Agency (EPA) in order to minimize the use of hazardous and toxic chemical substances or reagents [37]. Green chemistry, or sustainable chemistry, defines a set of actions linked to the life cycle of chemical reagents, which includes the amount of raw material, the process of transformation at reduced risk, the life of the product, the waste generated, and the use of reusable materials. In addition, it is based on the 12 principles of green chemistry proposed by Anastas and Warner [38], shown in figure 2.1, which are guidelines for carrying out an ecological process of elaboration [39] while reducing the environmental and health impacts of chemical production. The concept of sustainability is broader than the term 'green,' since it also includes social, environmental, and economic aspects, which are required for the design of innovative processes and the selection of chemicals with less impact on the environment, in order to conserve natural resources [40]. This section focuses on analyzing green synthesis methods in terms of their sustainability criteria for the elaboration of nanoparticles using natural extracts from different plants and biopolymers. Moreover, other common chemical synthesis routes employed for the synthesis of nanomaterials are also reviewed in the context of sustainability.

Among the ecological methods used for the synthesis of nanoparticles and nanostructures, some other perspectives are identified. These include the use of green raw materials such as plant extracts (stems, leaves, fruits, and flowers) and natural polymers (chitosan, alginate, and collagen) in electrochemical methods and ultrasound and microwave techniques.

2.3.1.1 Green synthesis using plant extracts

Plants contain a great variety of macromolecules, such as antioxidants and secondary metabolites (flavonoids, polyphenols, alkaloids, ascorbic acid, saponins, triterpenes, sterols, alcohols, polymers, β-phenylethylamines, sugars, cofactors, proteins, and enzymes). Their extracts have antioxidant activity and act as metal-ion-reducing agents when exposed to nanoparticles. Green synthesis can be performed using plant tissues, such as extracts of leaves, bark, roots, stems, flowers, and fruits [35]. Various researchers have suggested different mechanisms for the

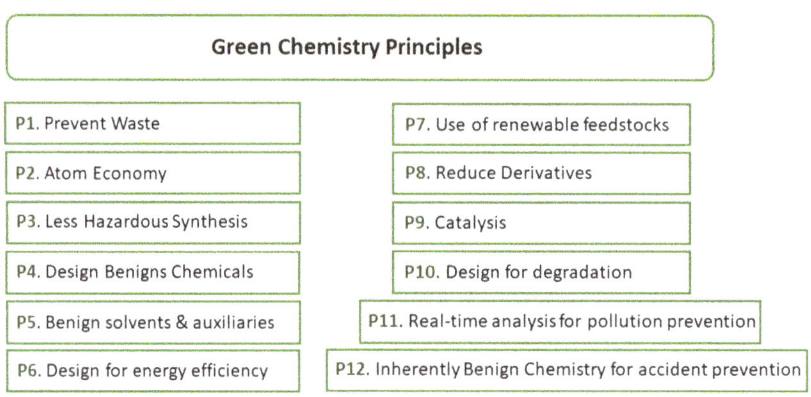

Figure 2.1. The 12 principles of green chemistry.

green synthesis of nanoparticles using plants, which depend on the species and the types of biomolecule present.

As an example, mango (*Mangifera indica* L.) is a plant that belongs to the Anacardiaceae family and is planted in many parts of the world, especially in tropical countries. It has alternate, smooth, dark green leaves with edible fruits. The extract of the leaves has been studied for the synthesis of silver nanoparticles (AgNPs) [41]. Phytochemical analysis of the extract has revealed the presence of several compounds, including alcohols, amines, phenols, fatty acids, terpenoids, flavonoids, and manguiferine, among others.

For the biological synthesis of nanoparticles (NPs), the metallic (Ag) salt is mixed at a fixed concentration with the plant extract; in this step, heterogeneous nanoparticles are produced at a low yield for the subsequent production of homogeneous, stable, and high-yield nanoparticles. The process must be improved by adjusting the following parameters: reaction time, the mixture ratio, temperature, and pH. To control the size, shape, and morphology of the nanoparticles, these variables can be modified and optimized during the process [42].

Some examples of green routes used in the synthesis of gold nanoparticles (AuNPs), AgNPs, nickel and NiO nanoparticles (NiNPs, NiONPs), iron and iron oxide nanoparticles (FeNPs, FeONPs), as well as TiO_2 nanoparticles using plant and natural extracts are reviewed below.

2.3.1.1.1 The green synthesis of AuNPs using plant-mediated extracts

AuNPs have gained much attention due to their optical properties, biocompatibility, thermal and chemical stability, and resistance to oxidation. They have various environmental applications, such as catalysis, sensors, and electrical conductivity, and also applications in the medical field. Moreover, AuNPs have economic importance due to the use of simple and fast green methods for the synthesis of AuNPs through plant extracts, since this process facilitates the reduction of their salts and offers excellent antimicrobial activity [43]. In 2017, Sundararajan *et al* used the leaf extract of *Artemisia vulgaris* L., which is an analgesic medicinal herb, for the preparation of gold nanoparticles using the phytochemicals contained in the terpenoid leaves, i.e. flavonoids and polyphenols, as substances that reduce Au^{3+} ions to Au^0 particles. Transmission electron microscopy (TEM) analysis showed that the synthesized nanoparticles presented triangular, hexagonal, and spherical forms with an average size between 50 and 100 nm. The reaction speed of the gold nanoparticles depends on the solution's pH, since it affects the size and shape of the particles [44]. The authors of another study synthesized gold nanoparticles (AuNPs) using *Cymbopogon citratus* leaf extract, which had a reducing and protective effect and worked as a stabilizing agent. TEM images showed that the synthesized nanoparticles had triangular, hexagonal, spherical, and elongated shapes, with sizes from 20 to 50 nm [45].

Different researchers have reported the synthesis of AuNPs using extracts of parts of *M. indica* L., and flavonoids have been identified as gold-salt-reducing agents in the green synthesis of gold nanoparticles. Fresh and dry leaf extracts of *M. indica* L. were used for the rapid synthesis of gold nanoparticles at room temperature, achieving spherical, almost monodisperse Au nanoparticle colloids at sizes of

~17 nm and 20 nm [46]. Another study used the aqueous seed extract of *M. indica* L. as a reducing agent for AuNPs through a simple and fast method, obtaining spherical nanoparticles 46.8 nm in size [47].

2.3.1.1.2 The green synthesis of AgNPs using plant extracts

AgNPs have received much attention due to their various applications in medicine, drug delivery, wound healing, household appliances, electronics, water disinfection, nanodevice manufacturing, cosmetics, and as antimicrobial agents, to name a few [48]. AgNPs are very popular due to their low toxicity to humans compared to other types of metallic particle. Today, AgNPs are synthesized by green or ecological methods from plant extracts, as such extracts include reducing agents that act as stabilizing substances [49]. Phytochemical analysis has revealed the presence of several compounds in *M. indica* L., including alcohols, amines, phenols, fatty acids, terpenoids, flavonoids, manguiferin, etc. which act as reducing agents in the synthesis of AgNPs [50].

One study has reported that a change of the temperature and the pH of the reaction mixture composed of an aqueous solution of $AgNO_3$ and the leaf extract of *M. indica* L. occurred during the easy and fast biosynthesis of $AgNO_3$ nanoparticles; at a pH of 8, a colloid of dispersed nanoparticles ~20 nm in size with a round shape and similar size was obtained [46]. Another study focused on the synthesis of AgNPs using mango flower extract; the resulting nanoparticles were spherical in shape and had a mean size of 10–20 nm [51]. Another study used aqueous leaf extracts from three plant species: *Brillantaisia patula*, *Crosspteryx febrifuga*, and *Senna siamea* to synthesize AgNPs. The synthesized nanoparticles were spherical in shape and well dispersed with sizes ranging from 45 to 110 nm. The modification of silver ions (Ag^+) to metallic silver (Ag^0) can be achieved because of the reducing power of some phytochemical compounds, such as terpenes, polyphenols, proteins, and alkaloids, which form spherical and crystalline silver colloids and nanoparticles [52]. The mechanism of synthesis in the biological pathway is due to binding between the silver ions and the active sites of the proteins; this occurs by electrostatic interactions between the Ag^+ ions and the reducing agents of the plant extracts. The reduction of the Ag^+ ions is consistent with the compounds present in the plant extracts that act as reducing substances [53]. The composition of the extract, the concentration of metallic ions, and the reaction time are parameters that influence the shape and size of AgNPs [42], hence these parameters should be controlled during the synthesis.

2.3.1.1.3 The green synthesis of NiNPs and NiONPs using plant extracts

Among the properties of NiNPs, the following stand out: a large surface area and good magnetic and chemical properties; NiNPs have applications in magnetic fluids, sensors, fuel cells, solar-cell systems, catalysts, and antimicrobial agents [54, 55]. Researchers have developed a novel and rapid method, employing green chemistry, for the synthesis of NiNPs using the aqueous extracts of pomegranate peel, which contains phenolic compounds, flavonoids, and flavonols [56], and orange peel, which contains citrus flavonoids such as polyphenols, which are reducing substances that can be used for the generation of Ni nanoparticles [57]. Morphological studies indicated that the particle sizes obtained were in the range between 100 nm and 1 μm for both extracts; the particles exhibited a rod shape with different crystal sizes and without agglomeration [58].

Another study investigated the production of fine NiONP powders from cactus extract; the particle sizes obtained were in the range of 20–35 nm and the average crystal size was 16 nm. In a morphological analysis, the particles presented a spherical shape and agglomerations due to the fact that NiO nanoparticles have increased surface tension and a high-energy surface. This study demonstrated that this natural extract is economical and environmentally friendly and has different applications for energy storage [59].

2.3.1.1.4 The green synthesis of FeNPs using plant extracts

A fast, easy, and ecological process used the leaves of *M. indica* L. for the green production of FeNPs, as the extract contains phytochemical agents such as polyphenols, flavonoids, and alkaloids, among other compounds. The size of the synthesized nanoparticles was 100–150 nm due to the concentration of polyphenols and their importance in the process of reduction and production of nanoparticles. The pH plays a key role in agglomeration and stability at acidic and basic pH values, respectively. FeNPs nanoparticles were efficiently produced in the decontamination processes for domestic wastewater with a high phosphorus content [60].

Other researchers conducted studies combining green chemistry and biogenesis to synthesize magnetic nanoparticles using *Moringa oleifera* and chitosan leaf extract. TEM analysis showed that *M. oleifera* and moringa–chitosan nanoparticles created spherical agglomerates at particle sizes of 4 ± 2 nm and 5 ± 2 nm, respectively. At the end of the process, the iron nanoparticles appeared as large, crystalline particles with a maximum saturation of 1.57 emu g^{-1}. The polyphenols, phenolic acids, and flavonoids present in *Moringa* leaves intervened in the nucleation of the nanoparticles due to the electrostatic attraction between the chitosan cations and the free radicals in the extract. Therefore, moringa and chitosan act as reducing, protective, and stabilizing agents for the synthesized nanoparticles [61]. This study reported the green synthesis of FeONPs using an extract of *C. citratus* as a reducing agent and sodium carbonate as a pH stabilizer. The average size of the magnetic nanoparticles corresponded to 17 ± 8 nm, and they exhibited an irregular cubic structure with polycrystalline agglomerates 27 nm in size due to strong dipole–dipole interactions and a superparamagnetic behavior around 28 emu g^{-1} [62]. In another study, $Fe_{55}Co_{45}$ nanoparticles were produced from the phytochemical compounds present in the aqueous extract of *M. indica* leaves. The nanoparticles synthesized by the green method had high magnetization values (92.44 emu g^{-1}) compared to those produced by the chemical method and their particle sizes were between 19.1 and 28.6 nm [63].

2.3.1.1.5 The green synthesis of TiO_2 nanoparticles using natural extracts

Titanium dioxide (TiO_2) is a semiconductor metal oxide recognized for having diverse applications. It has excellent properties such as chemical stability, considerable broadband energy (anatase 3.2 eV and rutile 3.0 eV), strong ultraviolet absorption, a high surface area, and photocatalytic and antibacterial activity; it is also nontoxic [64]. The ultrasonic method is a new green technique that is low-cost and clean and is applied in different areas of knowledge such as the chemical industry, pharmaceuticals, food, environmental bioremediation, and nanotechnology [65]. One study coupled two green methods, namely, natural aqueous extracts and the ultrasonic technique. TiO_2 nanoparticles were synthesized by the green chemistry method using ultrasound-assisted Fe^{3+}-doped lemongrass leaves. TEM analysis showed that the nanoparticles had a spherical shape and sizes in the range of 13.7–15.2 nm [66]. S-doped TiO_2 nanoparticles (TiO_2NPs) were prepared using an aqueous extract of the *Malva parviflora* plant (which had excellent reducing and protective characteristics) to favor S-doping and increase the size of anatase crystals. The synthesized nanoparticles presented a small particle size in a range between 20.3 and 29.5 nm and had a large surface area of 135 $m^2\ g^{-1}$ [67]. Researchers have used green synthesis for the elaboration of titanium dioxide nanoparticles by means of two natural plants, *Terminalia catappa* and *Carissa carandas*. The compositions of the extracts included reducing and protective agents for green synthesis. X-ray diffraction (XRD) analysis indicated the formation of the anatase phase in the TiO_2 nanoparticles; they had an average size of 10–20 nm and their morphology was homogeneous, spherical, and without agglomerates [68]. A scheme for the synthesis of TiO_2 nanoparticles following a green chemistry route based on a plant extract is shown in figure 2.2.

2.3.1.2 The biogenic synthesis of AgNPs using polymers

Chitosan is a natural polysaccharide derived from deacetylated chitin. It is low-cost and nontoxic, it has high permeability and balanced chemical resistance, and it is eco-friendly [69]. The synthesis of nanoparticles mediated by chitosan has attracted much attention due to chitosan's high degree of hydrophilicity, biodegradability, and biocompatibility, its molecular structure, and the presence of amino and hydroxyl groups; therefore, it is useful for various applications [70]. The main characteristic of chitosan is that it is an amino cationic polymer capable of interacting with polymers and negatively charged molecules [71]. In the synthesis of metallic nanoparticles, the primary amino groups are fixed to the surface of the

Figure 2.2. Synthesis of TiO_2 nanoparticles following a green chemistry route.

metal and function as protection and stabilization sites in the process of reducing silver ions to Ag^0 [72]. As a stabilizing agent, chitosan improves the stability of nanoparticles, provides good surface physical absorption, and generates electrostatic interaction [73, 74].

A group of researchers reported the synthesis of chitosan-mediated AgNPs [72]. In their synthesis route, the metallic salt was mixed with the chitosan solution at at a fixed concentration of 1%–2% (w/v), agitated for 10–12 h at room temperature, then mixed with a solution of metallic nanoparticles followed by agitation for 12 h at 95 °C. After a color change, the solution was centrifuged for 10 min, the supernatant was removed, and the precipitate was dried in an oven at 70 °C for 24 h. The UV–visible absorption spectrum revealed that the formation efficiency of AgNPs increased with the addition of chitosan and that higher concentrations of chitosan favored decreasing NP sizes.

Primary amino groups bind to the surface of the metal and act as protection and stabilization sites for Ag^+ NPs in the process of reducing silver ions to Ag^0. The resulting particles are spherical in shape with sizes ranging from 10 to 60 nm [72]. Another study used chitosan extracted from shrimp shells to synthesize biogenic AgNPs with a spherical and crystalline structure. TEM analysis revealed the formation of spherical particles at sizes from 17 to 50 nm [75].

Other researchers reported the isolation of polysaccharides from algae (alginate, focoidan, laminaran) and analyzed their activity as stabilizing and reducing agents in the synthesis of silver nanoparticles by a biogenic method. They proved their catalytic activity as reducing agents in the following order: laminaran > fucoidan > alginate. These polymers act as biocatalysts for the wet chemical reduction of silver ions and as stabilizing substances for Ag nanoparticles produced by biological methods. TEM analysis showed that the synthesized particles were spherical. The average crystalline sizes were 18, 24, and 29 nm for alginate, focidan, and laminar, respectively. This difference may be associated with variations in the hydrodynamic diameter, impurities, or the polycrystalline structure [76].

2.3.2 Other chemical synthesis routes

2.3.2.1 Hydrothermal methods

It is well known that hydrothermal methods are a promising alternative to the high temperatures and long reaction times usually employed by solid-state reactions. Hydrothermal synthesis can be described as a way of synthesizing single crystals that depends on the solubility of the precursors under high pressure. One of the main advantages of this technique is its ability to synthesize crystalline phases which are not stable at the melting point, avoiding the post-annealing process, and therefore reducing costs. From safety and sustainability points of view, hydrothermal methods allow the use of water as a solvent, avoiding the use of organic solvents which are usually toxic and/or explosive, limiting their practical application. Since hydrothermal methods operate at high pressures and low temperatures, this is one of the most employed routes for the synthesis of biomaterials. Moreover, this chemical process allows the synthesis of a wide variety of materials, from semiconducting oxides to metallic nanoparticles or carbon materials, often using natural precursors as reagents. For example, Ocsoy *et al* [77] synthesized AgNPs using an $AgNO_3$ aqueous solution as the precursor and a solution of red cabbage extract as a reducing and stabilizing agent. The resultant mixture was transferred to a Teflon-lined hydrothermal synthesis reactor and heated to 100 °C for different reaction times. To give an example of the use of this method, large quantities of AgNPs were obtained and a high degree of control over their size was exercised, as shown in figure 2.3.

Lam *et al* [78] reported a green hydrothermal synthesis method for producing ZnO nanotubes using ZnO powder and an H_2O_2 aqueous solution as starting materials. Using this approach, ZnO nanotubes were obtained at low temperature (130 °C) in a friendly environment, since there was no need to control the pH of the solution with ammonia, NaOH, or ethylenediamine and consequently no release of waste products. Green hydrothermal synthesis is also employed to produce high-quality carbon materials in a wide variety of morphologies. Kalamuei *et al* [79] synthesized high-quality single and few-layer graphene sheets via a hydrothermal method that used bread waste as the precursor. For this purpose, ground bread was added to a distilled water solution to produce a gel-like suspension, which was transferred to a Teflon-lined stainless autoclave and heated at 180 °C for 12 h. Since food (such as bread) is mainly formed of hydrocarbons, the use of waste materials as precursors to produce carbon materials has now become one of the most eco-friendly alternatives, reducing production costs and environmental impact.

2.3.2.2 Sol–gel methods

The sol–gel process is a wet chemical method which involves several steps: hydrolysis and polycondensation, gelation, drying, densification, and crystallization. This synthesis method usually employs alkoxides as precursors; it is one of the most

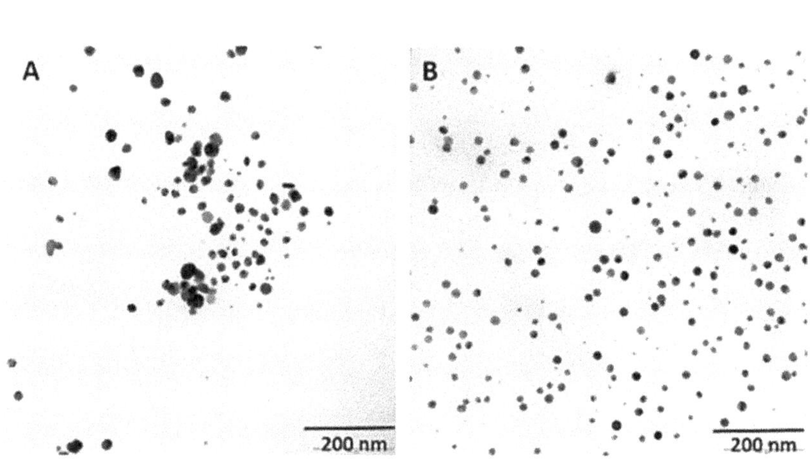

Figure 2.3. Scanning transmission electron microscopy (STEM) images of Ag nanoparticles synthesized by the addition of (a) 1 ml and (b) 5 ml of red cabbage extract to $AgNO_3$. Reprinted from [77], Copyright (2017), with permission from Elsevier.

versatile chemical routes, since metal oxides/non-oxides or hybrid organic–inorganic materials can be obtained. Moreover, when this method is used, a reduction in energy consumption is achieved because there is no need to reach the melting temperature, since the network structure can be achieved at relatively low temperatures. In contrast to the hydrothermal methods, sol–gel routes do not allow a high degree of control over the morphology of materials; however, the sol–gel process provides an ideal way to control the level and the homogeneity of doping, which are crucial in materials science. Since the term 'green chemistry' has become a hot topic in recent years, most of the synthesis methods are trying to use natural precursors, reducing the use of toxic and eco-unfriendly solvents. In this context, the use of sol–gel methods avoids the use of expensive precursors and does not need special or expensive equipment. Some examples are described in this section. For example, Fardood *et al* [80] reported the synthesis of MgO nanoparticles by the sol–gel method using gum arabic (a polysaccharide component) as a biotemplate and magnesium nitrate as the magnesium source. This method has many advantages such as absence of toxicity, economic viability, scalability, speed, and environmental friendliness, and it can be used for the synthesis of MgO nanoparticles without the need for any organic chemicals. Using this route, cubic-phase MgO nanoparticles with an average particle size of 14 nm were obtained. Many natural precursors are commonly used in sol–gel synthesis; they act as templating agents that reduce the use of organic compounds. Among these biocompatible precursors [81], the most common are the lotus root, green leaves, gelatin, and rice straw. Since the sol–gel method always employ a capping agent (usually an organic material), the use of natural and soft templates has become the most promising alternative to the use of hazardous chemicals.

2.3.2.3 Coprecipitation methods

Coprecipitation reactions involve the simultaneous use of nucleation, growth, coarsening, and/or agglomeration processes. They have been highlighted as simple, cost-effective, and fast processes that are easily scalable for industrial applications. The advantages of the co-precipitation method are the high yield, high product purity, and the absence of a requirement to use organic solvents. However, the presence of a precipitating agent (a basic or acidic solution) and a surfactant (an organic compound) is essential in order to cause the precipitation of the desired compound. NaOH, NH_4OH, HNO_3, HCl, and organic compounds such as dioctyl sodium sulfosuccinate or alkyl ether phosphates (usually corrosive products) are the most common precipitating agents and surfactants, respectively. Due to this use of corrosive products, a lot of effort is being made to find alternative routes to the traditional ones in order to reduce the environmental impact. For example, Nouri *et al* [82] synthesized Cu_3Se_2 nanoparticles using copper nitrate and selenium powder as the precursors and water as the solvent. To obtain the corresponding precipitate, $NaBH_4$ was used as a reducing agent, NaOH was used as a precipitating agent, and glycine amino acid was used as a green surfactant. Zhu *et al* [83] developed an energy-saving, environmentally friendly, and non-high-pressure route for the sustainable synthesis of 1D $Mg_2B_2O_5$ submicron whiskers via a facile byproduct-assisted thermal conversion. The resultant and residual byproduct NaCl within the room-temperature precipitate was employed as the flux needed in the subsequent thermal conversion, leading to submicron $Mg_2B_2O_5$ whiskers with uniform 1D morphology and high crystallinity (figure 2.4), easy product purification, and low energy consumption.

2.3.2.4 Sonochemical methods

Sonochemistry was one of the earliest techniques used to prepare nanomaterials. In this method, the molecules undergo a chemical reaction due to the application of powerful ultrasound radiation (20 kHz–10 MHz). The experimental setup usually consists of a sonic processor, a transducer, and a water cooler (figure 2.5). Different sonochemical reactions can be carried out in liquid systems in order to prepare

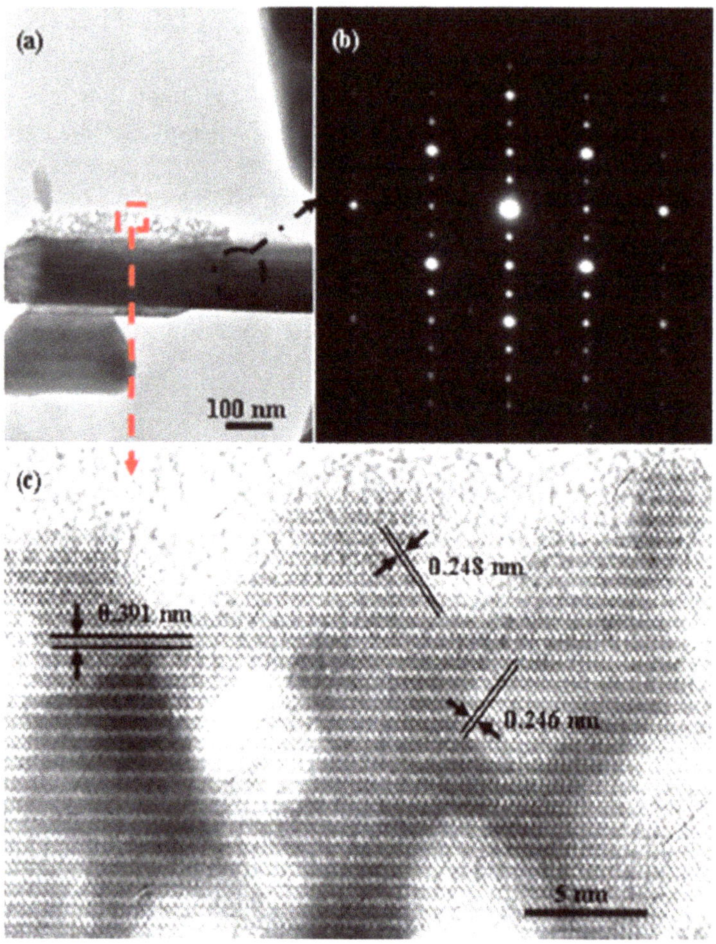

Figure 2.4. TEM image (a), selected area electron diffraction (SAED) pattern (b), and high-resolution transmission electron microscopy (HRTEM) image (c) of the calcined $Mg_2B_2O_5$ submicron whiskers treated at 800 °C. Reproduced from [83] with permission of The Royal Society of Chemistry.

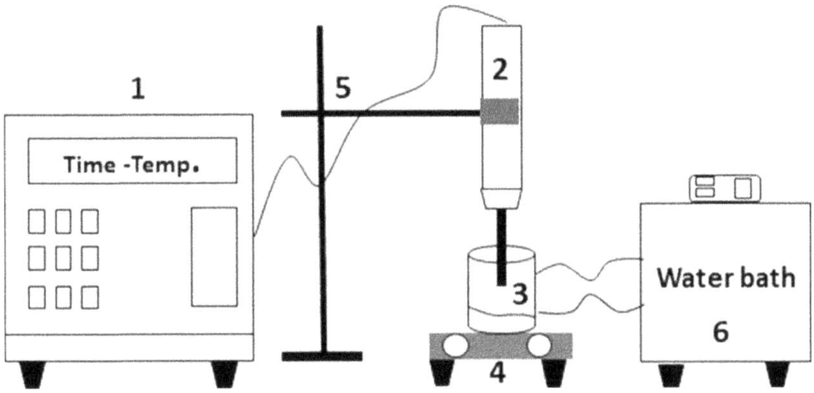

Figure 2.5. Experimental setup for sonochemical synthesis. (1) Sonic processor, (2) transducer, (3) reaction mixture, (4) hot plate, (5) stand, and (6) water cooler.

organic and inorganic functional materials. Moreover, this technique exhibits several advantages in comparison with those previously described. For example, this route allows the preparation of amorphous nanomaterials, the insertion of nanomaterials into mesoporous materials, or the deposition of nanoparticles on ceramic and polymeric surfaces. This chemical synthesis approach is a perfect fit with the concept of 'green chemistry,' since it does not require large amounts of energy or hazardous chemicals. In addition, the cost of the equipment is typically low in the case of a simple ultrasonic cleaning bath (frequencies range between 20 and 10 kHz), with the exception of high-power ultrasonic devices. Different nanomaterials can be synthesized by following this procedure. For example, Bozkurt [84] prepared Ag/graphene nanocomposites by a sonochemical method using silver nitrate and graphene oxide as precursors and sodium citrate as a green reducing agent. For this purpose, an aqueous solution of the precursors and sodium citrate was sonicated for 60 min and the nanocomposites obtained were finally dried at 60 °C. The mechanism responsible for the formation of the Ag/graphene nanocomposite was based on the dissociation of silver nitrate in the presence of graphene oxide by a sonochemical process. Ag particles were dispersed onto the surface of graphene oxide by ultrasonic irradiation, leading to an electrostatic interaction between the functional groups on the surface of the graphene oxide and the Ag^+ cations.

As has been previously mentioned, this method also allows organic nanomaterials to be obtained. For example, Siadatnasab *et al* [85] synthesized a novel inorganic–organic nanohybrid based on a copper dithiocarbamate by following a facile and single-step sonochemical method with a reaction time of around 30 min. A morphological analysis of the samples confirmed that this method allows nano-structures with homogeneous shapes and sizes to be obtained. This chemical method has been also employed to produce semiconductor oxides such as NiO [86] or ZnO [87], among others, yielding nanostructures with high crystallinity and a wide variety of morphologies.

2.3.2.5 Microwave-assisted methods

Microwave-assisted synthesis works by aligning dipoles of the materials in an external field via the excitation produced by microwave electromagnetic radiation. Stuerga and Galliard [88] concluded that, for chemical reactions, the energy of the microwave photon is insufficient to break chemical bonds and that the electric field strength cannot induce organization. Thus, the microwave heating effects are expected to be thermal effects only. Microwave energy is considered to be a more efficient source for selective heating in many processes. This source of energy has been employed in many recent chemical reaction studies and has been found to change the kinetics and selectivity of organic and inorganic synthesis. Moreover, this process is considered to be more environmentally friendly, requiring less energy than conventional processes. Microwave heating can provide different advantages in comparison to traditional heating for chemical synthesis [89], such as high heating rates, the absence of direct contact between the heating source and the reactants and/or solvents, excellent control of the reaction parameters, selective heating, higher yields, better selectivity, improved reproducibility and automatization, and high-throughput synthesis. Using this approach, various materials can be synthesized, from nanoporous oxides or metallic nanostructures to organic heterocycles. In the case of metallic nanostructures, Au, Ag, or Ni nanostructures [89] can be obtained via microwave-assisted processes, as shown in figure 2.6. The morphology and size of such nanostructures can be modulated not only by the molar ratio of the reactants but also by the physical parameters provided by the microwave reactor.

On the other hand, when metal oxide nanostructures are produced using a microwave method, this is usually accompanied by a hydrothermal method. Komarneni's group [90] was one of the pioneers of the microwave–hydrothermal synthesis of inorganic powders. In 1992 they reported the synthesis of TiO_2, ZrO_2, Fe_2O_3, $KNbO_3$, and $BaTiO_3$ using a combination of both techniques. They underlined that the short reaction times resulting from the use of microwave processes represented ideal conditions for scaling up the production of metal oxide nanostructures.

In summary, different modifications of chemical synthesis methods are being developed in order to realize the 'green chemistry' principles. As a function of the

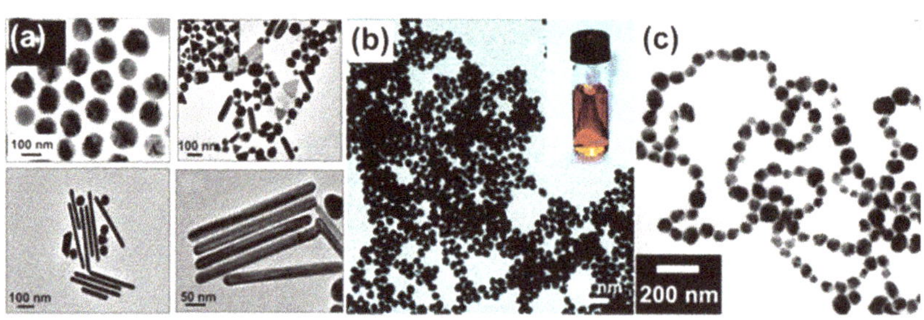

Figure 2.6. TEM images of various metallic nanostructures. (a) Au nanostructures with different morphologies, (b) Ag nanoparticles, and (c) Ni nanoparticles. Reproduced from [89] with permission of The Royal Society of Chemistry.

synthesis method, different materials with various compositions, shapes, and sizes can be obtained in a sustainable environment.

2.4 Fabrication techniques for sustainable thin layers

It is well known that material dimensions can create unique structural characteristics and improved chemical or physical properties. In fact, 2D nanomaterials have been widely exploited due to their optimized properties, and current research is focusing on their sustainability and adopting bio-inspired approaches [91]. A wide variety of materials can adopt 2D morphologies, such as metal oxides (MOs), graphene and its allotropes, transition-metal dichalcogenides (TMDCs), black phosphorus (BP) and organic polymers. The degree to which these materials can be regarded as sustainable depends on different considerations. As previously discussed, the use of so-called 'green synthesis' processes for the synthesis of these materials has the minimum environmental impact. In fact, in the previous section, we have discussed that fact that metal oxides are good candidates for sustainable materials due to their unique optoelectronic and electrochemical properties as well as their low toxicity. A wide variety of MOs have been synthesized in 2D, including both binary (e.g. TiO_2, SnO_2, and MnO) and ternary oxides [92], using similar techniques to those previously discussed (hydrothermal, CVD, etc). Graphene (currently a representative material in the materials science industry [92]) as well as its allotropes (graphene oxide, rGO) can be considered sustainable [93], especially when synthesized by low-impact synthesis methods [94]. As graphene attracts great interest, this topic has been covered in other works [95, 96], and in this chapter we will focus our attention on other potential sustainable materials. Regarding its synthesis routes (figure 2.7), graphene can be obtained via bottom-up approaches such as CVD (discussed in the previous section). However, the mechanical exfoliation of graphite via adhesive tips and atomic force microscopy (AFM) are easy processes by which to obtain graphene with minimum impact.

It is not only inorganic materials that can be used in 2D thin layers, but also organic materials. Organic materials are categorized as small molecules, and the most used are

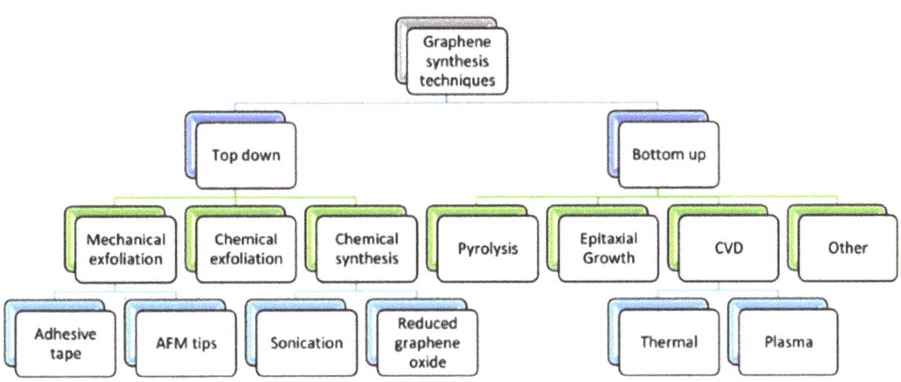

Figure 2.7. Graphene synthesis methods. Reproduced from [96]. CC BY 4.0.

polymers such as poly(3-hexylthiophene), polypropylene (PP), and polystyrene (PS), among others [97]. Polymers are fundamental for materials in our daily life; most of them are derived from petrochemicals (synthetics) [98], and their main disadvantage in terms of sustainability is the difficulty of their degradation; other polymers, however, can be obtained from nature. Natural biopolymers, obtained from biomass such as proteins (collagen) or polysaccharides (cellulose, chitosan), are gaining increasing attention, as they can be sourced from plants or animals [99], resulting in a reduced environmental impact. However, synthetic biopolymers such as polylactic acid (PLA) have achieved major importance in recent decades because of their use in current technologies such as food packaging, shopping bags, and 3D printing; thus, major efforts are being made to produce them via a more sustainable approach and to make them more easily disposable [100].

Combinations of the different materials previously described are called composites; these contain at least two constituents, often inorganic–inorganic or organic–inorganic; they offer superior mechanical and optoelectronic properties, as well as a more suitable balance between sustainability, low weight, and performance as compared to single-component materials, which can be observed in figure 2.8. Composites can be classified depending on their matrix material (metal, ceramic, or

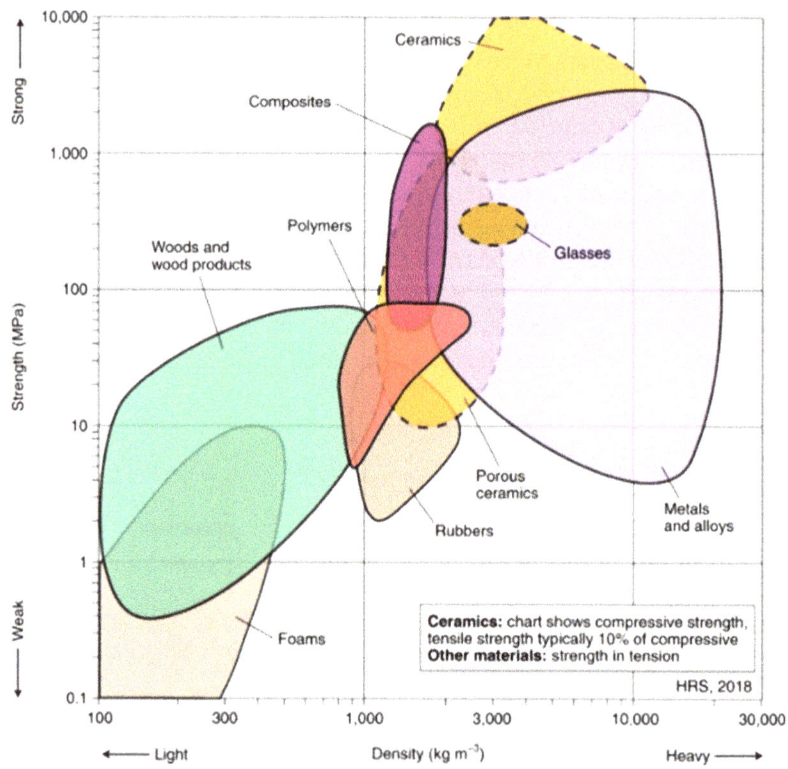

Figure 2.8. Comparison of properties such as the density and strength of different materials. Composites and polymers are perfect in-betweens [110] (2018) (Copyright © 2018, The Author(s)). With permission of Springer.

polymer) and at the same time, polymer-based materials can be composed either of biopolymers or synthetic polymers. Biopolymer-based composites, such as reinforced natural fibers, are suitable materials for aerospace engineering [101], the automobile industry [102], or construction, as they reduce weight and cost while reinforcing stability and strength. All green biocomposites represent a former objective which is still underdeveloped, but is gaining momentum [103]. Composites based on synthetic polymers, such as those composed of a conductive polymer, are more suitable for energy-related applications due to their superior optical, electrical, and mechanical properties [104]. They present certain limitations which can be overcome by combining them with a reinforcement material such as an MO; the resulting materials have found applications in several electronic and energy-related fields, such as sensors [105, 106], solar cells [107, 108], and energy storage [109].

It is clear that thin layers and 2D structures offer advantages due to their large surface-area-to-volume ratios; they exhibit unique physical, chemical, mechanical, electrical, optical, optoelectronic, and magnetic properties that cannot be provided by the bulk forms of these materials [97]. From an energy perspective, MOs, synthetic polymers, and hybrid composites present an interesting starting point for sustainability and efficiency. To help the reader understand the methods used to obtain these materials, this section presents the state of the art and discusses the synthesis and fabrication methods used for MOs, polymers, and hybrid composites with applications for energy. Some of these methods are schematically shown in figure 2.9.

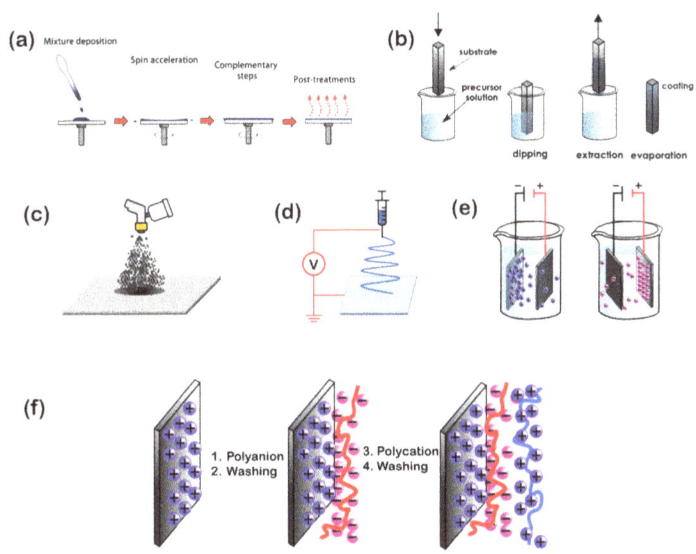

Figure 2.9. Various synthesis protocols used for the obtention of thin layers of MOs, polymers, or composites: (a) spin coating, (b) dip coating, (c) spray coating, (d) electrospinning, (e) cathodic electrophoretic deposition (EPD) and anodic EPD, and (f) layer-by-layer assembly.

2.4.1 Spin coating

Spin coating has been established as one of the predominant techniques for the easy production of uniform thin layers [111]. Its main advantage is the simplicity of the setup, which avoids the use of a high vacuum or large energy consumption. The basis of this technique is that a solution containing the material is deposited onto a spinning substrate. Different parameters are considered during the process, such as the spin speed, acceleration, and time, which lead to a certain degree of control over the final thickness and the appearance of the layer. Other parameters related to the mixture containing the target material, such as its viscosity and concentration, have to be controlled during the spin-coating process. The spin-coating process is depicted in figure 2.9(a). First, the solution is applied to the center of a substrate, which can be static or already spinning (often referred to as static and dynamic coating, respectively). The rotation is then accelerated to a high velocity (around 2000–6000 rpm) which, due to the action of the centrifugal force produced by the continuous rotation of the substrate, creates a thin layer; this process can be complemented by different steps with different spin and acceleration rates (depending on the desired outcome) and finalized by stopping the spinning of the substrate. Other post-treatments, such as thermal treatments, are often employed to remove the remaining solvent or moisture.

While this technique offers several advantages, mostly due its simplicity, it also has some drawbacks. The coating process starts with the solubilization and dissolution of the target material; therefore, a suitable solvent is needed, which can be toxic. Moreover, this technique is not suitable for large substrates, as they cannot be spun efficiently at a high rate. Nonetheless, spin coating can be employed without the use of any harmful or endangered material; it can be almost zero cost, as it uses a very small quantity of the solution, and it is sustainable. A great variety of materials, including organic, inorganic, and hybrid composites can be spin coated in the form of thin layers from tens to hundreds of nanometers.

To give some examples: spin coating is the one of the techniques most employed to obtain thin layers for many types of organic solar cell: dye-sensitized solar cells (DSSCs), perovskite solar cells (PSCs), and hybrid silicon solar cells [112]. In fact, water-soluble polymers such as poly(methyl methacrylate) (PMMA) [113] and poly (3,4-ethylenedioxythiophene) polystyrene sulfonate (PEDOT:PSS) [114] are good candidates for this technique, as they are easy to process and can be used to coat different substrates. This technique is also useful for the production of high-quality thin layers based on MOs (TiO_2, SnO_2) [115, 116], which employs a modified sol–gel process in order to achieve good liquid dispersion of the particles. Hybrid composites formed from combinations of MOs and polymers can be also deposited after the MOs are deposited on the polymer matrix [107, 108, 117]; this approach is particularly used for hybrid silicon solar cells, which are considered to be an alternative candidate for sustainable and efficient solar cells.

2.4.2 Dip coating

Dip coating is another fast, low-energy process used to deposit a thin film on a substrate after the substrate material has been introduced into a flask containing the solution of the coating material. The film's physical properties depend on parameters such as the viscosity, surface tension, and gravity. The dip-coating steps are depicted in figure 2.9(b): (1) immersion of the substrate in the coating material solution (for minutes or hours); (2) the pullout: the substrate is pulled out of the solution; thus, the remaining solution starts to precipitate (the thickness is controlled by the pull speed); (3) draining and evaporation: any remaining liquid is evaporated and the substrate is drained. The solution commonly contains volatile materials to boost drying. This method has often been used to obtain thin layers of polymer [118]. Using a modified approach based on sol–gel dip coating, MO layers can be obtained. This method is similar to sol–gel synthesis: the precursor conforms a colloidal solution named 'sol,' which gradually evolves into a gel-system containing a combination of liquid and solid phases, but in this case combined with dip coating.

This method had been recently employed to construct thin mesoporous layers of WO_3 [119] as photoanodes for water splitting and ceramic membranes of TiO_2/SiO_2 for the treatment of wastewater [120] as well as photoanodes [121] and photochromic devices made from WO_3 or TiO_2 [122].

2.4.3 Spray coating

Spray coating is a very useful deposition method for low-cost large-area deposition; it can easily be scaled. In this method (shown in figure 2.9(c)), an aerosol or liquid particles are introduced intro a closed funnel and are then ejected from a nozzle, commonly a spray gun, due to a carrier gas action which causes the liquid to be ejected. The different parameters that affect these coatings are factors such as the spread, the distance to the target, the carrier gas pressure, the nozzle angle, the flow rate to the nozzle, etc. The substrate also is important, as the liquid droplet contact angle and surface wettability are important factors in obtaining a good spread. This system is easily scaled and is used in industry. However, it presents drawbacks, as the dynamics of the gas and liquids, splashing, and other factors can affect the quality of the coating. One solution consists of fixing the nozzle to a rack or moving stage while applying a constant flow rate.

Spray coating can be used in different applications; for example, in perovskite solar cells, spray-coated microspheres allow high efficiencies of up to 17.08% to be reached [123]. Spray coating is an easy approach for reinforcement, as it allows easy surface modification, for instance, for modified chitosan.

2.4.4 Electrophoretic deposition

Another facile and green method for fabricating additive-free films is EPD. The underlying principle of EPD (figure 2.9(e)) is that charged particles suspended in a solution are deposited on a metal substrate due to the influence of a uniform electric field. This method allows the fabrication of dense and porous thin films at thicknesses from 1 μm to 50 μm. The ability of this method to control the thickness and to use different low-cost materials has increased the interest in EPD in recent decades [124]. This method has been used to form anodes such as SnO_2 for LIBs [125]. Ganaja *et al* [126] developed micrometer-thick films of $Li_4Ti_5O_{12}$ (LTO) as anodes for Li-ion storage, which had the advantages of high capacity and a long cycle life (over 10 000 cycles); they were also additive-free electrodes. Other materials such as $ZnFe_2O_4$ (ZFO)-based negative electrodes also showed promising results without any complex processing, with reversible specific capacities of 560 mAh g^{-1} [127].

2.4.5 Electrospinning

Electrospinning involves an electrohydrodynamic process in which a liquid droplet solution is extruded and elongated, forming fibers with aid of an electrostatic field, which are deposited over a substrate [128, 129] as shown in figure 2.9(d). As the jet is stretched into finer diameters, it solidifies quickly, leading to the deposition of solid fiber(s) on the grounded collector. An electrospinning setup consists of a high-voltage supply which generates the electrically-charged jet, a pipette or syringe which contains the solution and controls the diameter of the fibers (from nanometers to micrometers), and a substrate which collects the fibers. The power supply is connected to both the pipette and the substrate; therefore, when the electric field is applied, a charge is induced over the liquid. When the electric field is sufficiently strong, the liquid exits the needle. This method is simple and can be used for energy-related devices. Lui *et al* [130] summarized the application of electrospun one-dimensional (1D) nanomaterials, such as nanofibers (NFs), nanotubes (NTs), nanorods, and nanowires which are often employed in solar cells (SCs), fuel cells (FCs), LIBs, lithium–sulfur batteries (LSBs), sodium-ion batteries (SIBs), etc. However, these materials have drawbacks, as charged materials can facilitate this process, whereas uncharged polymers make it more difficult [118]. In addition, the outcome of this technique is dependent on several parameters, such as the flow rate and the distance between the needle tip and the substrate.

Electrospun fibers of metal oxides, such as SnO_2, ZnO, and TiO_2 have drawn increasing attention in different applications, such as DSSCs [130, 131], for which this technique can create photoanodes, counter electrodes, and the electrolyte. In LIBs and sodium batteries (NaBs), this technique is particularly useful, as it allows the synthesis of anodes, cathodes, electrolytes, and even separators [130, 132]. Almost all of the soluble polymers can generally be electrospun into nanofibers it the molecular weight is high enough; examples include poly(vinylidene fluoride) (PVDF), PLA, polymethacrylate (PMMA), PS, and polypyrrole (PPy) [130].

2.4.6 Layer-by-layer assembly

Layer-by-layer (LbL) deposition is a thin-film fabrication technique in which thin films are formed via the alternate deposition of layers of oppositely charged materials (polyanions/polycations), with washing steps in between depositions, as shown in figure 2.9(f).

This technique is particularly useful for ultrathin films (1–100 nm thick) [118]. The LbL technique can be used for the production of inorganic–organic assemblies [133] as well as the formation of metal–organic frameworks (MOFs) [134].

2.4.7 Other techniques

Several other techniques can be used to obtain thin layers of the described materials, such as doctor blade coating, drop casting, and screen printing. Doctor blade coating (DBC) (also known as tape casting) uses a well-mixed slurry made of a suspension of particles or additives which are deposited onto a substrate which has previously been fixed to a bench to determine the area of deposition. The paste formed by the slurry is then spread over the substrate using a glass rod. The thickness of the coating is related to the spreading process. DBC is considered to be a scalable method and is often used in the production of organic solar cells (OSCs) [135]. DBC is also employed to coat anode or cathode slurries in the production of Li-ion and solid-state batteries [136]. The underlying principle of drop casting is similar to that of spin coating: a mixture of the desired material is deposited on the substrate. However, in this case, either the mixture is left at room conditions or thermal or gas treatments are applied. Drawbacks, such as the formation of the so-called coffee ring [137] make the spin-coating technique dominant. However, DBC is still widely employed for some applications, such as the coating of electrodes. Drop casting has recently been used to coat the electrodes used in Li-ion batteries, for which $ZnFe_2O_4$ was used with rGO [138].

2.5 Questions and work assessments for the student

1. Describe the most challenging issues, in your opinion, among the 12 principles of green chemistry.
2. Indicate some advantages and drawbacks of the thermal evaporation methods in comparison with other physical methods for the synthesis of nanomaterials.
3. What are the main differences between the VS and VLS methods?
4. In your opinion, which properties of plant extracts support their use in the synthesis of metallic nanoparticles?
5. Apart from green chemistry, which chemical routes should permit higher stoichiometric control?
6. Compare the main advantages and drawbacks of two synthesis methods commonly used in the fabrication of sustainable thin layers.
7. List some technologies for which spin coating is considered to be a key synthesis method.
8. Select one metal oxide and search for information about the diverse synthesis routes used in the fabrication of its nanostructures with variable dimensions and morphologies, including 0D, 1D, and 2D structures.
9. Cite the most relevant weaknesses of the PLD and MBE techniques as compared to other physical synthesis routes.
10. Apart from plant extracts, search for information on the diverse biological precursors employed in the green synthesis of metal oxide nanoparticles.

References

[1] Gusev A I 2020 High-energy ball milling of nonstoichiometric compounds *Usp. Fiz. Nauk* **190** 371–95

[2] del Prado F, Andersen H F, Taeño M, Mæhlen J P, Ramírez-Castellanos J, Maestre D, Karazhanov S and Cremades A 2020 Comparative study of the implementation of tin and titanium oxide nanoparticles as electrodes materials in Li-ion batteries *Sci. Rep.* **10** 1–8

[3] Ulissi U, Ito S, Hosseini S M, Varzi A, Aihara Y and Passerini S 2018 High capacity all-solid-state lithium batteries enabled by pyrite-sulfur composites *Adv. Energy Mater* **8** 1–10

[4] Kahimbi H, Hong S B, Yang M H and Choi B G 2017 Simultaneous synthesis of NiO/reduced graphene oxide composites by ball milling using bulk Ni and graphite oxide for supercapacitor applications *J. Electroanal. Chem.* **786** 14–9

[5] Swihart M T 2003 Vapor-phase synthesis of nanoparticles *Curr. Opin. Colloid Interface Sci.* **8** 127–33

[6] Kaiser N 2002 Review of the fundamentals of thin-film growth *Appl. Opt* **41** 3053

[7] Huang H Y, Xiang Z, Ye W, Yun Guang Z, Yumeng S, Jen It W and Yang 2016 3D graphene supported MoO_2 for high performance binder-free lithium ion battery *Nanoscale* **6** 9839–45

[8] Gnanasekar P, Periyanagounder D, Nallathambi A, Subramani S, Palanisamy M and Kulandaivel J 2018 Promoter-free synthesis of monolayer MoS_2 by chemical vapour deposition *CrystEngComm* **20** 4249–57

[9] Kolasinski K W 2006 Catalytic growth of nanowires: vapor–liquid–solid, vapor–solid–solid, solution–liquid–solid and solid–liquid–solid growth *Curr. Opin. Solid State Mater. Sci.* **10** 182–91

[10] Alonso-Orts M, Sánchez A M, Nogales E and Méndez B 2020 Ge incorporation in gallium oxide nanostructures grown by thermal treatment *J. Mater. Sci.* **55** 11431–8

[11] Taeño M, Maestre D and Cremades A 2020 Formation of NiO/SnO_2 p–n heterostructures grown by a single-step process based on a vapor–solid method *Mater. Lett.* **271** 11–3

[12] Taeño M, Maestre D and Cremades A 2020 Fabrication and study of self-assembled NiO surface networks assisted by Sn doping *J. Alloys Compd* **827** 154172

[13] Zhu X, Ten Brink G H, De Graaf S, Kooi B J and Palasantzas G 2020 Gas-phase synthesis of tunable-size germanium nanocrystals by inert gas condensation *Chem. Mater* **32** 1627–35

[14] Xiang X, Xia F, Zhan L and Xie B 2015 Preparation of zinc nano structured particles from spent zinc manganese batteries by vacuum separation and inert gas condensation *Sep. Purif. Technol* **142** 227–33

[15] Zhan L, Xiang X, Xie B and Sun J 2016 A novel method of preparing highly dispersed spherical lead nanoparticles from solders of waste printed circuit boards *Chem. Eng.* J **303** 261–7

[16] Piqueras J and Hidalgo P 2021 Growth of metal oxide nanostructures by thermal oxidation of metals under influence of external electric fields and electric current flow *Phys. Status Solidi* A **218** 2100323

[17] Rodríguez B, Hidalgo P, Piqueras J and Méndez B 2020 Influence of an external electric field on the rapid synthesis of MoO_3 micro- and nanostructures by Joule heating of Mo wires *RSC Adv* **10** 11892–7

[18] Tang C, Liao X, Zhong W, Yu H and Liu Z 2017 Electric field assisted growth and field emission properties of thermally oxidized CuO nanowires *RSC Adv* **7** 6439–46

[19] Teoh W Y, Amal R and Mädler L 2010 Flame spray pyrolysis: an enabling technology for nanoparticles design and fabrication *Nanoscale* **2** 1324–47
[20] Liu G, Hall J, Nasiri N, Gengenbach T, Spiccia L, Cheah M H and Tricoli A 2015 Scalable synthesis of efficient water oxidation catalysts: insights into the activity of flame-made manganese oxide nanocrystals *ChemSusChem.* **8** 4162–71
[21] Liu G, Karuturi S K, Chen H, Spiccia L, Tan H H, Jagadish C, Wang D, Simonov A N and Tricoli A 2018 Tuning the morphology and structure of disordered hematite photoanodes for improved water oxidation: a physical and chemical synergistic approach *Nano Energy* **53** 745–52
[22] Lu Z, Raad R, Safaei F, Xi J, Liu Z and Foroughi J 2019 Carbon nanotube based fiber supercapacitor as wearable energy storage *Front. Mater* **6** 1–14
[23] Sun Y, Zhao C, Shen M, Pan Z and Liu X 2016 SnO_2 nanoparticles encapsulated by curved graphite layers as anode materials for Li-ion batteries with high performances *J. Alloys Compd* **683** 191–7
[24] Ashkarran A A and Mohammadi B 2015 ZnO nanoparticles decorated on graphene sheets through liquid arc discharge approach with enhanced photocatalytic performance under visible-light *Appl. Surf. Sci.* **342** 112–9
[25] Khamlich S, Nuru Z Y, Bello A, Fabiane M, Dangbegnon J K, Manyala N and Maaza M 2015 Pulsed laser deposited Cr_2O_3 nanostructured thin film on graphene as anode material for lithium-ion batteries *J. Alloys Compd* **637** 219–25
[26] Qiu Z, Gong H, Zheng G, Yuan S, Zhang H, Zhu X, Zhou H and Cao B 2017 Enhanced physical properties of pulsed laser deposited NiO films via annealing and lithium doping for improving perovskite solar cell efficiency *J. Mater. Chem.* C **5** 7084–94
[27] Wang H, Wu Y, Ma M, Dong S, Li Q, Du J, Zhang H and Xu Q 2019 Pulsed laser deposition of $CsPbBr_3$ films for application in perovskite solar cells *ACS Appl. Energy Mater* **2** 2305–12
[28] Tian G L, Zhao M Q, Yu D, Kong X Y, Huang J Q, Zhang Q and Wei F 2014 Nitrogen-doped graphene/carbon nanotube hybrids: *in situ* formation on bifunctional catalysts and their superior electrocatalytic activity for oxygen evolution/reduction reaction *Small* **10** 2251–9
[29] Tang C, Li B Q, Zhang Q, Zhu L, Wang H F, Le Shi J and Wei F 2016 CaO-templated growth of hierarchical porous graphene for high-power lithium–sulfur battery applications *Adv. Funct. Mater* **26** 577–85
[30] Qi G, Yang J, Bao R, Xia D, Cao M, Yang W, Yang M and Wei D 2017 Hierarchical graphene foam-based phase change materials with enhanced thermal conductivity and shape stability for efficient solar-to-thermal energy conversion and storage *Nano Res.* **10** 802–13
[31] Rioult M, Stanescu D, Fonda E, Barbier A and Magnan H 2016 Oxygen vacancies engineering of iron oxides films for solar water splitting *J. Phys. Chem.* C **120** 7482–90
[32] Zhuang J, Xu X, Peleckis G, Hao W, Dou S X and Du Y 2017 Silicene: a promising anode for lithium-ion batteries *Adv. Mater* **29** 1–8
[33] Asveld L 2019 Towards including social sustainability in green and sustainable chemistry *Curr. Opin. Green Sustain. Chem.* **19** 61–5
[34] To M H, Uisan K, Ok Y S, Pleissner D and Lin C S K 2019 Recent trends in green and sustainable chemistry: rethinking textile waste in a circular economy *Curr. Opin. Green Sustain. Chem.* **20** 1–10

[35] Rana A, Yadav K and Jagadevan S 2020 A comprehensive review on green synthesis of nature-inspired metal nanoparticles: mechanism, application and toxicity *J. Clean. Prod* **272** 122880
[36] Rawtani D, Rao P K and Hussain C M 2020 Recent advances in analytical, bioanalytical and miscellaneous applications of green nanomaterial *TrAC Trends Anal. Chem.* **133** 116109
[37] Waris A, Din M, Ali A, Ali M, Afridi S, Baset A and Ullah Khan A 2021 A comprehensive review of green synthesis of copper oxide nanoparticles and their diverse biomedical applications *Inorg. Chem. Commun* **123** 108369
[38] Anastas P T and Warner J C 1998 *Green Chemistry: Theory and Practice* (Oxford: Oxford University Press)
[39] Khalaj M, Kamali M, Costa M E V and Capela I 2020 Green synthesis of nanomaterials—a scientometric assessment *J. Clean. Prod* **267** 122036
[40] Dutta D and Das B M 2021 Scope of green nanotechnology towards amalgamation of green chemistry for cleaner environment: a review on synthesis and applications of green nanoparticles *Environ. Nanotechnol. Monit. Manag* **15** 100418
[41] Jahurul M H A, Zaidul I S M, Ghafoor K, Al-Juhaimi F Y, Nyam K L, Norulaini N A N, Sahena F and Omar A K M 2015 Mango (*Mangifera indica* L.) by-products and their valuable components: a review *Food Chem.* **183** 173–80
[42] Behzad F, Naghib S M, kouhbanani M A J, Tabatabaei S N, Zare Y and Rhee K Y 2021 An overview of the plant-mediated green synthesis of noble metal nanoparticles for antibacterial applications *J. Ind. Eng. Chem.* **94** 92–104
[43] Elemike E E, Onwudiwe D C, Nundkumar N, Singh M and Iyekowa O 2019 Green synthesis of Ag, Au and Ag–Au bimetallic nanoparticles using *Stigmaphyllon ovatum* leaf extract and their *in vitro* anticancer potential *Mater. Lett.* **243** 148–52
[44] Sundararajan B and Ranjitha Kumari B D 2017 Novel synthesis of gold nanoparticles using *Artemisia vulgaris* L. leaf extract and their efficacy of larvicidal activity against dengue fever vector *Aedes aegypti* L *J. Trace Elem. Med. Biol* **43** 187–96
[45] Murugan K, Benelli G, Panneerselvam C, Subramaniam J, Jeyalalitha T, Dinesh D, Nicoletti M, Hwang J S, Suresh U and Madhiyazhagan P 2015 *Cymbopogon citratus*-synthesized gold nanoparticles boost the predation efficiency of copepod Mesocyclops aspericornis against malaria and dengue mosquitoes *Exp. Parasitol* **153** 129–38
[46] Philip D 2010 Rapid green synthesis of spherical gold nanoparticles using *Mangifera indica* leaf *Spectrochim. Acta* A **77** 807–10
[47] Vimalraj S, Ashokkumar T and Saravanan S 2018 Biogenic gold nanoparticles synthesis mediated by *Mangifera indica* seed aqueous extracts exhibits antibacterial, anticancer and anti-angiogenic properties *Biomed. Pharmacother* **105** 440–8
[48] Al-Katib M, Al-Shahri Y and Al-Niemi A 2015 Biosynthesis of silver nanoparticles by cyanobacterium *Gloeocapsa* sp. *Int. J. Enhanc. Res. Sci. Technol. Eng* **4** 115–35 https://www.erpublications.com/sitepanel/download.php?filename=download/download_20_09_2015_17_09_16.pdf
[49] Shukla A K and Iravani S 2017 Metallic nanoparticles: green synthesis and spectroscopic characterization *Environ. Chem. Lett.* **15** 223–31
[50] Philip D 2011 *Mangifera indica* leaf-assisted biosynthesis of well-dispersed silver nanoparticles *Spectrochim. Acta* A **78** 327–31

[51] Ameen F, Srinivasan P, Selvankumar T, Kamala-Kannan S, Al Nadhari S, Almansob A, Dawoud T and Govarthanan M 2019 Phytosynthesis of silver nanoparticles using *Mangifera indica* flower extract as bioreductant and their broad-spectrum antibacterial activity *Bioorg. Chem.* **88** 102970

[52] Kambale E K, Nkanga C I, Mutonkole B P I, Bapolisi A M, Tassa D O, Liesse J M I, Krause R W M and Memvanga P B 2020 Green synthesis of antimicrobial silver nanoparticles using aqueous leaf extracts from three Congolese plant species (*Brillantaisia patula*, *Crossopteryx febrifuga* and *Senna siamea*) *Heliyon* **6** e04493

[53] Doughari J H 2012 Phytochemicals: extraction methods, basic structures and mode of action as potential chemotherapeutic agents *Phytochemicals – A Global Perspective of Their Role in Nutrition and Health* (London: IntechOpen Ltd.) 1–34

[54] Habtemariam A B 2020 Plant extract mediated synthesis of nickel oxide nanoparticles *Mater. Int* **2** 205–9

[55] Prasad C, Sreenivasulu K, Gangadhara S and Venkateswarlu P 2017 Bio inspired green synthesis of Ni/Fe_3O_4 magnetic nanoparticles using *Moringa oleifera* leaves extract: a magnetically recoverable catalyst for organic dye degradation in aqueous solution *J. Alloys Compd* **700** 252–8

[56] Derakhshan Z, Ferrante M, Tadi M, Ansari F, Heydari A, Hosseini M S, Conti G O and Sadrabad E K 2018 Antioxidant activity and total phenolic content of ethanolic extract of pomegranate peels, juice and seeds *Food Chem. Toxicol* **114** 108–11

[57] Sawalha S M S, Arráez-Román D, Segura-Carretero A and Fernández-Gutiérrez A 2009 Quantification of main phenolic compounds in sweet and bitter orange peel using CE-MS/MS *Food Chem.* **116** 567–74

[58] Al-Qasmi N, Al-Gethami W, Saleh D and Abuziad A 2020 A sustainable approach for the synthesis of metallic nanoparticles and its application as antimicrobial agents *J. Mater. Res. Technol* **9** 13036–42

[59] Gebretinsae H G, Tsegay M G and Nuru Z Y 2019 Biosynthesis of nickel oxide (NiO) nanoparticles from cactus plant extract *Mater. Today Proc.* **36** 566–70

[60] Devatha C P, Thalla A K and Katte S Y 2016 Green synthesis of iron nanoparticles using different leaf extracts for treatment of domestic waste water *J. Clean. Prod.* **139** 1425–35

[61] Tovar G I, Briceño S, Suarez J, Flores S and González G 2020 Biogenic synthesis of iron oxide nanoparticles using *Moringa oleifera* and chitosan and its evaluation on corn germination *Environ. Nanotechnol. Monit. Manag* **14** 100350

[62] Patiño-Ruiz D, Sánchez-Botero L, Tejeda-Benitez L, Hinestroza J and Herrera A 2020 Green synthesis of iron oxide nanoparticles using *Cymbopogon citratus* extract and sodium carbonate salt: nanotoxicological considerations for potential environmental applications *Environ. Nanotechnol. Monit. Manag* **14**

[63] Julio Jimenez L F, Herrera Barros A and Caamaño De Ávila Z I 2019 Synthesis and characterization of $Fe_{55}Co_{45}$ magnetic nanoparticles by polyol and green chemistry method *Results Phys.* **15** 102785

[64] Solano R A, Herrera A P, Maestre D and Cremades A 2019 Fe–TiO_2 nanoparticles synthesized by green chemistry for potential application in waste water photocatalytic treatment *J. Nanotechnol* **2019** 4571848

[65] Mallakpour S and Azadi E 2020 Sonochemical protocol for the organo-synthesis of TiO_2 and its hybrids: properties and applications *Green Sustainable Process for Chemical and Environmental Engineering and Science* (Amsterdam: Elsevier) 287–323

[66] Solano Pizarro R A and Herrera Barros A P 2020 Cypermethrin elimination using Fe–TiO_2 nanoparticles supported on coconut palm spathe in a solar flat plate photoreactor *Adv. Compos. Lett.* **28** 1–13
[67] Helmy E T, Abouellef E M, Soliman U A and Pan J H 2021 Novel green synthesis of S-doped TiO_2 nanoparticles using *Malva parviflora* plant extract and their photocatalytic, antimicrobial and antioxidant activities under sunlight illumination *Chemosphere* **271** 129524
[68] Rajendhiran R, Deivasigamani V, Palanisamy J, Masan S and Pitchaiya S 2021 *Terminalia catappa* and *Carissa carandas* assisted synthesis of TiO_2 nanoparticles—a green synthesis approach *Mater. Today Proc* **45** 2232–8
[69] Bonfante-Alvarez H, De Avila-Montiel G, Herrera-Barros A, Torrenegra-Alarcón M and González-Delgado Á D 2018 Evaluation of five chitosan production routes with astaxanthin recovery from shrimp exoskeletons *Chem. Eng. Trans* **70** 1969–74
[70] Aljebory A M, Alsalman T M, Aljebory A and Alsalman T M 2017 Chitosan nanoparticles: review article *Imp. J. Interdiscip. Res.* **3** 233–42 https://onlinejournal.in/IJIRV3I7/039.pdf
[71] Yu Z, Li B, Chu J and Zhang P 2018 Silica *in situ* enhanced PVA/chitosan biodegradable films for food packages *Carbohydr. Polym* **184** 214–20
[72] Kalaivani R, Maruthupandy M, Muneeswaran T, Hameedha Beevi A, Anand M, Ramakritinan C M and Kumaraguru A K 2018 Synthesis of chitosan mediated silver nanoparticles (Ag NPs) for potential antimicrobial applications *Front. Lab. Med* **2** 30–5
[73] Chouhan D and Mandal P 2021 Applications of chitosan and chitosan based metallic nanoparticles in agrosciences—a review *Int. J. Biol. Macromol* **166** 1554–69
[74] DeLong R K, Reynolds C M, Malcolm Y, Schaeffer A, Severs T and Wanekaya A 2010 Functionalized gold nanoparticles for the binding, stabilization, and delivery of therapeutic DNA, RNA, and other biological macromolecules *Nanotechnol. Sci. Appl* **3** 53–63
[75] Priya K, Vijayakumar M and Janani B 2020 Chitosan-mediated synthesis of biogenic silver nanoparticles (AgNPs), nanoparticle characterisation and *in vitro* assessment of anticancer activity in human hepatocellular carcinoma HepG2 cells *Int. J. Biol. Macromol.* **149** 844–52
[76] Yugay Y A, Usoltseva R V, Silant'ev V E, Egorova A E, Karabtsov A A, Kumeiko V V, Ermakova S P, Bulgakov V P and Shkryl Y N 2020 Synthesis of bioactive silver nanoparticles using alginate, fucoidan and laminaran from brown algae as a reducing and stabilizing agent *Carbohydr. Polym* **245** 116547
[77] Ocsoy I, Demirbas A, McLamore E S, Altinsoy B, Ildiz N and Baldemir A 2017 Green synthesis with incorporated hydrothermal approaches for silver nanoparticles formation and enhanced antimicrobial activity against bacterial and fungal pathogens *J. Mol. Liq* **238** 263–9
[78] Lam S M, Sin J C, Zuhairi Abdullah A and Rahman Mohamed A 2013 Green hydrothermal synthesis of ZnO nanotubes for photocatalytic degradation of methylparaben *Mater. Lett.* **93** 423–6
[79] Panahi-Kalamuei M, Amiri O and Salavati-Niasari M 2020 Green hydrothermal synthesis of high quality single and few layers graphene sheets by bread waste as precursor *J. Mater. Res. Technol* **9** 2679–90
[80] Taghavi Fardood S, Ramazani A and Woo Joo S 2018 Eco-friendly synthesis of magnesium oxide nanoparticles using Arabic gum *J. Appl. Chem. Res.* **12** 8–15

[81] Saikumari N, Preethi T, Abarna B and Rajarajeswari G R 2019 Ecofriendly, green tea extract directed sol–gel synthesis of nano titania for photocatalytic application *J. Mater. Sci., Mater. Electron* **30** 6820–31

[82] Nouri M, Azimi H R, Saray A M and Yousefi R 2017 S-doping effects on optical properties and highly enhanced photocatalytic performance of Cu_3Se_2 nanoparticles under solar-light irradiation *Ceram. Int* **43** 14983–8

[83] Zhu W, Zhang Q, Xiang L and Zhu S 2011 Green co-precipitation byproduct-assisted thermal conversion route to submicron $Mg_2B_2O_5$ whiskers *CrystEngComm* **13** 1654–63

[84] Bozkurt P A 2017 Sonochemical green synthesis of Ag/graphene nanocomposite *Ultrason. Sonochem* **35** 397–404

[85] Siadatnasab F, Farhadi S, Dusek M, Eigner V, Hoseini A A and Khataee A 2020 Sonochemical synthesis and structural characterization of an organic–inorganic nanohybrid based on a copper–dithiocarbamate complex and $PMo_{12}O_{40}{}^{3-}$ polyanion as a novel sonocatalyst *Ultrason. Sonochem* **64** 104727

[86] Duraisamy N, Numan A, Fatin S O, Ramesh K and Ramesh S 2016 Facile sonochemical synthesis of nanostructured NiO with different particle sizes and its electrochemical properties for supercapacitor application *J. Colloid Interface Sci.* **471** 136–44

[87] Bao Y, Gao L, Feng C, Ma J, Zhang W, Liu C and Simion D 2020 Sonochemical synthesis of flower-like ZnO assembled by hollow cones toward water vapor permeability and water resistance enhancement of waterborne film *J. Ind. Eng. Chem.* **82** 180–9

[88] Stuerga D A C and Gaillard P 1996 Microwave athermal effects in chemistry: a myth's autopsy: part I: historical background and fundamentals of wave-matter interaction *J. Microw. Power Electromagn. Energy* **21** 87–100

[89] Bilecka I and Niederberger M 2010 Microwave chemistry for inorganic nanomaterials synthesis *Nanoscale* **2** 1358–74

[90] Komarneni S, Roy R and Li Q H 1992 Microwave-hydrothermal synthesis of ceramic powders *Mater. Res. Bull* **27** 1393–405

[91] Zhang Y, Mei J, Yan C, Liao T, Bell J and Sun Z 2020 Bioinspired 2D nanomaterials for sustainable applications *Adv. Mater* **32** 1902806

[92] Guo T, Yao M S, Lin Y H and Nan C W 2015 A comprehensive review on synthesis methods for transition-metal oxide nanostructures *CrystEngComm* **17** 3551–85

[93] Titirici M M, White R J, Brun N, Budarin V L, Su D S, Del Monte F, Clark J H and MacLachlan M J 2015 Sustainable carbon materials *Chem. Soc. Rev.* **44** 250–90

[94] Kartick B, Srivastava S K and Srivastava I 2013 Green synthesis of graphene *J. Nanosci. Nanotechnol* **13** 4320–4

[95] Lee X J, Hiew B Y Z, Lai K C, Lee L Y, Gan S, Thangalazhy-Gopakumar S and Rigby S 2019 Review on graphene and its derivatives: synthesis methods and potential industrial implementation *J. Taiwan Inst. Chem. Eng* **98** 163–80

[96] Bhuyan M S A, Uddin M N, Islam M M, Bipasha F A and Hossain S S 2016 Synthesis of graphene *Int. Nano Lett.* **6** 65–83

[97] Neupane G P, Ma W, Yildirim T, Tang Y, Zhang L and Lu Y 2019 2D organic semiconductors, the future of green nanotechnology *Nano Mater. Sci.* **1** 246–59

[98] Zhu Y, Romain C and Williams C K 2016 Sustainable polymers from renewable resources *Nature* **540** 354–62

[99] Wang Z, Ganewatta M S and Tang C 2020 Sustainable polymers from biomass: bridging chemistry with materials and processing *Prog. Polym. Sci.* **101** 101197

[100] Thakur S, Chaudhary J, Sharma B, Verma A, Tamulevicius S and Thakur V K 2018 Sustainability of bioplastics: opportunities and challenges *Curr. Opin. Green Sustain. Chem.* **13** 68–75

[101] Puttegowda M, Rangappa S M, Jawaid M, Shivanna P, Basavegowda Y and Saba N 2018 Potential of natural/synthetic hybrid composites for aerospace applications *Sustainable Composites for Aerospace Applications* (Amsterdam: Elsevier) pp 315–51

[102] Ravishankar B, Nayak S K and Kader M A 2019 Hybrid composites for automotive applications—a review *J. Reinf. Plast. Compos* **38** 835–45

[103] Mohanty A K, Vivekanandhan S, Pin J M and Misra M 2018 Composites from renewable and sustainable resources: challenges and innovations *Science* **362** 536–42

[104] Namsheer K and Rout C S 2021 Conducting polymers: a comprehensive review on recent advances in synthesis, properties and applications *RSC Adv* **11** 5659–97

[105] Wang S, Kang Y, Wang L, Zhang H, Wang Y and Wang Y 2013 Organic/inorganic hybrid sensors: a review *Sensors Actuators* B **182** 467–81

[106] Vázquez-López A, Bartolomé J, Maestre D and Cremades A 2022 Gas sensing and thermoelectric properties of hybrid composite films based on PEDOT:PSS and SnO or SnO_2 nanostructures *Phys. Status Solidi* **219** 2100794

[107] Vázquez-López A *et al* 2021 Hybrid materials and nanoparticles for hybrid silicon solar cells and Li-ion batteries *J. Energy Power Technol* **3** 020

[108] Vázquez-López A, Yaseen A, Maestre D, Ramírez-Castellanos J, Karazhanov S Z, Marstein E S and Cremades A 2020 Improved silicon surface passivation by hybrid composites formed by PEDOT:PSS with anatase TiO_2 nanoparticles *Mater. Lett.* **271** 127802

[109] Louis H and Magu T O 2019 A review on conducting polymers-based composites for energy storage application *J. Chem. Rev.* **1** 19–34

[110] Babbitt C W and Moore E A 2018 Sustainable nanomaterials by design *Nat. Nanotechnol.* **13** 621–3

[111] Sahu N, Parija B and Panigrahi S 2009 Fundamental understanding and modeling of spin coating process: a review *Indian J. Phys.* **83** 493–502

[112] Khang D-Y 2019 Recent progress in Si-PEDOT:PSS inorganic–organic hybrid solar cells *J. Phys. D: Appl. Phys.* **52** 503002

[113] Shekar B C, Sathish S and Sengoden R 2013 Spin coated nano scale PMMA films for organic thin film transistors *Phys. Procedia* **49** 145–57

[114] Hu L, Song J, Yin X, Su Z and Li Z 2020 Research progress on polymer solar cells based on PEDOT:PSS electrodes *Polymers (Basel)* **12** 145

[115] Jiang Q, Zhang X, You J, Jiang Q, Zhang X and You J 2018 SnO_2: a wonderful electron transport layer for perovskite solar cells *Small* **14** 1801154

[116] Johari N D, Rosli Z M, Juoi J M and Yazid S A 2019 Comparison on the TiO_2 crystalline phases deposited via dip and spin coating using green sol–gel route *J. Mater. Res. Technol* **8** 2350–8

[117] García-Fernández J, García-Carrión M, Torres-Pardo A, Martínez-Casado R, Ramírez-Castellanos J, Nogales E, González-Calbet J and Méndez B 2020 New insights into the luminescence properties of a Na stabilized Ga–Ti oxide homologous series *J. Mater. Chem.* C **8** 2725–31

[118] Song J, Winkeljann B and Lieleg O 2020 Biopolymer-based coatings: promising strategies to improve the biocompatibility and functionality of materials used in biomedical engineering *Adv. Mater. Interfaces* **7** 2000850

[119] Hilliard S, Baldinozzi G, Friedrich D, Kressman S, Strub H, Artero V and Laberty-Robert C 2017 Mesoporous thin film WO_3 photoanode for photoelectrochemical water splitting: a sol–gel dip coating approach *Sustain. Energy Fuels* **1** 145–53

[120] Marzouk S S, Banat F and Hasan S W 2020 Preparation of TiO_2/SiO_2 ceramic membranes via solgel dip coating for the treatment of produced wastewater *Frontiers in Water-Energy-Nexus—Nature-Based Solutions, Advanced Technologies and Best Practices for Environmental SustainabilityAdvances in Science, Technology & Innovation* (Cham: Springer) 211–4 pp

[121] Kirchberg K and Marschall R 2019 Sol–gel synthesis of mesoporous $CaFe_2O_4$ photocathodes with hierarchical pore morphology *Sustain. Energy Fuels* **3** 1150–3

[122] Wu L, Yang D, Fei L, Huang Y, Wu F, Sun Y, Shi J and Xiang Y 2017 Dip-coating process engineering and performance optimization for three-state electrochromic devices *Nanoscale Res. Lett.* **12** 390

[123] Fan X, Rui Y, Han X, Yang J, Wang Y and Zhang Q 2020 Spray-coated monodispersed SnO_2 microsphere films as scaffold layers for efficient mesoscopic perovskite solar cells *J. Power Sources* **448** 227405

[124] Besra L and Liu M 2007 A review on fundamentals and applications of electrophoretic deposition (EPD) *Prog. Mater Sci.* **52** 1–61

[125] Ui K, Kawamura S and Kumagai N 2012 Fabrication of binder-free SnO_2 nanoparticle electrode for lithium secondary batteries by electrophoretic deposition method *Electrochim. Acta* **76** 383–8

[126] Gangaja B, Muralidharan H P, Nair S and Santhanagopalan D 2018 Ultralong (10K) cycle-life and high-power Li-ion storage in $Li_4Ti_5O_{12}$ films developed via sustainable electrophoretic deposition process *ACS Sustain. Chem. Eng.* **6** 4705–10

[127] Das D, Mitra A, Jena S, Majumder S B and Basu R N 2018 Electrophoretically deposited $ZnFe_2O_4$-carbon black porous film as a superior negative electrode for lithium-ion battery *ACS Sustain. Chem. Eng* **6** 17000–10

[128] Liu J, Zhang F, Hou L, Li S, Gao Y, Xin Z, Li Q, Xie S, Wang N and Zhao Y 2020 Synergistic engineering of 1D electrospun nanofibers and 2D nanosheets for sustainable applications *Sustain. Mater. Technol* **26** e00214

[129] Xue J, Wu T, Dai Y and Xia Y 2019 Electrospinning and electrospun nanofibers: methods, materials, and applications *Chem. Rev.* **119** 5298–415

[130] Liu Q, Zhu J, Zhang L and Qiu Y 2018 Recent advances in energy materials by electrospinning *Renew. Sustain. Energy Rev.* **81** 1825–58

[131] Wan T, Ramakrishna S and Liu Y 2018 Recent progress in electrospinning TiO_2 nanostructured photo-anode of dye-sensitized solar cells *J. Appl. Polym. Sci.* **135** 45649

[132] Wang L, Yang G, Peng S, Wang J, Yan W and Ramakrishna S 2020 One-dimensional nanomaterials toward electrochemical sodium-ion storage applications via electrospinning *Energy Storage Mater* **25** 443–76

[133] Richardson J J, Cui J, Björnmalm M, Braunger J A, Ejima H and Caruso F 2016 Innovation in layer-by-layer assembly *Chem. Rev.* **116** 14828–67

[134] Wang Z, Wöll C, Wang Z and Wöll C 2019 Fabrication of metal–organic framework thin films using programmed layer-by-layer assembly techniques *Adv. Mater. Technol.* **4** 1800413

[135] Ji G, Zhao W, Wei J, Yan L, Han Y, Luo Q, Yang S, Hou J and Ma C Q 2019 12.88% Efficiency in doctor-blade coated organic solar cells through optimizing the surface morphology of a ZnO cathode buffer layer *J. Mater. Chem.* A **7** 212–20

[136] Gurung A, Pokharel J, Baniya A, Pathak R, Chen K, Lamsal B S, Ghimire N, Zhang W H, Zhou Y and Qiao Q 2019 A review on strategies addressing interface incompatibilities in inorganic all-solid-state lithium batteries *Sustain. Energy Fuels* **3** 3279–309
[137] Kaliyaraj Selva Kumar A, Zhang Y, Li D and Compton R G 2020 A mini-review: how reliable is the drop casting technique? *Electrochem. Commun* **121** 106867
[138] Askari M B, Salarizadeh P, Seifi M, Ramezan zadeh M H and Di Bartolomeo A 2021 $ZnFe_2O_4$ nanorods on reduced graphene oxide as advanced supercapacitor electrodes *J. Alloys Compd* **860** 158497

Chapter 3

The fundamental properties of energy-technology-related nanomaterials

Francisco Domínguez-Adame, Bianchi Méndez and María Taeño

3.1 Introduction

Materials are ubiquitous in energy-related technologies, as they are involved in either a passive or an active way. The advances in materials research have significantly contributed to the achievement of scientific goals related to efficient energy management, in which advanced nanomaterials indeed played a crucial role. Herein, we will focus on the understanding of the physical properties of materials at the nanoscale that has contributed to improved designs and devices offering higher efficiencies and/or lower costs.

Sustainable energy generation, conversion, storage, or saving requires a fundamental knowledge of the properties of the so-called active materials and the nanostructures based on them. Some of the areas of interest are: photovoltaics, batteries, supercapacitors, fuel cells, thermoelectrics, efficient lighting, and hydrogen technologies. Although each particular device or application has particular needs, in most of them it is necessary to suitably manage the movement of carriers through the device (including the carrier generation and scattering processes) in order to get the desired performance. In other cases, light–matter interactions are also of special relevance to this purpose.

In this chapter, we present a glimpse of the fundamental physical properties of the nanomaterials involved in energy-related devices. The basics of the relationship between the crystalline structure and the electronic band structure are explained, including the roles of structural defects and impurities in the electronic and optical properties. One of the advantages of nanomaterials is their high aspect ratio, which makes their surface properties very important. In many energy applications, the active processes take place at the surface, so a brief review of surfaces and interfaces is also included. The particular features of two families of relevant materials, namely semiconducting oxides and graphene and layered materials, are then briefly presented.

doi:10.1088/978-0-7503-3531-7ch3

© IOP Publishing Ltd 2023

3.2 Physical properties relevant to energy applications

3.2.1 Bonding, crystal structure, and energy bands

Atoms combine together to form solids because the free energy is reduced for some ranges of temperature and pressure. In many situations, the lowest free-energy state is reached when atoms are arranged periodically in space to shape *crystalline solids*. Such a periodic ordering of atoms gives rise to emergent crystalline symmetries that greatly facilitate the study of the physical and chemical properties of solids [1–3]. Individual atoms are held together in solids by attractive and repulsive interatomic forces that depend on the electronic structure of the atoms concerned. From a fundamental point of view, forces in solids are basically electrostatic in origin, and therefore quantum mechanics plays a major role in our understanding of crystal binding. In this regard, when solids are compressed and the interatomic distance is thus reduced, the electron clouds of neighboring atoms begin to overlap. However, according to the Pauli principle, the electrons of one atom cannot occupy the orbitals of other atoms that are already occupied. Therefore, the only way to bring atoms together is to excite electrons to higher, unoccupied orbitals, increasing the energy of the crystal. This results in short-range repulsive interatomic forces that prevent atoms from being in close proximity.

A complete discussion of crystal binding is outside the scope of the chapter, and we will restrict ourselves to briefly presenting a classification of solids according to their binding forces. Broadly speaking, there are four classes of solids, according to the attractive forces between constituent atoms:

- *Molecular solids*. Chemically inert atoms, such as Ne and Ar, are known to condense to form crystalline solids at very low temperatures (below 24 K and 84 K for Ne and Ar, respectively). Such solids are characterized by low cohesive energies (20 meV/at and 80 meV/at for Ne and Ar, respectively). Chemically inert molecules can also bind together to form a solid. The attractive interaction between chemically inert atoms and molecules is described by the van der Waals force. Correlations in the quantum fluctuations of the charge distributions on atoms and molecules induce transitory dipole moments and give rise to weak short-range attractive forces (see, for example, references [2, 3]).
- *Ionic solids*. Group I or II elements are electropositive and can easily transfer outer electrons to group VI or VII electronegative elements, making both more stable due to their closed-shell electronic configurations. Consequently, atoms become ionized and can bind together to form crystalline solids (ionic bonding). Common examples of ionic crystals are the alkali halides LiF, NaCl, KBr, and some oxides (MgO, CuO, and CrO_2). As an example, consider the electronic configuration of Na and Cl atoms, namely [Na]=[Ne] $3s^1$ and [Cl]=[Ne] $3s^2$ $3p^5$. After one electron is transferred from the Na atom to the Cl atom, the closed-shell electronic configurations of the resulting ions are $[Na^+]$=[Ne] and $[Cl^-]$=[Ar]. The spherically symmetric charge distributions of these ions are only slightly distorted when they are brought together

to form a crystal. Hence, the point charge approximation works nicely and the calculation of the Coulomb potential energy is straightforward [4]. The cohesive energy and melting temperature are high (6.4 eV and 1073 K for NaCl), corresponding to strong bonding.

- *Covalent solids.* The energy required to reduce Group IV elements, among others, to a closed-shell electronic configuration is too large, and ionic bonding is therefore unlikely [5]. However, attractive interatomic interactions can still occur through electron sharing between neighboring atoms (covalent bonding). As an example, consider the electronic configuration of a C atom in the ground state [C]=$1s^2 2s^2 2p^2$ (Hund's rule). An excited state is obtained by promoting an electron from the $2s^2$ orbital to the empty $2p^2$ orbital, as shown in figure 3.1(a). The angular dependences of the wave functions concerned are given by $s(\theta, \phi) = \sqrt{1/4\pi}$, $p_x(\theta, \phi) = \sqrt{3/4\pi} \sin\theta \cos\phi$, $p_y(\theta, \phi) = \sqrt{3/4\pi} \sin\theta \sin\phi$ and $p_z(\theta, \phi) = \sqrt{3/4\pi} \cos\theta$. Linus Pauling argued that a convenient way of representing the electron states in the excited configuration is the following linear combination of atomic orbitals (four equivalent *hybrid orbitals*) [6]:

$$\begin{aligned} sp_1^3 &= \left(s + p_x + p_y + p_z\right)/2, \quad & sp_2^3 &= \left(s + p_x - p_y - p_z\right)/2, \\ sp_3^3 &= \left(s - p_x + p_y - p_z\right)/2, \quad & sp_4^3 &= \left(s - p_x - p_y + p_z\right)/2. \end{aligned} \tag{3.1}$$

These linear combinations are referred to as sp^3 orbitals and are directed towards the vertices of a tetrahedron, as shown in figure 3.1(b). The angle between half-filled sp^3 orbitals is 109.5°. The energy required to push an electron to a higher orbital in the excited state of the C atom is balanced by the energy released when an sp^3 orbital is filled by a second electron from another atom. Therefore, atoms can form covalent bonds by sharing electrons.

- *Metallic solids.* Metallic bonding can be viewed as the sharing of electrons among positively charged ions. As an illustrative example, let us consider the electronic configuration of Na atoms, [Na]=$1s^2\ 2s^2\ 2p^6\ 3s^1$=[Ne] $3s^2$. Core electrons (e.g. 2p electrons) are tightly bound to the atom and spatially localized, as shown in figure 3.2. In contrast, 3s valence electrons are only

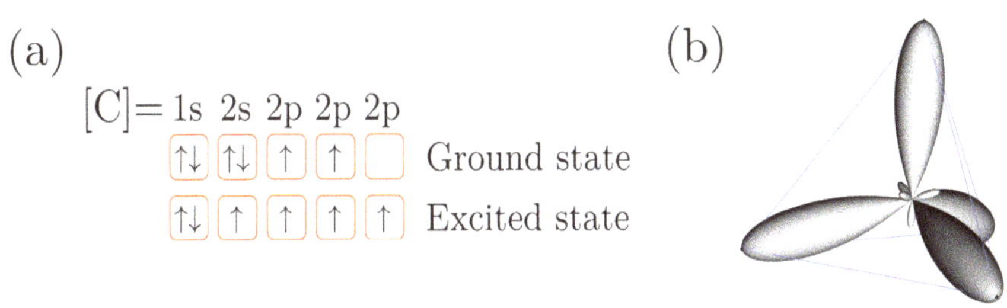

Figure 3.1. (a) Electronic configuration of carbon in the ground state and in an excited state. (b) Four sp^3 orbitals resulting from the hybridization of an s orbital and three p orbitals pointing towards the four vertices of a tetrahedron.

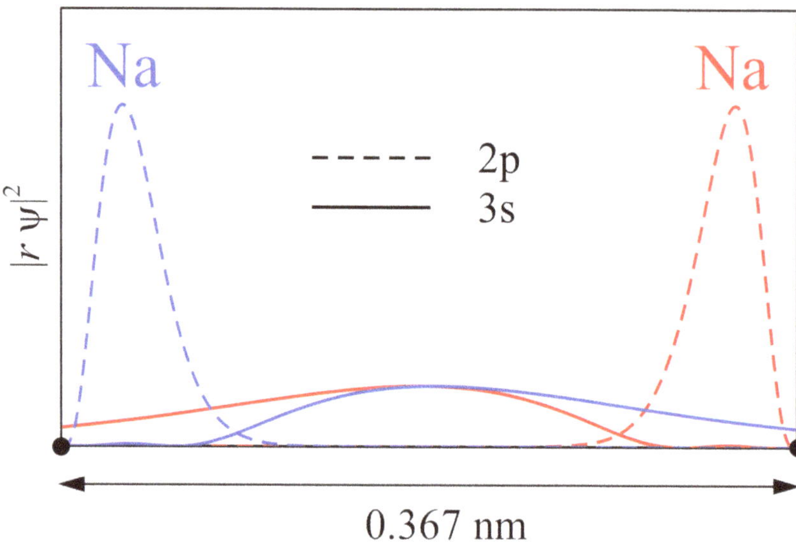

Figure 3.2. Probability densities of the 2p (dashed lines) and 3s (solid lines) electrons of two Na atoms separated by the internuclear distance of the nearest neighboring atoms in the solid metal. The 2p electrons are strongly localized around the atomic nuclei, but the spatial extent of the 3s electron wave function is larger than the internuclear distance.

loosely bound, and the corresponding wave function extends beyond the internuclear distance of the nearest neighboring Na atoms in the solid metal (0.367 nm), as seen in figure 3.2. Consequently, valence electrons move freely throughout the space between the atomic nuclei, and the free-electron model is a reasonable description of their dynamics. The delocalization of valence electrons leads to a decrease in both kinetic and potential energies, and this energy decrease is responsible for metallic bonding [5].

The type of bond determines to a large extent how atoms arrange periodically to form crystals. For instance, the metallic bond is nondirectional, resulting in compact structures. On the other hand, the covalent bond is highly directional, and atoms bind together while preserving the symmetries of the hybrid orbitals. An ideal crystal is constructed by periodically repeating an identical group of atoms, called the basis [4]. The lattice is a periodic set of mathematical points to which the basis is attached. The lattice is generated by three linearly independent vectors $\boldsymbol{a}_1$, $\boldsymbol{a}_2$, and $\boldsymbol{a}_3$, known as primitive vectors. The lattice points are given by $\boldsymbol{R} = n_1\boldsymbol{a}_1 + n_2\boldsymbol{a}_2 + n_3\boldsymbol{a}_3$, where n_i are arbitrary integers. Thus, it is clear that infinite lattices are invariant under discrete translations of the form $\boldsymbol{R} \rightarrow \boldsymbol{R} + \boldsymbol{R}'$. In addition to their translational symmetry, lattices have point symmetries and, as a consequence, only 14 different lattice types can be defined in three dimensions, known as the *Bravais lattices* (see figure 3.3).

The existence of the abovementioned discrete translational symmetry has profound consequences for the electronic states in crystals. Felix Bloch demonstrated that the single-electron wave function must be of the form [7]

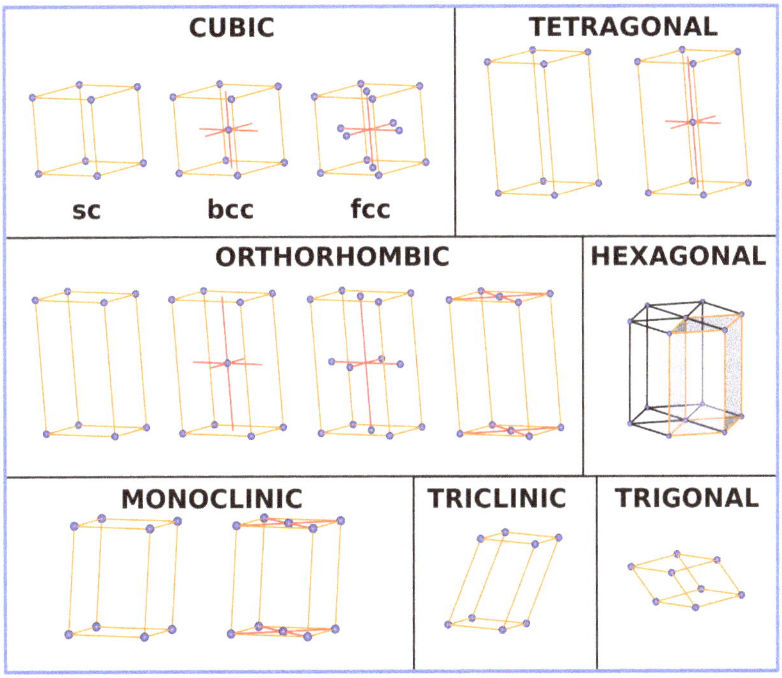

Figure 3.3. The unit cells of the 14 Bravais lattices. Lattice points are indicated by dots. Cubic lattices are named *simple cubic* (sc), *body-centered cubic* (bcc), and *face-centered cubic* (fcc).

$$\psi_k(\boldsymbol{r}) = \mathrm{e}^{i\boldsymbol{k}\cdot\boldsymbol{r}} u_k(\boldsymbol{r}). \tag{3.2}$$

Here, $u_k(\boldsymbol{r})$ is known as a *Bloch function* and has the same periodicity as the underlying lattice $u_k(\boldsymbol{r}) = u_k(\boldsymbol{r} + \boldsymbol{R})$. The wave vector $\boldsymbol{k}$ is a conserved quantity that characterizes the electronic state with energy E_k (the band structure). It is customary to restrict it to the so-called *Brillouin zone* (the Wigner–Seitz size of the reciprocal space). In this way, E_k becomes a multivalued function, each branch of which is called an *energy band.* Bands are separated by gaps, where the density of states (DOS) vanishes in the ideal crystal, i.e. in the absence of defects or surfaces. As an example, figure 3.4 displays the band structure of Ge, which has an fcc lattice with a basis of two Ge atoms, one at (0, 0, 0) and the other at (1/4, 1/4, 1/4) in units of the cube edge (sp^3 hybridization). The magnitude of the bandgap of Ge is $E_G = 0.66$eV at room temperature.

In a typical semiconductor, such as Ge, Si, or GaAs, the band structures of the carriers (electrons and holes) can generally be approximated by quadratic forms [2]. In the case of isotropic bands,

$$\begin{aligned} E_k &\simeq E_c + \frac{\hbar^2}{2m_e^*}(\boldsymbol{k} - \boldsymbol{k}_c)^2, \quad \text{(electrons)},\\ E_k &\simeq E_v - \frac{\hbar^2}{2m_h^*}(\boldsymbol{k} - \boldsymbol{k}_v)^2, \quad \text{(holes)}. \end{aligned} \tag{3.3}$$

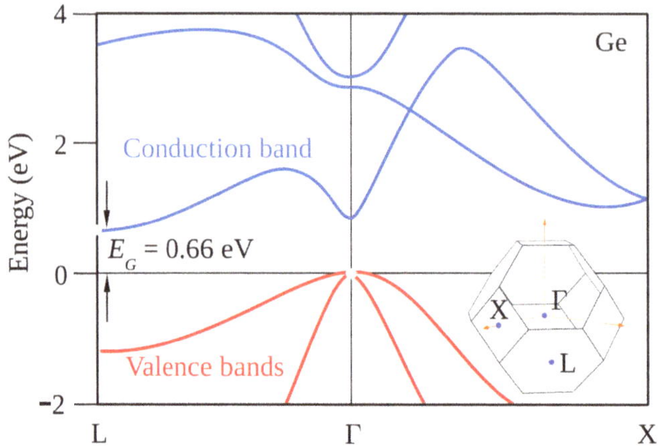

Figure 3.4. Band structure of Ge. The small circles mark the conduction band minimum and the valence band maximum, corresponding to an indirect-gap semiconductor. The inset shows the Brillouin zone along with high-symmetry points.

Here, E_c and E_v are the conduction and valence band edges, which occur at points $\boldsymbol{k}_c$ and $\boldsymbol{k}_v$ of the Brillouin zone, respectively, and m_e^* and m_h^* are the effective masses of electrons and holes. Since free electrons and holes have an energy close to the band edges, where the energy versus the wave vector takes the form shown in (3.3), they behave as free particles except for their effective masses of m_e^* or m_h^*. In the case of anisotropic bands, the effective mass turns out to be a second-rank tensor.

From the expansion (3.3) it can readily be concluded that electrons in the conduction band and holes in the valence band of semiconductors behave as free particles with effective masses of m_e^* and m_h^*. This approximation makes the calculation of spectral and transport properties particularly straightforward in the so-called *restricted geometries*, such as quantum wells, superlattices, quantum wires, and quantum dots [8]. As an example, figure 3.5 shows the DOSs in the conduction bands of bulk semiconductors, quantum wells, quantum wires, and quantum dots.

The spatial confinement of quantum particles, such as electrons or holes, can be studied by means of the steady-state Schrödinger equation, which for a particle of mass m and wave function $\psi(x)$, with a potential energy $U(x)$, in one-dimensional space is:

$$\frac{d^2\psi(x)}{dx^2} + k^2\psi(x) = 0, \tag{3.4}$$

where $k^2 = 2m(E - U)/\hbar^2$. As described in introductory quantum mechanics books, two major consequences of spatial confinement are the discretization of the energy levels in the system and the tunneling through potential barriers. These behaviors are quite relevant in the case of the restricted geometries displayed in figure 3.5 and affect the propagation of electrons through those simple nanostructures and also more complex nanostructures, such as superlattices or nanomaterials with mixed dimensionality.

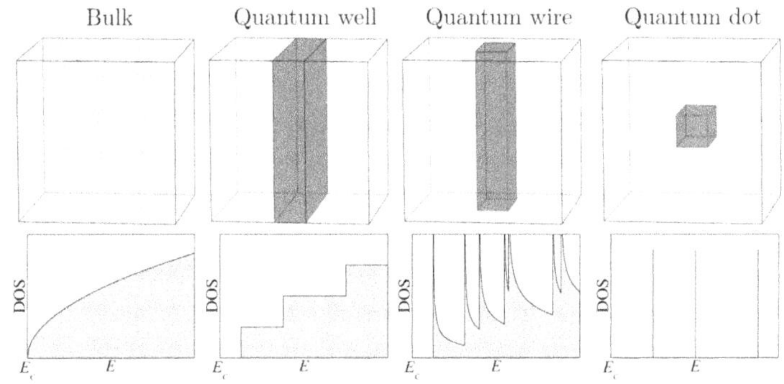

Figure 3.5. Schematic of the densities of states in bulk semiconductors, quantum wells, quantum wires, and quantum dots. E_C stands for the minimum energy of the conduction band. In nanostructures, the DOS vanishes well above E_C due to quantum confinement effects.

In the case of semiconductors, the electronic band structure in solids also includes electronic states related to the presence of crystalline defects, such as vacancies, interstitial atoms, and even impurities. Unlike electrons in the conduction band or holes in the valence band, these electronic states linked to defects are localized, which hinders carrier mobility in devices.

3.2.2 Electric conductivity

A material's crystal structure and the nature of its atomic/ionic bonding determine the way that charges can travel across it. Electric charges in solids arise from electrons and also from ions in the lattice. The movement of these charges in response to an electric field provokes an electric current, being the conductivity σ the link factor between them: $\boldsymbol{j} = \sigma\mathcal{E}$. The intriguing features of both electric and ionic conductivity are quite relevant to energy-related devices, since they affect their performance and efficiency, as occurs, for example, in solar cells or batteries.

- Electronic conductivity

In 1900, Paul Drude proposed a classical model of electron conduction in metals [9, 10]. Assuming that electrons are scattered by relatively immobile positive ions, Drude found a linear relationship between the current density $\boldsymbol{j}$ and the applied electric field $\mathcal{E}$:

$$\boldsymbol{j} = \frac{ne^2\tau}{m}\,\mathcal{E} \equiv \sigma_e\mathcal{E}, \tag{3.5}$$

where n is the number of electrons per unit volume, e is the elementary charge, τ is the momentum relaxation time (related to the mean time between collisions [11]), and σ_e is the electronic conductivity. Therefore, Drude provided a microscopic explanation of Ohm's law. However, Bloch's theorem (3.2) helped us to understand that immobile ions cannot scatter electrons in metals, but that the collective motion of ions (phonons), defects, and surfaces are responsible for the existence of finite relaxation times. Shortly after the advent of quantum mechanics, Arnold Sommerfeld [12] provided the first quantum mechanical treatment of the electrical conductivity of metals. Using Fermi–Dirac statistics, Sommerfeld concluded that the mean velocity of free electrons in metals, even at $T = 0\mathrm{K}$, is of the order of the Fermi velocity v_F. Therefore, the *mean free path* $\ell = v_F\tau$ gives an estimation of the mean distance between electron collisions with phonons, defects, or surfaces.

Many energy-related devices are based on p–n junctions that place p-type and n-type semiconductors in close contact, in such a way that an abrupt interface separates a region of hole majority from another of electron majority. This induces a built-in energy barrier due to the diffusion of carriers and its effect is spread over the so-called *depletion region* around the interface.

In the case of transport through a nanostructure of size L, it may happen that $\ell > L$. One can assume that charge carriers move through the active region without scattering in the so-called ballistic transport regime (see, for example, reference [13]). This approach, named the Landauer–Büttiker method, establishes the relation between the transmission coefficient and the electronic conductivity (see, for example, reference [14]). This approach is presented in more detail in chapter 7.

- Ionic conductivity

Ionic conductivity refers to an electric current derived from the movement of ions (cations or anions) in the material subjected to an electrical field as a driving force. In solids, the process can be imagined as similar to a diffusion process, in which an ion jumps from one lattice site to another, although in ionic conduction, the ions' movement is not random but preferentially aligned with the electric field. In short, the achievement of good ionic conductivity demands the presence of ions in an environment that allows high mobility. This environment can be achieved in several ways, such as through point defects (vacancies), tunnels, or the spaces between layers in crystalline materials.

The ionic conductivity, σ_{I}, can be estimated by calculating the velocity of ions passing through the crystal in a simple scenario. Unlike electrons, ions must overcome potential energy barriers, E, due to the other lattice ions. These energy barriers depend on the nature of the native defects that interact with mobile ions, so the E_{v} and E_{i} energies are referred to as vacancy and interstitial diffusion mechanisms, respectively. The barrier height is also modified by the applied electric field, ZeV/a, where Z is the atomic number of the ion, a the lattice spacing, and V the applied voltage. Assuming a low field strength in comparison with $k_{\mathrm{B}}T$, the mobility of the ions can be expressed as:

$$\mu_{\mathrm{I}} = \frac{\nu Z a^2 e}{k_{\mathrm{B}}T}\exp(-E/k_{\mathrm{B}}T), \tag{3.6}$$

where ν is the vibration frequency of the solid. Since the conductivity is defined as the product of the electric charge, the charge concentration, and the mobility, by taking account of (3.6) and defining $\sigma_0 = \frac{\nu Z a^2 e}{k_{\mathrm{B}}}$, the ionic conductivity can be generally expressed as:

$$\sigma_I(T) = \sigma_0 \exp(-E/k_B T). \tag{3.7}$$

The ionic conductivity is a key factor in the development of solid-state batteries, which require high conductivity values. The most studied systems are Li-ion batteries, although magnesium are calcium ions are also currently under study. The small size of alkaline cations eases the diffusion process through the material. However, one of the main drawbacks is the breakdown of the components after a finite number of charge–discharge cycles. In this sense, a number of materials systems have been proposed for the construction of efficient, long-lasting, and powerful Li batteries. Among these, the most promising system incorporate nanomaterials such as silicon fibers [15] or core–shell nanoparticles as electrodes [16]. In addition to the requirement for knowledge and deep understanding of ionic conductivity in cathode, anode, and electrolyte materials, the interfaces between the different elements add another obstacle to charge transport. To address this issue, post-annealing treatments have been reported as potential solutions for all solid solid-state batteries [17].

Finally, the ionic conductivity is also a key factor in liquid electrolytes, such those utilized as transport media in dye-sensitized solar cells (DSSCs). Recent developments in batteries and solar cells have incorporated polymers and ionic liquids, and high ionic conductivity is needed [18]. This has become a challenge, since these materials exhibit poor conduction.

3.2.3 Optical properties

Light–matter interactions produce a variety of phenomena of great importance in energy-related devices. One of the most shining examples is the photovoltaic effect, which converts solar radiation into electricity due to the opposite movements of excess electrons and holes generated by light absorption in the neighborhood of a p–n junction. In addition to absorption, light reflection and transmission are also quite relevant for the efficient performance of energy-related devices. For this reason, antireflective coatings and transparent conductive layers are used in solar cell layouts. Solar cells are an example of energy conversion from optical radiation into electric power. The reverse, i.e. devices that generate light, are key in other fields, such as optical sensors and solid-state lighting, which also aim to achieve a sustainable and environmentally friendly world. In this section, we summarize the fundamental optical properties relevant to the abovementioned energy applications, as outlined in figure 3.6. These properties refer to the optical phenomena (reflection, propagation, and transmission) that occur when light impinges on a material. In addition, scattering, absorption, and luminescence processes are also phenomena that can take place during light propagation through a material.

The optical properties of solids have been studied for a long time. The ancient Egyptians and Persians tried to explain the relationship governing the angles of refraction observed when light passed from one transparent medium to another. The formulation later derived by Snell in 1621 in the framework of ray optics is still valid in a general way; it determines the direction taken by light rays passing through refractive media with different refractive indexes: $n_1 \sin \theta_1 = n_2 \sin \theta_2$. The refractive

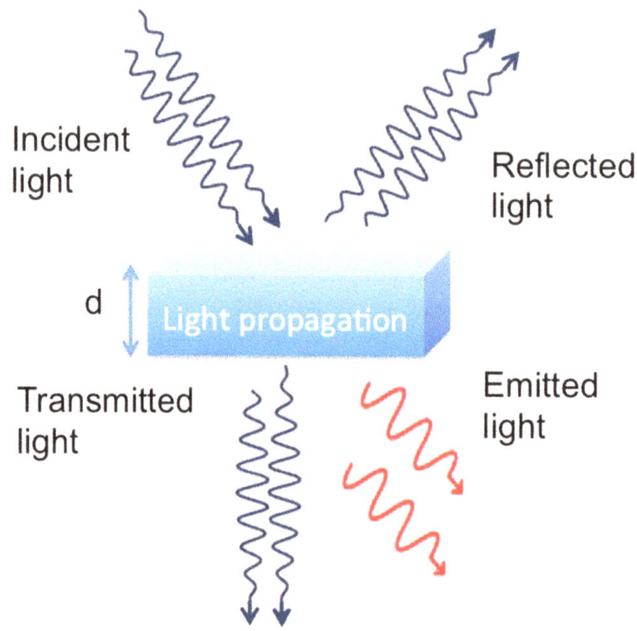

Figure 3.6. Outline of the fundamental optical properties.

index, n, represents the ratio between a light ray's speed in vacuum, c, and its velocity in a transparent medium. This parameter may depend on the frequency of the light beam, which is called dispersion.

The development of electromagnetic theory derived from Maxwell's equations provided a modern vision of light waves, which can be pictured as a self-propagating transverse oscillating wave of electric and magnetic fields. The classical theory of optical propagation considers light to be an electromagnetic wave and the medium as a collection of dipole oscillators, which represent the atoms and molecules of the material. This purely classical approach provides a good understanding of the basics of light propagation in materials. However, the most common way to tackle optical properties is the semi-classical model, in which light is considered to be a classical electromagnetic wave, while the atoms and molecules of solid materials are described by quantum mechanical rules. On the other hand, a fully quantum approach for both light and matter would lead us into the field of quantum optics. Nevertheless, even in the semi-classical models, light is often considered to be a beam of quantum particles, i.e. photons. The dual wave–particle concept developed in quantum mechanics can also be applied to the nature of light. The oscillation frequency determines the energy of the zero-mass quantum particles $E = \hbar\omega$ when light is considered to be a beam of photons.

3.2.3.1 Light reflection and transmission

The reflectivity or coefficient of reflection represents the ratio between the powers of light incident on and reflected by a surface. It is usually denoted by R. Likewise, the transmissivity or coefficient of transmission, T, denotes the ratio between the transmitted power and the incident power. If no absorption or scattering occurs during the propagation, the equality $R + T = 1$ should be fulfilled by the conservation of energy law. This is the case for transparent materials with small dispersion effects in the visible range, for which a single refractive-index value can be given. In this case, only light refraction occurs, and the expression for R of the surface of a material of refractive index n_2 embedded into a medium of refractive index n_1 is given by:

$$R = \left(\frac{n_2 - n_1}{n_2 + n_1}\right)^2. \tag{3.8}$$

Let us consider a photovoltaic solar cell based on silicon, which has a refractive index of 3.6 for $\lambda = 600$ nm. The reflectivity at the front surface is 31%, according to equation (3.8) and taking $n_1 = 1$ for air. This is indeed a high value that hampers improvements in solar efficiency. The reflection coefficient can be reduced by covering the top of the solar cell surface in antireflective coatings. To this end, single or double layers of several materials have been proposed as reflective coatings for current solar cells, lowering the reflectivity to 3%. Some examples are oxide layers such as TiO_2 ($n = 2.3$), Al_2O_3 ($n = 1.8$–1.9), SiO_2 ($n = 1.4$–1.5), and Ta_2O_5 ($n = 2.1$–2.3) [19] or other silicon-based compounds, such as amorphous a-SiNx:H [20], or phosphorous silicate glass (PSG) [21], to name a few.

However, the refractive indexes of electronic materials are not constant over the whole solar electromagnetic range due to light absorption effects, which generally speaking occur when the frequency of the light matches a transition frequency of atoms of the material. The absorption coefficient α quantifies this phenomenon and is strongly dependent on the frequency of the light. The absorption coefficient is defined as the fraction of power absorbed by a unit length of the material. The relationship between the intensity of the incident light I_0 and its intensity after traveling a distance z inside a medium is given by Beer's law:

$$I(z) = I_0 e^{-\alpha z}. \tag{3.9}$$

Therefore, absorption effects have to be considered when evaluating the transmission coefficient T, since only unabsorbed light is transmitted. Considering the slab shown in figure 3.6, if scattering effects are negligible, the transmissivity is described by:

$$T = (1 - R_t)e^{-\alpha d}(1 - R_b), \tag{3.10}$$

where R_t and R_b are the reflectivities of the top and bottom surfaces, respectively, and d is the slab thickness.

Absorption effects can be incorporated into the refractive index of a medium with the aid of the complex refractive index, usually denoted by $\tilde{n}$ and defined as: $\tilde{n} = n + i\kappa$. The real part, n, is the usual refractive index, $n = c/v$, while the imaginary part, κ, is called the extinction coefficient.

The relationship between the absorption coefficient, α, and the extinction factor, κ, can be calculated from an analysis of the propagation of plane electromagnetic waves through a material with a complex refractive index. For a light wave propagating in the z direction, the electric field component can described by: $\mathcal{E}(z, t) = \mathcal{E}_0 e^{i(kz-\omega t)}$, where k is the wave vector of the light and ω is the angular frequency. If there is no absorption, these quantities are related each other via $k = n\omega/v$. This formula can be generalized to take absorption effects into account by using the complex refractive index:

$$k = \tilde{n}\frac{\omega}{c} = \left(n + \mathrm{i}\kappa\right)\frac{\omega}{c}.$$

After some mathematics, taking into account that the optical intensity is proportional to the square of the electric field and by comparison with Beer's law (3.9), it can be concluded that:

$$\alpha = \frac{2\kappa\omega}{c} = \frac{4\pi\kappa}{\lambda}, \tag{3.11}$$

where λ is the wavelength of light in free space. This expression shows the linear relationship between the absorption coefficient and the imaginary part of the refractive index.

3.2.3.2 Light propagation

The propagation of light is commonly described in the framework of classical electromagnetism in optics textbooks [22]. This classical model applies Maxwell's theory of electromagnetic waves and the concept of the dipole oscillator to simulate the solid or medium that light is traveling through. To a first approximation, it is rather easy to identify several oscillators in a solid, together with their own characteristic resonant frequencies. The optical frequencies of near-infrared, visible, and ultraviolet light are in the range of 10^{14}–10^{15} Hz, and these are precisely the resonant frequencies of bound electrons in solids. On the other hand, lattice vibrations have resonances at lower frequencies in the infrared range (10^{12}–10^{13} Hz) according to the harmonic crystal model. Finally, collective oscillations of free electrons in metals also resonate at optical frequencies in the ultraviolet and visible ranges, depending on the material.

The dipole oscillator model allows the frequency dependence of the refractive index and the absorption coefficient to be calculated for solids. This model can be applied to the interaction between light and a single oscillator or multiple resonators. The frequency dependence of the complex dielectric function, defined as $\tilde{\epsilon}_r = \epsilon_1 + \mathrm{i}\epsilon_2$, is first derived, and then the real and imaginary parts of the complex refractive index, $\tilde{n}$, can be calculated from $\tilde{\epsilon}_r$.

An example of the dispersion relations of the refractive index, n, and the extinction coefficient, κ, of SiO_2 is shown in figure 3.7 [23]. Resonant peaks in the extinction curve correspond to light absorption at particular wavelengths in the infrared (lattice vibrations) and in the ultraviolet (electrons in the band structure).

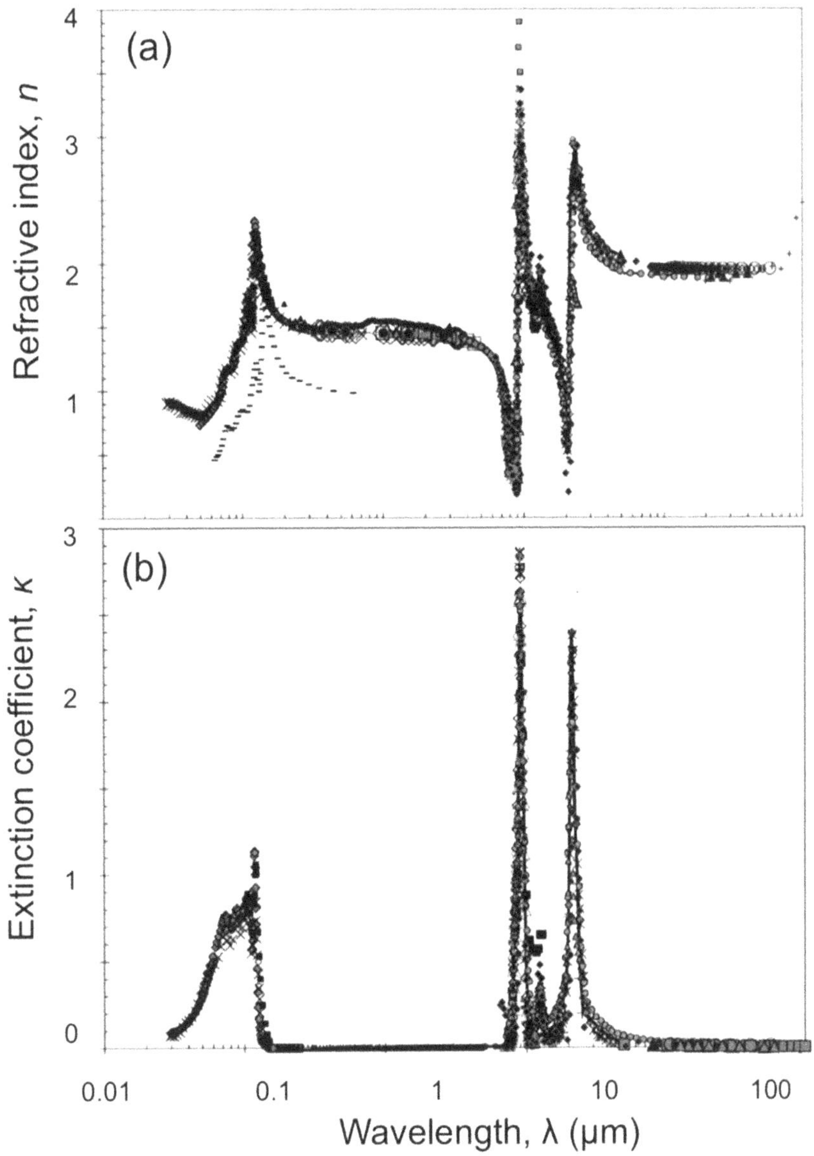

Figure 3.7. The real (a) and imaginary (b) parts of the complex refractive index of SiO_2. Adapted with permission from [23] © The Optical Society.

On the other hand, the refractive index shows a rather constant value over the visible transparency range, with no absorption effects.

Finally, let us consider the propagation of light through microstructured materials such as a system of stacked materials with different indices of refraction. In this case, it is very useful to consider the scalar Helmholtz equation that describes the space distribution of an electric field when an electromagnetic wave of frequency ω propagates in a medium with refractive index $n(\boldsymbol{r})$:

$$\nabla^2 E(\boldsymbol{r}) + k^2 E(\boldsymbol{r}) = 0, \quad k = \frac{n(\boldsymbol{r})}{c}\,\omega. \tag{3.12}$$

This Helmholtz equation for waves is formally equivalent to the steady-state Schrödinger equation for particles (equation 3.4) under a potential profile $U(x)$. This fact allows some analogies to be found between light transmission/reflection through a medium with a spatially variable refractive index $n(x)$ and electrons under equivalent $U(x)$ potential profiles [24]. In particular, new insights into our understanding of optical propagation at the micro- and nanoscales can be gained by extrapolating the behavior of electronic states in quantum wells, multiple quantum wells, and superlattices.

3.2.3.3 Absorption and luminescence

An accurate description of light absorption and emission in solids requires a quantum mechanical approach. Light can be absorbed by transitions across the bandgap in a semiconductor or by absorption by shallow or deep electronic levels in gaps that originate from impurities or defects in the crystal. In an absorption experiment, a solid experiences a perturbation in the form of electromagnetic radiation with a vector potential, $\boldsymbol{A}(x, t)$, which for small fields can be expressed in an operator form as: $\boldsymbol{H}' = -\boldsymbol{pA}e/m$, where $\boldsymbol{p}$ is the momentum of the electron. The absorption is then determined by the quantum mechanical probability of the transition between an initial (usually ground) state and a final (excited) electronic state in the solid. The transition probability is proportional to the square of the magnitude of the matrix element, $M = \langle f|H'|i\rangle$. According to the golden rule of quantum mechanics, the probability of a transition per unit of time is:

$$W_{fi} = \frac{2\pi}{\hbar}|M|^2 g(\hbar\omega). \tag{3.13}$$

The absorption coefficient may then be calculated as the ratio between the rate at which energy is absorbed per volume, V, and the rate at which energy is incident per unit area:

$$\alpha = \frac{\hbar\omega W_{fi}}{I(\omega)V}. \tag{3.14}$$

Light absorption depends on two factors: the matrix element, M, and the density of states, $g(\hbar\omega)$, which have to be evaluated at the energy of the incident photons.

The specific shape of the absorption coefficient as a function of the photon energy depends on the electronic states involved in the optical transitions. The main expected contribution arises from absorption by extended states, e.g. from the valence band to the conduction band. This is called the absorption edge; at this point, an abrupt increase in the absorption coefficient occurs at a radiation energy, $\hbar\omega$, that exceeds the energy bandgap, E_G, of the semiconductor. Light absorption curves are useful in determining the optical bandgap in semiconductors, which refers to the photon energy that excites electrons from the valence band to the conduction band. On the other hand, optical absorption is also possible via localized states such as impurity levels or lattice-defect-related levels, which are reflected in the absorption coefficient spectrum as small features below the absorption edge.

Optical absorption is of particular interest in solar cell applications. New generations of photovoltaic devices use quantum dots and/or hybrid organic–inorganic structures to achieve bandgap tuning. Figure 3.8(a) shows the absorption coefficients of a series of PbS quantum dots in a planar hybrid heterojunction synthesized at different temperatures, in which E_G variations from 0.86 to 2.1 eV are achieved by a suitable thermal treatment of the quantum dot–polymer assembly [25]. The tunability of the bandgap can also be achieved by adjusting the size of the quantum dot. Figure 3.8(b) shows an example of the absorption of CdSe particles with sizes in the range of 2.3–3.7 nm in which a clear blueshift in the excitonic peak is observed as the particle diameter decreases [26].

Once a solid is excited in some way, e.g. by light irradiation, the subsequent relaxation process allows electron recombination to occur from excited states to the ground state after a certain time. This recombination can be radiative (spontaneous luminescence) or non-radiative, the characteristic lifetimes of which are denoted by τ_r and τ_{nr}, respectively. Spontaneous emission is also a quantum process, formulated by Einstein in the description of transition processes in a two-level energy system, along with absorption and stimulated emission. The quantum efficiency η is defined as:

$$\eta = \frac{1/\tau_r}{1/\tau_r + 1/\tau_{nr}}. \tag{3.15}$$

If the radiative lifetime is shorter than the non-radiative lifetime, luminescence is the main de-excitation mechanism and strong luminescence can be achieved. Otherwise, non-radiative processes cause heating of the crystal.

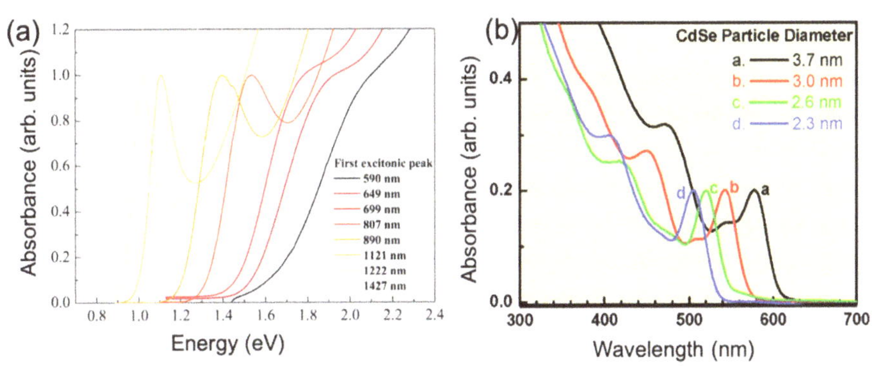

Figure 3.8. (a) Absorption spectra of PbS quantum dots synthesized at temperatures between 45 °C and 185 ° C. Reprinted from [25], Copyright (2019), with permission from Elsevier. (b) Absorption spectra of CdSe quantum dots of different sizes. Reprinted with permission from [26]. Copyright (2008) American Chemical Society.

Optical absorption and photoluminescence (PL) measurements are usually combined; these have become powerful techniques for the study the dynamic of excited electronic states in electronic materials without the need for electrical contacts. They can provide information about electronic states in the bandgap as well as their interaction with other localized states or the valence and conduction bands. In addition, they can also give hints about the eventual electron–phonon coupling due to the interaction between vibronic bands and electronic states in the solid.

3.2.4 Surfaces and interfaces

Every material is delimited by a surface, which can be considered to be an abrupt interface between the solid and the surrounding medium. A simple approach to the creation of surfaces is to break a solid into two parts, which leads to two new free surfaces. As a consequence, the atoms in the outermost layers may rearrange due to energy minimization considerations. Therefore, in a rigorous way, the surfaces of real solids are composed of a few atomic layers which may exhibit different physical/chemical properties than those of the solids they separate. Figure 3.9 illustrates the creation of a surface of thickness δ which is achieved by breaking atomic bonds. This concept can also be extended to solid interfaces such as grain boundaries, interfaces in multilayers, or nanocrystalline bulk materials.

If we consider a small sphere of diameter d, its surface-to-volume ratio is $A/V = 6/d$. This means that when it comes to nanomaterials, the atomic percentage that lies at the surface increases considerably in comparison to bulk materials, in which most of the atoms are in the interior of the solid. Hence, the peculiar properties of the surface layer can lead to remarkable phenomena due to 'size effects' that are not observed in bulk materials. These size effects are different from the quantum confinement effects described above for the luminescent properties of nanomaterials. It is quite straightforward to demonstrate that the smaller the particle size, the higher the relevance of the surface layer can be in the physico-chemical properties of a material. In fact, atoms within a small thickness, δ, close to the geometric surface, suffer displacements and/or the rearrangement of atomic bonds due to the abrupt ending of the material. This parameter is often taken to be

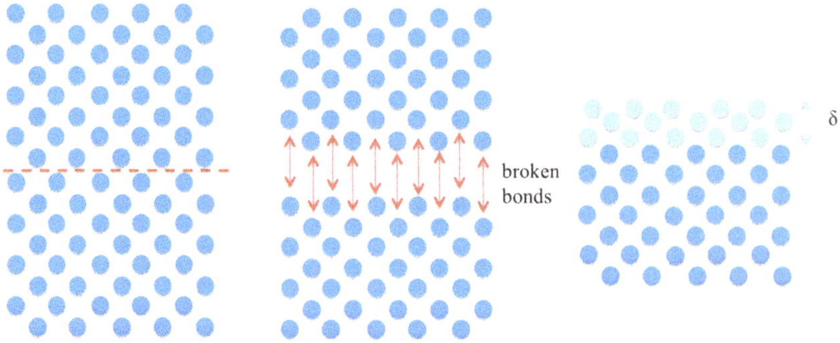

Figure 3.9. Schematic of the creation of two free surfaces by breaking chemical bonds in the solid, leading to a free surface of thickness δ.

the thickness of a free surface. It is a straightforward exercise to calculate the surface-to-volume ratio versus the particle diameter for spherical and cubic particles.

Studying the surface properties is of paramount importance when dealing with applications in which the functionality of the material is manifested at the surface. Chemical sensors are a clear example. Chemical species attach to the surface of the sensor and an output signal proportional to the species concentration is provided via the corresponding transducer mechanism. It is obvious that the sensor efficiency is higher when the surface-to-volume ratio is increased as much as possible. Other examples are also found in photocatalysis and photovoltaic devices, in which light capture should be maximized to achieve the best performance.

3.2.4.1 Surface energy

The concept of surface tension or surface forces was developed by Gibbs and explains the formation of particular surfaces or interfaces according to thermodynamic considerations, which lead to the shapes that solids may adopt.

The specific surface energy per unit area, $\gamma = (\partial G/\partial A)_{n,T,p}$, is usually defined as the energy needed to divide a solid into two pieces, and hence to create new surfaces. Let us consider a spherical particle of diameter d and specific surface energy γ. This particle thus has an energy surface equivalent to a geometrical surface of $u_{\mathrm{surf}} = \gamma \pi d^2$. The energy surface per mol is usually used for thermodynamic considerations. Taking into account the number of particles per mol, $N = M/\rho v$, where v is the volume of the particle, the surface energy per mol is $U_{\mathrm{surf}} = Nu_{\mathrm{surf}} = \gamma 6M/\rho d$. This shows that the surface energy per mol is inversely proportional to the particle diameter. In the case of very small particle sizes, this energy cost is a major issue that has to be addressed.

There are several contributions to γ that help us to understand the origin of the surface energy [27]. As a first approximation, γ_0 can be estimated from the energy needed to break chemical bonds in the material. The specific surface energy, γ_0, can be expressed as $\gamma_0 = N^*\epsilon_b/2$, where ϵ_b is the bonding energy and N^* the number of broken bonds per meter square. However, the neighbor's loss of surface atoms creates a force, f, perpendicular to the surface that pulls atoms down toward the interior of the solid, which causes some surface stretching in the outermost atomic layers. The surface stress, $\sigma = f/a$, contributes to the specific surface energy, γ_s. Thus, $\gamma = \gamma_0 + \gamma_s$. Finally, in the case of spherical particles, hydrostatic pressure due to their curvature also affects the surface energy considerations.

It is also important to realize that the orientation of bonds in the solid leads to anisotropic effects in the surface energy. This implies that in order to minimize the surface energy, preferential crystal orientations occur, leading crystals to adopt shapes such as rods, nanowires, or plates. The addition of functional groups to a surface also modifies the surface energy, allowing one- or two-dimensional nanomaterials to be produced.

3.2.4.2 Surface states

The broken bonds at the surface of a solid not only lead to slight atomic/ionic displacements from their equilibrium state and may provoke atomic rearrangements, as described above, but also have consequences for the electronic structure of the solid via the creation of eventual dangling bonds at the surface. Most energy-related applications involve the movement of carriers (electrons or holes) through particular devices or systems, in which either free surfaces or interfaces are present. Part of the success in the design of energy-related devices results from the proper management of carriers interacting with these surfaces. In the case of semiconductors, it is well known that structural defects and impurities introduce localized energy states in the bandgap, due to the disruption of lattice periodicity (figure 3.10(a)). These energy levels usually act as electronic traps for carriers via recombination processes, which is detrimental to device performance. This is critical in solar cells, for example, in which high surface recombination increases the rate at which minority carriers are depleted, affecting the cells' efficiency.

When it comes to nanomaterials, for which the surface-to-volume ratio is highly relevant, surface states are needed to explain the optical and electrical properties of the nanomaterial. An example of the impact of surface hole or electron traps on the carrier dynamics of ZnO nanowires is shown in figures 3.10(b) and (c).

Surfaces and interfaces can be considered to be planar defects in a solid that interrupt the periodicity of the lattice, so an increase in recombination processes is to be expected, leading to carrier depletion in the vicinity of a surface. This depletion region, which has a low carrier concentration, causes a flow of carriers from the bulk into this region. Therefore, the surface recombination rate can be limited by the rate at which minority carriers move towards the surface. The parameter that quantifies the recombination processes at surfaces is known as the surface recombination velocity, v_s. A surface with no carrier recombination would have $v_s = 0$, whereas for real surfaces or interfaces, this parameter can reach values as high as $v_s = 10^7$ cm s^{-1}. The higher the value of v_s, the higher the difference in recombination properties between the bulk and the surface. In many applications, a low v_s is required, for example in light-emitting devices or solar devices, in order to avoid a severe loss of the carriers needed for optimum performance. The dangling bonds at semiconductor surfaces cause a high local recombination rate. One way to avoid this is to fill such dangling bonds with functional groups. In later chapters, several such strategies are shown for energy-related devices.

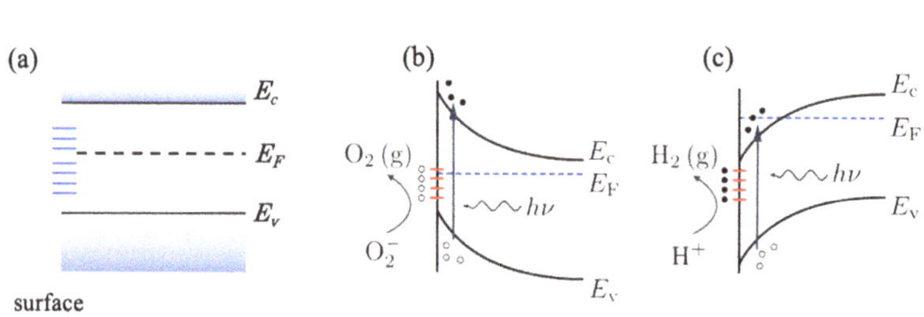

Figure 3.10. (a) Schematic showing the distribution of surface states (short blue lines) within the bandgap. (b) Energy band bending due to electron depletion and (c) electron accumulation at the surface of a nanowire; (b) and (c) reproduced from [28] with permission of The Royal Society of Chemistry.

3.2.4.3 Interfaces and energy band alignment

A surface can be considered to be a particular case of an interface. Most electronic devices incorporate several electronic materials in their design in a way that forms heterojunctions. The role of interfaces in physical properties has been a long-studied issue in materials science in general and in electronic materials in particular since the development of the first solid-state transistors. Rectifying junctions, such as those made of p-type/n-type semiconductors or metal–semiconductor junctions, are exploited in many electronic devices [29]. Under equilibrium conditions, an intimate atomic-level junction between materials that have different bandgaps provokes changes in the layout of the energy bands (taking the vacuum level as the reference energy level).

Figure 3.11 summarizes the three types of energy band alignment of A–B–A heterojunctions in a flat band scheme. Here, ΔE_c and ΔE_v stand for the conduction and valence band offsets, respectively. Type I heterostructures are the best choice for optoelectronic devices, since they confine electrons and holes to the same spatial region. However, type II heterostructures are best suited for photodetectors and sensors due to the effective charge separation which is required in these devices. Finally, type III heterojunctions, also called broken-gap junctions, are suitable for use in tunnel diodes and zero-gap configurations.

The band alignment offsets of heterojunctions composed of MoS_2 (which has an $E_G = 1.35$ eV) and several nitrides (AlN, GaN, and InN) have recently been experimentally determined [30]. Figure 3.12 is a schematic of the bandgap alignments of MoS_2–AlN, MoS_2–GaN, and MoS_2–InN. The results show that they follow the type I, type II, and type III alignments, respectively. The precise shape in each case determines the way in which electrons and holes can pass through these devices, which is critical for the design of future photonic and electronic bipolar devices.

3.3 Semiconducting oxides

Oxide materials can exhibit a wide variety of physical properties due to the huge number of possibilities available in the mixing of metal cations and oxygen. In particular, some transition-metal oxides and basic metal oxides exhibit semiconducting properties, leading to a well-recognized material class. Recently, several roadmaps have promoted the relevance of oxide materials for electronics in the post-

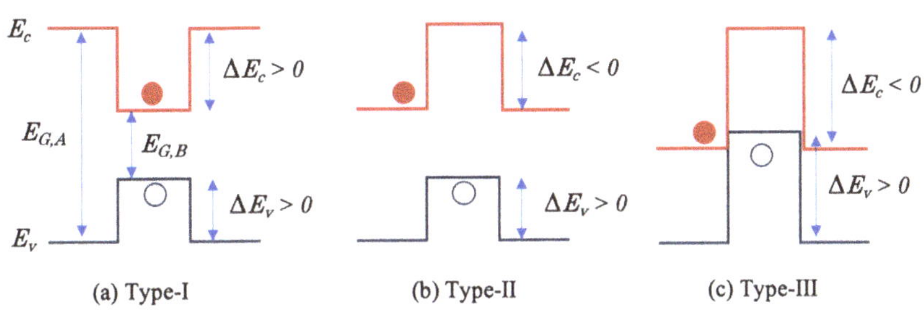

Figure 3.11. Sketch of the three types of band alignment in heterojunctions. Red and black lines stand for conduction and valence bands, respectively.

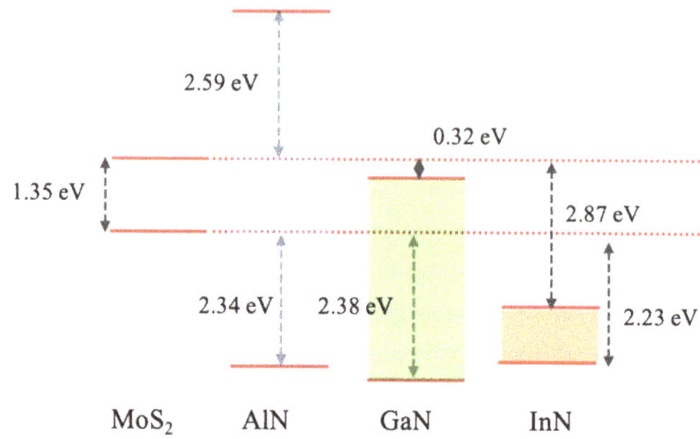

Figure 3.12. Energy band offsets in MoS_2–nitride heterojunctions. Reprinted from [30], with the permission of AIP Publishing.

silicon era [31]. In contrast to mainstream microelectronics, in which Si and some selected III–V semiconductors take the lead, oxide electronics are open to a big part of the periodic table. This fact offers plenty of room for research from the fundamental and technological points of view. Binary and ternary semiconducting oxides, such as ZnO, SnO_2, and indium tin oxide (ITO) and insulators such as $SrTiO_3$ have been the subjects of research in recent decades, with the aim of understanding their electronic, mechanical, thermal, optical, and magnetic properties. The advances in this knowledge have favored the implementation of oxide materials in electronic and energy-related devices. Semiconducting oxides are bursting into, for example, areas such as photovoltaics, smart windows, solid-state fuel cells, neuromorphic computing, oxide microelectromechanical systems (MEMS), and transparent conductors, making a huge impact on today's electronics. In addition, the wide bandgaps that some of these metal oxides exhibit allow optical transparency in the visible range while retaining adequate electronic conductivity; such oxides form the transparent conducting oxide (TCO) materials family.

The variety of electronic properties that oxides present arises from the O^{2-} bonds with metal-ion *d* orbitals that form tetrahedra and octahedra, from the variability of oxygen occupancy and doping, and from the enormous polarizability of oxygen ions. Therefore, oxides may display several structural phases, which determine their physical properties. Table 3.1 summarizes the basic properties of some binary semiconducting oxides. The improvements achieved in recent years have allowed better control over the crystalline quality, carrier mobility, impurity doping, and bandgap engineering and the development of ternary oxide alloys.

In addition to the multiple advantages of semiconducting oxides in bulk or thin-film form, low dimensionality adds an extra trait to their appeal. Oxide nanoparticles, nanowires, and nanomembranes are currently a hot research topic, which aims to achieve synergy between the material properties and the selected nanoarchitecture. Unlike semiconductor nanomaterials such as quantum dots or nanowires, which exploit strong quantum confinement effects, in the case of

Table 3.1. A comparison of the structural and physical properties of some semiconducting oxides.

Oxide	Stable structure	Melting point (°C)	Bandgap (eV)	Debye temperature (K)
Ga_2O_3	$C2/m$	1725	4.9	738
TiO_2	$P4_2/mnm$	1840	3.05	600
SnO_2	$P4_2/mnm$	1650	3.6	570
ZnO	$P4_3/mc$	1975	3.4	305
NiO	$Fm\bar{3}m$	1955	3.5–4.5	495
Cu_2O	$Pn\bar{3}m$	1235	2.2	540
SnO	$P4/nmm$	1080	2.5–3.4	245

semiconducting oxide nanomaterials, the role of the surface is of the utmost interest due to the considerable increase in the surface-to-volume ratio.

Some oxide nanomaterials have been inspired by carbon nanostructures, such as carbon nanotubes and graphene, as paradigmatic 1D and 2D materials, respectively. Carbon nanotubes have become the seed of further work on many semiconductor nanowires, including oxide nanowires. Nanowires and nanobelts of ZnO, SnO_2, or Ga_2O_3 can be fabricated via low-cost physical or chemical routes in many cases, which is a clear advantage in comparison to the epitaxial techniques used for other semiconductors [32]. On the other hand, the post-graphene era has prompted the incorporation of metal oxides as novel members of the two-dimensional materials family, in particular, those that exhibit a layered structure, such as SnO and MoO_3.

The electronic states (deep and/or shallow) associated with oxygen vacancies in semiconducting oxides, which to a large extent determine the electronic properties of such oxides, have been theoretically studied for a quite good selection of binary oxides [33]. In particular, it is worth mentioning that most of the semiconducting oxides exhibit intrinsic n-type conductivity due to the presence of oxygen vacancies that act as shallow donors. This is the case for TiO_2, ZnO, SnO_2, Ga_2O_3, and In_2O_3. In contrast, NiO, SnO, and Cu_2O are natural p-type conductors which have been heavily investigated in recent years. Some highlights of the semiconducting oxides displayed in table 3.1 are summarized below.

3.3.1 n-type oxides

The energy bandgaps of metal oxides range from the visible to the ultraviolet. Some of them are considered to be wide bandgap semiconductors, i.e. $E_g > 3$ eV. In spite of the large bandgap, conductivity arises due to the presence of native defects in the oxide; oxygen vacancies are the main defects that generate electronic levels in the bandgap close to the conduction band, providing electrons for conduction and making most of the undoped oxides behave as n-type semiconductors.

Tin oxide, SnO_2

SnO_2 is a transparent direct-bandgap semiconductor ($E_G \simeq 3.6$ eV) with applications in gas sensors, solar cells, and glass coatings as well as in electronic devices.

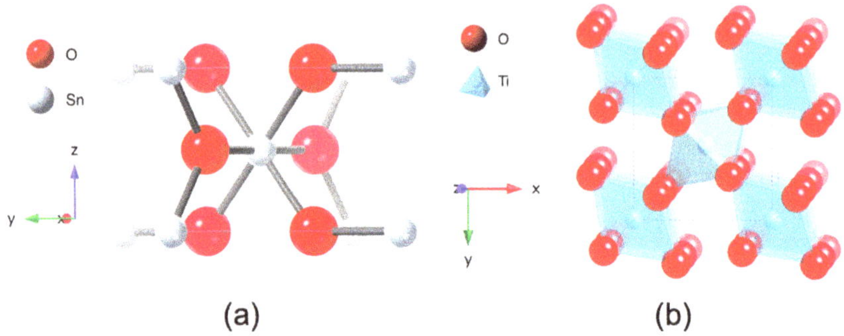

Figure 3.13. The crystalline structures of the most stable phases of (a) TiO_2 and (b) SnO_2.

The stable phase is the rutile phase, which has a tetragonal structure, (figure 3.13(a)) with six ions in the unit cell. Each tin ion is surrounded by six oxygen ions in an octahedral configuration.

The electronic band structure is complex due to the large number of valence electrons that occupy each unit cell. As in the case of other oxides, the conduction band is the result of the s and p states of Sn, while the upper valence band results from O 2p and Sn p. The intrinsic defects and their consequences in terms of physical properties have been thoroughly investigated. In particular, the defects resulting in a net oxygen deficiency are favored, since it is the oxygen vacancy and not the tin interstitial which has the lowest formation energy and is facilitated by the ease of reducing Sn(IV) to Sn(II). These oxygen vacancies are responsible for the shallow donor levels at 0.03 and 0.15 eV below the conduction band minimum (CBM), which enhance the conductivity by introducing carrier electrons into the conduction band [34]. Experimental evidence for the existence of intrinsic defect-related states has been provided by the cathodoluminescence (CL) technique in SnO_2 microtubes [35]. In addition to identifying the oxygen vacancy state levels, the results also showed a luminescence dependence on the surface planes in SnO_2.

Titanium oxide, TiO_2

Titanium dioxide, TiO_2, crystallizes in three structural forms: rutile and anatase (tetragonal) and brookite (orthorhombic). Although rutile is the thermodynamically stable phase (see figure 3.13(b)), the anatase structure is rather easy to obtain, depending on the synthesis route. The anatase phase is most suitable, for example, for photocatalytic applications. The bandgap of TiO_2 rutile is 3.0 eV. The electronic conductivity of this oxide strongly depends on the oxygen partial pressure, as expected, due to the role of oxygen vacancies. However, Ti interstitials also behave as electron donors under low oxygen partial pressures.

The nanostructured form of TiO_2 has also been found indispensable in numerous applications [36], many of which are related to its capability for light absorption, such as photocatalysis, dye-sensitized solar cells, pigments, UV photodetectors, or biomedical uses, to name a few. The material properties of this environmentally friendly and chemically stable oxide are strongly influenced by external factors such as temperature, pressure, and the surrounding medium, particularly in the nanoparticle form. A review of the optical and structural properties of TiO_2 nanoparticles can be found in reference [37].

Zinc oxide, ZnO

ZnO is a direct-bandgap semiconductor which has been well known for many years; its properties are similar to those of GaN. It possesses a wide bandgap of 3.4 eV and n-type conductivity due to its native defects. The stable phase is the hexagonal one known as wurtzite, which has four atoms per unit cell (figure 3.14(a)), although cubic phases (zinc blende and rocksalt) have also been stabilized. ZnO exhibits piezoelectricity and has spontaneous interface charges. The piezoelectric effect occurs upon strain parallel to the *c*-axis, and the spontaneous charge is the result of the lattice distortion from the ideal structure at room temperature.

The natural donor behavior of undoped ZnO is due to intrinsic defects, zinc-on-oxide antisites, oxygen vacancies, and zinc interstitials. However, dopants such as aluminum, indium, and gallium have also been used to increase its n-type conductivity. As an alternative, p-type behavior has been obtained by locating Li or Na on the zinc site, where they behave as shallow acceptors.

The literature about this oxide is huge, and excellent reviews and books have been published about it, e.g. [38, 39]. Therein, its thermal, phonon, electrical, and optical properties are reviewed, including their relevant parameters. Of particular interest is the anisotropy of its physical properties due to its hexagonal structure, which is also revealed in the variety of nanomaterials based on ZnO. In summary, ZnO is a valuable candidate for use in optoelectronics, gas sensors, thin-film transistors, and solar cells.

Gallium oxide, Ga_2O_3

Ga_2O_3 can be stabilized in five polymorphs, α, β, γ, δ, and ϵ, among which the β (monoclinic) form is the thermodynamically stable phase at room temperature [40].

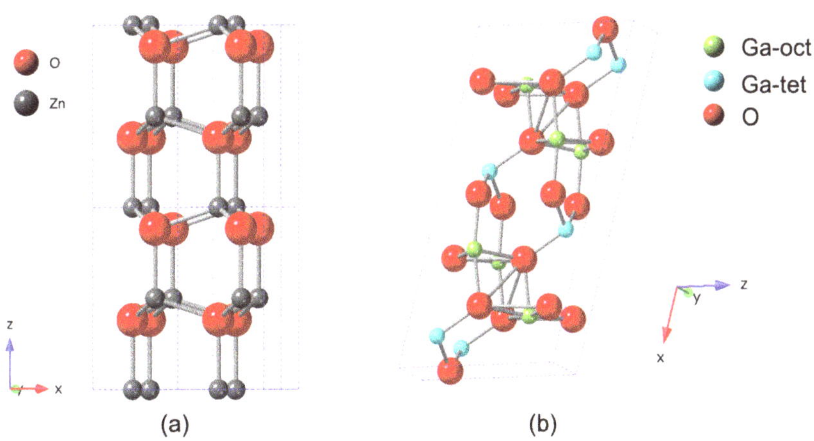

Figure 3.14. The crystalline structures of (a) ZnO and (b) Ga_2O_3.

The material properties of β-Ga_2O_3 exhibit strong anisotropy, due to the monoclinic angle, β, of 103°. Ga ions occupy tetrahedral and octahedral sites in the lattice, whereas O ions occupy three different positions in the lattice. The crystalline structure of β-Ga_2O_3 is shown in figure 3.14(b).

The electronic band structure of β-Ga_2O_3 is quite complicated and still under investigation. Several theoretical calculations have produced different results, and the more recent works consider the anisotropic features of its structure. Nevertheless, experimental work has reported an ultrawide band gap in the range between 4.7 and 5.4 eV, depending on the transition dipole orientation within the a–c-plane, α, given with respect to a-axis [41]. In spite of its ultrawide bandgap, undoped β-Ga_2O_3 shows a decent electron mobility of around 100 $cm^2\ V^{-1}\ s^{-1}$ due to intrinsic defects and unintentional impurities. In addition, its electron conductivity can be adjusted by suitable doping (10^{13}–10^{20} cm^{-3}), which is a valuable feature for device applications.

The structural and electronic properties of Ga_2O_3 have recently boosted the interest in this oxide. The most promising applications are in the fields of high-power electronics [42] and UV optoelectronic devices, such as solar-blind photodetectors [43]. The ultrawide bandgap allows it to be doped with optically active impurities, such as transition-metal or rare-earth ions, which display characteristic emission lines due to their intraionic transitions. This allows films and nanomaterials based on Ga_2O_3 to be efficient emitters [44]. Finally, its chemical stability allows its use under extreme conditions, such as radiation, high temperatures, and high electric fields. A new trends in the exploitation of Ga_2O_3 is the development of ternary oxides, which are obtained by alloying it with Al, In, or Zn, with the aim of achieving bandgap engineering in a similar way to that used for group III–V semiconductors. Regarding the design of novel devices, the band alignment of some Ga_2O_3-based heterostructures allows self-powered devices to be produced, which is currently also a hot research topic.

3.3.2 p-type oxides

In contrast to the previously described and well-known n-type oxides, further in-depth knowledge is still required for p-type oxides. NiO, Cu_2O, and SnO are some of the few known intrinsic p-type binary oxides, in which interest has resurged in recent decades for diverse technological applications. Such applications include sensors, batteries, supercapacitors, electronics, photovoltaics, and electrochromic devices, among others [45–47]. In contrast to n-type oxides, in which negative charges (electrons) are responsible for the n-type behavior, positive charges (holes) are responsible for p-type behavior.

The difficulty in achieving p-type oxides with similar performances to n-type ones originates from the specific electronic configuration of oxide materials. In the case of n-type oxides, the presence of oxygen vacancies produces enough electrons, and the electron-transport path around the CBM is mainly composed of metal s orbitals. The highly dispersed and delocalized CBM facilitates a low electron effective mass and thus high mobility [48]. On the other hand, in p-type oxides, the creation of holes (positive carriers) is limited by two main factors: the low formation energy of native acceptors that produce holes, such as cation vacancies, and the low formation energy of native donors that annihilate holes, such as anion vacancies; these result in a large hole effective mass and low mobility [49]. Taking this into account, it is more evident why the fabrication of high-performance p-type oxides is still a challenging task. In this context, different strategies have been developed in recent years in order to achieve one goal: to reduce the localization of the valence band edge and thereby increase the carrier mobility.

In this section, we present a brief description of some of the most interesting properties of the p-types oxides, such as NiO, SnO, and Cu_2O, as well as the recent advances in their implementation in relevant technological applications.

3.3.2.1 Nickel oxide, NiO

This oxide has an intrinsic p-type conductivity with an energy bandgap of between 3.6 and 4.3 eV [50]. Overall, NiO exhibits a non-stoichiometric form with native Ni deficiency that is responsible for the p-type conductivity, which is usually associated with the formation of Ni^{3+} to maintain charge neutrality [51]. NiO adopts the NaCl structure, which has octahedral Ni^{2+} and O^{2-} sites, as shown in figure 3.15(a). This p-type oxide is one of the most promising materials due to its high thermal and chemical stability, low toxicity, and abundance. NiO is currently being extensively studied for implementation as the p-type side of p–n junctions based on oxides.

3.3.2.2 Tin monoxide, SnO

In contrast to n-type SnO_2, SnO possesses p-type conductivity and an energy bandgap of between 2.5 and 3.4 eV. [52, 53]. The relatively higher mobility in SnO is largely due to the low defect-formation energy of Sn vacancies, which serve as acceptors, and to the creation of a suitable dispersed valence band maximum by the hybridization of Sn 5s and oxygen 2p orbitals [49]. Tin monoxide often exhibits a layered structure with a tetragonal crystal structure, as shown in figure 3.15(b). Although this material has generated increasing interest in recent years due to its low toxicity and high abundance, its stability over time is still one of its main drawbacks.

3.3.2.3 Copper oxide (Cu_2O) and Cu-based compounds ($CuMO_2$)

Cu_2O, which has a direct bandgap of around 2.5 eV [54], exhibits a simple cubic crystal structure in which the Cu atoms are arranged in an fcc sublattice and oxygen

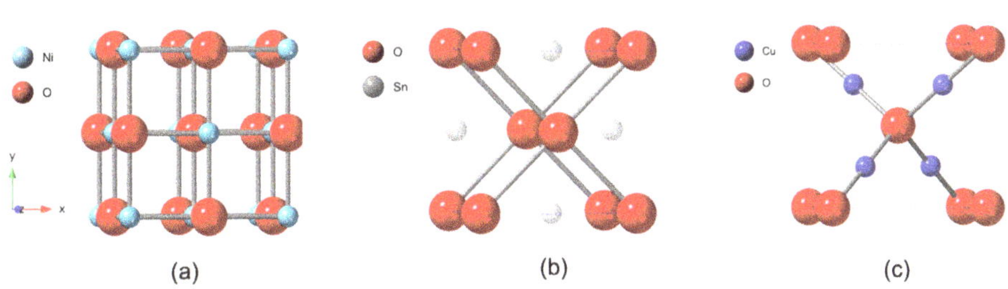

Figure 3.15. Crystalline structure from (a) NiO, (b) SnO and (c) Cu_2O.

atoms occupy the tetrahedral sites [55], as shown in figure 3.15(c). This oxide has been considered to be a promising p-type oxide for a long time and still has the best p-type Hall mobility value [56], which exceeds 100 $cm^2 V^{-1} s^{-1}$. The p-type character of Cu_2O has been theoretically studied and is largely attributed to the formation of copper vacancies. Cu_2O has a unique band structure in which the valence band is composed of fully occupied hybridized levels from the Cu 3d and O 2p orbitals. The top of the valence band is composed of Cu-d states, which creates a less-localized hole-transport pathway [57, 58]. Among Cu-based compounds, a novel family of materials, namely delafossites, with the formula $CuMO_2$ (M = Al, Ga, and In) has been widely studied in recent years. However, due to the presence of a high valence band maximum tail state, these Cu-based oxides suffer from either low hole mobility or unsuitable carrier concentrations [49].

3.4 2D materials

Fueled by the advent of graphene—the first truly two-dimensional (2D) material [59]—in 2004, the effort to understand the fundamental properties and to seek novel applications of single-layer materials and van der Waals solids has been growing at an ever-increasing rate [60, 61]. Graphene and (more recently) transition-metal dichalcogenides (TMDs) stand out among 2D materials due to their unique electronic and mechanical properties. A number of other 2D materials and their applications in energy-related devices are presented in chapter 7. Here, we focus on the basic properties of graphene and TMDs.

3.4.1 Graphene

Graphene is a 2D crystalline allotrope of carbon. The carbon atoms in graphene are tightly bound in a hexagonal honeycomb lattice [see figure 3.16(a)]. Three of the four unpaired electrons in the excited state [see figure 3.1(a)] form sp^2 hybrid orbitals pointing towards the three vertices of an equilateral triangle and give rise to the threefold coordination of C atoms in graphene (σ bonds). The remaining electron is distributed throughout the whole crystal lattice and forms π bonds [62]. π electrons are responsible for the remarkable electronic properties of graphene, which has an electron mobility of more than 2×10^5 $cm^2 V^{-1} s^{-1}$. Its band structure can be calculated with reasonable accuracy using the nearest-neighbor tight-binding approximation [63]:

$$E_k = \pm t \left[1 + 4\cos\left(\frac{3}{2}k_x a\right)\cos\left(\frac{\sqrt{3}}{2}k_y a\right) + 4\cos^2\left(\frac{\sqrt{3}}{2}k_y a\right) \right]^{1/2}, \tag{3.16}$$

where $t \simeq 2.8$eV is the hopping parameter and $a = 0.142$nm is the C–C distance. In undoped graphene, the two bands lie symmetrically above and below the Fermi energy $E_F = 0$, as shown in figure 3.16(b). The Fermi energy, however, can be shifted upwards or downwards, inducing an excess of charge carriers (either electrons or holes) by the application of a gate voltage or by chemical doping [64–66]. Novoselov *et al* demonstrated that the density of carriers induced by a gate voltage reaches

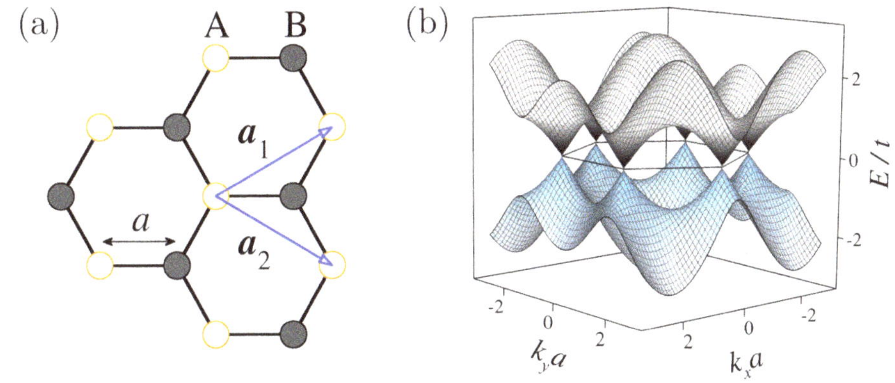

Figure 3.16. (a) A honeycomb lattice of graphene, formed by two interpenetrated sublattices, A and B. $\boldsymbol{a}_1$ and $\boldsymbol{a}_2$ are primitive vectors and the C–C distance is $a = 0.142nm$. (b) π (lower surface) and π^* (upper surface) bands of graphene over the whole (hexagonal) Brillouin zone in the nearest-neighbor approximation. Energy is expressed in units of the hopping parameter $t \simeq 2.8eV$. The two bands touch each other at the corners of the Brillouin zone (K and K' points), where the dispersion is linear in momentum (Dirac cones).

values up to 10^{17} m^{-2} [59]. Bilayer graphene is a material consisting of two layers of graphene bonded by van der Waals forces. Interestingly, an electric field perpendicular to the layers can induce a tunable gap, opening possibilities for applications in electronics. Furthermore, when two graphene sheets have been stacked and twisted close to the *magic angle* of 1.1°, the system can host an unconventional superconducting phase [67].

In addition to its remarkable electronic properties, graphene is the stiffest material ever discovered, with an ultimate tensile strength of 130 GPa [68]. First-principles calculations have found that the Young's moduli of ordered graphene oxides vary from 380 to 470 GPa as the coverage of oxygen groups changes, whereas the Young's moduli of amorphous graphene oxides with comparable coverage are smaller at 290–430 GPa [69]. In addition, due to the strong σ bonding, graphene is an outstanding heat conductor, with a thermal conductivity as high as 2000–4000 W $(mK)^{-1}$ [70].

In energy-related technologies, graphene is widely used as the material of choice in batteries. Graphene-based batteries can handle up to 1500 charge cycles and generate less heat than normal ones at an equal charge power. Furthermore, due to its remarkable mechanical properties, it is more resilient to known issues in normal batteries, such as puncturing.

3.4.2 Transition-metal dichalcogenides

TMDs are semiconductors with a composition given by MX_2, where M is a transition-metal atom (such as an atom of Mo, W, or Ti) and X is a chalcogen atom (such as and atom of S, Se, or Te) (see references [71, 72] for a review of the main properties of TMDs). One of the most studied layered materials in this family is MoS_2 (see figure 3.17). The absence of inversion symmetry results in a spin split in the electronic bands. TMDs have emerged as a promising alternative to graphene

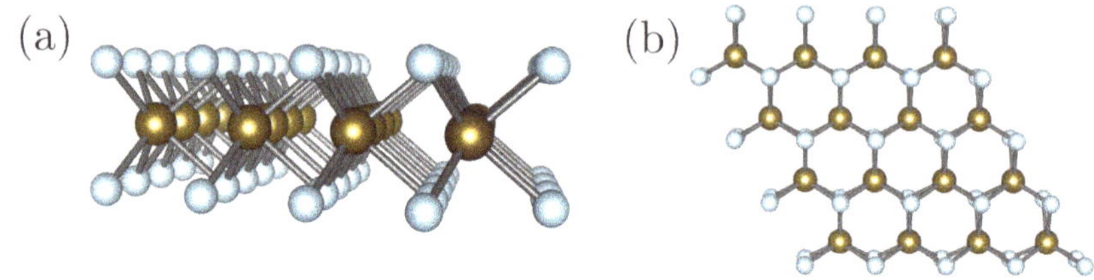

Figure 3.17. The crystal structure of a MoS_2 monolayer. (a) Side view and (b) top view.

because of their unique combination of physical and chemical properties (a direct bandgap and a strong spin–orbit interaction, to name some properties not shared with graphene). In particular, the relatively large separation between stacked MoS_2 layers and their high specific surface area make this material highly attractive for capacitive energy storage (e.g. supercapacitors and batteries) [73] and heat-to-electricity conversion (see chapter 7).

References

[1] Seitz F 1940 *The Modern Theory of Solids* (New York: McGraw-Hill)

[2] Ashcroft N W and Mermin N D 1976 *Solid State Physics* (Philadelphia, PA: Saunders College)

[3] Patterson J D and Bailey B C 2019 *Solid-State Physics: Introduction to the Theory* (Berlin: Springer International Publishing)

[4] Kittel C 1986 *Introduction to Solid State Physics* (New York: Wiley)

[5] Lovell M C, Avery A J and Vernon M W 2012 *Physical Properties of Materials* (Heidelberg: Springer)

[6] Pauling L 1931 The nature of the chemical bond. Application of results obtained from the quantum mechanics and from a theory of paramagnetic susceptibility to the structure of molecules *J. Am. Chem. Soc.* **53** 1367–400

[7] Bloch F 1929 Über die Quantenmechanik der Elektronen in Kristallgittern *Z. Phys.* **52** 555–600

[8] Harriwon P and Valavanis A 2016 *Quantum Wells, Wires and Dots* (Chichester: Wiley)

[9] Drude P 1900 Zur Elektronentheorie der Metalle; I. Teil. *Ann. Phys.* **306** 566–613

[10] Drude P 1900 Zur Elektronentheorie der Metalle; II. Teil. Galvanomagnetische und thermomagnetische Effecte *Ann. Phys.* **308** 369–402

[11] Datta S 1997 *Electronic Transport in Mesoscopic Systems* (Cambridge: Cambridge University Press)

[12] Sommerfeld A 1928 Zur Elektronentheorie der Metalle auf Grund der Fermischen Statistik *Z. Phys.* **47** 1–32

[13] Ferry D K, Goodnick S and Bird J 2009 *Transport in Nanostructures* (Cambridge: Cambridge University Press)

[14] Ryndyk D A 2016 *Theory of Quantum Transport at Nanoscale: An Introduction* (Berlin: Springer International Publishing)

[15] Boaretto N, Rana M, Marcilla R and Vilatela J 2020 Revealing the mechanism of electrochemical lithiation of carbon nanotube fibers *ACS Appl. Energy Mater.* **3** 8695–705

[16] Kunduraci M and Çağlayan U 2022 Spherical and core-shell-structured $LiMn_{1.5}Ni_{0.5}O_4$ lithium-ion battery cathode with enhanced cyclability *J. Appl. Electrochem.* **52** 477–86

[17] Wang C *et al* 2020 Unveiling the critical role of interfacial ionic conductivity in all-solid-state lithium batteries *Nano Energy* **72** 104686

[18] Lin X *et al* 2021 A high-performance, solution-processable polymer/ceramic/ionic liquid electrolyte for room temperature solid-state Li metal batteries *Nano Energy* **89** 106351

[19] Green M A 1986 *Solar Cells: Operating Principles, Technology, and System Applications* (Kensington N.S.W: University of New South Wales)

[20] Swatowska B and Stapinski T 2008 Amorphous hydrogenated silicon-nitride films for applications in solar cells *Vacuum* **82** 942–6

[21] Chakravarty B C, Vinod P N, Singh S N and Chakraborty B R 2002 Design and simulation of antireflection coating for application to silicon solar cells *Sol. Energy Mater. Sol. Cells* **73** 59–66

[22] Fox M 2007 *Optical Properties of Solids* (Oxford: Oxford University Press)

[23] Kitamura R, Pilon L and Jonasz M 2007 Optical constants of silica glass from extreme ultraviolet to far infrared at near room temperature *Appl. Opt.* **46** 8118–33

[24] Gaponenko S V 2010 *Introduction to Nanophotonics* (Cambridge: Cambridge University Press)

[25] Xie Q, Ming S, Chen L, Wu Y, Zhang W, Liu X, Cao M, Wang H-Q and Fang J 2019 Parameters in planar quantum dot-polymer solar cell: tuned by QD Eg, ligand exchange and fabrication process *Org. Electron.* **69** 1–6

[26] Kongkanand A, Tvrdy K, Takechi K, Kuno M and Kamat P V 2008 Quantum dot solar cells. Tuning photoresponse through size and shape control of CdSe-TiO_2 architecture *J. Am. Chem. Soc.* **130** 4007–15

[27] Fischer F D, Waitz T, Vollath D and Simha N K 2008 On the role of surface energy and surface stress in phase-transforming nanoparticles *Prog. Mater. Sci.* **53** 481–527

[28] Li M, Xing G, Nien Ah Qune L F, Xing G, Wu T, Huan C H A, Zhang X and Sum T C 2012 Tailoring the charge carrier dynamics in ZnO nanowires: the role of surface hole/electron traps *Phys. Chem. Chem. Phys.* **14** 3075–82

[29] Bardeen J 1947 Surface states and rectification at a metal semi-conductor contact *Phys. Rev.* **71** 717–27

[30] Singh D K, Roul B, Pant R, Chowdhury A M, Nanda K K and Krupanidhi S B 2020 Different types of band alignment at MoS_2/(Al, Ga, In)N heterointerfaces *Appl. Phys. Lett.* **116** 252102

[31] Coll M *et al* 2019 Towards oxide electronics: a roadmap *Appl. Surf. Sci.* **482** 1–93 Roadmap for Oxide Electronics, pushed forward by the TO-BE EU Cost action

[32] Wang Z L 2003 Nanobelts, nanowires, and nanodiskettes of semiconducting oxides–from materials to nanodevices *Adv. Mater.* **15** 432–6

[33] Linderälv C, Lindman A and Erhart P 2018 A unifying perspective on oxygen vacancies in wide band gap oxides *J. Phys. Chem. Lett.* **9** 222–8

[34] Godinho K G, Walsh A and Watson G W 2009 Energetic and electronic structure analysis of intrinsic defects in SnO_2 *J. Phys. Chem.* C **113** 439–48

[35] Maestre D, Cremades A and Piqueras J 2005 Growth and luminescence properties of micro- and nanotubes in sintered tin oxide *J. Appl. Phys.* **97** 044316

[36] Haider A J, Jameel Z N and Al-Hussaini I H M 2019 Review on: Titanium dioxide applications *Energy Procedia* **157** 17–29

[37] Zhang H and Banfield J F 2014 Structural characteristics and mechanical and thermodynamic properties of nanocrystalline TiO_2 *Chem. Rev.* **114** 9613–44

[38] Kołodziejczak-Radzimska A and Jesionowski T 2014 Zinc oxide–from synthesis to application: a review *Materials* **7** 2833–81

[39] Jagadish G and Pearton S 2006 *Zinc Oxide Bulk, Thin Films and Nanostructures* (Oxford: Elsevier Science)

[40] Pearton S J, Yang J, Cary P H, Ren F, Kim J, Tadjer M J and Mastro M A 2018 A review of Ga_2O_3 materials, processing, and devices *Appl. Phys. Rev.* **5** 011301

[41] Spencer J A, Mock A L, Jacobs A G, Schubert M, Zhang Y and Tadjer M J 2022 A review of band structure and material properties of transparent conducting and semiconducting oxides: Ga_2O_3, Al_2O_3, In_2O_3, ZnO, SnO_2, CdO, NiO, CuO, and Sc_2O_3 *Appl. Phys. Rev.* **9** 011315

[42] Pearton S J, Ren F, Tadjer M and Kim J 2018 Perspective: Ga_2O_3 for ultra-high power rectifiers and MOSFETS *J. Appl. Phys.* **124** 220901

[43] Chen X, Ren F-F, Ye J and Gu S 2020 Gallium oxide-based solar-blind ultraviolet photodetectors *Semicond. Sci. Technol.* **35** 023001

[44] Alonso-Orts M, Nogales E and Méndez B 2020 Optical properties of gallium oxide nanostructures *Wide Bandgap Semiconductor-Based Electronics* (Bristol: IOP Publishing), 8 8.1–8.26

[45] Zhang Y, Dong Z and Li P 2021 Construction of electron and grain boundary barrier in quantum dots light-emitting diodes: the role of nio interface coating *Opt. Mater.* **117** 111204

[46] Vajda M, Ursu D, Mosoarca C, Duteanu N and Miclau M 2021 Experimental investigation of hydrogen insertion in copper oxide on photovoltaic performance of p-type dye-sensitized solar cell *Int. J. Energy Res.* **45** 5309–17

[47] Yang X, Liang H-J, Zhao X-X, Yu H-Y, Wang M-Y, Nie X-J and Wu X-L 2021 A sandwich nanocomposite composed of commercially available SnO and reduced graphene oxide as advanced anode materials for sodium-ion full batteries *Inorg. Chem. Front.* **8** 396–404

[48] Nomura K, Ohta H, Takagi A, Kamiya T, Hirano M and Hosono H 2004 Room-temperature fabrication of transparent flexible thin-film transistors using amorphous oxide semiconductors *Nature* **432** 488–92

[49] Wang Z, Nayak P K, Caraveo-Frescas J A and Alshareef H N 2016 Recent developments in p-type oxide semiconductor materials and devices *Adv. Mater.* **28** 3831–92

[50] Yang H, Tao Q, Zhang X, Tang A and Ouyang J 2008 Solid-state synthesis and electrochemical property of SnO_2/NiO nanomaterials *J. Alloys Compd.* **459** 98–102

[51] Enke C G 1974 Nonstoichiometry, diffusion, and electrical conductivity in binary metal oxides *Mater. Corros.* **25** 801–2

[52] Dias J S, Batista F R M, Bacani R and Triboni E R 2020 Structural characterization of SnO nanoparticles synthesized by the hydrothermal and microwave routes *Sci. Rep.* **10** 9446

[53] Batzill M and Diebold U 2005 The surface and materials science of tin oxide *Prog. Surf. Sci.* **79** 47–154

[54] Murali D S, Kumar S, Choudhary R J, Wadikar A D, Jain M K and Subrahmanyam A 2015 Synthesis of Cu_2O from CuO thin films: optical and electrical properties *AIP Adv.* **5** 047143

[55] Gao L, Pang C, He D, Shen L, Gupta A and Bao N 2015 Synthesis of hierarchical nanoporous microstructures via the Kirkendall effect in chemical reduction process *Sci. Rep.* **5** 16061

[56] Li B S, Akimoto K and Shen A 2009 Growth of Cu2O thin films with high hole mobility by introducing a low-temperature buffer layer *J. Cryst. Growth* **311** 1102–5

[57] Nolan M and Elliott S D 2006 The p-type conduction mechanism in Cu_2O: a first principles study *Phys. Chem. Chem. Phys.* **8** 5350–8
[58] Greiner M T, Helander M G, Tang W-M, Wang Z-B, Qiu J and Lu Z-H 2012 universal energy-level alignment of molecules on metal oxides *Nat. Mater.* **11** 76–81
[59] Novoselov K S, Geim A K, Morozov S V, Jiang D, Zhang Y, Dubonos S V, Grigorieva I V and Firsov A A 2004 Electric field effect in atomically thin carbon films *Science* **306** 666–9
[60] Das S, Robinson J A, Dubey M, Terrones H and Terrones M 2015 Beyond graphene: progress in novel two-dimensional materials and van der Waals solids *Annu. Rev. Mater. Res.* **45** 1–27
[61] Khan K, Tareen A K, Aslam M, Wang R, Zhang Y, Mahmood A, Ouyang Z, Zhang H and Guo Z 2020 Recent developments in emerging two-dimensional materials and their applications *J. Mater. Chem.* C **8** 387–440
[62] Katsnelson M I 2012 *Graphene: Carbon in Two Dimensions* (Cambridge: Cambridge University Press)
[63] Wallace P R 1947 The band theory of graphite *Phys. Rev.* **71** 622–34
[64] Liu H, Liu Y and Zhu D 2011 Chemical doping of graphene *J. Mater. Chem.* **21** 3335–45
[65] Guo B, Fang L, Zhang B and Gong J R 2011 Graphene doping: a review *Insciences* J **1** 80–9
[66] Pinto H and Markevich A 2014 Electronic and electrochemical doping of graphene by surface adsorbates *Beilstein J. Nanotechnol* **5** 1842–8
[67] Cao Y, Fatemi V, Fang S, Watanabe K, Taniguchi T, Kaxiras E and Jarillo-Herrero P 2018 Unconventional superconductivity in magic-angle graphene superlattices *Nature* **556** 43–50
[68] Papageorgiou D G, Kinloch I A and Young R J 2017 Mechanical properties of graphene and graphene-based nanocomposites *Prog. Mater. Sci.* **90** 75–127
[69] Liu L, Zhang J, Zhao J and Liu F 2012 Mechanical properties of graphene oxides *Nanoscale* **4** 5910–6
[70] Pop E, Varshney V and Roy A K 2012 Thermal properties of graphene: fundamentals and applications *MRS Bull.* **37** 1273–81
[71] Manzeli S, Ovchinnikov D, Pasquier D, Yazyev O V and Kis A 2017 2D transition metal dichalcogenides *Nat. Rev. Mater.* **2** 17033
[72] Choi W, Choudhary N, Han G H, Park J, Akinwande D and Lee Y H 2017 Recent development of two-dimensional transition metal dichalcogenides and their applications *Mater. Today* **20** 116–30
[73] Choudhary N, Patel M D, Park J, Sirota B and Choi W 2016 Synthesis of large scale MoS_2 for electronics and energy applications *J. Mater. Res.* **31** 824–31

IOP Publishing

Sustainable Nanomaterials for Energy Applications

Ana Cremades and David Maestre

Chapter 4

Introducing nanoconcepts used in solar cell development

Aleksandra B Djurišić, Annie Ng, Gaukhar Nigmetova, Fangzhou Liu, Zhilin Ren and Yi Zhang

In this chapter, we discuss the applications of nanomaterials for clean energy generation using solar cells. We focus on oxide nanomaterials, as they are the most commonly used nanomaterials in this type of device, although other nanomaterials are also discussed where relevant. We discuss the functions of nanomaterials in different types of solar cell and discuss how their properties (material choice, material morphology, and material doping and/or defect engineering) affect solar cell performance. For each type of solar cell, we provide an overview of the main challenges and problems to be solved. We mainly focus on emerging photovoltaic technologies, though we briefly mention the relevant functions of metal oxides and other nanomaterials for other types of solar cell. In the case of the most common application of metal oxide nanomaterials, namely as charge transport layers, we provide a broad overview of both electron transport and hole transport materials in organic, dye-sensitized, quantum dot, and perovskite solar cells, with a particular focus on perovskite solar cells, as these are most promising of the emerging photovoltaic technologies. Our discussion of the challenges and future outlook includes a review of the different types of photovoltaic technology and the identification of today's most promising materials for different types of application.

4.1 Introduction

Nanomaterials and nanostructuring play important roles in solar cell development, although the degree of importance, the number of nanomaterials used, and their functions in the device are dependent on the type of solar cell. In this chapter, we primarily focus on emerging photovoltaic technologies, namely organic, dye-sensitized, quantum-dot-sensitized, and perovskite solar cells. This is due to the fact that these technologies commonly employ nanomaterials/nanostructures and

doi:10.1088/978-0-7503-3531-7ch4

© IOP Publishing Ltd 2023

thus they are highly relevant to the topic of this chapter. The record efficiencies for different types of solar cell are summarized in the National Renewable Energy Laboratory (NREL) efficiency chart shown in figure 4.1. It can be observed that the efficiencies for emerging photovoltaics have improved significantly, and that several of these technologies are outperforming amorphous silicon solar cells. In some cases, for example, that of perovskite solar cells, the record efficiencies exceed those achieved for different conventional thin-film technologies, such as CdTe, multi-crystalline Si, copper indium gallium selenide (CIGS), etc. The problem of device stability remains, in addition to some technology-specific issues (lead toxicity for perovskite solar cells, for example), but the efficiencies have achieved the necessary minimum for potential commercialization. In addition, these emerging technologies offer not only the potential for lower-cost fabrication, but also significantly lower energy payback times and lower environmental impacts. For example, flexible tandem organic modules were found to have 3%–10% of the environmental impact of multi-Si or CdTe modules for the same device area [1]. In addition to discussing the use of nanomaterials in emerging photovoltaics, we also mention some applications of nanostructures in light management, as well as transparent contacts and recombination layers for tandem devices in more conventional technologies.

In terms of the materials covered, the major part of the chapter is devoted to metal oxide nanomaterials, since they can serve multiple functions in devices, including charge transport, electrodes, and encapsulation (for a recent short review of functional metal oxides in perovskite solar cells, see reference [2]). The most commonly used metal oxide materials in solar cells are various charge transport layers, namely TiO_2, ZnO, and SnO_2 among the n-type oxides and NiO_x and copper-based oxides among the p-type oxides. A common function of these materials is to serve as electron or hole charge transport layers, but they can also be used as parts of the electrode, light management structures, or recombination layers in tandem cells.

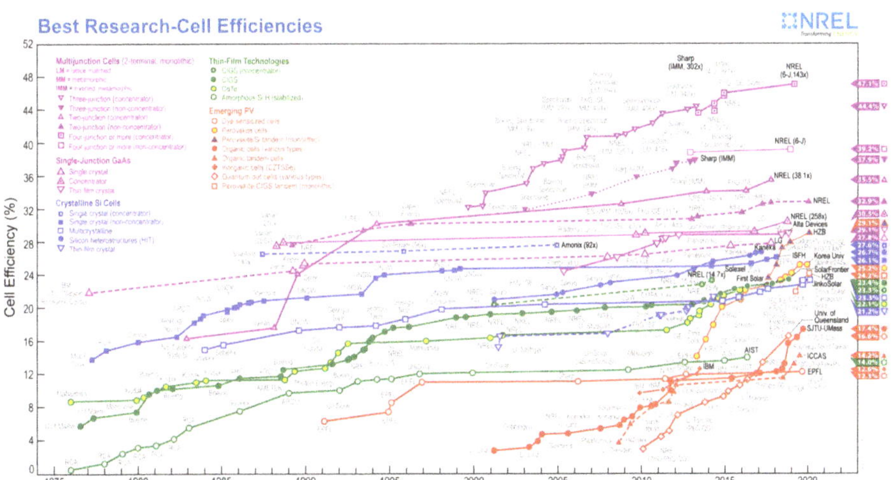

Figure 4.1. Efficiency chart of different solar cell technologies. This plot is provided by courtesy of the National Renewable Energy Laboratory, Golden, CO, USA.

There have been a number of comprehensive reviews of different types of metal oxide nanomaterial for solar cell applications, including TiO_2 [3], ZnO [4], etc. We particularly focus on low-temperature deposition methods for nanomaterials, since these are of interest for flexible solar cells. Flexible devices are particularly suitable for low-cost large-scale manufacturing, for example using the roll-to-roll technique, slot-die coating, doctor blading, and spray coating; they have excellent power-to-weight ratios and lower transportation and installation costs [5]. In addition, they are of interest for some niche applications, such as wearable devices and self-powered electronics [6, 7]. While the efficiency of flexible devices typically lags behind that of rigid devices, the record efficiencies achieved for some technologies (19.11% for conventional and 18.3% for inverted perovskite solar cells, for example) is sufficient for commercialization, provided that the stability issues are addressed [5].

Here, we mainly focus on recent literature (the majority of the references were published in the last six years), and we discuss the problems which need to be solved for different types of device. Some of these issues are common to all types of device, namely surface modifications/interfacial layers used to achieve favorable energy level alignment and/or passivate interfacial defects, while others are material- or technology-specific (for example, the instability of ZnO/perovskite interfaces). Due to the high level of interest in perovskite solar cells, a major part of the discussion is devoted to these devices, but we also discuss recent progress and applications of various nanomaterials in organic (small molecule and polymer) photovoltaics, as well as quantum dot (QD)-sensitized solar cells. Dye-sensitized solar cells (DSSCs), as devices which share a basic architecture with QD-sensitized solar cells (QDSSCs) and perovskite solar cells (PSCs) are briefly discussed as well. This is because interest in DSSCs has been declining with the rise of perovskite solar cells. DSSCs with liquid electrolytes exhibit higher efficiency than solid-state DSSCs, while the reverse is true for perovskite solar cells. The use of liquid electrolytes results in difficulties for cheap and reliable mass production, combined with significantly lower efficiencies compared to perovskite devices; liquid electrolyte efficiencies have not improved significantly in decades, making this type of device less competitive than other emerging photovoltaic technologies. It is thus anticipated that the research into these devices will continue to decline, although they remain potentially interesting for some niche applications, for example, in fiber-shaped devices [6].

In addition to a comprehensive discussion of the use of nanomaterials as charge transport layers, we also provide an overview of the use of nanomaterials in active/adsorber layers of different types of solar cell, followed by a discussion of their applications in electrodes and tandem cells. For convenience, we have also summarized the commercial nanomaterial sources for various applications in different types of solar cell, as shown in table 4.1. While the highest-performing devices are commonly made using nanomaterials synthesized by researchers, commercial nanomaterials (especially nanoparticle dispersions) offer the advantage of convenience and possibly improved reproducibility, since nanomaterial properties are notoriously sensitive to small variations in synthesis conditions (including ambient environment factors). Finally, we also briefly discuss the use of

Table 4.1. Commercially available nanomaterials for solar cell applications. The cited references are examples, not an exhaustive list, due to the large number of papers using materials from popular suppliers, especially for DSSC applications and mesoporous devices in general. Acronyms used: aluminum-doped ZnO (AZO), solar cell (SC), organic photovoltaic (OPV).

NPs	Solar cell	Supplier	Reference
P25	PSC	Degussa (Evonik)	[8]
TiO_2	PSC	GreatCell Solar	[9]
TiO_2	PSC	Solaronix	[10]
TiO_2	PSC	Sharif Solar	[11]
TiO_2	DSSC	Sigma-Aldrich	[12]
TiO_2	QDSSC	Alfa Aesar	[13]
NiO	PSC	Sigma-Aldrich	[10]
NiO	DSSC	Solaronix	[14]
NiO	DSSC	Inframat	[14]
NiO	DSSC	Dyenamo	[14]
ZnO	DSSC	Aladdin	[15]
ZnO	DSSC	Alfa Aesar	[16]
ZnO	DSSC	Sigma-Aldrich	[17]
AZO	PSC	Sigma-Aldrich	[18]
AZO	Polymer SC	US Research Nanomaterials	[19]
Al_2O_3	PSC	Aladdin	[20]
Al_2O_3	PSC	Sigma-Aldrich	[21]
Al_2O_3	QDSSC	Degussa (Evonik)	[22]
CuO	DSSC	Iolitec	[23]
SnO_2	PSC	Alfa Aesar	[24]
ZrO_2	PSC	Solaronix	[25]
WO_{3-x}	PSC	Sigma-Aldrich	[26]
Au	OPV	Sigma-Aldrich	[27]
Au@SiO_2	OPV	nanoComposix	[28]
$CuAlO_2$/Cu–O	PSC	Avantama Ltd	[29]
NiO	PSC	High Purity Chemicals, Japan	[30]

nanomaterials in encapsulation and then provide a future outlook and discuss some material- and device-specific challenges. It should be noted that here we cover a number of broad topics, many of which (electrodes, different charge transport materials, for example) could be the topics of an entire chapter on their own. Hence, we focus more on breath rather than depth of coverage, and where relevant we refer readers to a recent comprehensive review paper for deeper coverage of a particular topic.

4.2 Nanomaterials as charge transport layers in solar cells

In this section, we discuss the use of nanomaterials as charge transport layers in different types of solar cell. Charge transport layers generally serve as charge-selective contacts in a device, and the interfaces between different layers play a key function in device performance (this is particularly important for PSCs as well as QDSSCs) [31]. The chapter is organized by the type of material, since the same materials may be used as electron transport layers (ETLs) or hole transport layers (HTLs) in different types of solar cell. While some of the challenges are shared among different types of solar cell, for example, the need to passivate interfacial defects, some issues (specific processing limitations) are applicable only to a specific technology or specific application (for example, when the charge transport layer is deposited on top of the active material), and we will outline those issues when discussing the application of a certain material in a certain type of solar cell.

We will mainly consider emerging photovoltaics, namely dye-sensitized, quantum-dot-sensitized, perovskite, and organic solar cells. Dye-sensitized solar cells typically consist of a thick mesoporous layer, an organic dye, a liquid electrolyte, and a counter electrode [3]. The organic dye serves as the light absorber, and the photoexcited electrons are transferred from the dye to the mesoporous charge transport layer (usually titania), which conducts electrons to the electrode, and the dye is regenerated by the electrolyte [3]. To enhance compatibility with more traditional deposition and encapsulation processes, solid-state DSSCs have been investigated using polymers rather than liquid electrolytes, but the efficiencies achieved were typically less than those achieved using liquid electrolytes [3]. As can be observed from the NREL efficiency chart shown in figure 4.1, since the initial breakthrough in 1991, progress on DSSCs has been quite slow and their efficiency has stagnated at around 12% for a number of years.

From the basic idea of DSSCs, two different classes of device have been developed, namely QDSSCs and PSCs. Both of these currently outperform DSSCs, reaching record efficiencies 16.6% and 25.2% for QDSSCs and PSCs, respectively. In QD-based cells, inorganic quantum dots, typically composed of CdS, CdSe, PbS, Bi_2S_3, or InP have replaced organic dyes and sensitizers, providing increased light harvesting due to better bandgap tunability and higher extinction coefficients [3]. The surfaces and interfaces used as contacts between different types of nanomaterial play even more significant roles in this type of device than in other types of solar cell [3]. In PSCs, organic–inorganic metal halide perovskite materials have been used as sensitizers to replace organic dyes [3]. Interestingly, PSCs exhibit significantly improved efficiency and stability in solid-state architectures, in contrast to DSSCs, which is attributed to the dissolution of the perovskite in an iodide-containing electrolyte [3]. The device architecture can involve mesoscopic and planar devices, as illustrated in figures 4.2(a) and (b). Devices with an ETL at the bottom are commonly described as having a conventional architecture, while those with an HTL at the bottom are commonly known as inverted-architecture devices.

Organic photovoltaics (OPVs) are another highly promising emerging photovoltaic technology. OPVs have recently achieved a record efficiency of 17.4%, as shown in figure 4.1. This type of device can be based on polymers, and in this case typically involves solution processing or small molecules, which can be both solution processed or vapor deposited. While the use of nanomaterials as charge transport layers is very common in DSSCs, QDSSCs, and PSCs and the construction of such

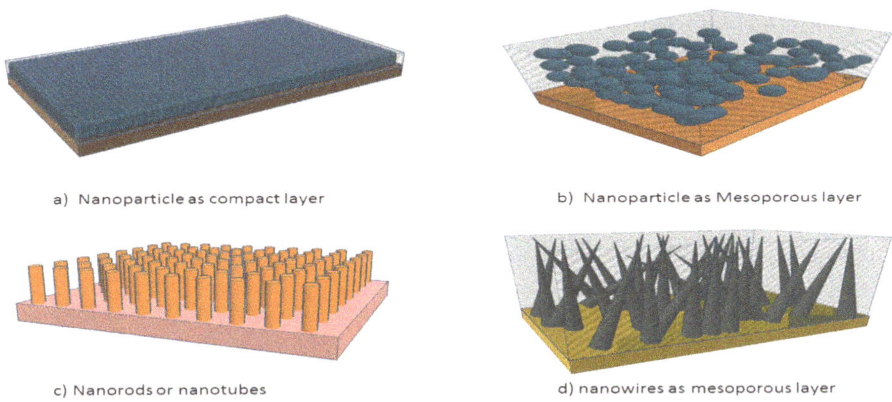

Figure 4.2. An illustration of different charge transport layer morphologies.

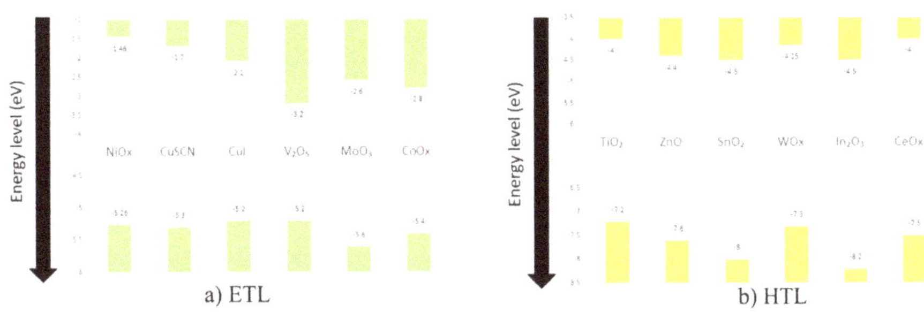

Figure 4.3. The energy levels of different charge transport materials. The energy levels were extracted from references [34–44].

devices without nanomaterials is rare, the situation is reversed for OPVs since thin-film organic charge transport layers are more commonly used. In addition, the common usage of architectural labels, i.e. conventional and inverted, is swapped: in the case of OPVs, 'conventional architecture' refers to the architecture with the HTL at the bottom of the film stack [32]. Nevertheless, there is still an interest in the use of inorganic nanomaterials as charge transport layers in OPVs, due to their potential for improved lifetimes and environmental stability, as well as possible advantages in energy level alignment (more efficient hole collection for HTLs, more efficient hole blocking for ETLs). The energy levels of common inorganic charge transport materials are summarized in figure 4.3. It should be noted that for this type of device, there are somewhat different considerations than for DSSCs, QDSSCs, and PSCs; due primarily to the low thickness of the organic layers, planar architectures or short nanorod arrays are typically preferred. The limiting factors for the performance of OPVs are the low conductivities of organic materials and short exciton diffusion lengths (excitons instead of free carriers are generated upon photoexcitation due to high exciton binding energy, which are then dissociated at the interfaces between donor and acceptor materials), which necessitate the use of bulk heterojunctions and relatively low overall device thicknesses [33].

Furthermore, for the majority of the devices reported in the literature, charge transport layers are commonly deposited on the substrate, which is typically a transparent conductive-oxide-coated glass, such as fluorine-doped tin oxide (FTO)/glass or indium-doped tin oxide (ITO)/glass. While flexible devices prepared on ITO/plastic substrates are subject to processing condition limitations in order to avoid damage to the plastic substrates, the use of nanomaterials as charge transport layers located below the active layer is generally simpler and subject to fewer restrictions compared to nanomaterial deposition as a top charge transport layer. The deposition of a nanostructured charge transport layer on top of the active layer, rather than below the active layer, can result in significant processing challenges, since the solvents used and/or deposition temperatures are limited by the lower layers. Nevertheless, the successful deposition of ETLs and HTLs as top charge transport layers has been demonstrated in different types of device. For example, in OPVs, the use of metal oxide materials as top charge transport layers has been reported for ZnO [45], graphene oxide/NiO_x [34], MoO_x [34], TiO_x [46], WO_3 [47], etc. In perovskite solar cells, there have also been examples of the successful deposition of different charge transport layers on top of the active layer, for example ZnO [48], SnO_2 [49], NiO_x [50–52], a $CuCrO_2$-based hole transport layer [53], etc.

4.2.1 Electron transport layers

The general requirement for electron transport materials is that they should have low parasitic absorption, high electron mobility, and a suitable energy level alignment with the absorbing layer for efficient electron collection and low recombination losses. The most commonly used ETLs are TiO_2, ZnO, and SnO_2, although other materials have also been reported for this purpose. As mentioned, in terms of morphology, we can distinguish between devices with mesoporous charge transport layers and planar devices using flat and compact charge transport layers, as illustrated in figures 4.2(a) and (b). In addition to these types of device based on nanoparticles, there are also devices that use 1D nanostructures (nanorods, nanowires, and nanotubes) either as ordered arrays (Figure 4.2(c)) or as randomly oriented 1D nanomaterials (figure 4.2(d)) in an attempt to improve charge transport (although this typically results in a reduced overall surface area compared to the nanoparticle-based mesoporous layers, which hinders overall efficiency despite improving the transport). Nanoparticle films can be deposited by spin coating nanoparticle dispersions (commonly used in planar devices), or by doctor blading or screen printing nanoparticle-containing pastes, followed by annealing, which are commonly used for mesoscopic charge transport layers. Different methods have been reported for the particle synthesis; the main foci have been control of the particle size and the optimization of synthesis conditions to achieve a low-cost, low-temperature, scalable synthesis process [54]. For some of the synthesis procedures, small particle sizes (<5 nm) were achieved, which is of interest for very thin compact ETLs [54]. For a recent review of synthesis methods that produce particles suitable for solar cell charge transport layer applications, see reference [34].

In the following, we discuss the use of different ETLs in various types of solar cell. However, here we would like to highlight some common issues which are shared among different types of device. A common issue when using metal oxide charge transport layers in organic solar cells is the need to use interfacial layers to passivate surface/interface defects and/or adjust the energy level alignment across the metal oxide/absorber layer interface. Oxygen vacancies have been found to be a common type of surface trap in oxide ETLs, and their negative influence on recombination losses has been shown to be mitigated by the deposition of interfacial layers by atomic layer deposition (ALD) [55]. ALD is a particularly convenient technique for this purpose, since it allows a large degree of control over layer thickness and produces conformal coatings. This is a general feature of ALD, and it applies to interfacial layers in other types of solar cell as well. Interfacial layers can also be deposited by other deposition methods, and they can have different chemical compositions. For example, different interfacial layer/charge transport layer combinations proposed in the literature include ZnO/Bi_2S_2O [56], TiO_2/alumina or zirconia [55], ZnO/TiO_2 [57], etc. Improvements in device performance can also be achieved using oxides with high surface hydrogenation, which has been found to affect the morphology and self-organization of organic layers deposited on top of oxide HTLs or ETLs deposited below the organic layer [58]. In addition to defect passivation and energy level adjustment, interfacial layers and surface passivation play important roles in solution-processed devices, since they can alter the wetting of the surface by the solution used for the deposition of the next layer and consequently affect the film quality. This is a concern for both planar devices, for which poor wetting can result in incomplete coverage, and also for mesoscopic devices, for which poor wetting results in incomplete infiltration. The crystallinity of the active layer can also be affected by the interfacial layer [56].

4.2.2 Titanium dioxide

Titanium dioxide remains the most commonly used metal oxide nanomaterial in emerging photovoltaics. For a recent comprehensive review of TiO_2 applications in solar cells, see reference [3]. Titanium dioxide has been used in OPVs, mainly as a compact layer [58], but the early origins of its use in emerging photovoltaics can be traced back to DSSCs, for which a mesoporous layer of titania nanoparticles is commonly used. This mesoporous layer is commonly prepared by annealing a film deposited as a paste containing small titania nanoparticles. Scattering layers containing larger particles are also often included to improve light management [59]. Scattering layers can consist of titania nanostructures or can be based on other oxide materials [59]. Titania is still commonly used in DSSCs and QDSSCs [3]. In addition to nanoparticle-based mesoporous layers that result in high-efficiency DSSCs, other titania morphologies have been used, but mesoscopic thick layers generally tend to work better than nanorod and nanotube arrays due to their larger surface area for dye adsorption. For a recent review of DSSC applications, see reference [60].

TiO_2-based ETLs are also still commonly used in PSCs due to the high versatility of their preparation techniques; they yield good efficiency in PSCs [3]. The earliest types of ETL used in PSCs were single layers of compact TiO_2 (c-TiO_2) and bilayers of c-TiO_2 and mesoporous TiO_2 (mp-TiO_2) nanoparticles, due to the evolution of this architecture from DSSCs. Despite its common use, the use of TiO_2 as an ETL has a number of shortcomings related to manufacturing processes and device stability, such as the high sintering temperature required to achieve the desired semiconductor properties [3] and the strong photocatalytic activity of TiO_2 that leads to PSC degradation after long-term illumination [61]. Nevertheless, these drawbacks have gradually been improved by the progress of research and development [3]. Mesoporous TiO_2 has been regarded as an efficient electron acceptor, since it possesses a large surface area which can facilitate the interfacial charge transfer between the photovoltaic absorber and the photoanode. However, there is considerable interest in exploring other nanostructural morphologies of TiO_2 [62–83] to enhance charge carrier transport efficiency and thus improve the performance of PSCs. In addition to the commonly used TiO_2 nanoparticles, TiO_2 nanorods [62–64, 66, 67–76, 81] and nanowires [65, 77–79] are the most dominant types of nanostructure among the various nanostructural morphologies used in PSCs [80, 82, 83]. Among the different nanostructural morphologies which can result in power conversion efficiencies (PCEs) of more than 18%, TiO_2 nanorods in particular have been identified as very promising compared to other morphologies [78]. In general, the morphologies of nanostructured ETLs can play a significant role in the efficiencies obtained. For example, the growth of closely packed rutile titania nanowires parallel to FTO facets has resulted in a significant improvement in device efficiency due to improved electron collection and reduced recombination losses [84].

It is necessary to further optimize TiO_2 nanostructures, the interfaces of nanostructured ETL/perovskite, and the crystallization of perovskite on nanostructured ETLs. It is noteworthy that the high performances of nanostructured TiO_2-based PSCs (PCE > 19%) are usually attributed to the synergistic effects arising from the increased interfacial area of nanostructures for carrier transport as well as the multiple positive effects induced by optimization strategies [62, 73, 78, 81]. The doping of TiO_2 nanostructures is a common strategy for enhancing the electron conductivity of the ETL and tuning the band alignment at the interface between TiO_2 and perovskite for efficient carrier transport. Different types of dopant, such as Mg [64], Ti [66], N [69], Cd [71], Ta [73], Y [74], Nb [60, 75, 76, 85, 86], F, Cr, Sn, W, and Ni [60] used for nanostructured TiO_2 ETL have been investigated for both DSSC and PSC applications. It is noteworthy that the doping amounts should be carefully controlled, as the presence of dopants in the precursor solution can alter the growth dynamics of TiO_2 nanostructures, leading to a change in the morphology of the nanostructured ETL as well as the crystallization of perovskite atop of the ETL [71, 75]. It has been demonstrated that a similar nanostructural morphology can be still obtained if a certain small amount of dopant within a narrow concentration range is adopted during preparation of ETL [64, 73].

Controlling the interface between the TiO_2 nanostructure and the perovskite is also an important strategy for optimizing the performance of PSCs. This can be achieved by: (i) adjusting the parameters of the growth techniques [3], such as the composition of the precursors, concentration, temperature, time, etc. [63, 68, 72, 79, 80] used to prepare the desired morphology of TiO_2 nanostructure; (ii) applying post-treatments to the as-deposited TiO_2 nanostructures, such as UV ozone treatment [63], $TiCl_4$ treatment, etching, etc. [83]; (iii) incorporating an interlayer for interface modification [62, 67, 78]. An ultrathin TiO_2 interlayer prepared by ALD inserted between FTO and a TiO_2 ETL [62] and another deposited at the interface between TiO_2 nanorods and perovskite [67] have shown their effectiveness in surface passivation. Wu *et al* [78] demonstrated the insertion of a fullerene layer between amorphous TiO_2 nanowires and perovskite to improve the surface work function, which is desirable because it extracts electrons more efficiently at the interface; the resulting low-temperature-processed PSCs achieved PCEs of 20.3% and 19.0% for active areas of 0.16 cm^2 and 1.00 cm^2, respectively.

The quality of the perovskite layer deposited on the nanostructured ETL is another concern. Since the surface morphology of a nanostructured ETL is substantially different from that of a conventional planar ETL, the established protocols for perovskite deposition on conventional substrates should be adjusted to suit nanostructured ETL-coated substrates in order to ensure good infiltration of the precursor solution into the gaps of the nanostructures and uniform coverage of the perovskite film on the nanostructured ETL. In general, the effective approaches are to use multiple spinning steps during spin coating [79, 80], to vary the concentrations of the precursor solutions [70], to adjust the compositions of the precursor solutions [81], etc. A scanning electron microscopy (SEM) top view and cross-sectional images of perovskite layers deposited on TiO_2 nanorods using different precursor solutions are shown in figure 4.4.

Finally, in addition to pure and doped titania in various morphologies, nanocomposites of TiO_2 with other materials have also been reported, such as WO_3 with TiO_2 [87] and TiO_2 with different forms of nanostructured carbon in DSSCs [60] and CeO_2 with TiO_2 [88] in PSCs.

4.2.3 Zinc oxide

ZnO is generally considered to be an attractive choice for ETLs due to its low-cost deposition, variety of available morphologies, and excellent electron transport properties. ZnO also exhibits a large variety of nanostructured morphologies, such as nanorods, nanowires, microrods, nanoflowers, tetrapods, etc. that can readily be synthesized by solution-based methods and/or vapor-based methods [89, 90]. It has a wide direct bandgap of 3.37eV at room temperature and a high mobility in the range of around 130–170 $cm^2\ V^{-1}\ s^{-1}$ for thin films, which is higher than that of titania [91]. It has been found that processing ZnO thin film composed of nanoparticles at temperatures in the range of 100 °C–200 °C yields reasonably good device performance for optoelectronics, which is of significant importance in the development of flexible PSCs for future portable and wearable electronics [92]. Furthermore, ZnO is known to have a rich defect chemistry and as a result, undoped ZnO displays a wide range of carrier concentrations [93]. The properties of ZnO are significantly affected by its surface, which in turn is significantly affected by terminating crystal facets, since different facets have different crystal structures [93]. The properties of ZnO surfaces and consequently the interfaces of

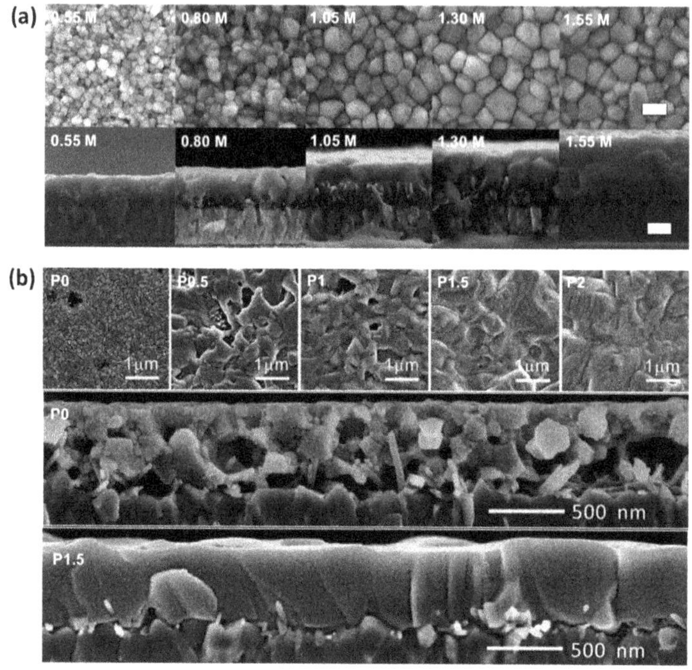

Figure 4.4. (a) The surface and cross-sectional morphologies of perovskite films prepared using different concentrations of precursor solution ($Cs_{0.1}(FA_{0.83}MA_{0.17})_{0.9}Pb(I_{0.83}Br_{0.17})_3$) deposited on TiO_2 nanorods. The optimal concentration is 1.3 M. The scale bar is 200 nm. Reprinted from [70], Copyright (2018), with permission from Elsevier. Acronyms used: methylammonium (MA), formamidinium (FA). (b) The surface and cross-sectional morphologies of perovskite films prepared using different compositions of precursor solution deposited on TiO_2 nanorods. The labels P0–P2 represent the ratio of MACl to MAI. The optimal condition is P1.5. Reproduced from [81] with permission of The Royal Society of Chemistry.

ZnO-containing devices can be altered by doping or by interface modifications using organic or inorganic layers [93]. This is particularly important for perovskite solar cells due to the instability of perovskite deposited on ZnO by annealing and for organic photovoltaics due to surface dipole formation and interfacial recombination losses. Here, we will not discuss all the reported molecules that can be used to modify the ZnO surface, but instead we focus on some selected examples to explain the relevant mechanisms. For a detailed review of the surface/interface modifications of ZnO, see reference [93]. It is, however, worth reiterating that doping in general is highly desirable for metal oxides, in particular those with rich defect chemistries. One of the important reasons why the defects in undoped ZnO have such an important influence on the resulting properties is the fact that in emerging photovoltaics, the focus is commonly on the use of ZnO prepared at low temperatures. As a result of the low-temperature preparation (regardless of whether the material is processed as a film or a nanoparticle layer), surface defects, surface adsorbates, and dangling bonds are abundant [94]. Consequently, there are intergap states which can serve as recombination centers in solar cells, and their passivation can result in performance improvements for these devices [94]. The introduction of dopants results in better control over the material's properties and improved synthesis control and reproducibility, since material prepared without intentional doping can be very sensitive to environmental factors during synthesis, especially for synthesis performed under ambient conditions.

In general, ZnO exhibits inferior chemical stability compared to TiO_2, which is problematic for applications in some types of solar cell, such as perovskite solar cells and DSSCs [4]. Nevertheless, a number of studies have demonstrated the opportunities to use ZnO nanostructures to replace the mesoscopic TiO_2 nanostructures in conventional PSCs [48, 57, 95–102]. However, the PCEs of ZnO-based PSCs are usually lower than those of PSCs employing other metal oxides such as TiO_2 and SnO_2 for the ETL. To date, the reported PCEs associated with ETLs made from pure ZnO nanostructures in PSCs are significantly fewer than those of ETLs made from titania and SnO_2. The low performance of ZnO-based PSCs is attributed to a number of factors, such as large surface defect concentrations [103] and the deprotonation of methylammonium in contact with ZnO, leading to the decomposition of perovskites on top of the ZnO ETL during thermal annealing [103]. Nevertheless, ZnO could be more suitable for use as the ETL in all-inorganic perovskite solar cells [104, 105].

The use of ZnO nanostructures as the ETLs of PSCs has been reported for both n–i–p [96] and p–i–n [48] structures. ZnO nanorod arrays have also been demonstrated to grow vertically on the flexible plastic substrates of PSCs [101] with good mechanical robustness as exhibited in a bending cycle test [98]. As is the case for other metal oxides, doping ZnO nanostructures with extrinsic materials such as nitrogen [97], cobalt [101], etc. has been demonstrated to be effective in enhancing the electron conductivity of ETLs. However, the major concern raised by using ZnO in PSCs is its surface properties, since the presence of –OH and –COOH groups at the surface of ZnO could easily induce the decomposition of perovskite [102]; therefore, the processing temperatures used for device fabrication are usually kept at low temperatures for a short duration, most likely causing incomplete reaction of the perovskite precursors and low crystallinity of the perovskite thin films, leading to poor device efficiency and instability in the early reports of ZnO-based PSCs. A substantial improvement in PCE was usually observed for $CH_3NH_3PbI_3$-based PSCs after interfacial engineering via the incorporation of additional materials with ZnO nanostructures [95, 97, 99]. For example, a PSC based on ZnO nanorods passivated by the optimal thickness of Al_2O_3 monolayers exhibited an enhancement in PCE from 10.33% to 16.08% (figures 4.5(a) and (b)) [95]. An interfacial coating of polyethyleneimine (PEI) was used in ZnO nanorod array-based PSCs, yielding an enhanced PCE of 16.1% due to the dual effects of PEI on the nanostructured ETL, namely, reducing the work function of ZnO and forming optimal high-aspect-ratio (HAR) nanorods for efficient charge extraction (figures 4.5(c)–(f)) [97]. In addition, a quasi-core–shell structure of ZnO/reduced graphene oxide (rGO) quantum dots was shown to be important in protecting ZnO from reacting with $CH_3NH_3PbI_3$ as well as promoting faster electron extraction and transfer in PSCs due to the presence of the desirable electrical properties of rGO, resulting in a PCE of 15.2% and enhanced device stability [99].

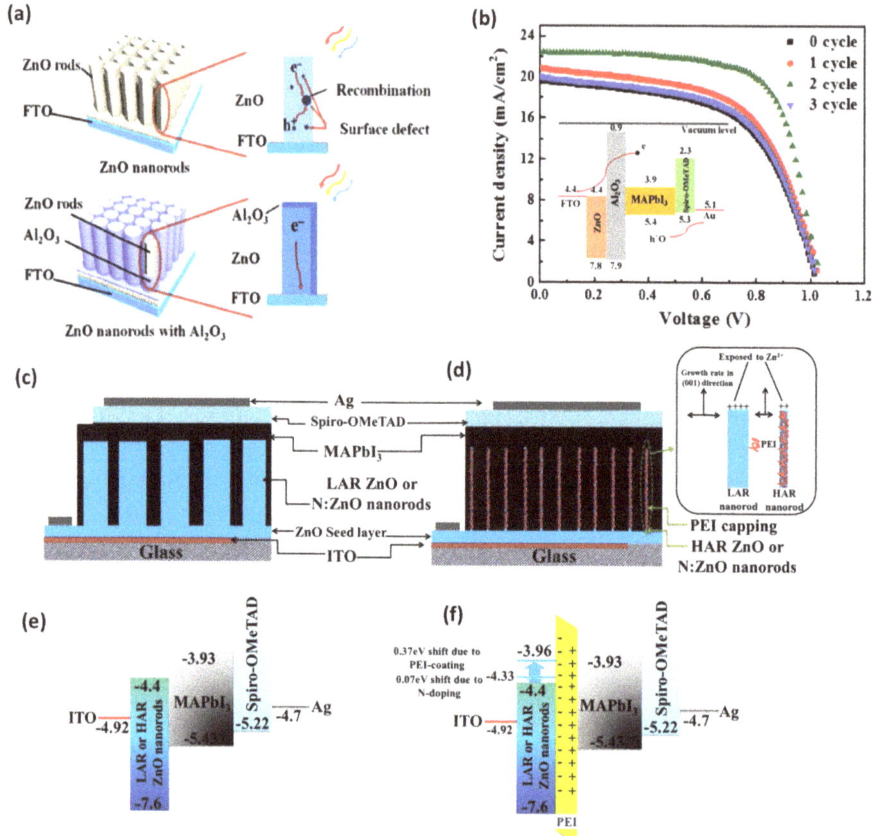

Figure 4.5. (a) A schematic illustration of the passivation mechanism of Al_2O_3 in the ZnO-nanorod-based PSC. (b) The *I–V* characteristics of ZnO-nanorod-based PSCs with different numbers of ALD cycles for the Al_2O_3 interfacial layer (inset: an energy band diagram of the ZnO-nanorod-based PSC with an Al_2O_3 interfacial layer). (a) and (b) reproduced from [95] (2016) (Copyright © 2016, Tsinghua University Press and Springer-Verlag Berlin Heidelberg). With permission of Springer. (c) A schematic illustration of the PSC with low-aspect-ratio (LAR) nanorod arrays. (d) A schematic illustration of the PSC with HAR nanorod arrays with the PEI coating (inset: a representation of PEI as a capping agent to control nanorod growth). (e) and (f) Energetic diagrams of PSCs based on nanorod arrays without and with PEI coatings, respectively. The conduction band energy of ZnO was upshifted by 0.07 eV, which was attributed to nitrogen doping; it was upshifted by 0.37 eV, which was attributed to dipole formation by the PEI layer. (c)–(f) reproduced from [97] John Wiley & Sons. [© 2015 WILEY-VCH Verlag GmbH & Co. KGaA, Weinheim].

Recently, the application of ZnO ETLs in all-inorganic perovskite-based PSCs has received increasing interest, since the interfacial problem that appears in methylammonium-based PSCs has not been observed when all-inorganic perovskites are used as the photovoltaic absorbers. Meanwhile, the higher-lying conduction band minimum (CBM) of ZnO is a good match for all-inorganic perovskites such as $CsPbI_2Br$, facilitating better carrier transport than those of the all-inorganic perovskite-based PSCs that use ETLs composed of other types of metal oxide alone [104, 105]. Li *et al* [100] recently prepared an ETL composed of core–shell

ZnO@SnO_2 nanoparticles for $CsPbI_2Br$-based PSCs. The synergetic effect attributed to the material advantages of SnO_2 and ZnO yielded a champion PCE of 14.35%, compared to a control PCE of 11.66% obtained through the use of SnO_2 nanoparticles as the ETL.

For DSSC applications, which are similar to those of PSCs, the reported efficiencies for ZnO-based DSSCs have been lower than those of titania-based devices despite the advantage of the higher electron mobility of ZnO. For a recent review of the use of ZnO in DSSCs, see reference [106], which provides quite a comprehensive discussion of the use of ZnO nanostructures in DSSCs. Here, we just briefly summarize the main challenges and provide updates on more recent progress instead of providing a detailed discussion of ZnO nanomaterials for DSSC applications. The main problem with the use of ZnO in DSSCs is the fact that ZnO is less chemically stable than TiO_2, especially for low pH values, so that acidic dye can etch ZnO, and ZnO can also react with the iodide in the electrolyte, which also deteriorates the performance [106]. Thus, the efficiencies achieved are generally lower than those of TiO_2, which is most pronounced in devices with liquid electrolytes. To clarify the reasons for the lower efficiency of devices that use a ZnO ETL, a comparative study used ZnO and TiO_2 deposited by ALD on an insulating mesoporous template to ensure comparable thicknesses of the two materials [107]. In this type of device, comparable performances of the two materials were observed. However, it was found that while ZnO exhibited improved electron transport as expected, titania exhibited higher dye loading, faster electron transfer, and a lower recombination rate, which likely contribute to the high performance of titania-based devices with thick mesoporous layers [107]. ZnO has also been used in combination with other materials, for example cupric and cuprous oxides, and while increased efficiency was obtained, these devices still significantly lagged behind titania-based DSSCs [108]. Similarly, increased efficiency has been reported for Cu-modified ZnO compared to pure ZnO [109], but its maximum efficiency of 7.27% was still significantly lower than the record performances of titania-based cells.

ZnO has also been used in QDSSCs and OPVs [58, 94, 110]. As for other types of solar cell, surface/interface defects at the oxide ETL/active layer interface play a significant role in the device performance. These defects can be addressed either by changes in the deposition conditions or by surface passivation. Different types of molecules can be used for this purpose. For example, $MgCl_2$-passivated ZnO resulted in a significant increase in the efficiency of PbS QD-based cells compared to unmodified ZnO [110]. Surface passivation has also been shown to be effective method for improving the performance of OPVs that use ZnO ETLs. For example, ethanedithiol treatment was shown to be a simple way to passivate surface defects in ZnO for OPV applications [94]. Such defect passivation was found to lead not only to an efficiency increase but also to a substantial improvement in solar cell stability [94]. In addition, doped ZnO, such as AZO has been used in organic solar cells. It has been demonstrated that an ammonia treatment of AZO nanoparticles results in defect passivation by nitrogen, the production of smooth, uniform, compact nanoparticle films, and consequently improved solar cell performance compared to that of unmodified AZO; this was attributed to the reduction of defects and traps [19].

4.2.4 Tin oxide

SnO_2 has a wide bandgap of 3.6 eV at 300 K [111] with a high electron mobility of around 100–250 $cm^2 V^{-1} s^{-1}$ [112, 113], compared to that of the TiO_2 ETL prevalent in emerging solar cells, which is around 0.1–1 $cm^2 V^{-1} s^{-1}$ [35]. SnO_2 has a deep conduction band which aligns well with the organic–inorganic hybrid perovskite materials, resulting in an enhanced electron extraction efficiency and a higher open-circuit voltage (V_{OC}) in PSCs [114]. In addition, unlike ZnO and TiO_2, tin oxide has low photocatalytic activity, while in common with TiO_2 but in contrast to ZnO, it has good chemical stability, and in common with ZnO but in contrast to titania, it can readily be deposited by low-temperature methods, which makes it suitable for flexible devices. Thus, tin oxide is a particularly promising candidate for inorganic ETLs in high-efficiency solar cells. Among the different types of solar cell, tin oxide has been most commonly and most successfully used in perovskite solar cells, while its use in DSSCs and OPVs has been less common.

Therefore, we mainly discuss the applications of SnO_2 nanostructures in high-efficiency PSCs (for a recent review, see reference [115]). A variety of morphologies of SnO_2 nanostructures have been reported, such as zero-dimensional (0D) nanoparticles [116] and quantum dots; [117, 118] one-dimensional (1D) nanorods [119], nanowires [120], nanotubes [121], nanobelts [122], etc.; two-dimensional (2D) nanosheets and three-dimensional (3D) hierarchical structures prepared from low-dimensional SnO_2 nanostructures [123]. In general, in PCSs with very high efficiencies (>20%) which are based on nanostructured SnO_2, the ETLs are usually take the form of nanoparticles [124–126] or quantum dots [127, 128] prepared at low temperatures. Jiang *et al* [124–126] demonstrated a straightforward process for the preparation of the SnO_2 nanoparticle films by spin coating a diluted commercially available SnO_2 colloid dispersion in ambient air, followed by thermal annealing at 150 °C for 30min. PSCs based on the mixed perovskite absorber and employing ETLs composed of SnO_2 nanoparticles as demonstrated by Jiang *et al* exhibited PCEs from 20.54% to 23.2% with negligible hysteresis. SnO_2 quantum dots can be prepared via the sol–gel method by continuously stirring a solution containing $SnCl_2 \cdot 2H_2O$ as the precursor dissolved in high-polarity solvents such as alcohol, water, etc [127, 128]. Yang *et al* [127] introduced a two-step low-temperature method for the preparation of ETLs composed of SnO_2 quantum dots (3–5 nm in size). To suppress charge recombination at the interface between the ETL and the perovskite due to the excess electron density associated with n-type SnO_2, highly resistive colloidal SnO_2 quantum dots were synthesized at room temperature by the incorporation of thiourea into water containing an $SnCl_2{\cdot}2H_2O$ precursor (figure 4.6(a)). The ETL was then prepared by spin coating the colloidal SnO_2 solution, followed by thermal annealing at various temperatures from 100 °C to 250 °C to remove different amounts of thiol and amino groups from the thiourea in order to control the carrier concentration of SnO_2 quantum dots. PSCs with ETLs prepared using this approach yielded stabilized PCEs of 20.32% and 16.97% for rigid and flexible devices, respectively. SnO_2 nanoparticles can be also synthesized by a hydrothermal method [49, 129] or a microwave-assisted method [130]. Transmission

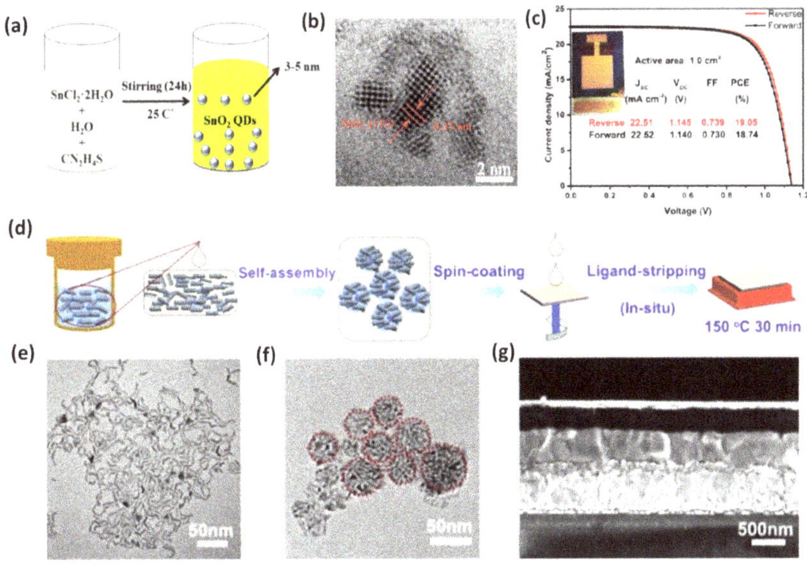

Figure 4.6. (a) A schematic diagram of the synthesis of a colloidal SnO_2 quantum dot solution at room temperature. (b) A high-resolution TEM image of the SnO_2 quantum dots obtained. (c) The *I–V* characteristics of the PSCs based on the ETL composed of SnO_2 quantum dots. (a)–(c) reproduced with permission from reference [131] John Wiley & Sons. [© 2020 WILEY-VCH Verlag GmbH & Co. KGaA, Weinheim]. (d) A schematic illustration of the processes used to prepare a SnO_2 ETL from well-dispersed SnO_2 nanorods self-assembled into SnO_2 microspheres. (e) A TEM image of the SnO_2 nanorods obtained. (f) A TEM image of the SnO_2 emulsified microspheres obtained. (g) A SEM cross-sectional image of the mesoscopic PSC based on a SnO_2 microsphere ETL. (d)–(g) reproduced with permission from reference [127].

electron microscopy (TEM) images of the obtained SnO_2 quantum dots and the *I–V* characteristics of the PSCs based on SnO_2 quantum dots are shown in figures 4.6(b) and (c), respectively.

In addition, PSCs with ETLs based on other SnO_2 nanostructures [131–141] have also been reported, for example using nanorods [131–136], nanosheets [139–141], nanowires [137], and nanotubes [138]. However, the reported PCEs of these types of PSC were generally lower than those of PSCs whose ETLs were composed of 0D nanostructures, with the exception of a few cases based on SnO_2 nanorods that showed PCEs of more than 20% [131, 132]. This is likely due to the poor penetration of the perovskite precursor solution into the nanostructures during the spin-coating process, leading to the formation of a nonuniform perovskite layer on the nanostructured ETL. The quality of the interface between the nanostructured ETL and the perovskite should be also considered, as the increased surface area of the nanostructured ETL may be associated with an increasing defect concentration at the material interface induced during the growth of the ETL and the perovskite. Sun *et al* [132] introduced $CH_3NH_3PbI_3$-based PSCs with a mesoporous ETL composed of antimony-doped SnO_2 nanorod arrays, yielding a champion PCE of 20.1%. This work demonstrated two important strategies for tackling the common problems that appear in PSCs with nanostructured ETL. The nonuniformity of perovskite on antimony-doped SnO_2

nanorods can be solved using a dynamic two-step spin-coating method (which consists of dropping the CH_3NH_3I solution onto the spinning sample with a freshly deposited PbI_2 film). Carrier recombination between the ETL and the perovskite can be significantly suppressed by modifying the surface of antimony-doped SnO_2 nanorods with a thin layer of TiO_2 (~ 10 nm) prepared by ALD. It should also be noted that Sb-doped tin oxide nanostructures have been used, and, as discussed, doped metal oxides are more likely to yield reproducible properties than undoped ones.

Other ways to further improve the efficiency of devices using tin oxide nanorods have also been demonstrated. Recently, Sun *et al* [131] demonstrated the use of a low-temperature process to produce an ETL composed of SnO_2 nanorods synthesized by a solvothermal method, which were horizontally oriented on an FTO-coated substrate; SnO_2 microspheres were deposited on the top to form a mesoporous structure. The microspheres were formed by the aggregation of SnO_2 nanorods into spherical shapes using the surfactant effect, leading to the emulsification of nanorods in the precursor solution. Low-temperature procedures for removing the organic surfactant and ligand stripping were also demonstrated in this work. It was also found that the replacement of oleic acid (OA) ligands by $[BF_4]^-$ ligands on SnO_2 nanorods and microspheres could enhance electron extraction due to better charge carrier transport dynamics and lower trap-state densities compared to the OA-capped nanostructures, yielding a PCE of 21.35% with a low hysteresis index for the champion PSC. Schematic illustrations of the processes used to prepare microsphere ETLs from SnO_2 nanorods in a PSC and the corresponding images obtained via electron microscopy are shown in figures 4.6(d)–(g).

It should be noted that the majority of the reports on the application of SnO_2 are for conventional-architecture PSCs, in which a SnO_2 ETL is located below the perovskite layer. Nevertheless, the use of an inverted PSC has been reported [49]. In this case, an interfacial fullerene layer was used between the perovskite and SnO_2, and the use of SnO_2 resulted in a significant enhancement in device stability due to the protective effect of a compact SnO_2 layer against ambient humidity [49].

Tin oxide has also been used in DSSCs; for a recent review, see reference [142]. It should be noted that SnO_2 has a significant advantage over ZnO in terms of stability, while it also exhibits superior charge transport properties compared to titania [142]. However, fast recombination has been observed in DSSCs with SnO_2 layers, and the resulting efficiencies have been low, generally below 5% [142]. One exception is devices that utilize TiO_2/SnO_2, for which higher efficiencies of up to 8.4% have been obtained [142], but even in these cases, the performance was worse than the reported results for titania-based DSSCs. Thus, SnO_2 is highly promising for PSC applications, but TiO_2 remains the most promising material for DSSCs and QDSSCs. The progress made in the low-temperature processing of SnO_2 for PSCs could likely be translated to performance improvements in OPVs as well.

4.2.5 Other electron transport materials

Other nanomaterials used as ETLs in various solar cells include WO_3, Nb_2O_5, Zn_2SnO_4, $SrTiO_3$, etc [60]. While some of these materials are promising, at present the most commonly used TiO_2, ZnO, and SnO_2 represent the best candidates for efficient devices. ZnO is likely to be the most suitable for OPV applications due to its processing temperature, while for other types of solar cell, SnO_2 is likely the best candidate for the possible replacement of TiO_2, which is currently the commonest ETL in high-performance devices.

4.2.6 Hole transport materials

In this section, we discuss commonly used HTLs. The discussion is mainly devoted to applications in PSCs and OPVs. This is because the use of p-type oxides in DSSCs has been rare. The materials which have been reported for this purpose include NiO_x, CuO, and $CuGaO_2$ (undoped and Mg-doped) [143]. p-type DSSCs are of interest for potential tandem devices, but their efficiency is typically significantly lower than those of their n-type counterparts. Hence, there has been limited progress in DSSC tandem devices.

4.2.6.1 Nickel oxide

Nickel oxide (NiO_x) has become a popular HTL in PSCs, where it is used to replace the widely used organic hole transport materials (HTMs) such as poly(3,4-ethylenedioxythiophene)-poly (styrenesulfonate) (PEDOT:PSS), 2,2',7,7'-tetrakis[*N*,*N*-di (4-methoxyphenyl)amino]-9,9'-spirobifluorene (spiro-MeOTAD) and poly[bis(4-phenyl)(2,4,6-trimethylphenyl)amine] (PTAA) [144–147]. The exploration of efficient metal oxide HTLs is of significant importance for the future commercialization of PSCs for practical use, as it provides opportunities to fabricate PSCs consisting of all-inorganic materials using a combination of the long-established metal oxide ETLs and the recently developed all-inorganic perovskites (e.g. ABX_3, where A is Cs). In this form, the metal oxide HTL and ETL can also serve as protective layers, defending the perovskite absorber from unfavorable environmental conditions [146]. Nevertheless, the aforementioned organic materials currently still dominate as the materials of choice for the HTL because they usually yield PSCs with higher efficiency. However, these organic materials have poor ambient and thermal stability; in addition, the hygroscopic additives used in these organic materials in an attempt to increase their carrier conductivity limit long-term device stability [148]. The acquisition of a high-performance metal-oxide-based HTL would be an important breakthrough in the photovoltaic community, accelerating progress toward the commercial applications of PSCs. To date, different morphologies of NiO_x nanostructure used for hole transport in PSCs have been reported [30, 50–52, 149, 150–153] with promising results.

Among the different types of inorganic HTM, the application of NiO_x as the HTL in PSCs has been the most extensively investigated by research. Promising PCEs of more than 20% have been achieved for NiO_x-based PSCs in a planar structure [144, 145], while NiO_x HTLs composed of nanostructures still yield lower PCEs than those of PSCs that use the optimized planar NiO_x HTL. Nevertheless, the nanostructures of carrier transport layers have enlarged surface areas for efficient charge extraction, which is beneficial to device efficiency and stability, whereas the technical challenges—such as the negative impacts due to defect generation—associated with the increased interfacial area between the nanostructured HTL and the perovskite absorber as well as the variation in the quality of perovskite thin films grown on nanostructured HTL can be addressed by the precise control of nanomorphology via the continuous development of material growth methods and surface engineering techniques [146, 147].

Nowadays, the spin coating of a NiO_x nanoparticle dispersion solution on the top of perovskite is the dominant approach to the production of nanostructured HTL for n–i–p PSCs [50–52] due to multiple fabrication constraints, such as the compatibility of the solvents used, the material growth, the annealing temperatures, etc. used to deposit HTL on the deposited perovskite layer, which narrow the options for other material growth techniques. The PCEs reported for this type of nanostructured NiO_x-based n–i–p PSC are generally lower (usually <16%) than those of their counterparts based on a p–i–n structure. Yang *et al* [51] demonstrated a novel concept that incorporated a carbon electrode in the form of multiwall carbon nanotubes on a submonolayer of NiO_x nanoparticles deposited on perovskite, in which the NiO_x caused favorable energy band bending at the material interfaces for efficient hole extraction and reduced nonradiative recombination, leading to a champion PCE of 15.8% with good stability for a PSC based on the carbon electrode.

In contrast to n–i–p PSCs, the nanomorphology of NiO_x HTLs in p–i–n PSCs is easier to control via adjusting the growth parameters used in different growth techniques. Since the oxide is deposited below the perovskite layer and is commonly directly deposited on the ITO, there are fewer processing limitations. In addition, the use of prefabricated NiO_x particles that are either undoped or doped with different dopants has been shown to be a promising way to obtain efficient flexible perovskite solar cells [154]. For example, the surface morphology of NiO_x nanostructures deposited on conductive substrates can be controlled by varying the concentration of the dispersion solution used for spin coating [151], by regulating the deposition current in the electrochemical deposition process [152], by adjusting the oxygen partial pressure used for pulsed laser deposition [30], or by varying the growth time of the chemical bath deposition [153]. The surface morphologies of HTMs composed of NiO_x nanostructures prepared by different methods are shown in figure 4.7. Furthermore, Park *et al* [30] demonstrated that the (111) plane is the preferred orientation of nanostructured NiO_x prepared by pulsed laser deposition in ~200 mTorr of oxygen partial pressure; it results in better HTL charge extraction ability and enhanced device performance. Yin *et al* [149] showed that an HTL composed of 1D NiO_x nanotubes exhibited a more pronounced improvement in its charge transport properties for PSCs compared those of its counterpart based on NiO_x nanocrystals. A champion PCE of 18.77% was demonstrated for a p–i–n PSC based on NiO_x nanotubes, which was an impressive result among those reported so far for NiO_x-nanostructure-based PSCs. NiO_x HTMs exhibit p-type properties due to Ni vacancies (or the self-doping of Ni^{3+}). However, the hole carrier concentration is relatively low in undoped NiO_x materials due to the large ionization energy required to form Ni vacancies, resulting in HTLs with low conductivity [147].

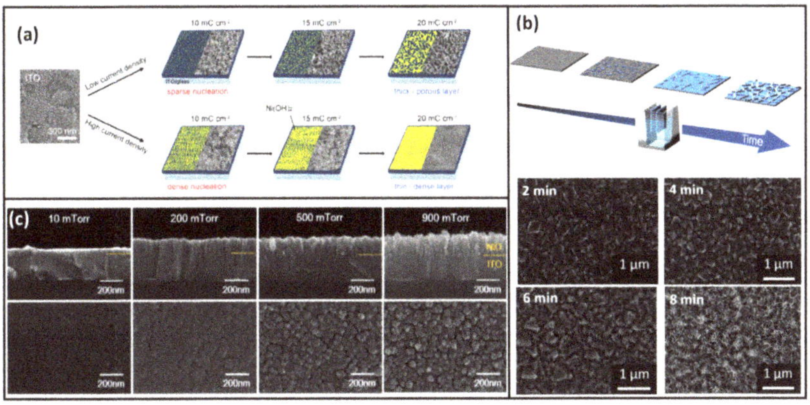

Figure 4.7. The morphology of NiO_x deposited by: (a) electrochemical deposition at different current densities [152] John Wiley & Sons. [© 2017 Wiley-VCH Verlag GmbH & Co. KGaA, Weinheim]; (b) the chemical bath process with different growth times [153] Copyright (2018), with permission from Elsevier; (c) pulsed laser deposition at different oxygen partial pressures [30] John Wiley & Sons. [© 2015 WILEY-VCH Verlag GmbH & Co. KGaA, Weinheim].

Extrinsic dopants, such as Cu, Cs, Li, Mg, Zn, Fe, etc. in planar NiO_x-based PSCs have been widely explored in order to enhance the HTL conductivity and improve the energy band alignment between the HTM and the perovskite [147, 154], while additional investigations should examine the doping of NiO_x nanostructures, as the incorporation of dopants into NiO_x nanostructures will also affect their nanomorphologies [150].

4.2.6.2 Copper-based HTLs

Copper (Cu)-based inorganic p-type semiconductors such as copper iodide (CuI), copper thiocyanate (CuSCN), copper oxides (CuO, Cu_2O) and Cu-containing ternary oxides have also been shown to be the promising alternatives for the replacement of organic HTLs in PSCs [155], among which, CuO- and Cu ternary oxide-based HTLs have been demonstrated in different nanostructures. This class of materials can be combined with a wide range of elements to form effective HTLs [156], which allows more flexibility in choosing an appropriate HTM with the desired carrier transport properties that can be prepared at low temperatures using low-cost methods. Nanoparticles/quantum dots based on CuO [157, 158] or its ternary oxides [159–165] are the approaches most often used to produce HTLs for PSCs, while other nanomorphologies such as nanoplates and nanopowders have been also reported [29, 166]. The p-type copper oxides and their ternary oxides have substantially higher hole mobility (on the order of 10^{-2}–$10\ cm^2\ V^{-1}\ s^{-1}$) [166] than the widely used organic HTM spiro-OMeTAD in the presence of additives (on the order of $10^{-3}\ cm^2\ V^{-1}\ s^{-1}$) [155]. The existing reports have shown that the replacement of organic HTMs with Cu-based materials in PSCs inhibits the UV-induced decomposition of the perovskite [165] and leads to significantly enhanced device stability [156]. As in the case of mesoscopic ETLs, it has been found that HTLs composed of mesoscopic Cu-based ternary oxides exhibit more efficient carrier extraction due to the enlarged interfacial area between the perovskite absorber and the HTL, yielding higher device efficiencies and longer lifetimes than their planar counterparts [160]. Despite the advantages of using nanostructured Cu oxides as HTLs in PSCs, most of their early reported PCEs were lower than those of conventional PSCs that employed spiro-OMeTAD as the HTL. The quality of the interface between the enlarged surface area of the nanostructured HTL and the perovskite is a major concern. The nanomorphology of the Cu-oxide-based HTLs, which significantly affects the crystallization of perovskite in p–i–n devices, should be optimized. The effective approaches demonstrated for Cu-oxide- based HTLs in PSCs are physical treatments for dispersion solutions, such as high-energy sonication and centrifugation [29], as well as the incorporation of a surfactant during the growth of the nanostructure [159], which can control the size, shape, and uniformity of nanoparticles in dispersion solutions, resulting in different HTL surface morphologies for the subsequent perovskite crystallization. In a p–i–n device structure, Cu-oxide-based nanoparticles have been directly deposited onto conductive metal-oxide-coated substrates [29, 158, 159, 163, 165], yielding PSC PCEs from 12% to 19%, whereas Chen *et al* [160] demonstrated a bilayer inorganic HTL composed of compact NiO_x and mesoporous $CuGaO_2$ (figures 4.8(a) and (b)). The bilayer inorganic HTL was shown to be efficient in charge carrier transfer and collection (see the energy band diagram shown in figure 4.8(c)) and to reduce interfacial trap states, yielding a high PSC efficiency of 19.95%. The same research group further improved the stabilized efficiency of PSCs in the same device structure to 20.15% by adding Zn doping to the $CuGaO_2$. A high stable efficiency of 20.53% PSC has been also reported for an In-doped $CuCrO_2$ HTL in a p–i–n structure [162]. Optimized dopants in Cu-oxide-based HTMs have been shown to be beneficial in enhancing the hole conductivity and engineering the band alignment of HTL with that of perovskite. It is noteworthy that the application of a Cu-oxide-based HTM in an n–i–p-structured PSC is more challenging, since the preparation process used for this type of thin film requires a high annealing temperature to convert the precursors into the final product or to eliminate the residual high-boiling-point solvents (e.g. water) [158], which denatures the underlying temperature-sensitive perovskite layer. A common approach used to avoid this problem is to pre-synthesize Cu-oxide-based nanostructures and disperse them in a solvent, which does not destroy the underlying perovskite layer. However, the coverage of HTL on the perovskite layer can be poor due to the wettability mismatch between the solvent used and the perovskite surface as well as the nonuniformity of the dispersion solution. A thicker HTL can be deposited on perovskite to ensure that it is well covered by the HTM. However, it is noteworthy that the thickness of the HTL is a critical factor that affects the carrier transport and hence the device performance [166]. The series resistance increases with the thickness of the HTL, resulting in a reduced fill factor (FF) and thus a poor PCE [166]. Liu *et al* [157] demonstrated a surface modification strategy that incorporated silane-based surfactants in a nonpolar solvent, which was able to yield a homogenous HTL composed of Cu_2O quantum dots because the surfactant in the nonpolar solvent allowed the Cu_2O quantum dots to become hydrophobic, thereby

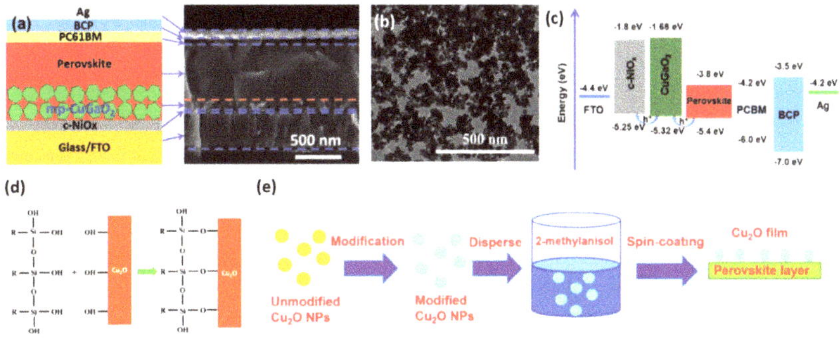

Figure 4.8. (a) A schematic illustration of the p–i–n PSC based on the bilayer metal oxide HTL composed of the compact NiO_x and mesoscopic $CuGaO_2$ and the corresponding cross-sectional SEM image; (b) a TEM image of the $CuGaO_2$ nanoparticles used in the bilayer HTL; (c) an energy band diagram depicting the hole carrier transport process within the p–i–n PSC based on the bilayer HTL. (a)–(c) reproduced from [160] John Wiley & Sons. [© 2018 WILEY-VCH Verlag GmbH & Co. KGaA, Weinheim]. (d) The chemical structure of the silane-based surfactant used to modify the surface properties of Cu_2O quantum dots to become hydrophobic; and (e) a schematic illustration of the preparation processes used for the HTL composed of Cu_2O quantum dots deposited on the top of a perovskite layer via a surface modification strategy. (d) and (e) reproduced from [157] John Wiley & Sons. [© 2019 The Authors. Published by WILEY-VCH Verlag GmbH & Co. KGaA, Weinheim].

improving the dispersion solution. A PCE of 18.9% with negligible hysteresis can be achieved for PSCs based on this surface modification technique, whose working mechanism is illustrated in figures 4.8(d) and (e). This surface engineering approach offers a straightforward procedure for the deposition of metal oxide nanostructures on the top of perovskite, leading to the formation of a high-quality carrier transport layer. In addition, nanoparticles can be incorporated into a polymer layer, as demonstrated for $CuCrO_2$ incorporated into PTAA, which resulted in improved charge extraction and transport as well as a significant improvement in stability [53].

4.2.6.3 Other HTLs

In addition to charge transport layers based on copper oxides (binary and ternary), the use of other nanomaterials as hole transport layers has also been reported. For example, copper-based chalcogenide nanoparticles incorporated in PEDOT: PSS resulted not only in improved hole transport, but also in enhanced PSC stability under illumination due to the absorption of the UV light by the nanoparticles [167]. Other examples of HTLs include the use of tungsten oxide (nanoparticles and nanosheets) as a top HTL in OPVs [47]. In addition to WO_x, MoO_x is another commonly used HTL in OPVs [58]. Examples of other HTL materials used in OPVs include BiOCl nanoplates [168]. It can be observed that this approach can result in high-efficiency devices, resulting in an improvement in performance compared to the PEDOT:PSS commonly used with several organic materials in the active layer; the best efficiency achieved was 16.6% [168]. Another option for the HTL of an OPV is graphene oxide (GO). It has been shown that chlorinated GO (GO-Cl) can result in higher solar cell efficiency due to its higher work function, which results in a more favorable energy level alignment across the interface and consequently improved hole extraction [169]. Another approach explored in OPVs is the incorporation of metal oxide nanoparticles into a polymer charge transport layer to enhance the hole transport and change the work function. An example of this approach was an HTL consisting of WO_x in PEDOT:PSS [170].

QDSSCs are based on the same concept as DSSCs, except that quantum dots are used as sensitizers instead of organic dyes [171]. As in the case of DSSCs, different materials and n-type oxide morphologies have been used in these devices. These include ZnO hollow spheres [171].

4.2.7 Other nanomaterials as charge transport layers in solar cells

Carbon-based nanomaterials such as carbon nanotubes can also be used in charge transport layers in different types of solar cell [172]. It should be noted, however, that in devices which are commonly thin, such as OPVs, there are processing challenges in integrating long carbon nanotubes into the devices in such a way as to avoid defects and short circuits.

Conductive nanomaterials can also be incorporated into charge transport layers to obtain plasmonic enhancement of the solar cell performance [173, 174]. For a detailed review of the enhancement mechanisms and the recent progress of their application to solar cells, see reference [174]. Briefly, plasmons represent collective oscillations of electrons in a conductive material (typically a metal, although ITO or graphene can also be used), and they can contribute to the enhancement of photovoltaic performance through far-field scattering, near-field enhancement, and charge carrier or resonant energy transfer [174]. The use of different metals (Ag, Au, Al, etc.) in different sizes and shapes, with or without coatings, as well as carbon-based materials (graphene oxide) has been reported for this purpose [173, 174]. In addition to the incorporation of pre-synthesized nanostructures into solution-processed layers, various nanolithographic techniques (e.g. e-beam lithography, nanoimprinting) can be used to prepare plasmonic nanostructures as part of a solar cell [173]. It should be noted that in cells containing liquid electrolyte, such as DSSCs, it may be necessary to coat the plasmonic particles with a protective shell to prevent their corrosion [175]. Due to the reactivity of halide perovskites with a wide range of metals, core–shell structures with metallic cores would likely be of interest for this type of device as well, although graphene-based or ITO-based plasmonic enhancement is expected to be possible without encountering chemical stability issues.

4.3 Nanomaterials as electrodes and buffer layers

Nanomaterials can be used as a part of the bottom or the top electrode in a solar cell. In thin-film devices on glass substrates, the bottom electrode is typically a transparent conductive oxide (TCO), most commonly ITO [6]. ITO can also be deposited on flexible plastic substrates such as polyethylene terephthalate (PET) and polyethylene naphthalate (PEN), but the ITO layer is not sufficiently mechanically robust enough to withstand a sufficient number of bending cycles to achieve the required stability for flexible photovoltaics [6]. Thus, a common objective in the field of flexible devices is to develop a conductive electrode which is more mechanically robust than ITO while retaining high conductivity and high transmission [6]. Commonly used transparent electrodes include doped metal oxides, thin metals (usually combined with wide bandgap films for light management and improved mechanical integrity), conductive polymers, and nanomaterials (metal nanowire or carbon based) [6, 176]. While different metals could be used for metal nanowire electrodes, the majority of the reported studies have used Ag nanowires due to their excellent scalability [176]. It should be noted that while this is of significant interest for flexible devices in order to address the problem of their mechanical stability, such alternative electrodes are also of interest for devices produced on rigid substrates to reduce their cost and environmental impact and also to reduce the use of indium (which is in limited supply) by eliminating ITO film.

Composites of conductive polymers and different types of nanomaterial have been extensively used as electrodes or electrode buffer layers in different types of device. For a recent review, see reference [177]. Carbon nanotubes [172] and graphene oxide are particularly promising nanomaterials for electrode applications. However, carbon-based conductive materials have the drawback of lower conductivities compared to those of metals, resulting in limitations in the device area to avoid a significant drop in fill factor due to the increased series resistance of the devices that use them. Nevertheless, carbon-based conductive materials are non-reactive, which makes them interesting for applications in which metal corrosion could occur, such as PSC electrodes. In addition to avoiding the degradation of performance due to electrode corrosion, carbon-based electrodes also serve as a protective layer by reducing the penetration of moisture into the sensitive perovskite layer when used as a top electrode in PSCs [178]. Carbon-based nanomaterials are also of interest for counter electrodes in DSSCs, as a possible replacement for platinum nanoparticles (or other nanostructures) or other metal nanostructures [60, 179]. CuO nanoparticles have also been used as counter electrodes in DSSCs, but the efficiency of 3.4% achieved was less than the record efficiencies reported for DSSCs with Pt-based counter electrodes [180]. In addition to their use in the counter electrodes of DSSCs, nanomaterials have also been used in the counter electrodes of QDSSCs. As in the case of DSSCs, the objective is to increase the surface area of the electrode, thereby improving the efficiency by increasing the number of catalytic sites and improving charge collection. Consequently, nanorods or nanowires are commonly used. For example, ITO nanowires with Cu_2S nanocrystal shells have been reported to be effective counter electrodes [181].

A common electrode application in solid-state devices, such as PSCs, OPVs, etc. involves mixing a conductive nanomaterial (metal nanowires or carbon-based nanomaterials, such as carbon nanotubes, graphene oxide, etc.) into a conductive polymer [177]. When used as a cathode buffer layer, for example in polymer solar cells, metal oxide nanoparticles are typically mixed into a conductive polymer [177]. PEDOT:PSS remains a very common conductive polymer host for different conductive nanomaterials [6]. However, it should be noted that its hydrophilicity and processing in aqueous solution (although there are also alternative PEDOT:PSS formulations using organic solvents) limit its applications in devices containing moisture-sensitive materials, since the use of PEDOT:PSS typically results in inferior stability compared to its inorganic alternatives.

In addition to various applications in rigid devices, nanomaterial-containing electrodes have commonly been used in different combinations (in polymers, sandwiched between oxide layers, in combinations of metal and carbon nanomaterials, etc.) in flexible PSCs, DSSCs, and OPVs [6]. Furthermore, nanomaterial-containing electrodes, such as multilayer electrodes consisting of Al:ZnO/Ag nanowires/Al:ZnO/ZnO, have also been used in more established solar cell technologies, such as CIGS solar cells [182]. Among various nanomaterials, metal nanowire networks are of particular interest, since metal nanowires (for some metals at least) can be inexpensively produced and they are solution processable [183]. While different metals such as Cu, Au, and CuNi have been reported for this purpose [183], Ag nanowires remain most the frequently used in different types of solar cell [183, 184]. In metal nanowire electrodes, the dispersion and density of the nanowires is a significant consideration in ensuring the best compromise between low resistance and high transmission of the electrode [183]. For photovoltaic applications, light scattering by the nanowire network also needs to be optimized [183]. For a detailed overview of metal nanowire-based transparent electrodes and their various applications, see reference [183].

In addition to careful optimization of the processing conditions to obtain a good compromise between the conductivity and transparency of the electrode, there are additional considerations for some specific technologies. For example, while Ag nanowires are commercially available and also simple to prepare and typically result in transparent electrodes with a good compromise between optical transmission and sheet resistance, silver is readily corroded by halide ions; this creates a difficulty for the application of this type of electrode in perovskite solar cells. In addition, it is necessary to improve the ambient and thermal stability of Ag nanowire networks, which can be achieved by sandwiching such networks in a protective conductive coating, such as graphene oxide, which protects the silver from corrosion [185]. Furthermore, the use of sacrificial layers such as cellulose acetate enables conductive GO/Ag/GO film to be transferred to an arbitrary substrate [185]. The deposition process is illustrated in figure 4.9. Similarly, sandwiching the Ag nanowires between layers which do not react with the perovskite or diffusing halide ions can address this problem. For example, it has been shown that the use of rGO/Ag nanowire/rGO results in reduced Ag corrosion due to the blocking of iodine diffusion by the rGO [186]. Although corrosion is not a concern in organic solar cells (individual junctions and tandems), graphene/Ag nanowire electrodes have been used in this type of device as well [187]. The main advantage of the approach was the use of laminated electrodes (graphene mesh/Ag nanowire) without causing damage to the underlying organic layers [187]. An alternative choice of material for the protection of Ag nanowire networks is AZO [184].

Nanomaterials are also used as interfacial layers, to adjust energy level alignments, or for protection from the environment or additional processing. For example, nanoparticles can also be used as a part of a multilayer electrode to protect the softer organic layers below from sputtering damage, as illustrated for

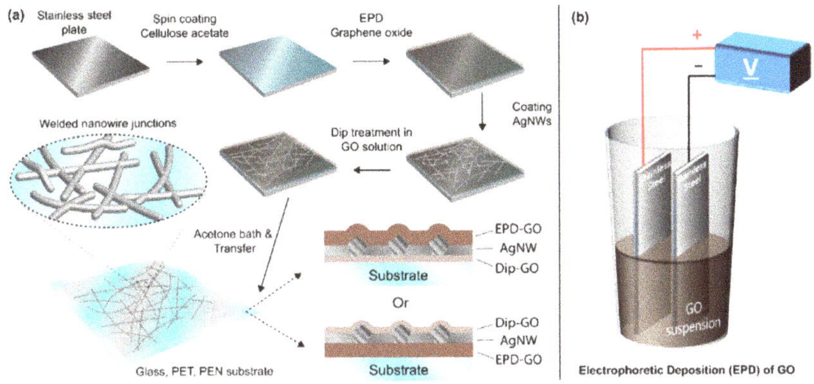

Figure 4.9. (a) A process flow for the fabrication of EPD-GO/AgNW/GO conductive films. (b) A schematic of the electrophoretic deposition (EPD) setup. (a) and (b) reprinted with permission from [185]. Copyright (2020) American Chemical Society.

ZnO nanoparticle/AZO/Ag grid electrodes [188]. In addition, nanostructured interfacial layers can be used to enhance PSC stability, as demonstrated for PbS quantum dots inserted between the active perovskite layer and the top hole transport layer, which was attributed to the hydrophobicity of PbS [189].

While the majority of cases involving the use of nanomaterials as part of the electrodes involve solution processing of the prefabricated nanomaterial, it is worth mentioning that the vacuum deposition of very thin layer can result in the deposition of nanoparticles on a surface. Such nanoparticles can then serve a variety of functions in the device, for example, for plasmonic enhancement or as seed layers in the multilayer electrode structure for dielectric/metal/dielectric electrodes [190].

It should also be noted that nanostructures are sometimes deposited on top of the solar cell for the purpose of light-trapping enhancement [191]. Insulating or semiconducting materials with high refractive indices are commonly used for this purpose [191]. Nanostructured antireflection coatings are also a common feature in high-efficiency devices, where they are used to minimize optical losses [191]. By choosing a specific material (according to its refractive index) and tuning the shape and size of nanostructures, device performance can be optimized to minimize optical losses. It should be noted that while the principles for light management are applicable to all types of solar cell, nanostructuring of the front or back surfaces and the use of antireflective coatings is more common for the more established photovoltaic technologies, such as silicon-based solar cells. This is partly due to the need to study the relationships between material properties, device architecture, and device performance in emerging technologies, in which such relationships are not yet fully understood and partly due to the processing limitations of softer, more sensitive materials such as organics and halide perovskites.

4.4 Nanomaterials in the absorber layers of solar cells

Nanomaterials have been used as absorbers or parts of the absorber layer in different types of solar cell. For example, QDSSCs use nanomaterials (quantum dots) as absorbers. These nanomaterials include PbS, PbSe, CdS, CdSe, CdTe, CdS_xSe_{1-x}, and $Zn_xCd_{1-x}Se$ [171]. While PSCs and OPVs most commonly use thin-film absorber layers (the use of perovskite QDs is possible but results in lower efficiencies than those of champion thin-film devices), nanostructures in the absorber layer can still play a significant role in these devices. For example, nanomaterials can also be added to the absorber layer to enhance solar cell performance either via scattering, nonlinear optical effects (upconversion), or plasmonic effects. Plasmonic enhancement of performance is typically achieved using conductive nanomaterials, as discussed in the section on charge transport materials. In addition to incorporation into ETLs and HTLs, different plasmonic nanostructures have also been incorporated into the active layers of different polymer solar cells, as reported in [173]. These approaches can also be used within charge transport layers, and the use of metallic nanomaterials is more common in charge transport layers or at the charge transport layer/absorber layer interface due to the need to avoid undesirable exciton quenching, increased recombination losses, and phase separation [173]. It should also be noted that plasmonic nanostructures can be used to generate hot electrons, which are then captured by the semiconductor material [192]. However, this type of solar cell needs further development to become competitive with conventional devices [192].

The application of scattering for performance enhancement can use any type of insulating particle, while upconversion typically relies on materials containing rare-earth elements. For example, scattering by nanoparticles and upconversion by rare-earth dopants using a $NaYF_4:Yb^{3+}$, Er^{3+} nanocomposite absorber was found to increase the efficiency of bulk heterojunction solar cells [177]. The same material has also been used as a mesoporous layer in perovskite solar cells, resulting in an increase in efficiency due to upconversion [193]. For further details on the use of nonlinear optical effects (upconversion, downconversion, and luminescent down-shifting (LDS)) for performance enhancement, see reference [194]. It should also be noted that nonlinear effects such as LDS can result not only in an efficiency enhancement due to improved light harvesting, but can also contribute to the enhanced stability of devices that are sensitive to UV light exposure; thus, they are of interest for PSC applications.

The nanostructuring of the absorber layer can also be performed as a part of light management for different types of solar cell [4]. Nanostructuring can also increase the charge extraction by increasing the interfacial area, provided that the interfacial traps are addressed, which can be achieved using post-deposition treatments, annealing, and passivation layers [4]. The nanostructuring of the absorber layer can be performed directly or by the deposition of a nanostructured layer below the adsorber layer. For a review of nanowire-based solar cells for conventional inorganic semiconductors, see references [4, 195, 196]. The use of ZnO for this purpose, which is motivated by the ease and low cost of deposition of ZnO nanorod or nanowire arrays, has been recently reviewed in detail [4]. For emerging photovoltaic technologies, such as organic and perovskite solar cells, nanostructuring can also be performed for the purpose of light management, but due to the softer nature of the absorber materials, different techniques are used than those used for inorganic semiconductors; these include nanoimprinting the active layer with different structures to improve light trapping [197]. Various structures that can enhance light trapping in devices include diffraction gratings, photonics crystal structures, light-trapping whispering gallery structures, moth-patterned structures, etc [197]. It is also possible to pattern the both the underlying nanoparticle layer and the active organic layer through sequential molding using polydimethylsiloxane (PDMS) to prepare a grating pattern [7].

Finally, in OPVs, the nanostructuring of the active layer also serves the purpose of increasing the interface area and shortening the distance between the donor's molecular domain and the acceptor's molecular domain, which is critical in order to achieve efficient exciton dissociation [198]. This concept is utilized in bulk heterojunctions, which involve simple mixing of the two materials. However, it is expected that higher efficiencies could be achieved by the use of more ordered nanostructured interfaces between the donor's domain and the acceptor's domain [198]. A possible way to achieve this is to use ordered bulk heterojunctions that are fabricated using templates. An example of this strategy is the use of alumina nanotube arrays as templates [198]. Organic nanomaterials can then be either directly deposited inside the template, followed by template removal, or the template can be used for nanoimprinting [198]. Finally, in hybrid organic–inorganic solar cells, nanomaterials are in some cases mixed with the polymer material to form an organic–inorganic bulk heterojunction absorber layer, with or without the use of an organic adsorber [177]. The materials used for this purpose include n-type metal oxides such as TiO_2 and ZnO [45, 177], as well as various carbon-based nanomaterials [177].

4.5 Metal oxide nanomaterials in tandem solar cell applications

Tandem cells are commonly considered to be a possible way to increase overall power conversion efficiency and overcome the Shockley–Queisser efficiency limit on single-junction devices. They have been explored for many different combinations of active materials, including silicon/perovskite, perovskite/organic, quantum dot/organic, organic/organic, etc [199]. In a tandem cell, the top (or front) cell (which typically has a higher-bandgap absorber) and the bottom (or back) cell (which typically has a lower-bandgap absorber) are connected in parallel or in series, so that the short-circuit current and the open-circuit voltage of the tandem cell are the sums of the respective values for the individual cells [200]. Perovskite solar cells have been identified as particularly suitable for use as the top (front) cells of tandem devices due to their suitable bandgap, high open-circuit voltage, and mild deposition process [199]. However, since the focus of this chapter is the use of nanomaterials in solar cells, we will mainly discuss the connection between the top and bottom cells rather than the material choices for the top and bottom cells, since the connecting layers between the devices are the typical sites where nanomaterials are used.

Tandem cells are typically fabricated in four-terminal and two-terminal configurations, as illustrated in figure 4.10 [201]. It should be noted that a three-terminal configuration (one terminal in front, two terminals at the back) is also possible, but is rarely considered [202]. In addition, optically coupled four-terminal cells with a

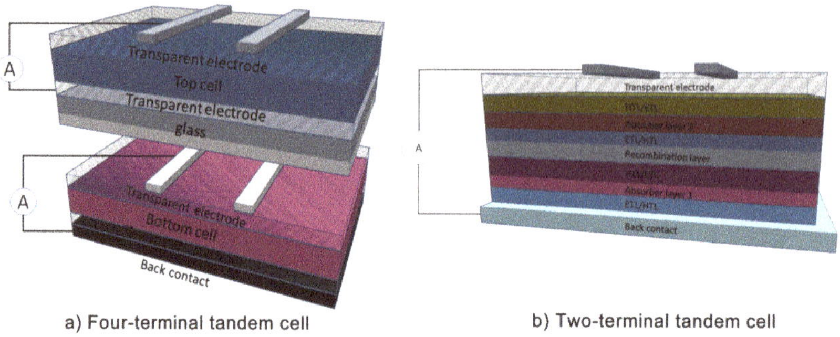

Figure 4.10. Schematic diagrams of (a) four-terminal and (b) two-terminal tandem solar cells.

beam splitter between the top and bottom cells are not a common architecture (primarily due to the high cost of dichroic mirrors), whereas mechanically stacked four-terminal cells are inexpensive and quite common [199]. Therefore, we will limit our discussion to the more common four-terminal (mechanically stacked) and two-terminal configurations. In a four-terminal configuration, in which the top contact of the bottom cell and the bottom contact of the top cell are separated by an insulating interlayer, the primary design consideration is the transmittance of the contacts and the interlayer [201]. The electrode choice can include metal nanowire networks or transparent conductive oxides, while any non-absorbing insulator can be used as the interlayer [201]. Tandem solar cells, in particular those based on Si, can also employ an intermediate reflector, which also serves as a planarization layer for the deposition of the top cell [203]. SiO_2 nanoparticles, due to their low conductivity and low refractive index, are particularly suitable for use as vertical interconnects in tandem solar cells [203].

In contrast to four-terminal solar cells, two-terminal tandem cells have a more stringent requirement for the recombination layer between the two cells. Recombination layers in two-terminal tandem solar cells need to have low absorption, enable the recombination of electrons and holes to take place, and be sufficiently mechanically robust [204]. In addition, they need to perform efficient extraction of their respective charges, i.e. they need to exhibit favorable energy level alignments and have good conductivity [205]. The fact that the recombination layer allows the recombination of electrons and holes to take place implies that the contact between the electron transport part and the hole transport part of the recombination layer should be ohmic [205]. Thus, the achievement of high efficiency is inevitably more difficult in two-terminal configurations of tandem cells than in four-terminal configurations. The use of different materials has been reported for the recombination layers in different types of tandem cell; such materials include Ca/Ag/MoO_3 [200], ZnO/organic materials [204, 205], TiO_2/organic materials [187, 205], MoO_3/Au/poly[(9,9-bis(30-(*N*,*N*-dimethylamino)propyl)-2,7-fluorene)-alt-2,7-(9,9-dioctylfluorene)] (PFN) [206], Ag/Al/Ca/MoO_3 [187], ZnO/Al [187], nanocrystalline silicon [207], etc. It can be observed that the interlayer commonly contains thin metal films sandwiched between the hole transport layer and the electron transport layer [200, 206]. However, the presence of a thin metal layer, which inevitably results in some absorption losses, is not mandatory. For example, the polymer PEDOT:PSS is commonly used for the recombination layer in tandem devices with organic or perovskite cells [204, 205]. In addition, an efficient recombination layer based on ZnO/modified PEDOT:PSS (M-PEDOT:PSS)/phosphomolybdic acid hydrate (PMA) (or PMA/M-PEDOT:PSS/ZnO), which can serve as the recombination layer in both conventional and inverted architectures, has been demonstrated to be effective for organic photovoltaic cells [204]. It should also be noted that parasitic absorption needs to be minimized for all the layers of tandem solar cells, including the various charge transport layers. Thus, the replacement of PEDOT:PSS with CuSCN can be beneficial for overall device performance, and tandem solar cells need to be carefully optimized to minimize parasitic absorption [184]. Furthermore, to minimize the optical losses in monolithic two-terminal tandem devices, it is necessary to consider not only the absorption within the recombination layer but also refractive index matching with the bottom cell, since a significant refractive index difference can cause reflection losses at the interface with the recombination layer as well as lateral conductivity that provides shunt paths through the top cell [207].

It should be noted that although the general design of tandem devices implies the use of a higher-bandgap material for the front cell and a lower-bandgap material for the back cell, in devices with very thin absorbing layers, such as organic solar cells, this rule may not necessarily apply [200]. Furthermore, for any tandem devices containing back (bottom) cells with solution-processable materials (polymers, small molecules, or perovskites), it is necessary to consider processing limitations (imposed by the deposition or annealing temperatures, or the solvents used) on the deposition of the recombination layer, the top (front) cell, and the top electrode.

Finally, other uses of metal oxides in tandem cells have also been reported, e.g. light management [208]. When a transparent conductive oxide is deposited on a ZnO nanowire array, followed by the conformal growth of absorbent Si layers, increased efficiency can be obtained, compared to the efficiencies of devices deposited on flat substrates, due to increased light trapping in the devices that use nanorods to achieve a rough surface [208]. In addition to metal oxide nanorods, other materials can be used to obtain nanostructured electrodes, such as polymer nanopillars [209]. Nanostructuring can be achieved through a variety of patterning techniques based on the use of polymeric materials, such as polymer blend lithography [209] and nanoimprint lithography, in addition to conventional patterning techniques such as e-beam lithography. Light management can also be achieved by spin coating metal oxide particles onto the tops of the devices [210].

4.6 Nanomaterials in solar cell encapsulation and packaging

In addition to their roles in charge transport, charge collection, light management, and performance enhancement, nanomaterials can also play an important role in device packaging, especially for emerging technologies, where there is a significant interest in hydrophobic coatings due to the fact that the materials used are often moisture sensitive. The incorporation of silica nanoparticles into polymers has been shown to result in superhydrophobic coatings [211], and similar approaches would obviously be of interest for encapsulation. Superhydrophobic coatings can also be achieved by nano- or microstructuring the polymer without the incorporation of oxide nanoparticles [212]. This approach has the advantage that superhydrophobicity can be achieved simultaneously with the excellent antireflection and light trapping provided by the textured surface [212]. Furthermore, superhydrophobicity is of interest not only for emerging moisture-sensitive technologies, but it is also relevant for conventional Si-based cells, since it endows them with anti-soiling (self-cleaning) functionality [212]. An additional self-cleaning function can be provided by coatings containing titania nanoparticles, as demonstrated by a PDMS–TiO_2/SiO_2/CeO_2 coating on the glass substrate of a PSC with a superhydrophobic carbon electrode [213].

The encapsulation of a nanomaterial by a polymer can also enhance the stability of a device against photodegradation, as demonstrated by the encapsulation of GO in an ethylene vinyl acetate co-polymer for use in Si-based solar cells [214]. It should be noted, however, that the concentration of carbon-based nanomaterials in the encapsulating material needs to be carefully optimized in order to retain good optical transmission [214]. Furthermore, for graphene-based nanomaterials, the orientation can make a significant difference to the barrier properties of the composite, as demonstrated for graphene nanoplatelets, for which a horizontal orientation resulted in enhanced barrier properties compared to a vertical orientation in the encapsulation of OPVs [215].

In addition, appropriately designed packaging also addresses one of the significant challenges for the commercialization of perovskite solar cells, namely the use of lead in a water-soluble form. Safe-by-design lead-trapping encapsulation has been demonstrated using several approaches. While the majority of these have been based on functional organic materials such as a cation-exchange resin [216] and a self-healing polymer [217], it is likely that lead trapping can be further enhanced using composites that include carbon-based nanomaterials, which are capable of both increasing moisture resistance and trapping lead.

Finally, another possible use of nanomaterials is in coating the substrate or cover glass for the purpose of light management. For example, it has been demonstrated that hollow MgF_2 nanoparticles can function as an effective omnidirectional and broadband antireflective coating for solar cells [218].

4.7 Conclusions and future outlook

Nanomaterials and nanostructuring have been used in a variety of different types of solar cell. In the more established technologies, the biggest challenges are to incorporate the use of materials and/or nanostructuring without a significant cost increase and to obtain a reliable estimate of the added cost and environmental impact for the purpose of life cycle assessment (LCA). Even for the use of nanomaterials in established technologies (silicon-based solar cells, including tandem devices), such as the use of nanowires either by direct growth or transfer,

LCA can contain significant uncertainties due to differences caused by upscaling from the lab to an industrial process, and by an inability to account for nano-specific toxicity, which is often unknown [219]. The environmental impacts of nanomaterials in different types of solar cell are a significant concern for their widespread application in general. In the field of emerging photovoltaics, the significant environmental concerns are the solvents used for processing, metal electrodes, any additional nanomaterials used, and in some cases, toxic elements (for example, for lead-based perovskites). To reduce their environmental impact, considerable efforts have been made to recover the metals. However, this typically involves a large quantity of acids, so that the recovery process also has significant environmental impacts [220]. The environmental impact of metal recovery for flexible devices can be reduced by combustion, followed by metal recovery from the ashes [220]. However, work on the environmental impact of the use of nanomaterials and the possibility of their recovery and reuse has been scarce compared to other parts of solar cells. It is known that the manufacture of nanomaterials is more energy intensive than the manufacture of bulk materials [221], and thus it is expected that the environmental impact of solar cell manufacture will be dependent on the nanomaterials used, if any. While LCAs have been performed for both photovoltaics and nanomaterials, further work is needed on LCAs of the use of nanomaterials in photovoltaics, especially emerging photovoltaic technologies, and investigations of end-of-life impacts and recyclability are of particular importance [221].

In addition to these challenges, emerging photovoltaic technologies face significant challenges in demonstrating adequate stability and sufficiently long lifetimes for practical applications. This is a particularly huge challenge for PSCs. While significant progress has been made (for a recent review, see reference [222]), their stability still requires further improvements for practical applications. In addition, it is necessary to use standardized testing conditions when reporting stability data. The International Summit on Organic and Hybrid Photovoltaics Stability (ISOS) protocols which have been developed for organic photovoltaics have recently been updated for PSCs [223]. It is expected that the wider adoption of stability testing protocols will result in faster progress in device stability improvement, which would bring these devices closer to practical applications. However, a determination of the acceleration factors is needed to establish the relationship between accelerated ageing lifetimes and actual outdoor lifetimes.

Concerning the advancements in the efficiency and stability of various types of device, we have emphasized the importance of surface passivation/modification and/or interfacial layers. While significant success in this area has been achieved in the lab, the scalability and manufacturability of these processes needs to be considered to establish their relevance for possible wide-scale application. Among various methods of surface modification, ALD is particularly promising due to the degree of control it offers over layer thickness and the achievement of uniform, conformal coatings [224]. For an overview of the technique and its various practical applications, including photovoltaics, see reference [224]. However, despite the promising results obtained through the use of ALD in lab-scale device fabrication, this process is generally regarded as slow and consequently unsuitable for industrial manufacturing. However, advances in this technology have recently been made, and high-throughput ALD has been realized [225]. The availability of high-throughput ALD is also promising for encapsulation applications [225].

Finally, nanomaterials are also expected to play a significant role in the integration of different types of energy harvesting and storage device, since nanomaterials represent key components not only for solar cells, but also for many of other types of device, such as Li-ion batteries, nanowire-based piezoelectric energy harvesters, nanocomposite-based thermoelectrics, etc. For recent reviews of the integration of different types of device, in particular the integration of energy generation and energy storage devices, see references [226–228]. The integration of different devices and improvements in flexible devices are of significant interest for potential applications in smart textiles and ultimately in smart textile-integrated microelectronics and technology (SIMTES), which has a number of very promising applications and will likely play a significant role in technological advancement in the future [229].

Acknowledgments

The authors acknowledge the financial support of the Seed Funding for Strategic Interdisciplinary Research Scheme of the University of Hong Kong, Shenzhen Science and Technology Commission Project JCYJ20170818141216288, and Scientific Research Grant AP08856931 funded by the Ministry of Education and Science of the Republic of Kazakhstan.

References

[1] Hengevoss D, Baumgartner C, Nisato G and Hugi C 2016 Life cycle assessment and eco-efficiency of prospective, flexible, tandem organic photovoltaic module *Sol. Energy* **137** 317–27

[2] Wang J, Liu Y, Chen X, Chen C, Chen P, Wang Z and Duan Y 2019 Functional metal oxides in perovskite solar cells *Chem. Phys. Chem.* **20** 2580–6

[3] Bai Y, Mora-Sero I, De Angelis F, Bisquert J and Wang P 2014 Titanium dioxide nanomaterials for photovoltaic applications *Chem. Rev.* **114** 10095–130

[4] Consonni V, Briscoe J, Karber E, Li X and Cossuet T 2019 ZnO nanowires for solar cells: a comprehensive review *Nanotechnology* **30** 362001

[5] Mujahid M, Chen C, Hu W, Wang Z-K and Duan Y 2020 Progress of high-throughput and low-cost flexible perovskite solar cells *Sol. RRL* **4** 1900556

[6] Hashemi S A, Ramakrishna S and Aberle A G 2020 Recent progress in flexible-wearable solar cells for self-powered electronic devices *Energy Environ. Sci.* **13** 685–743

[7] Park S *et al* 2018 Self-powered ultra-flexible electronics via nano-grating-patterned organic photovoltaics *Nature* **561** 516

[8] Liu P, Yu Z, Cheng N, Wang C, Gong Y, Bai S and Zhao X-Z 2016 Low-cost and efficient hole-transport-materialfree perovskite solar cells employing controllable electron-transport layer based on p25 nanoparticles *Electrochim. Acta* **213** 83–8

[9] Anuratha K S, Peng H-S, Xiao Y, Su T-S, Wei T-C and Lin J-Y 2019 Electrodeposition of nanostructured TiO_2 thin film as an efficient bifunctional layer for perovskite solar cells *Electrochim. Acta* **295** 662–7

[10] Wang Y, Rho W-Y, Yang H-Y, Mahmoudi T, Seo S, Lee D-H and Hahn Y-B 2016 Airstable, hole-conductor-free high photocurrent perovskite solar cells with $CH_3NH_3PbI_3$–NiO nanoparticles composite *Nano Energy* **27** 535–44

[11] Amini M, Keshavarzi R, Mirkhani V, Moghadam M, Tangestaninejad S, Mohammadpoor-Baltork I and Sadegh F 2018 From dense blocking layers to different templated films in dye sensitized and perovskite solar cells: toward light transmittance management and efficiency enhancement *J. Mater. Chem.* A **6** 2632–42

[12] Chougala L S, Kadadevarmath J S, Kamble A A, Bayannavar P K, Yatnatti M S, Linganagoudar R K, Nirupama J, Kamble R R and Qiao Q 2017 Effect of TiO_2 nanoparticles on newly synthesized phenothiazine derivative-CPTA dye and its applications as dye sensitized solar cell *J. Mol. Liq.* **244** 97–102

[13] Tyagi J, Gupta H and Purohit L 2020 Cascade structured $ZnO/TiO_2/CdS$ quantum dot sensitized solar cell *Solid State Sci.* **102** 106176

[14] Wood C J *et al* 2016 A comprehensive comparison of dye-sensitized NiO photocathodes for solar energy conversion *Phys. Chem. Chem. Phys.* **18** 10727–38

[15] Ge Z, Wang C, Chen T, Chen Z, Wang T, Guo L, Qi G and Liu J 2020 Preparation of Cu-doped ZnO nanoparticles via layered double hydroxide and application for dyesensitized solar cells *J. Phys. Chem. Solids* **109833**

[16] Nguyen Q and Kwon J W 2018 Silver nanowire-based transparent electrode as FTO replacement for dye-sensitized solar cell *Int. Nano Lett.* **9** 83–7

[17] Zamiri G and Bagheri S 2018 Fabrication of green dye-sensitized solar cell based on ZnO nanoparticles as a photoanode and graphene quantum dots as a photo-sensitizer *J. Colloid Interface Sci.* **511** 318–24

[18] Bush K A, Bailie C D, Chen Y, Bowring A R, Wang W, Ma W, Leijtens T, Moghadam F and McGehee M D 2016 Thermal and environmental stability of semi-transparent perovskite solar cells for tandems enabled by a solution-processed nanoparticle buffer layer and sputtered ITO electrode *Adv. Mater.* **28** 3937–43

[19] Jagadamma L K *et al* 2015 Polymer solar cells with efficiency $< 10\%$ enabled via a facile solution-processed Al-doped ZnO electron transporting layer *Adv. Energy Mater.* **5** 1500204

[20] Liu S, Cao K, Li H, Song J, Han J, Shen Y and Wang M 2017 Full printable perovskite solar cells based on mesoscopic $TiO_2/al_2o_3/NiO$ (carbon nanotubes) architecture *Sol. Energy* **144** 158–65

[21] Chen W, Wu Y, Liu J, Qin C, Yang X, Islam A, Cheng Y-B and Han L 2015 Hybrid interfacial layer leads to solid performance improvement of inverted perovskite solar cells *Energy Environ. Sci.* **8** 629–40

[22] Lee J-W, Makuta S, Sukarasep S, Bo J, Suzuki T, Nakayama T, Suematsu H, Niihara K and Tachibana Y 2016 Electron injection from a CdS quantum dot to a TiO_2 conduction band as an efficiency limiting process: comparison of qd depositions between silar and linker assisted attachment *J. Photopolym. Sci. Technol.* **29** 357–62

[23] Langmar O, Ganivet C R, Schol P, Scharl T, de la Torre G, Torres T, Costa R D and Guldi D M 2018 Improving charge injection and charge transport in CuO-based p-type DSSCs – a quick and simple precipitation method for small CuO nanoparticles *J. Mater. Chem.* C **6** 5176–80

[24] Yun A J, Kim J, Hwang T and Park B 2019 Origins of efficient perovskite solar cells with low-temperature processed SnO_2 electron transport layer *ACS Appl. Energy Mater.* **2** 3554–60
[25] Bashir A *et al* 2018 Spinel CO_3O_4 nanomaterials for efficient and stable large area carbon-based printed perovskite solar cells *Nanoscale* **10** 2341–50
[26] Bhandari S, Roy A, Ghosh A, Mallick T K and Sundaram S 2019 Performance of WO_3-incorporated carbon electrodes for ambient mesoscopic perovskite solar cells *ACS Omega* **5** 422–9
[27] Canto-Reyes D, Soberanis-Ortiz R A, Riech I, Mendez-Gamboa J A and Acosta M 2020 Photocurrent enhancement estimation of P3HT:PCBM:au films as a function of gold nanoparticles concentration *Gold Bull.* **53** 141–5
[28] Lee Y H, Abdu H A E, Kim D H and Kim T W 2019 Enhancement of the power conversion efficiency of organic photovoltaic cells due to Au@SiO_2 core shell nanoparticles embedded into a WO_3 hole transport layer *Org. Electron.* **68** 182–6
[29] Savva A *et al* 2019 Inverted perovskite photovoltaics using flame spray pyrolysis solution based CuAlO2/Cu-O hole-selective contact *ACS Appl. Energy Mater.* **2** 2276–87
[30] Park J H *et al* 2015 Efficient $CH_3NH_3PbI_3$ perovskite solar cells employing nanostructured p-type NiO electrode formed by a pulsed laser deposition *Adv. Mater.* **27** 4013–9
[31] Fakharuddin A, Schmidt-Mende L, Garcia-Belmonte G, Jose R and Mora-Sero I 2017 Interfaces in perovskite solar cells *Adv. Energy Mater.* **7** 1700623
[32] Jiang B-H, Chan P-H, Su Y-W, Hsu H-L, Jeng R-J and Chen C-P 2020 Surface properties of buffer layers affect the performance of PM6:Y6-based organic photovoltaics *Org. Electron.* **87** 105944
[33] Collins S D, Ran N A, Heiber M C and Nguyen T-Q 2017 Small is powerful: recent progress in solution-processed small molecule solar cells *Adv. Energy Mater.* **7** 1602242
[34] Ouyang D, Huang Z and Choy W C H 2019 Solution-processed metal oxide nanocrystals as carrier transport layers in organic and perovskite solar cells *Adv. Funct. Mater.* **29** 1804660
[35] Hendry E, Koeberg M, O'Regan B and Bonn M 2006 Local field effects on electron transport in nanostructured TiO_2 revealed by terahertz spectroscopy *Nano Lett.* **6** 755–9
[36] Liu D and Kelly T L 2014 Perovskite solar cells with a planar heterojunction structure prepared using room-temperature solution processing techniques *Nat. Photon.* **8** 133–8
[37] Xi H, Tang S, Ma X, Chang J, Chen D, Lin Z, Zhong P, Wang H and Zhang C 2017 Performance enhancement of planar heterojunction perovskite solar cells through tuning the doping properties of hole-transporting materials *ACS Omega* **2** 326–36
[38] Dong X, Hu H, Lin B, Ding J and Yuan N 2014 The effect of ALD-ZnO layers on the formation of $CH_3NH_3PbI_3$ with different perovskite precursors and sintering temperatures *Chem. Commun.* **50** 14405–8
[39] Leijtens T, Eperon G E, Pathak S, Abate A, Lee M M and Snaith H J 2013 Overcoming ultraviolet light instability of sensitized TiO_2 with meso-superstructured organometal tri-halide perovskite solar cells *Nat. Commun.* **4** 3885
[40] Wei Y, Hao Z, Zhang F and Li H 2015 A functionalized graphene oxide and nano-zeolitic imidazolate framework composite as a highly active and reusable catalyst for [3 3] formal cycloaddition reactions *J. Mater. Chem.* A **3** 14779–85
[41] Habisreutinger S N, Noel N K, Snaith H J and Nicholas R J 2016 Investigating the role of 4-TertButylpyridine in perovskite solar cells *Adv. Energy Mater.* **7** 1601079

[42] Wijeyasinghe N *et al* 2017 Copper(i) thiocyanate (CuSCN) hole-transport layers processed from aqueous precursor solutions and their application in thin-film transistors and highly efficient organic and organometal halide perovskite solar cells *Adv. Funct. Mater.* **27** 1701818

[43] Lee Y-J, Yi J, Gao G F, Koerner H, Park K, Wang J, Luo K, Vaia R A and Hsu J W P 2012 Low-temperature solution-processed molybdenum oxide nanoparticle hole transport layers for organic photovoltaic devices *Adv. Energy Mater.* **2** 1193–7

[44] Khan M, Al-Oufi M, Tossef A, Al-Salik Y and Idriss H 2015 On the role of CoO in CoO_x/TiO_2 for the photocatalytic hydrogen production from water in the presence of glycerol *Catal. Struct. React.* **1** 192–200

[45] Kim J Y, Vincent P, Jang J, Jang M S, Choi M, Bae J-H, Lee C and Kim H 2020 Versatile use of ZnO interlayer in hybrid solar cells for self-powered near infra-red photo-detecting application *J. Alloys Compd.* **813** 152202

[46] Wang D H, Kyaw A K K, Pouliot J-R, Leclerc M and Heeger A J 2014 Enhanced power conversion efficiency of low bandgap polymer solar cells by insertion of optimized binary processing additives *Adv. Energy Mater.* **4** 1300835

[47] Remya R, Gayathri P T G and Deb B 2020 Studies on solution-processed tungsten oxide nanostructures for efficient hole transport in the inverted polymer solar cells *Mater. Chem. Phys.* **255** 123584

[48] You J *et al* 2016 Improved air stability of perovskite solar cells via solution-processed metal oxide transport layers *Nat. Nanotechnol.* **11** 75–81

[49] Zhu Z, Bai Y, Liu X, Chueh C-C, Yang S and Jen A K Y 2016 Enhanced efficiency and stability of inverted perovskite solar cells using highly crystalline SnO_2 nanocrystals as the robust electron-transporting layer *Adv. Mater.* **28** 6478–84

[50] Liu Z, Zhu A, Cai F, Tao L, Zhou Y, Zhao Z, Chen Q, Cheng Y-B and Zhou H 2017 Nickel oxide nanoparticles for efficient hole transport in p-i-n and n-i-p perovskite solar cells *J. Mater. Chem.* A **5** 6597–605

[51] Yang Y, Chen H, Zheng X, Meng X, Zhang T, Hu C, Bai Y, Xiao S and Yang S 2017 Ultrasoundspray deposition of multi-walled carbon nanotubes on NiO nanoparticlesembedded perovskite layers for high-performance carbon-based perovskite solar cells *Nano Energy* **42** 322–33

[52] Icli K C and Ozenbas M 2018 Fully metal oxide charge selective layers for n-i-p perovskite solar cells employing nickel oxide nanoparticles *Electrochim. Acta* **263** 338–45

[53] Gil B, Kim J, Yun A J, Park K, Cho J, Park M and Park B 2020 $CuCrO_2$ Nanoparticles incorporated into PTAA as a hole transport layer for 85 degrees c and light stabilities in perovskite solar cells *Nanomaterials* **10** 1669

[54] El Kass M, Brohan L, Gautier N, Bechu S, David C, Lemaitre N, Berson S and Richard-Plouet M 2017 TiO_2 Anatase solutions for electron transporting layers in organic photovoltaic cells *Chem. Phys. Chem.* **18** 2390–6

[55] Vasilopoulou M *et al* 2014 Atomic-layer-deposited aluminum and zirconium oxides for surface passivation of TiO_2 in high-efficiency organic photovoltaics *Adv. Energy Mater.* **4** 1400214

[56] Wu Z, Yu H, Shi S and Li Y 2019 Bismuth oxysulfide modified ZnO nanorod arrays as an efficient electron transport layer for inverted polymer solar cells *J. Mater. Chem.* A **7** 14776–89

[57] Xu X, Zhang H, Shi J, Dong J, Luo Y, Li D and Meng Q 2015 Highly efficient planar perovskite solar cells with a TiO_2/ZnO electron transport bilayer *J. Mater. Chem.* A **3** 19288–93

[58] Vasilopoulou M 2014 The effect of surface hydrogenation of metal oxides on the nanomorphology and the charge generation efficiency of polymer blend solar cells *Nanoscale* **6** 13726–39

[59] Deepak T G, Anjusree G S, Thomas S, Arun T A, Nair S V and Nair A S 2014 A review on materials for light scattering in dye-sensitized solar cells *RSC Adv.* **4** 17615–38

[60] Karim N A, Mehmood U, Zahid H F and Asif T 2019 Nanostructured photoanode and counter electrode materials for efficient dye-sensitized solar cells (DSSCs) *Sol. Energy* **185** 165–88

[61] Lee S-W *et al* 2016 UV degradation and recovery of perovskite solar cells *Sci. Rep.* **6** 38150

[62] Cao B *et al* 2019 Interfacial engineering for high-efficiency nanorod array-structured perovskite solar cells *ACS Appl. Mater. Interfaces* **11** 33770–80

[63] Li X, Dai S-M, Zhu P, Deng L-L, Xie S-Y, Cui Q, Chen H, Wang N and Lin H 2016 Efficient perovskite solar cells depending on TiO_2 nanorod arrays *ACS Appl. Mater. Interfaces* **8** 21358–65

[64] Manseki K, Ikeya T, Tamura A, Ban T, Sugiura T and Yoshida T 2014 Mg-doped TiO_2 nanorods improving open-circuit voltages of ammonium lead halide perovskite solar cells *RSC Adv.* **4** 9652–5

[65] Yu Y, Li J, Geng D, Wang J, Zhang L, Andrew T L, Arnold M S and Wang X 2015 Development of lead iodide perovskite solar cells using three-dimensional titanium dioxide nanowire architectures *ACS Nano* **9** 564–72

[66] Zhang X, Bao Z, Tao X, Sun H, Chen W and Zhou X 2014 Sn-doped TiO_2 nanorod arrays and application in perovskite solar cells *RSC Adv.* **4** 64001–5

[67] Mali S S, Shim C S, Park H K, Heo J, Patil P S and Hong C K 2015 Ultrathin atomic layer deposited TiO_2 for surface passivation of hydrothermally grown 1D TiO_2 nanorod arrays for efficient solid-state perovskite solar cells *Chem. Mater.* **27** 1541–51

[68] Thakur U K, Askar A M, Kisslinger R, Wiltshire B D, Kar P and Shankar K 2017 Halide perovskite solar cells using monocrystalline TiO_2 nanorod arrays as electron transport layers: impact of nanorod morphology *Nanotechnology* **28** 274001

[69] Zhang Z-L, Li J-F, Wang X-L, Qin J-Q, Shi W-J, Liu Y-F, Gao H-P and Mao Y-L 2017 Enhancement of perovskite solar cells efficiency using N-doped TiO_2 nanorod arrays as electron transfer layer *Nanoscale Res. Lett.* **12** 43

[70] Yang L, Wang X, Mai X, Wang T, Wang C, Li X, Murugadoss V, Shao Q, Angaiah S and Guo Z 2019 Constructing efficient mixed-ion perovskite solar cells based on TiO_2 nanorod array *J. Colloid Interface Sci.* **534** 459–68

[71] Li Y, Guo Y, Li Y and Zhou X 2016 Fabrication of Cd-doped TiO_2 Nanorod arrays and photovoltaic property in perovskite solar cell *Electrochim. Acta* **200** 29–36

[72] Wu S, Chen C, Wang J, Xiao J and Peng T 2018 Controllable preparation of rutile TiO_2 nanorod array for enhanced photovoltaic performance of perovskite solar cells *ACS Appl. Energy Mater.* **1** 1649–57

[73] Cui Q, Zhao X, Lin H, Yang L, Chen H, Zhang Y and Li X 2017 Improved efficient perovskite solar cells based on Ta-doped TiO_2 nanorod arrays *Nanoscale* **9** 18897–907

[74] Lv Y, Li Y, Sun H, Guo Y, Li Y, Tan J and Zhou X 2018 Yttrium-doped TiO_2 nanorod arrays and application in perovskite solar cells for enhanced photocurrent density *Thin Solid Films* **651** 117–23
[75] Xiao G, Shi C, Lv K, Ying C and Wang Y 2018 Nb-doping TiO_2 electron transporting layer for efficient perovskite solar cells *ACS Appl. Energy Mater.* **1** 2576–81
[76] Yang M, Guo R, Kadel K, Liu Y, O'Shea K, Bone R, Wang X, He J and Li W 2014 Improved charge transport of Nb-doped TiO_2 nanorods in methylammonium lead iodide bromide perovskite solar cells *J. Mater. Chem.* A **2** 19616–22
[77] Tao H, Ke W, Wang J, Liu Q, Wan J, Yang G and Fang G 2015 Perovskite solar cell based on network nanoporous layer consisted of TiO_2 nanowires and its interface optimization *J. Power Sources* **290** 144–52
[78] Wu W-Q, Chen D, Cheng Y-B and Caruso R A 2020 Low-temperature solution-processed amorphous titania nanowire thin films for 1 cm^2 perovskite solar cells *ACS Appl. Mater. Interfaces* **12** 11450–8
[79] Wu W-Q, Huang F, Chen D, Cheng Y-B and Caruso R A 2015 Thin films of dendritic anatase titania nanowires enable effective hole-blocking and efficient light-harvesting for high-performance mesoscopic perovskite solar cells *Adv. Funct. Mater.* **25** 3264–72
[80] Wu S, Cheng C, Jin J, Wang J and Peng T 2018 Low-temperature processed nanostructured rutile TiO_2 array films for perovskite solar cells with high efficiency and stability *Sol. RRL* **2** 1700164
[81] Liu C *et al* 2017 Investigation of high performance TiO_2 nanorod array perovskite solar cells *J. Mater. Chem.* A **5** 15970–80
[82] Yang H-Y, Rho W-Y, Lee S K, Kim S H and Hahn Y-B 2019 TiO_2 Nanoparticles/nanotubes for efficient light harvesting in perovskite solar cells *Nanomaterials* **9** 326
[83] Mali S S, Shim C S, Kim H, Betty C A, Patil P S and Hong C K 2017 Secondary hydrothermally processed engineered titanium dioxide nanostructures for efficient perovskite solar cells *Energy Technol.* **5** 1775–87
[84] Wu W-Q, Liao J-F, Jiang Y, Wang L and Kuang D-B 2019 Bifacial contact junction engineering for high-performance perovskite solar cells with efficiency exceeding 21% *Small* **15** 1900606
[85] Kim D H, Han G S, Seong W M, Lee J-W, Kim B J, Park N-G, Hong K S, Lee S and Jung H S 2015 Niobium doping effects on TiO_2 mesoscopic electron transport layer-based perovskite solar cells *Chem. Sus. Chem.* **8** 2392–8
[86] Chen B-X, Rao H-S, Li W-G, Xu Y-F, Chen H-Y, Kuang D-B and Su C-Y 2016 Achieving high-performance planar perovskite solar cell with Nb-doped TiO_2 compact layer by enhanced electron injection and efficient charge extraction *J. Mater. Chem.* A **4** 5647–53
[87] Younas M, Gondal M A, Dastageer M A and Baig U 2019 Fabrication of cost effective and efficient dye sensitized solar cells with WO_3-TiO_2 nanocomposites as photoanode and MWCNT as Pt-free counter electrode *Ceram. Int.* **45** 936–47
[88] Manibalan G, Murugadoss G, Thangamuthu R, Kumar R M and Jayavel R 2019 Facile synthesis of heterostructure CeO_2-TiO_2 nanocomposites for enhanced electrochemical sensor and solar cell applications *J. Alloys Compd.* **773** 449–61
[89] Rodrigues J, Fernandes A J S, Monteiro T and Costa F M 2019 A review on the laser-assisted flow deposition method: growth of ZnO micro and nanostructures *Cryst. Eng. Comm.* **21** 1071–90

[90] Umar A, Akhtar M S, Almas T, Ibrahim A A, Al-Assiri M S, Masuda Y, Rahman Q I and Baskoutas S 2019 Direct growth of flower-shaped ZnO nanostructures on FTO substrate for dye-sensitized solar cells *Crystals* **9** 405

[91] Jayah N A, Yahaya H, Mahmood M R, Terasako T, Yasui K and Hashim A M 2015 High electron mobility and low carrier concentration of hydrothermally grown ZnO thin films on seeded alpha-plane sapphire at low temperature *Nanoscale Res. Lett.* **10** 7

[92] Zhang J, Yang H, Zhang X, Morbidoni M, Burgess C H, Kilmurray R, Feng S and McLachlan M A 2002 Effect of processing temperature on film properties of ZnO prepared by the aqueous method and related organic photovoltaics and LEDs *Inorg. Chem. Front.* **7** 2809–17

[93] Hewlett R M and McLachlan M A 2016 Surface structure modification of ZnO and the impact on electronic properties *Adv. Mater.* **28** 3893–921

[94] Bai S *et al* 2015 Ethanedithiol treatment of solution-processed ZnO thin films: controlling the intragap states of electron transporting interlayers for efficient and stable inverted organic photovoltaics *Adv. Energy Mater.* **5** 1401606

[95] Li S, Zhang P, Wang Y, Sarvari H, Liu D, Wu J, Yang Y, Wang Z and Chen Z D 2017 Interface engineering of high efficiency perovskite solar ref2cells based on ZnO nanorods using atomic layer deposition *Nano Res.* **10** 1092–103

[96] Liang L, Huang Z, Cai L, Chen W, Wang B, Chen K, Bai H, Tian Q and Fan B 2014 Magnetron sputtered zinc oxide nanorods as thickness-insensitive cathode inter layer for perovskite planar-heterojunction solar cells *ACS Appl. Mater. Interfaces* **6** 20585–9

[97] Mahmood K, Swain B S and Amassian A 2015 16.1% Efficient hysteresis-free mesostructured perovskite solar cells based on synergistically improved ZnO nanorod arrays *Adv. Energy Mater.* **5** 1500568

[98] Zhao X, Shen H, Sun R, Luo Q, Li X, Zhou Y, Tai M, Li J, Gao Y and Lin H 2018 Bending durable and recyclable mesostructured perovskite solar cells based on superaligned ZnO nanorod electrode *Sol. RRL* **2** 1700194

[99] Tavakoli M M, Tavakoli R, Nourbakhsh Z, Waleed A, Virk U S and Fan Z 2016 High efficiency and stable perovskite solar cell using ZnO/rGO QDs as an electron transfer layer *Adv. Mater. Interfaces* **3** 1500790

[100] Li Z *et al* 2019 Core-Shell ZnO@SnO_2 Nanoparticles for efficient inorganic perovskite solar cells *J. Am. Chem. Soc.* **141** 17610–6

[101] Bouhjar F, Derbali L and Mari B 2020 High performance novel flexible perovskite solar cell based on a low-cost-processed ZnO:Co electron transport layer *Nano Res.* **13** 2546–55

[102] Luo J, Wang Y and Zhang Q 2018 Progress in perovskite solar cells based on ZnO nanostructures *Sol. Energy* **163** 289–306

[103] Zhang P *et al* 2018 Perovskite solar cells with ZnO electron-transporting materials *Adv. Mater.* **30** 1703737

[104] Yan L *et al* 2018 Interface engineering for all-inorganic $CsPbI_2Br$ perovskite solar cells with Efficiency over 14% *Adv. Mater.* **30** 1802509

[105] Ma J, Su J, Lin Z, Zhou L, He J, Zhang J, Liu S, Chang J and Hao Y 2020 Improve the oxide/perovskite heterojunction contact for low temperature high efficiency and stable all-inorganic $CsPbI_2Br$ perovskite solar cells *Nano Energy* **67** 104241

[106] Vittal R and Ho K-C 2017 Zinc oxide based dye-sensitized solar cells: A review *Renew. Sustain. Energy Rev.* **70** 920–35

[107] Chandiran A K, Abdi-Jalebi M, Nazeeruddin M K and Graetzel M 2014 Analysis of Electron Transfer Properties of ZnO and TiO_2 Photoanodes for Dye-Sensitized Solar Cells *ACS Nano* **8** 2261–8
[108] Wongratanaphisan D, Kaewyai K, Choopun S, Gardchareon A, Ruankham P and Phadungdhitidhada S 2019 CuO-Cu_2O nanocomposite layer for light-harvesting enhancement in ZnO dye-sensitized solar cells *Appl. Surf. Sci.* **474** 85–90
[109] Zhuang S, Lu M, Zhou N, Zhou L, Lin D, Peng Z and Wu Q 2019 Cu modified ZnO nanoflowers as photoanode material for highly efficient dye sensitized solar cells *Electrochim. Acta* **294** 28–37
[110] Gao Y, Patterson R, Hu L, Yuan L, Zhang Z, Hu Y, Chen Z, Teh Z L, Conibeer G and Huang S 2019 $MgCl_2$ passivated ZnO electron transporting layer to improve PbS quantum dot solar cells *Nanotechnology* **30** 085403
[111] Zhou W, Liu Y, Yang Y and Wu P 2014 Band gap engineering of SnO_2 by epitaxial strain: experimental and theoretical investigations *J. Phys. Chem.* C **118** 6448–53
[112] Hu Y, Hwang J, Lee Y, Conlin P, Schlom D G, Datta S and Cho K 2019 First principles calculations of intrinsic mobilities in tin-based oxide semiconductors SnO, SnO_2, and Ta_2SnO_6 *J. Appl. Phys.* **126** 185701
[113] Mun H, Yang H, Park J, Ju C and Char K 2015 High electron mobility in epitaxial SnO_2-x in semiconducting regime *APL Mater.* **3** 076107
[114] Jiang Q, Zhang X and You J 2018 SnO_2: A wonderful electron transport layer for perovskite solar cells *Small* **14** 1801154
[115] Xiong L, Guo Y, Wen J, Liu H, Yang G, Qin P and Fang G 2018 Review on the application of SnO_2 in perovskite solar cells *Adv. Funct. Mater.* **28** 1802757
[116] Han X, Jin M, Xie S, Kuang Q, Jiang Z, Jiang Y, Xie Z and Zheng L 2009 Synthesis of tin dioxide octahedral nanoparticles with exposed high-energy {221} facets and enhanced gas-sensing properties *Angew. Chem. Int. Ed.* **48** 9180–3
[117] Zhu H, Yang D, Yu G, Zhang H and Yao K 2006 A simple hydrothermal route for synthesizing SnO_2 quantum dots *Nanotechnology* **17** 2386–9
[118] Xu X, Zhuang J and Wang X 2008 SnO_2 quantum dots and quantum wires: controllable synthesis, self-assembled 2D architectures, and gas-sensing properties *J. Am. Chem. Soc.* **130** 12527–35
[119] Zhang D, Sun L, Yin J and Yan C 2003 Low-temperature fabrication of highly crystalline SnO_2 nanorods *Adv. Mater.* **15** 1022
[120] Wang Y, Jiang X and Xia Y 2003 A solution-phase, precursor route to polycrystalline SnO_2 nanowires that can be used for gas sensing under ambient conditions *J. Am. Chem. Soc.* **125** 16176–7
[121] Wang Y, Zeng H and Lee J 2006 Highly reversible lithium storage in porous SnO_2 nanotubes with coaxially grown carbon nanotube overlayers *Adv. Mater.* **18** 645
[122] Pan Z, Dai Z and Wang Z 2001 Nanobelts of semiconducting oxides *Science* **291** 1947–9
[123] Wang H and Rogach A L 2014 Hierarchical SnO_2 nanostructures: recent advances in design, synthesis, and applications *Chem. Mater.* **26** 123–33
[124] Jiang Q, Zhang L, Wang H, Yang X, Meng J, Liu H, Yin Z, Wu J, Zhang X and You J 2017 Enhanced electron extraction using SnO_2 for high-efficiency planar-structure HC (NH2)(2)PbI3-based perovskite solar cells *Nat. Energy* **2** 16177
[125] Jiang Q, Chu Z, Wang P, Yang X, Liu H, Wang Y, Yin Z, Wu J, Zhang X and You J 2017 Planar-structure perovskite solar cells with efficiency beyond 21% *Adv. Mater.* **29** 1703852

[126] Jiang Q, Zhao Y, Zhang X, Yang X, Chen Y, Chu Z, Ye Q, Li X, Yin Z and You J 2019 Surface passivation of perovskite film for efficient solar cells *Nat. Photon.* **13** 460
[127] Yang G *et al* 2018 Effective carrier-concentration tuning of SnO_2 quantum dot electron-selective layers for high-performance planar perovskite solar cells *Adv. Mater.* **30** 1706023
[128] Liu H, Chen Z, Wang H, Ye F, Ma J, Zheng X, Gui P, Xiong L, Wen J and Fang G 2019 A facile room temperature solution synthesis of SnO_2 quantum dots for perovskite solar cells *J. Mater. Chem.* A **7** 10636–43
[129] Zhu Z, Zheng X, Bai Y, Zhang T, Wang Z, Xiao S and Yang S 2015 Mesoporous SnO_2 single crystals as an effective electron collector for perovskite solar cells *Phys. Chem. Chem. Phys.* **17** 18265–8
[130] Abulikemu M, Neophytou M, Barbe J M, Tietze M L, El Labban A, Anjum D H, Amassian A, McCulloch I and Del Gobbo S 2017 Microwave-synthesized tin oxide nanocrystals for low-temperature solution-processed planar junction organo-halide perovskite solar cells *J. Mater. Chem.* A **5** 7759–63
[131] Song J, Li G, Wang D, Sun W, Wu J and Lan Z 2020 High-efficiency low-temperature-processed mesoscopic perovskite solar cells from SnO_2 nanorod self-assembled microspheres *Sol. RRL* **4** 1900558
[132] Sun H, Deng K, Zhu Y, Liao M, Xiong J, Li Y and Li L 2018 A novel conductive mesoporous layer with a dynamic two-step deposition strategy boosts efficiency of perovskite solar cells to 20% *Adv. Mater.* **30** 1801935
[133] Zhang X, Rui Y, Wang Y, Xu J, Wang H, Zhang Q and Mueller-Buschbaum P 2018 SnO_2 nanorod arrays with tailored area density as efficient electron transport layers for perovskite solar cells *J. Power Sources* **402** 460–7
[134] Lv Y, Wang P, Cai B, Ma Q, Zheng X, Wu Y, Jiang Q, Liu J J and Zhang W-H 2018 Facile fabrication of SnO_2 nanorod arrays films as electron transporting layer for perovskite solar cells *Sol. RRL* **2** 1800133
[135] Yeom E J, Shin S S, Yang W S, Lee S J, Yin W, Kim D, Noh J H, Ahn T K and Seok S I 2017 Controllable synthesis of single crystalline Sn-based oxides and their application in perovskite solar cells *J. Mater. Chem.* A **5** 79–86
[136] Xu X, Xu Z, Tang J, Zhang X, Zhang L, Wu J and Lan Z 2018 High-performance planar perovskite solar cells based on lowtemperature solution-processed well-crystalline SnO_2 nanorods electrontransporting layers *Chem. Eng. J.* **351** 391–8
[137] Han G S, Chung H S, Kim D H, Kim B J, Lee J-W, Park N-G, Cho I S, Lee J-K, Lee S and Jung H S 2015 Epitaxial 1D electron transport layers for highperformance perovskite solar cells *Nanoscale* **7** 15284–90
[138] Gao C, Yuan S, Cao B and Yu J 2017 SnO_2 nanotube arrays grown via an *in situ* template-etching strategy for effective and stable perovskite solar cells *Chem. Eng. J.* **325** 378–85
[139] Zhou P, Wu J, Tu Y, Zhen M, Huo J, Wei Y and Lan Z 2016 Tin oxide nanosheets as efficient electron transporting materials for perovskite solar cells *Sol. Energy* **137** 579–84
[140] Liu Q *et al* 2016 Enhanced stability of perovskite solar cells with low-temperature hydrothermally grown SnO_2 electron transport layers *Adv. Funct. Mater.* **26** 6069–75
[141] Yang G *et al* 2017 Reducing hysteresis and enhancing performance of perovskite solar cells using low-temperature processed Y-doped SnO_2 nanosheets as electron selective layers *Small* **13** 1601769
[142] Zainudin S N F, Abdullah H and Markom M 2019 Electrochemical studies of tin oxide based-dye-sensitized solar cells (DSSC): a review *J. Mater. Sci.: Mater. Electron.* **30** 5342–56

[143] Renaud A, Cario L, Deniard P, Gautron E, Rocquefelte X, Pellegrin Y, Blart E, Odobel F and Jobic S 2014 Impact of Mg doping on performances of $CuGaO_2$ based p-type dye-sensitized solar cells *J. Phys. Chem.* C **118** 54–9
[144] Chen W *et al* 2018 Understanding the doping effect on NiO: toward high-performance inverted perovskite solar cells *Adv. Energy Mater.* **8** 1703519
[145] Chen W, Pang G, Zhou Y, Sun Y, Liu F-Z, Chen R, Chen S, Djurišić A B and He Z 2020 Stabilizing n-type hetero-junctions for NiO_x based inverted planar perovskite solar cells with an efficiency of 21.6% *J. Mater. Chem.* A **8** 1865–74
[146] Yin X, Guo Y, Xie H, Que W and Kong L B 2019 Nickel oxide as efficient hole transport materials for perovskite solar cells *Sol. RRL* **3** 1900001
[147] Ma F, Zhao Y, Li J, Zhang X, Gu H and You J 2021 Nickel oxide for inverted structure perovskite solar cells *J. Energy Chem.* **52** 393–411
[148] Djurišić A B, Liu F Z, Tam H W, Wong M K, Ng A, Surya C, Chen W and He Z B 2017 Perovskite solar cells - An overview of critical issues *Prog. Quantum Electron.* **53** 1–37
[149] Yin X, Zhai J, Song L, Du P, Li N, Yang Y, Xiong J and Ko F 2019 Novel NiO nanoforest architecture for efficient inverted mesoporous perovskite solar cells *ACS Appl. Mater. Interfaces* **11** 44308–14
[150] He Q, Yao K, Wang X, Xia X, Leng S and Li F 2017 Room-temperature and solution-processable Cu-doped Nickel oxide nanoparticles for efficient hole-transport layers of flexible large-area perovskite solar cells *ACS Appl. Mater. Interfaces* **9** 41887–97
[151] Zhang H, Cheng J, Lin F, He H, Mao J, Wong K S, Jen A K-Y and Choy W C H 2016 Pinhole-free and surface-nanostructured NiO_x film by room-temperature solution process for high-performance flexible perovskite solar cells with good stability and reproducibility *ACS Nano* **10** 1503–11
[152] Park I J, Kang G, Park M A, Kim J S, Seo S W, Kim D H, Zhu K, Park T and Kim J Y 2017 Highly efficient and uniform 1 cm(2) perovskite solar cells with an electrochemically deposited NiO_x hole-extraction layer *Chem. Sus. Chem.* **10** 2660–7
[153] Sun J, Lu J, Li B, Jiang L, Chesman A S R, Scully A D, Gengenbach T R, Cheng Y-B and Jasieniak J J 2018 Inverted perovskite solar cells with high fill-factors featuring chemical bath deposited mesoporous NiO hole transporting layers *Nano Energy* **49** 163–71
[154] Chandrasekhar P S, Seo Y-H, Noh Y-J and Na S-I 2019 Room temperature solution-processed Fe doped NiO_x as a novel hole transport layer for high efficient perovskite solar cells *Appl. Surf. Sci.* **481** 588–96
[155] Bidikoudi M and Kymakis E 2019 Novel approaches and scalability prospects of copper based hole transporting materials for planar perovskite solar cells *J. Mater. Chem.* C **7** 13680–708
[156] Pitchaiya S, Natarajan M, Santhanam A, Asokan V, Yuvapragasam A, Ramakrishnan V M, Palanisamy S E, Sundaram S and Velauthapillai D 2020 A review on the classification of organic/inorganic/carbonaceous hole transporting materials for perovskite solar cell application *Arab. J. Chem.* **13** 2526–57
[157] Liu C, Zhou X, Chen S, Zhao X, Dai S and Xu B 2019 Hydrophobic Cu_2O quantum dots enabled by surfactant modification as top hole-transport materials for efficient perovskite solar cells *Adv. Sci.* **6** 1801169
[158] Zuo C and Ding L 2015 Solution-processed Cu_2O and CuO as hole transport materials for efficient perovskite solar cells *Small* **11** 5528–32

[159] Papadas I T, Savva A, Ioakeimidis A, Eleftheriou P, Armatas G S and Choulis S A 2018 Employing surfactant-assisted hydrothermal synthesis to control $CuGaO_2$ nanoparticle formation and improved carrier selectivity of perovskite solar cells *Mater. Today Energy* **8** 57–64
[160] Chen Y, Yang Z, Wang S, Zheng X, Wu Y, Yuan N, Zhang W-H and Liu S F 2018 Design of an inorganic mesoporous hole-transporting layer for highly efficient and stable inverted perovskite solar cells *Adv. Mater.* **30** 1805660
[161] Chen Y, Yang Z, Jia X, Wu Y, Yuan N, Ding J, Zhang W-H and Liu S F 2019 Thermally stable methylammonium-free inverted perovskite solar cells with Zn^{2+} doped $CuGaO_2$ as efficient mesoporous hole-transporting layer *Nano Energy* **61** 148–57
[162] Yang B, Ouyang D, Huang Z, Ren X, Zhang H and Choy W C H 2019 Multifunctional synthesis approach of In:$CuCrO_2$ nanoparticles for hole transport layer in high-performance perovskite solar cells *Adv. Funct. Mater.* **29** 1902600
[163] Dunlap-Shohl W A, Daunis T B, Wang X, Wang J, Zhang B, Barrera D, Yan Y, Hsu J W P and Mitzi D B 2018 Room-temperature fabrication of a delafossite $CuCrO_2$ hole transport layer for perovskite solar cells *J. Mater. Chem.* A **6** 469–77
[164] Akin S, Sadegh F, Turan S and Sonmezoglu S 2019 Inorganic $CuFeO_2$ delafossite nanoparticles as effective hole transport materials for highly efficient and long-term stable perovskite solar cells *ACS Appl. Mater. Interfaces* **11** 45142–9
[165] Zhang H, Wang H, Zhu H, Chueh C-C, Chen W, Yang S and Jen A K-Y 2018 Low-temperature solution-processed $CuCrO_2$ hole-transporting layer for efficient and photo-stable perovskite solar cells *Adv. Energy Mater.* **8** 1702762
[166] Zhang H, Wang H, Chen W and Jen A K Y 2017 $CuGaO_2$: A promising inorganic hole-transporting material for highly efficient and stable perovskite solar cells *Adv. Mater.* **29** 1604984
[167] Kim B G, Jang W, Choa Y-H and Wang D H 2020 Selective UV absorbance of copper chalcogenide nanoparticles for enhanced illumination durability in perovskite photovoltaics *ACS Sustain. Chem. Eng.* **8** 7617–27
[168] Liu B *et al* 2019 Boosting efficiency and stability of organic solar cells using ultralow-cost BiOCI nanoplates as hole transporting layers *ACS Appl. Mater. Interfaces* **11** 33505–14
[169] Stratakis E, Savva K, Konios D, Petridis C and Kymakis E 2014 Improving the efficiency of organic photovoltaics by tuning the work function of graphene oxide hole transporting layers *Nanoscale* **6** 6925–31
[170] Kim H P, Lee S J, Yusoff A R b M and Jang J 2015 Improvement of conversion efficiency of inverted organic photovoltaic with PEDOT: PSS:WOx by thermal annealing *IEEE J. Photovolt.* **5** 897–902
[171] Yu L and Li Z 2019 Synthesis of ZnxCd1-xSe@ZnO hollow spheres in different sizes for quantum dots sensitized solar cells Application *Nanomaterials* **9** 132
[172] Wieland L, Li H, Rust C, Chen J and Flavel B S 2020 Carbon nanotubes for photovoltaics: from lab to industry *Adv. Energy Mater.* **10** 2002880
[173] Chou C-H and Chen F-C 2014 Plasmonic nanostructures for light trapping in organic photovoltaic devices *Nanoscale* **6** 8444–58
[174] Jang Y H, Jang Y J, Kim S, Quan L N, Chung K and Kim D H 2016 Plasmonic solar cells: from rational design to mechanism overview *Chem. Rev.* **116** 14982–5034
[175] Rai P 2019 Plasmonic noble metal@metal oxide core-shell nanoparticles for dye-sensitized solar cell applications *Sustain. Energy Fuels* **3** 63–91

[176] Cao W, Li J, Chen H and Xue J 2014 Transparent electrodes for organic optoelectronic devices: a review *J. Photonics Energy* **4** 040990
[177] Zhan C, Yu G, Lu Y, Wang L, Wujcik E and Wei S 2017 Conductive polymer nanocomposites: a critical review of modern advanced devices *J. Mater. Chem.* C **5** 1569–85
[178] Shi T, Lin S, Fang M, Kong D, Yuan Y, Gao Y, Yang B, Han H and Zhou C 2020 Low-temperature fabrication of carbon-electrode based, hole-conductor-free and mesoscopic perovskite solar cells with power conversion efficiency < 12% and storagestability < 220 days *Appl. Phys. Lett.* **117** 163501
[179] Wu J, Lan Z, Lin J, Huang M, Huang Y, Fan L, Luo G, Lin Y, Xie Y and Wei Y 2017 Counter electrodes in dye-sensitized solar cells *Chem. Soc. Rev.* **46** 5975–6023
[180] Sharma J K, Akhtar M S, Ameen S, Srivastava P and Singh G 2015 Green synthesis of CuO nanoparticles with leaf extract of Calotropis gigantea and its dye-sensitized solar cells applications *J. Alloys Compd.* **632** 321–5
[181] Jiang Y, Zhang X, Ge Q-Q, Yu B-B, Zou Y-G, Jiang W-J, Song W-G, Wan L-J and Hu J-S 2014 ITO@Cu_2S Tunnel Junction Nanowire Arrays as Efficient Counter Electrode for Quantum-Dot-Sensitized Solar Cells *Nano Lett.* **14** 365–72
[182] Kim A, Won Y, Woo K, Jeong S and Moon J 2014 All-Solution-Processed Indium-Free Transparent Composite Electrodes based on Ag Nanowire and Metal Oxide for Thin- Film Solar Cells *Adv. Funct. Mater.* **24** 2462–71
[183] Sannicolo T, Lagrange M, Cabos A, Celle C, Simonato J-P and Bellet D 2016 Metallic Nanowire-Based Transparent Electrodes for Next Generation Flexible Devices: a Review *Small* **12** 6052–75
[184] Quiroz C O R *et al* 2018 Balancing electrical and optical losses for efficient 4-terminal Si-perovskite solar cells with solution processed percolation electrodes *J. Mater. Chem.* A **6** 3583–92
[185] Chae W H, Sannicolo T and Grossman J C 2020 Double-Sided Graphene Oxide Encapsulated Silver Nanowire Transparent Electrode with Improved Chemical and Electrical Stability *ACS Appl. Mater. Interfaces* **12** 17909–20
[186] Dou Y *et al* 2002 Self-augmented ion blocking of sandwiched 2D/1D/2D electrode for solution processed high efficiency semitransparent perovskite solar cell *Nano Energy* **71** 104567
[187] Yusoff A R b M, Lee S J, Shneider F K, da Silva W J and Jang J 2014 High-performance semitransparent tandem solar cell of 8.02% conversion efficiency with solution-processed graphene mesh and laminated Ag nanowire top electrodes *Adv. Energy Mater.* **4** 1301989
[188] Jang Y H, Lee J M, Seo J W, Kim I and Lee D-K 2017 Monolithic tandem solar cells comprising electrodeposited $CuInSe_2$ and perovskite solar cells with a nanoparticulate ZnO buffer layer *J. Mater. Chem.* A **5** 19439–46
[189] Sidhik S, Rosiles Perez C, Serrano Estrada M A, Lopez-Luke T, Torres A and De la Rosa E 2020 Improving the stability of perovskite solar cells under harsh environmental conditions *Sol. Energy* **202** 438–45
[190] Giuliano G, Cataldo S, Scopelliti M, Principato F, Martino D C, Fiore T and Pignataro B 2019 Nonprecious copper-based transparent top electrode via seed layer-assisted thermal evaporation for high-performance semitransparent ni-p perovskite solar cells *Adv. Mater. Technol.* **4** 1800688

[191] Brongersma M L, Cui Y and Fan S 2014 Light management for photovoltaics using high-index nanostructures *Nat. Mater.* **13** 451–60
[192] Clavero C 2014 Plasmon-induced hot-electron generation at nanoparticle/metal-oxide interfaces for photovoltaic and photocatalytic devices *Nat. Photon.* **8** 95–103
[193] He M, Pang X, Liu X, Jiang B, He Y, Snaith H and Lin Z 2016 Monodisperse dual-functional upconversion nanoparticles enabled near-infrared organolead halide perovskite solar cells *Angew. Chem. Int. Ed.* **55** 4280–4
[194] Day J, Senthilarasu S and Mallick T K 2019 Improving spectral modification for applications in solar cells: A review *Renew. Energy* **132** 186–205
[195] Chen W, Roca i and Cabarrocas P 2019 Rational design of nanowire solar cells: from single nanowire to nanowire arrays *Nanotechnology* **30** 194002
[196] Dasgupta N P, Sun J, Liu C, Brittman S, Andrews S C, Lim J, Gao H, Yan R and Yang P 2014 25th Anniversary article: semiconductor nanowires synthesis, characterization, and applications *Adv. Mater.* **26** 2137–84
[197] Wang W and Qi L 2019 Light management with patterned micro and nanostructure arrays for photocatalysis, photovoltaics, and optoelectronic and optical devices *Adv. Funct. Mater.* **29** 1807275
[198] Wu H, Yang J, Cao S, Huang L and Chen L 2014 Ordered organic nanostructures fabricated from anodic alumina oxide templates for organic bulk- heterojunction photovoltaics *Macromol. Chem. Phys.* **215** 584–96
[199] Wang Z, Song Z, Yan Y, Liu S F and Yang D 2019 Perovskite-a perfect top cell for tandem devices to break the S-Q limit *Adv. Sci.* **6** 1801704
[200] Jiang Z, Gholamkhass B and Servati P 2019 Effects of interlayer properties on the performance of tandem organic solar cells with low and high band gap polymers *J. Mater. Res.* **34** 2407–15
[201] Chapa M, Alexandre M F, Mendes M J, Aguas H, Fortunato E and Martins R 2019 All-thin-film perovskite/C-Si four-terminal tandems: interlayer and intermediate contacts optimization *ACS Appl. Energy Mater.* **2** 3979
[202] Santbergen R, Uzu H, Yamamoto K and Zeman M 2019 Optimization of three-terminal perovskite/silicon tandem solar cells *IEEE J. Photovolt.* **9** 446–51
[203] Niesen B, Blondiaux N, Boccard M, Stuckelberger M, Pugin R, Scolan E, Meillaud F, Haug F-J, Hessler-Wyser A and Ballif C 2014 Self-patterned nanoparticle layers for vertical interconnects: application in tandem solar cells *Nano Lett.* **14** 5085–91
[204] Lu S, Lin H, Zhang S, Hou J and Choy W C H 2017 A switchable interconnecting layer for high performance tandem organic solar cell *Adv. Energy Mater.* **7** 1701164
[205] Lechêne B, Perrier G, Emmanouil K, Kennou S, Bouthinon B and de Bettignies R 2014 Design of intermediate layers for solution-processed tandem organic solar cells: Guidelines from a case study on TiOx and ZnO *Sol. Energy Mater. Sol. Cells* **120** 709–15
[206] Kim T *et al* 2017 Hybrid tandem quantum dot/organic photovoltaic cells with complementary near infrared absorption *Appl. Phys. Lett.* **110** 223903
[207] Sahli F *et al* 2018 Improved optics in monolithic perovskite/silicon tandem solar cells with a nanocrystalline silicon recombination junction *Adv. Energy Mater.* **8** 1701609
[208] Chin A H, Keshavarz M, Wang G, Yu R, Niu X and Yang L 2017 Light management of tandem solar cells on nanostructured substrates *J. Photonics Energy* **7** 027001
[209] Hossain I M *et al* 2020 Nanostructured front electrodes for perovskite/c-Si tandem photovoltaics *Opt. Express* **28** 8878–97

[210] Das M, Hossain M J, Leung S-F, Lenox A, Jung Y, Davis K, He J-H and Roy T 2019 A leaf-inspired photon management scheme using optically tuned bilayer nanoparticles for ultra-thin and highly efficient photovoltaic devices *Nano Energy* **58** 47–56
[211] Ibrahim S and Sultan M 2020 Superhydrophobic coating polymer/silica nanocomposites: part I synthesis and characterization as eco-friendly coating *Silicon* **12** 805–11
[212] Centeno P, Alexandre M F, Chapa M, Pinto J, Deuermeier V J, Mateus T, Fortunato E, Martins R, Aguas H and Mendes M J 2020 Self-cleaned photonic-enhanced solar cells with nanostructured parylene-C *Adv. Mater. Interfaces* **7** 2000264
[213] Keshavarzi R, Molabahrami N, Afzali N and Omrani M 2020 Improving efficiency and stability of carbon-based perovskite solar cells by a multifunctional triple-layer system: antireflective, UV-protective, superhydrophobic, and self-cleaning *Sol. RRL* **4** 2000491
[214] Carvalho de Oliveira M C, Soares L G, Viana M M, Alves Cardoso Diniz A S and Cunha Lins V d F 2020 Development of ethylene-vinyl acetate copolymer/graphene oxide nanocomposites for crystalline silicon photovoltaic modules *Int. J. Adhes. Adhes.* **100** 102595
[215] Yuwawech K, Wootthikanokkhan J and Tanpichai S 2020 Functionalized graphene nanoplatelets as a barrier enhancing filler in organic photovoltaic encapsulant *J. Appl. Polym. Sci.* **138** e50351
[216] Chen S, Deng Y, Gu H, Xu S, Wang S, Yu Z, Blum V and Huang J 2002 Trapping lead in perovskite solar modules with abundant and low-cost cation-exchange resins *Nat. Energy* **5** 1003
[217] Jiang Y, Qiu L, Juarez-Perez E J, Ono L K, Hu Z, Liu Z, Wu Z, Meng L, Wang Q and Qi Y 2019 Reduction of lead leakage from damaged lead halide perovskite solar modules using self-healing polymer-based encapsulation *Nat. Energy* **4** 585–93
[218] Chundi N, Das B, Kolli C S R, Madiwala S P, Koppoju S, Ramasamy E and Shanmugasundaram S 2002 Single layer hollow MgF_2 nanoparticles as high-performance omnidirectional broadband antireflective coating for solar application *Sol. Energy Mater. Sol. Cells* **215** 110680
[219] Pallas G, Vijver M G, Peijnenburg W J G and Guinee J 2002 Life cycle assessment of emerging technologies at the lab scale: The case of nanowire-based solar cells *J. Ind. Ecol.* **24** 193
[220] Søndergaard R R, Zimmermann Y-S, Espinosa N, Lenz M and Krebs F 2016 Incineration of organic solar cells: efficient end of life management by quantitative silver recovery *Energy Environ. Sci.* **9** 857–61
[221] Kim J, Rivera J L, Meng T Y, Laratte B and Chen S 2016 Review of life cycle assessment of nanomaterials in photovoltaics *Sol. Energy* **133** 249–58
[222] Wali Q, Iftikhar F J, Khan M E, Ullah A, Iqbal Y and Jose R 2020 Advances in stability of perovskite solar cells *Org. Electron.* **78** 105590
[223] Khenkin M V *et al* 2020 Consensus statement for stability assessment and reporting for perovskite photovoltaics based on ISOS procedures *Nat. Energy* **5** 35–49
[224] Meng X, Wang X, Geng D, Ozgit-Akgun C, Schneider N and Elam J W 2017 Atomic layer deposition for nanomaterial synthesis and functionalization in energy technology *Mater. Horiz.* **4** 133–54
[225] Muñoz Rojas D, Maindron T, Esteve A, Piallat E, Kools J C S and Decams J-M 2019 Speeding up the unique assets of atomic layer deposition *Mater. Today Chem.* **12** 96–120
[226] Pan S, Ren J, Fang X and Peng H 2016 Integration: an effective strategy to develop multifunctional energy storage devices *Adv. Energy Mater.* **6** 1501867

[227] Yun S, Qin Y, Uhl A R, Vlachopoulos N, Yin M, Li D, Han X and Hagfeldt A 2018 New-generation integrated devices based on dye-sensitized and perovskite solar cells *Energy Environ. Sci.* **11** 476–526
[228] Pu X, Hu W and Wang Z L 2018 Toward wearable self-charging power systems: the integration of energy-harvesting and storage devices *Small* **14** 1702817
[229] Shi J *et al* 2020 smart textile-integrated microelectronic systems for wearable applications *Adv. Mater.* **32** 1901958

Chapter 5

Achieving batteries and supercapacitors with improved performances

Neslihan Yuca, Ömer Suat Taşkin, Sebahat Topal, Ozan Karakaya, Sema Topal and Gao Liu

5.1 Energy storage systems

Energy storage is defined as the storage of energy in different forms such as chemical, electrical, or thermal within a device or storage medium to be used at a later time. Energy storage systems first eliminate any difference that may arise between the time of generation and the time of demand; second, they store unneeded energy in locations where energy is used; finally, they provide sustainable energy by storing renewable energy that can only be produced at certain times.

An energy storage system should have:

- high storage capacity
- high charge/discharge efficiency
- no self-discharge
- long life
- low cost
- high energy density

Table 5.1. Energy storage technologies grouped according to power and storage capacity.

Large power (>50 MW) and storage (100 MWh) capacities	Medium power (1–50 MW) and storage (5–100 MWh) capacities	Large power or storage capacities	Energy storage systems
PHES	BES	FES	HESS
UPHES	FBES	SCES	TESS
CAES		SMES	EVs

PHES (Pumped hydroelectric energy storage), UPHES (Underground pumped-hydro energy storage), CAES (Compressed-air energy storage), BES (Battery energy storage), FBES (Flow battery energy storage), FES (Flywheel energy storage), SCES (Supercapacitor energy storage), SMES (Superconducting magnetic energy storage), HESS (Hydrogen energy storage), TES (Thermal energy storage), EVs (Electric vehicles)

doi:10.1088/978-0-7503-3531-7ch5

© IOP Publishing Ltd 2023

An energy storage system may, according to its nature, be suitable for specific applications. These can be grouped based on their powers and storage capacities, as shown in table 5.1.

This chapter mainly focuses on electrochemical energy storage devices, but brief information about all energy storage systems is given in the text. A comparison of the many electrochemical system specifications is summarized in table 5.2.

Energy storage systems can be categorized according to their working properties, as follows (figure 5.0):

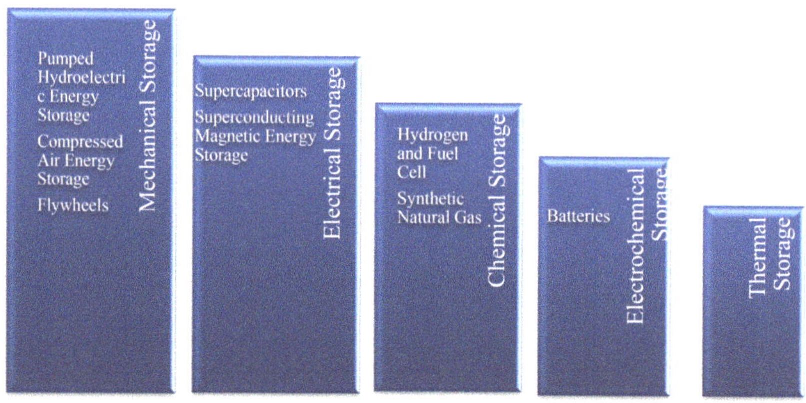

Figure 5.0: Categorization of the energy storage systems

5.1.1 Mechanical energy storage systems

5.1.1.1 Pumped hydroelectric energy storage

Pumped hydro storage (PHS) systems require lower and upper reservoirs so that water can be pumped from the lower to the upper reservoir during off-peak hours (charging). When the energy is needed, the water flows back from the upper to the lower reservoir. This flow powers a turbine with a generator to produce electricity (discharging). Around the world, there are many examples of this kind of system. The first PHS plants were used in Italy and Switzerland in the 1890s. The largest is in Virginia, USA; it has a peak power of 2100 MW. Typical discharge times range from several hours to a few days. The efficiency of PHS plants is in the range of 70%–85%. Table 5.3 shows the advantages and disadvantages of these systems. Their main applications are for energy management via time shifting, namely nonspinning reserve and supply reserve. PHS plants can have capacities from MWh to GWh, depending on the availability of reservoirs and the location of the plant.

PHS plants can be used as hybrid systems with renewable energy generation plants. The energy generated by the renewable systems is used to pump water during the day from a lower reservoir to an upper reservoir during the day. If there are no available reservoirs near a renewable energy production plant, PHS plants can be artificially established close to the plant to make the system highly efficient and sustainable.

Table 5.2. A comparison of the many energy storage system specifications.

	Lead–acid battery	Li-ion battery	Super capacitor	Superconducting magnetic energy storage	Flywheel	Pumped hydroelectric energy storage	Compressed-air energy storage
Energy storage capacity (kWh)	<100	<10	<10	<10	1–25	<150	<10
Energy density (Wh L^{-1})	50–80	200–500	2–10	0.2–2.5	20–80	0.5–1.5	3–6
Power density (W L^{-1})	10–400	0	100 000	1000–4000	1000–2000	0	0.5–2
Discharge time	Hours	Minutes–hours	Seconds	Hours	Seconds–minutes	A few hours	Hours
Charge time	Hours	Minutes–hours	Seconds	Subsecond	Minutes	A few hours	Hours
Response time	Subsecond	Seconds	Seconds	Submillisecond	Seconds	Seconds–minutes	Minutes
Lifetime (cycles)	500–800	1000–3000	50 000	<50 000	<100 000	>50 000	>10 000
Lifetime (years)	3–10	10–15	5–20	5–20	20	25+	20+

Table 5.3. The advantages and disadvantages of PHS systems.

Advantages	Disadvantages
Wide operational experience	Suitable geographic site
High maturity for establishment	Low energy density
Low self-discharge	High initial investment costs
Storage volume can be large	Environmental concerns
Storage period can be long	
Start/stop flexibility	
Long life	
Low storage cost	

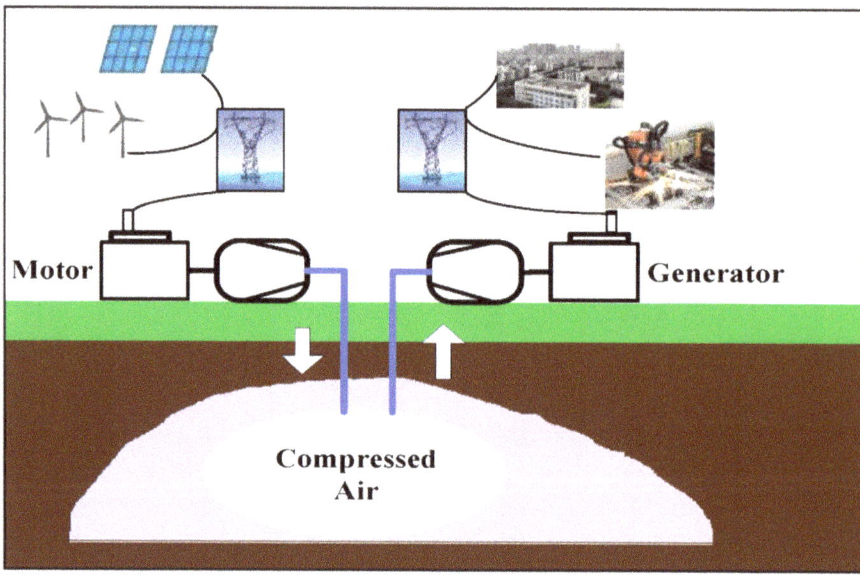

Figure 5.1. A schematic of a compressed-air energy storage plant. Reproduced with permission from Ref. 186. Copyright (2017) MDPI.

5.1.1.2 Compressed-air energy storage

Energy storage systems require a medium that can store the energy. In compressed-air energy systems, the storage medium is air. In this system, air is compressed in a container such as an underground structure or an overground system of vessels or pipes. When the energy is needed, the compressed air is used to burn natural gas, and it is then expanded in a modified gas turbine. A schematic illustration of the system is shown in figure 5.1. This system, which is called a diabatic CAES (Compressed Air Energy Storage), has an efficiency of <50%. These systems are reliable and well-proven; they do not require extraneous power. The ideal sites for air storage are old salt caverns, aquifers, or pore storage sites. Although this system can store huge amounts of energy, the storage medium has to be well sealed, and the compression/

Table 5.4. The advantages and disadvantages of CAES.

Advantages	Disadvantages
Capable of storing huge amounts of energy, similar to those stored by PSH	Requires sealed storage caverns
AA-CAES is capable of efficiencies nearly as good as that of PSH (around 70%)	Economical only up to a day of storage (for AA-CAES)
Fast response times	Competes with other storage needs (natural gas, hydrogen)
Inexpensive way to store energy	Not yet fully developed

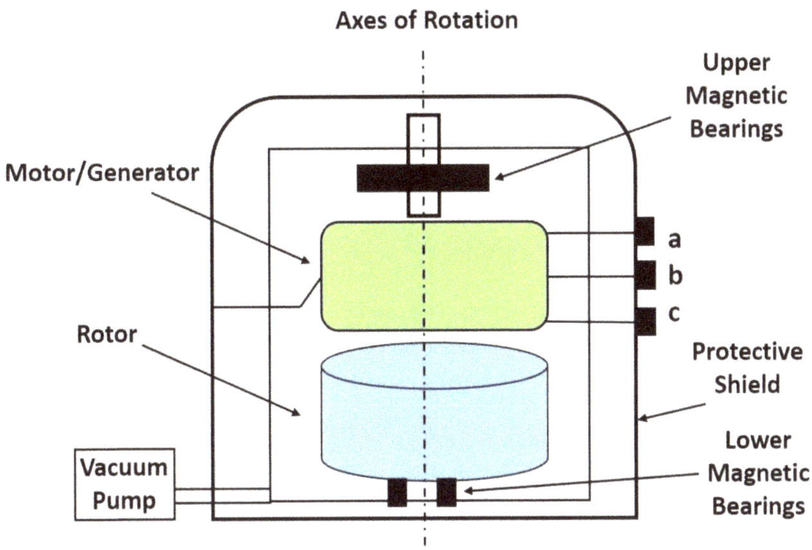

Figure 5.2. The system components of flywheel energy storage.

decompression processes can create many problems. Table 5.4 shows the advantages and disadvantages of this system.

5.1.1.3 Flywheels

Flywheels are mechanical energy storage systems that use a massive rotating cylinder. Such cylinders have a lot of angular momentum, which means they can store a great deal of kinetic energy. If a flywheel spins at high speed, it stores much more energy than those that spin more slowly. The main components of a flywheel are the rotating body/cylinder (comprised of a rim attached to a shaft) housed in a compartment, the bearings, and the transmission device (a motor/generator mounted onto the stator) (figure 5.2).

The flywheel storage system is a very good system that has excellent cycle stability and a long life. However, flywheels have a high level of self-discharge due to air

Table 5.5. The advantages and disadvantages of flywheels.

Advantages	Disadvantages
Fast charge capabilities	Low energy density
Long cycle life	High self-discharge rates
High power density	Need power to energize magnetic bearings
Low maintenance	Unexpected dynamic loads
Wide operational experience	Unexpected external shocks

resistance and bearing losses and suffer from low current efficiency. Table 5.5 shows the advantages and disadvantages of this system.

5.1.2 Electrical energy storage systems

5.1.2.1 Supercapacitors

Supercapacitors are used to store energy at high currents and for short durations; they store electrical energy at an electrode–electrolyte interface. They consist of two metal plates, which only are coated with a porous material known as activated carbon (figure 5.3). When a supercapacitor is charged, energy is stored in the dielectric material in an electrostatic field [1–3]. More information about supercapacitors can be found in section 5.2. Table 5.6 shows the advantages and disadvantages of this system.

Batteries and supercapacitors are complementary systems: supercapacitors have high power density and batteries have high energy density. These two technologies work together to increase the efficiency of applications such as electric vehicles.

5.1.2.2 Superconducting magnetic energy storage

Superconducting magnetic energy storage (SMES) systems use a magnetic field created by the flow of direct current in a superconducting coil. The components of this storage system are a coil made of superconducting material, power-conditioning equipment, and a cryogenic cooling system (figure 5.4). The principle of the system is to create an electromagnetic field surrounding the coil that stores energy. Practical systems can reach efficiencies of 90%–95%.

The system has the advantages of a very quick response time and the absence of moving parts in its main portion. Table 5.7 shows the advantages and disadvantages of this system.

5.1.3 Chemical energy storage systems

5.1.3.1 Fuel cells

A fuel cell (FC) is a system which uses the oxidation reaction to convert chemical energy into electrical energy. This system produces electricity and heat as long as fuel is supplied. It consists of two electrodes—a negative electrode (or anode) and a positive electrode (or cathode)—sandwiched around an electrolyte (figure 5.5). If hydrogen is fed to the anode as a fuel, a catalyst at the anode separates hydrogen molecules into protons and electrons. The electrons move to an external circuit to

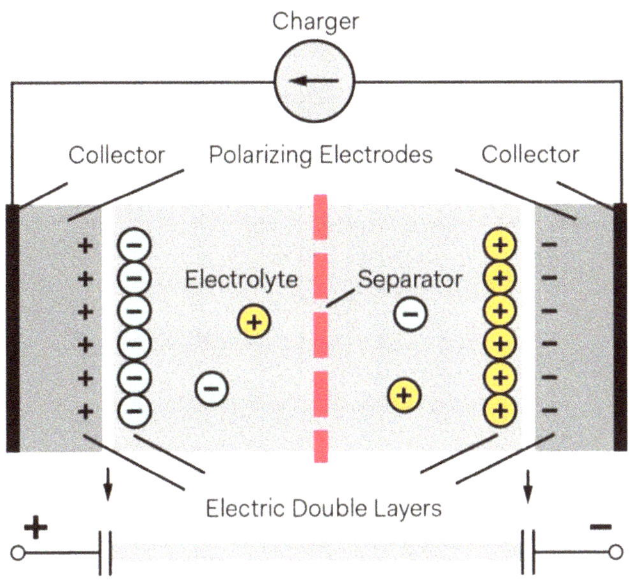

Figure 5.3. Schematic of a supercapacitor system. © 2022 Power Systems Design. All rights reserved. https://www.powersystemsdesign.com/articles/the-silent-power-of-supercapacitors/135/13510.

Table 5.6. Advantages and disadvantages of supercapacitors.

Advantages	Disadvantages
High power density	Limited capacity
Shorter charging times than those of conventional batteries	Relatively low energy density
	High self-discharge losses
Long life	Storing small quantities of electrical energy

complete their flow, while the protons migrate through the electrolyte to the cathode. Finally, water and heat are the products [4].

FCs are commonly classified according to their electrolytes. The fuel cell types are proton exchange/polymer electrolyte membrane fuel cells (PEMFCs), direct methanol fuel cells (DMFCs), alkaline fuel cells (AFCs), phosphoric acid fuel cells (PAFCs), molten carbonate fuel cells (MCFCs), solid oxide fuel cells (SOFCs), zinc–air fuel cells (ZAFCs), and photonic ceramic fuel cells (PCFCs). Table 5.8 summarizes the properties of these fuel cells.

5.1.3.2 Synthetic natural gas

Synthetic natural gas (SNG), which is artificially synthesized, can be used as another chemical energy storage system. This system is based on the reaction of hydrogen and carbon dioxide in a methanation reactor. This is system is feasible if SNG production can be performed at locations where CO_2 and excess electricity are both available. This system has relatively low efficiency due to the conversion losses in electrolysis,

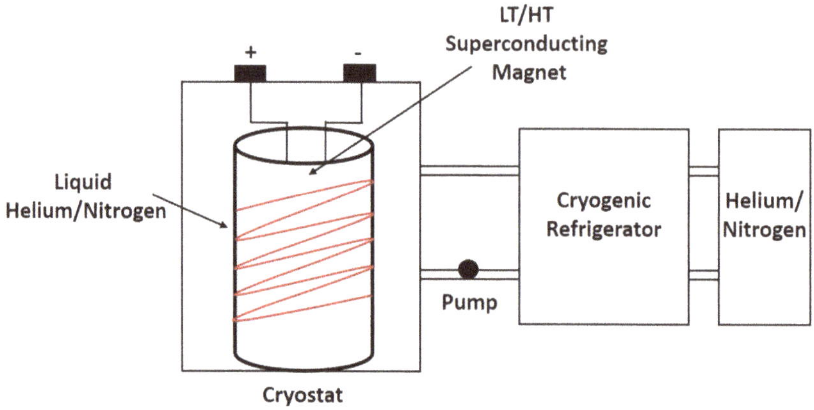

Figure 5.4. A schematic of an SMES.

Table 5.7. The advantages and disadvantages of SMES.

Advantages	Disadvantages
Fast response times	High energy losses (~12% per day)
Capable of partial and deep discharges	Very expensive to produce and maintain
No environmental hazard	Reduced efficiency due to the cooling process required

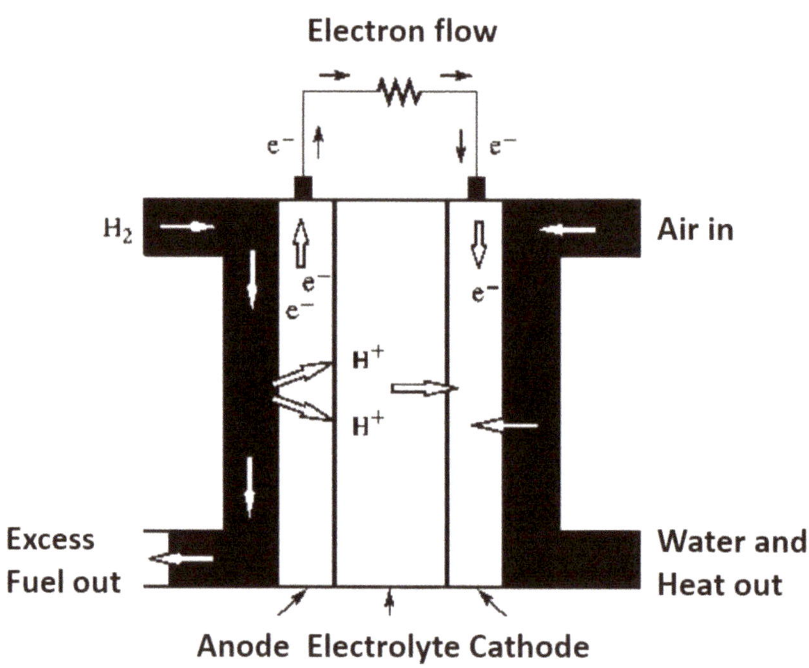

Figure 5.5. Schematic of a fuel cell.

Table 5.8. The properties of the various fuel cell types.

FC type	Mobile ion	Electrolyte	Electrical efficiency	Operating temperature
AFC	OH^-	KOH solution	60%–70%	Room temperature to 90 °C
PEMFC	H^+	Proton exchange membrane	40%–60%	Room temperature to 80 °C
PAFC	H^+	Phosphoric acid	55%	160 °C–220 °C
MCFC	CO_3^{-2}	Molten mixture of alkali metal carbonates	65%	620 °C–660 °C
SOFC	O^{-2}	Oxide-ion-conducting ceramic	60%–65%	1000 °C

methanation, storage, transport, and subsequent power generation. The overall AC–AC efficiency is around <35%. Therefore, this storage system should be integrated with an existing plant that already produces CO_2 and excess electricity.

5.1.4 Electrochemical energy storage systems

Electrochemical energy storage devices are batteries which convert chemical energy to electrical energy. This system contains an electrolyte medium which helps the ion transfer between electrodes. Electrodes are positive (+) and negative (−) terminals called the cathode and the anode, respectively. Batteries can be categorized into primary (non-rechargeable) and secondary (rechargeable) batteries.

5.1.4.1 Primary batteries

Primary batteries, also known as non-rechargeable batteries, are galvanic cells that are used once. This kind of electrochemical energy storage device stores energy for convenient use and has a long shelf life. These disposable batteries undergo irreversible changes that cause them to degrade and slowly die. The types of primary battery are zinc-based batteries (zinc–carbon, zinc–bromine), alkaline cells, metal–air-depolarized batteries, and lithium-based batteries (lithium iron disulfide, lithium manganese dioxide, and lithium sulfur dioxide).

5.1.4.2 Secondary batteries

Secondary batteries, also known as rechargeable batteries, can be charged and discharged many times. For many applications (electric vehicles, mobile phones etc), the cycle life of a battery is usually defined as the number of discharge–charge steps that can take place before the capacity decreases to 20% of a specified amount. The principle of a secondary battery is to store energy through a reversible chemical reaction. The types of secondary battery are described below.

5.1.4.2.1 Lead–acid batteries

Lead–acid batteries mainly contain lead components (figure 5.6). A lead–acid cell consists of two electrodes submerged in an electrolyte of sulfuric acid. The negative

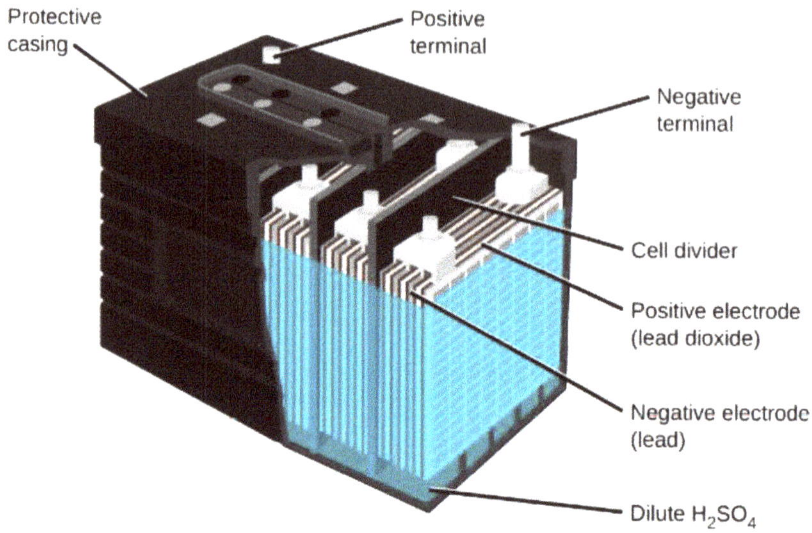

Figure 5.6. A schematic of a lead–acid battery. © LibreTexts. CC BY 4.0. https://chem.libretexts.org/Bookshelves/General_Chemistry/Chemistry_1e_(OpenSTAX)/17%3A_Electrochemistry/17.5%3A_Batteries_and_Fuel_Cells.

Table 5.9. The advantages and disadvantages of lead–acid batteries.

Advantages	Disadvantages
Low-cost manufacture	Low specific energy
Low self-discharge	Slow charging
High specific power	Sulfation problem
Wide temperature range	Limited cycle life
	Not environmentally friendly
	Transportation restrictions on the flooded type

plates are made of metallic lead that also contains minor additives such as calcium, tin, and antimony. The positive plates are made of lead oxides in various compositions. The grid structure of the lead–acid battery is made from a lead alloy. The advantages and disadvantages of lead–acid batteries are given in table 5.9.

There are two different types of lead–acid battery: flooded and valve-regulated. Flooded lead–acid batteries are less expensive, but require maintenance and ventilation. They can be operated safely within their whole temperature range with no need for temperature control. Valve-regulated lead–acid (VRLA) batteries require less maintenance and create less gaseous emissions. There are two primary types of VRLA battery: absorbent glass mat (AGM) and gel cell. In AGM cells, the electrolyte is absorbed into separators between the plates consisting of sponge-like fine glass fiber mats, while gel cells contain a jellified electrolyte, which is sulfuric acid mixed with fumed silica.

5.1.4.2.2 Nickel-based batteries

Nickel-based batteries can be categorized as nickel–cadmium (Ni–Cd), nickel–metal hydride (Ni–MH), nickel–iron (Ni–Fe), nickel–zinc (Ni–Zn) and nickel–hydrogen (Ni–H) batteries. Following the wide usage of Ni–Cd batteries in portable devices, Ni–MH batteries have been discovered to be environmentally friendly and the most readily available rechargeable batteries for consumer use. Table 5.10 summarizes the advantages and disadvantages of this system.

5.1.4.2.3 Lithium-ion batteries

The working principle of a Li-ion cell can be explained as follows: Li ions move between the anode and cathode. During discharge, Li ions de-intercalate from the anode and intercalate into the cathode. Here, the chemical potential difference between the Li ions at the two electrodes results in an electron simultaneously leaving the anode and flowing through the external circuit, which provides electrical energy to the load. If the process takes place in the opposite direction, a charging step occurs. A brief overview of a Li-ion battery (LIB) cell is shown in figure 5.7.

5.1.5 Thermal energy storage systems

Thermal energy storage (TES) systems store heat or cold under varying conditions for later use. They are mainly associated with solar thermal systems. In these systems, a material stores energy when its temperature increases and provides energy when its temperature decreases. Thus, many materials with different thermal properties can be used for these systems. The scale and the method of storage can be varied, and the important parameter is the length of the energy storage period. This directly affects the scale of the storage and the selected application. For example, a buffer tank can be selected for daily peak adjustment applications.

TES systems can be classified according to energy transferred between a body and its surroundings. When water or rock are used, they are called sensible systems; when phase-change materials such as molten salt are used, they are called latent systems; finally, when thermochemical materials such as silica gel/water or NaOH/water are used, they are called thermochemical energy storage systems. When these three systems are compared, it is well known that thermochemical energy storage can yield the highest heat storage capacity without suffering any thermal losses during the storage period [5].

Table 5.10. The advantages and disadvantages of nickel-based batteries.

Advantages	Disadvantages
Simple storage and transportation	High self-discharge
Wide temperature range	Memory effect
Nickel content makes recycling profitable	Low cell voltage
	Low coulombic efficiency

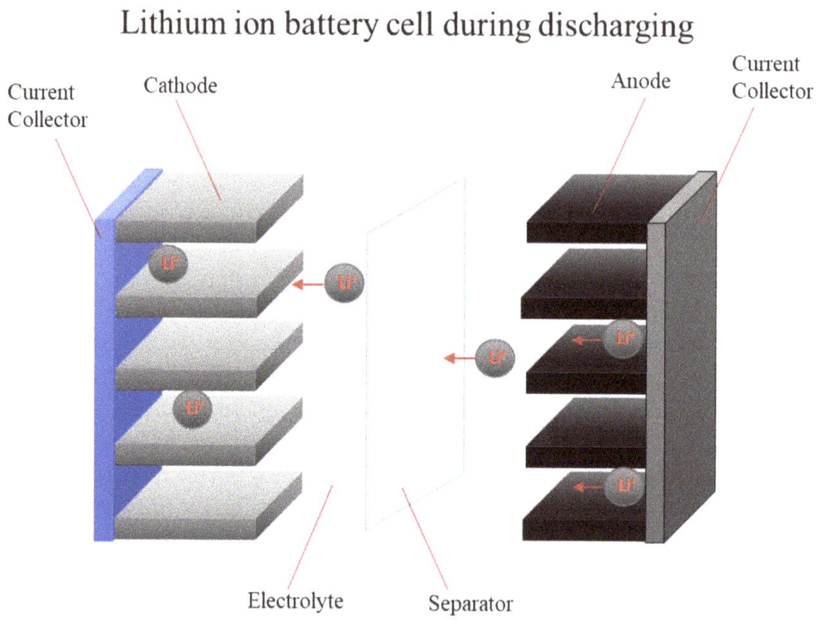

Figure 5.7. A schematic of a Li-ion cell.

5.2 Next-generation devices

5.2.1 Organic batteries and supercapacitors

5.2.1.1 Supercapacitors

Fossil fuels have limitations, leading to the need for alternative energy resources. Nowadays, batteries and supercapacitors are two attractive types of device which can be used for energy storage applications. Supercapacitors possess a fast charge–discharge ability which enhances their power density; they can be used alone or in combination with a battery in electronics. The most important key parameter for supercapacitors is the electrode material. Supercapacitors can use one of two different mechanisms, i.e. they can be electrical double-layer capacitors (EDLCs) or pseudocapacitors (redox capacitors). In general, EDLC-type electrodes involve carbon-based materials which have a high active surface area and a porous structure. Their mechanism is based on the electrostatic interactions between the large surface area and the electrolyte [6]. This fast charge–discharge process occurs within just seconds, and an EDLC-type electrode can work for more than 100 000 cycles. EDLC systems have some significant components: (i) a space charge layer, (ii) a diffusion layer, (iii) an electrolyte system and (iv) a Helmholtz layer, which defines the polarization layer when a voltage is applied. To achieve a high capacitance value, the pore distribution of the electrode and the ion size of electrolyte should be compatible with each other [7].

In pseudocapacitors, the redox-active electrode is composed of transition metals and conductive organic and polymeric materials [8]. It is well known that redox capacitors have high energy and power densities compared to those of EDLCs due to the electrochemical performance of their multiple components [9, 10]. In pseudocapacitors, the energy storage mechanism operates through a fast and reversible faradaic redox reaction which occurs on the electrode surface and between the electrode and the electrolyte ions. Pseudocapacitors can store more capacitance per gram than EDLCs, while EDLCs are kinetically faster than pseudocapacitors due to the fact that the EDLC charge–discharge process only occurs on the electrode surface [11]. Conductive polymers (CPs) and transition-metal oxides are common materials used in the design of pseudocapacitive materials, and much attention has been paid to these materials in the last few decades. In addition, hybrid-type devices which involve both charge storage mechanisms, i.e. the EDLC and pseudocapacitive mechanisms, offer great opportunities to obtain high energy and high power density. This review mainly focuses on the usage of organic compounds due to their sustainability, easy functionality, abundance in nature, and low cost. These organic compounds can be combined with carbon materials and can be used as composites or as unmixed electrode materials.

Sustainable, clean, and cheap materials are one of the major inputs in the creation of energy storage devices. Nowadays, many researchers are focusing on carbon materials due to their abundance in nature and high functionality. Carbon materials are well known for their use as an electrode material in supercapacitors due to their high cyclic stability. In particular, graphene and carbon nanoproducts have attracted great attention for use in the fabrication of supercapacitors with high electrochemical performance [12, 13]. Nanomaterials exhibit high surface area and low weight, enabling the manufacture of flexible devices with high energy and power. There are a number of ways to produce carbon materials, including synthetic organic chemistry and the use of biomass [14]. These materials are described below.

Graphene

Graphene is a monolayer of graphite; it can be prepared using several techniques, and it has become popular in many applications due to its superior electrical and optical properties, such as its tunable bandgap, electronic transport behavior, stability, thermal conductivity, large surface area, etc [15, 16]. In recent years, researchers have focused on graphene-based electrodes, which can be categorized according to their different dimensions (one dimension: 1D (fiber and yarn-type structures); two dimensions: 2D (films); three dimensions: 3D (graphene foam and composites)) [17–19].

The surface area of graphene is much larger than those of carbon black ($<900\ m^2\ g^{-1}$) and carbon nanotubes (from 100 to 1000 $m^2\ g^{-1}$) but similar to that of activated carbon [20]. High-quality graphene sheets with minimal defects have been observed following chemical vapor deposition on copper or nickel foil, so the production of graphene is too expensive compared to that of activated carbon [21]. Today, most studies based on graphene use different graphene derivatives: graphene oxide (GO), reduced graphene oxide (rGO), graphene nanopowder, etc. due to their easy and cheap production processes [22].

Biomass

Biomass-derived porous activated carbons offer superior performance due to their rich elemental structure, and they are very promising materials for the new field of energy storage materials. However, their capability is not efficient at high charge–discharge current densities. Hence, some development strategies are used to improve the performance of this storage mechanism, such as increasing its wettability using different types of electrolyte (e.g. organic, ionic liquids which show high stability at high voltage), increasing its electrical conductivity, etc.

The carbonization of biomaterials is the most popular way of obtaining activated carbon with high porosity and capacity. Cellulose is a most abundant, cheap, and suitable natural material, which consists of linear chain D-glucose. Cellulose fibers exhibit a high surface area, excellent flexibility and mechanical stretching, and a modification capability [23, 24]. A cellulose aerogel structure can be used as an active carbon material with high capacitance values (170–350 F g^{-1}) after several modifications: chemical modifications, carbonization, and doping procedures [25–29]. Other natural polysaccharides such as alginate have attracted great attention due to their ability to chelate metal ions in a so-called egg-box structure, which is useful technique for enhancing the capacity of carbon materials [30, 31]. The macro–meso porous structure of the egg box exhibits a large capacitance of 450 F g^{-1} and highly stable behavior [32]. Lignin is another natural polymer, which exhibits high capacitance values of 200–350 F g^{-1}. N-rich carbon materials are very promising due to their significant electrochemical performance [32, 33].

Conductive polymers

CPs have been regarded as promising pseudocapacitive materials due to their unique characteristics. Polyanilines (PAnis), polypyroles (PPys) and polythiophenes (PThs) are the most well-known CPs used in the preparation of supercapacitor electrodes due to their good conductivity, flexibility, cheap and easy synthesis, and excellent electrochemical performance [34, 35]. Their synthesis can be achieved in numerous ways, such as chemical oxidation, electrochemical polymerization, photopolymerization, different coupling reactions, etc [36–42]. As a result of the development of nanoengineering techniques, 3D porous-structure CPs have gained a lot of attention for use in the assembly of high-energy materials with high specific surface areas.
Conjugated polymers: PAnis, PPys, PThs, and PEDOTConjugated polymers are very attractive materials for energy storage applications due to their good conductivity in the doped state [43]. They possess relatively fast charge–discharge kinetics, fast doping–dedoping processes, and suitable morphologies, and their plastic properties allow flexible electronics to be designed [19, 44].

During the charge and discharge processes, CPs can be p-doped with counter anions in the oxidized state and conversely, they can be n-doped with counter cations in the reduced state. The mechanisms of p and n doping are are described by the equations below:

$$CP(p) \rightarrow CP(p)^{+} A^{-} + n.\ e_{-} \qquad \text{(p-doping process)}$$

$$CP(p) + n.\ e^{-} \rightarrow CP(n)^{-} C^{+} \qquad \text{(p-doping process).}$$

Many CPs, such as PTh, PAni, PPy, etc. can only be used as p-type materials because of their high negative potentials required for n-doping. Some studies have used ionic liquid electrolytes to reduce the negative potential of such CPs and they exhibited a wide potential window [36, 45, 46]. However, the potential of PTh and its derivatives for n doping was found to be more negatively shifted; thus, they are commonly used as p-type materials with a negative electrode made from another material such as carbon [47]. In generally, many CPs can be used as cathode materials to create asymmetric devices in combination with activated carbon anode electrodes.

PAnis have been used in many different energy storage application studies, and relatively high capacitance values of between 400 and 500 F g^{-1} have been obtained [34, 48–51]. Nevertheless, the swelling and shrinkage of PAni over repeated charge–discharge cycles caused a degradation in performance. Combinations of PAni with carbon materials are promising materials with which to improve stability along with maximum capacitance values [52, 53].

PPy is another important CP material due to its great flexibility in electrochemical systems [54–56]. In general, the cyclic stability of PPy film is worse than those of other CPs such as PAni or poly(3,4-ethylenedioxythiophene) (PEDOT) due to its poor porosity, which is caused by large particles on the electrode surface. However, its conductivity is higher than that of PAni, which can be significant in energy storage applications [57]. To solve the stability problem, several strategies can be used: (i) morphological designs based on different synthesis methods, (ii) combination with a nanomaterial or some metal oxides, (iii) a new organic configuration based on Py can be designed in order to enhance its capacity [58–62].

For energy storage applications, PEDOT shows great promise as an energy storage electrode material due to its electron-rich structure, low oxidation potential, wide operational window (1.2–1.5 V wide), low bandgap, and high conductivity in the p-doped state, good thermal and chemical stability, and high charge mobility, which results in fast electrochemical kinetics [63–68]. The expected capacitance values of PEDOT electrodes are around 80–200 F g^{-1} and they are expected to support up to 50 000 cycles. The main reasons for these values are good film formation, a higher surface area, and a long cycle life. In addition, PEDOT has been used in symmetric and asymmetric supercapacitors (with a PEDOT cathode and an activated carbon anode) because it can be doped with both p- and n-type materials [69].

New research topics involving the creation of new organic molecules which can be doped with both p- and n-type dopants (i.e. bipolar materials) have attracted great attention. For this purpose, donor and acceptor molecules were used to design an electrode which could be applied as both anode and cathode. Donor molecules such as Py, Th, 3,4-ethylenedioxythiophene (EDOT), etc. are involved in the cathode reactions, while the acceptor moiety stores the negative charge and is involved in the anode reactions [70]. This type of electrode material enhances the capacity, energy, and power density by nearly four times for symmetric supercapacitor applications [71].

Recently, combinations of CPs with the high specific surface areas of carbon materials, such as porous organic polymers (POPs), covalent organic frameworks (COFs), and conjugated micro-/mesoporous polymers (CMPs) have attracted great attention for use as active electrode materials in energy storage systems [72–74]. These types of material have become more attractive because they work through a combination of the faradaic and capacitive energy storage mechanisms and permit the manufacture of hybrid electrode devices with porous structures and redox-active moieties.

Covalent organic frameworksCrystalline covalent organic frameworks (COFs) are building-block materials that depend on the covalent bonds between boron and oxygen, carbon and nitrogen, boron and nitrogen, or boron, oxygen, and silicon [75]. A microporous structure is most important property of COF materials, because their large surface area makes them good candidates for use as EDLC electrode materials. Liu *et al* designed two different conjugated aromatic polymers, CAP1 and CAP2, based on phenazine [76]. CAP1 had bigger surface area of 704 m^2 g^{-1}, compared to that of CAP2 (594 m^2 g^{-1}); however, CAP1 showed a lower capacitance of 81 F g^{-1} at a current density of 1 A g^{-1}, while that of CAP2 was 240 A g^{-1}. The higher capacitance of CAP2 can be explained by its nanoporous crystalline structure, which was different than that of amorphous CAP1. In addition, CAP2 has a N-enriched skeleton, and this improves the wettability of the surface, enabling faster ion transportation between the electrolyte and the electrode surface. Another example of a new porous polymer was prepared by a solvothermal Schiff reaction, and this material displayed a large surface area (545 m^2 g^{-1}) with an excellent specific capacitance of 469 F g^{-1}. Another new material, N,N,N·,N·-tetrakis (4-aminophenyl)-1,4-benzenediamne (TPDA), displayed good stability after 1000 cycles with a capacitance retention of 75% at a current density of 5 A g^{-1}.

Research into traditional COF materials has focused on increasing the surface area to enhance EDLC performance, however this did not result in good electrochemical performance. Redox-active moieties should be added to the COF structure to improve the electrochemical reaction rate. In 2013, diaminoantroquinone and amine units were used as redox-active sites in the design of a new COF material, which exhibited the high surface area of 1280 m^2 g^{-1}. Electrochemical measurements showed that the maximum capacitance was only 40 F g^{-1}, which remained stable for 5000 charge–discharge cycles due to a poor electrical contact [77]. To overcome this issue, a 2D thin film COF was prepared by direct Au subtraction and the film thickness was controlled through the monomer concentration. The capacitance of an electrode made from oriented COF film increased by 400% compared to that of one made from COF powder [78]. The same method increased the conductivity and improved the capacitance when a new material, PEDOT, was added on the basis that it could effectively enhance the charge transfer. The new PEDOT-modified COF film exhibited excellent stability for 10 000 cycles, and this result demonstrates that modified redox-active COFs are promising materials for use as electrodes in supercapacitors [79]. This strategy has been used in various studies to improve the capacity and stability of COF electrode materials. The best capacitance value among COF supercapacitors was again observed using a PEDOT–COF nanocomposite material; the capacitance reached as much as 1663 F g^{-1} at a current density of 1 A g^{-1} [80]. In the research literature, there are many examples of COFs containing redox-active moieties, which have been prepared in different ways to improve the capacitance [81–83]. Composite COF materials have been prepared using graphene, carbon nanotubes (CNTs), single-walled carbon nanotubes (SWCNTs), rGO, graphite, etc. and their electrochemical performances have been investigated [84–87].

Amorphous porous organic polymers and conjugated microporous polymersA novel family of amorphous porous organic polymers (POPs) and conjugated microporous polymers (CMPs) which display high surface areas, low densities, and strong covalent frameworks could be widely used for applications such as gas storage, sensors, luminescence, and energy conversion and storage [73, 74]. However, their conductivity and oxidative stability show limited suitability for their potential use as electrode materials for supercapacitors. There are a couple of ways to solve this problem: (i) increase the conductivity using chemical or doping approaches and (ii) enhance the stability with the addition of rigid bridge molecular units.

The first supercapacitor electrode was prepared in 2009 by a pyrolytic reaction; it had a high specific surface area (900 m^2 g^{-1}) but a capacitance of only 110 F g^{-1} [88]. The low capacitance value and low stability of CMP were improved by increasing the conductivity. This was achieved through the use of ionothermal condensation to explore aza-CMP [89]. The authors of this paper explained that increasing the capacitance was accomplished by incorporating redox-active units into the CMP via aza moieties.

Recent studies have shown that various heteroatoms (nitrogen, oxygen, boron, and sulfur) can enhance the capacity of energy storage devices. The nitrogen hybridization of materials improves their conductivity, wettability, and surface area as well as increasing their electrochemical performance. Nitrogen-rich redox- -active triazatruxene moieties were used to design new CMPs (TAT-CMP1 and TAT-CMP2) and their capacitances were found to be 141 and 183 F g^{-1}, respectively [90]. This difference between the two molecules was explained by the fact that TAT-CMP2 had a higher nitrogen concentration which led to a larger surface area and a more suitable pore size distribution. A triphenylamine unit improved the electrochemical stability and capacitance of thienothiophene due to its porous structure, and the new compound exhibited a high capacitance of 447 F g^{-1} [91]. Thus, chemical modification with nitrogen enhances the performance of supercapacitors. Similarly, N-rich CMPs based on carbazole exhibited a high capacitance of 558 F g^{-1} with excellent stability, and their nitrogen concentration was found to be 6.07 wt% [92]. Further studies focused on increasing the capacitance of CMPs combined with inorganic nanocarbon materials: graphene, CNT, rGO, etc [93, 94]. Other studies have investigated flexible microsupercapacitors by preparing fiber electrodes with POPs [95].

In summary, POPs and CMPs have great potential to function as electrode materials for supercapacitors due to their high surface area, regular pore structure, and easy modification. However, the capacitance values of POPs are lower than those of MOFs, which also have higher conductivity and charge transfer. These are very significant parameters to consider in the design of new POPs with higher capacitance and more stable supercapacitor performance.

5.2.1.2 Rechargeable organic batteries

Rechargeable batteries are mainly composed of a cathode (the positive electrode), an anode (the negative electrode) and an electrolyte which possesses a cation and an anion in its structure. The cathode is the electrode that is oxidized during the charging process; it readily accepts an anion from the electrolyte to stabilize the positive charge formed. Conversely, the anode is the electrode that is reduced during the charging process and is willing to accept a cation from the electrolyte to stabilize the negative charge formed. During discharge, the cathode accepts electrons that were donated during charging and reduces itself, while the anode loses the electrons and oxidizes itself [96]. The cathode and/or anode materials used for electrochemical energy storage systems are traditionally metal-based inorganic compounds, such as cobalt, iron, tin, or manganese-based materials due to their high specific capacity, fast charge transfer, high working potential, and acceptable lifetime.

Unlike traditional inorganics, organic compounds are favorable electrode materials in batteries due to their recyclability, sustainability, safety, nonuse of costly heavy metals, and their easy modification for different purposes such as flexibility, biodegradability, wearability, and multiple functions [97]. There are infinite possibilities for producing organic electrode materials, since they consist of abundant elements such as C, H, N, and S. The diversification of organic materials and their structural modifications provide many opportunities for the storage of ions and reversible reduction–oxidation processes in targeted applications. However, as other materials, their commercialization and research have some challenges and drawbacks.

Their easy dissolution in liquid electrolytes is their first major disadvantage, which causes electrodes made from them to be intrinsically unstable. However, as a result of recent advances, the use of ionic liquids or organic-gel electrolytes can overcome this issue. Another drawback is the poor conductivity of organic compounds, with the exception of the CPs. Better conductivity could be achieved using some additives such as activated carbon, carbon nanomaterials, etc. thereby recovering the capacity.

5.2.1.2.1 Electroactive cathode materials

Electroactive materials can be commonly classified as p-type, n-type, or bipolar materials. P-type materials are electron rich and willing to donate their electrons to oxidize themselves and form a positive charge. By contrast, n-type materials are electron deficient and willing to accept electrons to reduce themselves and form a negative charge. Bipolar materials possess both electron-donating and -accepting active sites in their scaffold; thus, they can be oxidized and reduced in their neutral state. All these materials are suitable for use in cathodes.

(a) The use of p-type materials in cathodes

Azine- and amine-based p-type materials

An *N*,*N*′-diphenyl-5,10-dihydrophenazine oligomer (p-DPPZ) was synthesized via a coupling polymerization method and used as a p-type cathode in organic LIB applications. The phenazine-based electrode exhibited two sequential reversible electron transfers with redox potentials of 3.3 and 4.1 V vs Li/Li^+ while achieving an energy density of up to 530 Wh kg^{-1} [98]. A p-type material, a phenoxazine-based trimer (3PXZ), was synthesized through a one-step Buchwald–Hartwig reaction and investigated for use as an organic cathode material in rechargeable Li-ion batteries. A negligible structural reorganization of this trimer was facilitated a kinetically faster electrochemical pathway, which led to more precise anodic/cathodic peak separation. The 3PXZ cathode displayed a specific capacity of 112 mAh g^{-1} at 1 C with a high average discharge voltage of 3.7 V vs Li/Li^+ [99]. Other p-type cathodes, namely three N–N-linked diamine azine-based polymers displayed two reversible redox oxidations at 2.9 and 3.3V vs Li/Li^+, demonstrating their suitability for rechargeable battery applications. Moreover, among these three materials, a cross-linked and end-capped polymer exhibited specific capacities of up to 133 mAh g^{-1} and a high rate performance of up to 100 C in Li-based half cells [100].

A triphenylamine (TPA) derivative, 1,3,5-tris(4-diphenylamino-phenyl)benzene (TTPAB) was polymerized through oxidative polymerization to obtain a star-shaped polymer. A TPA-based cathode delivered a specific capacity of 86.7 mAh g^{-1} at a current density of 20 mA g^{-1} and a precise voltage of 3.7 V. The cathode remained stable, and showed a specific capacity of 86.5 mAh g^{-1} after the 50th cycle at the same current density [101]. Poly-hyperbranched triphenylamine (PHTPA) was obtained in one step by a Buchwald–Hartwig reaction and exhibited a high discharge voltage of 4 V, an ultrafast charge–discharge rate of 100–300 C, and a long cycle life of more than 5000 cycles [102].

Organosulfur-based p-type materials

A 3,4-ethylenedioxythiophene (EDOT) monomer was polymerized with $FeCl_3$ with the aim of exploring its use as an organic cathode in a $LiN(CF_3SO_2)_2$/1,2-dimethoxyethane/1,3-dioxopentane electrolyte system. A poly(3,4-ethylenedioxythiophene) (PEDOT) cathode exhibited a discharge capacity of 691 mAh g^{-1} at the first cycle and the capacity remained greater than 330 mAh g^{-1} after 44 cycles [103]. In another study, the polymerization of EDOT was performed through an oxidative method via the direct mechanical mixing of lithium iron phosphate $LiFePO_4$/C. The resulting $LiFePO_4$/C/PEDOT composite delivered a specific capacity of 70 mAh g^{-1} at a current density of 1600 mA g^{-1} [104]. A PEDOT polymer backbone attached to a TEMPO group exhibited a reversible redox potential of 3.5 V vs Li/Li^+ due to the presence of TEMPO radicals and had a good reversible capacity of 47 mAh g^{-1}, which was 62% of its theoretical capacity when used as a cathode [105].

(b) The use of n-type materials in cathodes

In one approach, a dicarbazole-benzothiadiazole (PDCzBT) based n-type conjugated polymer was obtained via oxidation polymerization and applied as a cathode material in Li/Na-ion battery applications due to its utilization of both ionic doping and dedoping processes. A reversible capacity of 404 mA h g^{-1} was obtained after 100 cycles at a current density of 100 mA g^{-1} with a specific energy of 224.4 Wh kg^{-1} [106].

Carbonyl-based organic compounds are conspicuous in rechargeable battery applications due to simplicity and abundance of the carbonyl group, which can easily be reduced without bond breaking. Since simple one-electron reduction of a carbonyl group causes dimerization and inhibits the reversibility of the redox processes, compounds in which two carbonyl groups are connected by a covalent bond between the carbon atoms (1,2-diketones) are more favored in rechargeable battery applications.

Pyrene-4,5,9,10-tetraone (PYT) was combined with polymethacrylate to enhance charge–discharge capability and recyclability (83% after 500 cycles). A cathode made using this material delivered an initial discharge capacity of 231 mAh g^{-1}, which was 88% of its theoretical capacity. The similarity between the theoretical value and the experimental results implies that this cathode can undergo four-electron reduction [107].

Anthraquinone (AQ), phenanthrenequinone (PQ), PYT, and their two derivative $-CO_2Li$ units were electrochemically investigated as organic cathode materials. Low-molecular-weight organic materials have a significant drawback of reduced cyclability due to their easy solubility in the electrolyte. For this reason, the $-CO_2Li$ units were attached to these materials to decrease their solubility. The $-CO_2Li$-coordinated materials LCAQ, LCPQ, and LCPYT exhibited very low initial capacities of 85, 90, and 217 mAh g^{-1}, respectively. However, their discharge capacities after 20 cycles were calculated to be 82, 105, and 86% of their initial capacities, respectively, while those of the non $-CO_2Li$-coordinated quinones had a significant drop after 20 cycles [108].

Poly(benzoquinonyl sulfide) (PBQS) was synthesized through its constructive units dichlorobenzoquinone and Li_2S via a polymerization–oxidation reaction process for use as an organic cathode material for rechargeable Li- and Na-ion batteries. In a Li battery, a PBQS cathode achieved an energy density of 734 Wh kg^{-1} (275 mAh g^{-1} × 2.67 V) and a stable cycling performance (1000 cycles, 86%), while a PBQS-Na battery exhibited an energy density of 557 Wh kg^{-1} (268 mAh g^{-1} × 2.08 V). The significant negative shift at the redox potential of the Na-PBQS battery can be explained by the larger ionic radius of Na^+, which causes a greater energy barrier to appear during the reduction of PBQS [109].

Poly(pyrene-4,5,9,10-tetraone) (PPYT) and poly(9,10-phenanthrenequinone) (PPQ) were obtained through their constructive units via cross-coupling polymerization and introduced into carbon nanotubes to obtain hybrid films, which showed initial capacities of 146.3 mAh g^{-1} and 73.5 mAh g^{-1}, respectively. The discharge capacities at 500 mA g^{-1} after 300 cycles were calculated to be 87.2 mAh g^{-1} for the PPYT-based nanotube film and 70.9 mAh g^{-1} for the PPQ-based nanotube film. The apparent difference in discharge/charge capacities can easily be explained by the multiple carbonyl centers of PPYT [110].

The capacitive behavior of poly(pentacenetetrone sulfide) (PPTS) was investigated as an organic cathode material in a rechargeable potassium-ion battery. The well-known drawbacks of potassium-ion-based batteries are limited capacity, low cyclability, and poor rate performance, which are attributed to the bulky size of K ions. The polymeric cathode showed a reversible capacity of 260 mAh g^{-1} at a current rate of 0.1 A g^{-1} over 100 cycles while achieving a discharge energy density of 450 Wh kg^{-1}. Moreover, after 3000 cycles, the electrode kept delivering a reversible capacity of more than 190 mAh g^{-1} at a high current density of 5 A g^{-1} [111].

Naphthalenetetracarboxylic dianhydride with four nitrogen atoms and perylenetetracarboxylic dianhydride with four nitrogen atoms were synthesized and explored for use as n-type cathodes in rechargeable batteries. Both materials were investigated at voltages between 2.0 and 3.9 V (vs Li^+/Li) and the naphthalene-based cathode exhibited an initial specific capacity of 141.8 mAh g^{-1}, while the perylene-based cathode exhibited an initial specific capacity of 116.3 mAh g^{-1} [112].

(c) The use of bipolar materials in cathodes

The redox-active bipolar polymer polyparaphenylene (PPP) was investigated for use as both a cathode material and an anode material. The material showed reversible capacities of 80 mAh g^{-1} and 350 mAh g^{-1} when used as a cathode and an anode, respectively. The results revealed that such PPP-based electrodes can be operated at 3.0 V, hence they are good alternatives to conventional metal-based batteries [113].

A triphenylamine and anthraquinone-based D–A–D polymer was obtained through oxidation polymerization and its capacitive behavior investigated in a bipolar cathode. The cathode displayed a discharge capacity of 132.7 mAh g^{-1}, which was 86.2% of its theoretical capacity, and it outperformed polyTPA (101 mAh g^{-1}). The electrode also demonstrated its cyclic stability by delivering a discharge capacity of 118.6 mAh g^{-1} for 100 cycles [114].

(d) The use of radical polymers in cathodes

The capacitive behavior of a nonconjugated radical polymer, poly(4-glycidyloxy-2,2,6,6-tetramethylpiperidine-1-oxyl) (PTEO), was investigated for its use as an organic cathode material in both Li- and Na-ion batteries at high oxidation potentials of ~3.6 V vs Li^+/Li and 3.3 V vs Na^+/Na, respectively. It was observed that the reaction mechanism of the polymer developed as the neutral-state nitroxide radical ($N–O^{\cdot}$) turned into oxoammonium cations ($^+N=O$) and aminooxy anions ($N–O^-$). With only 30 wt% of a carbon additive, a PTEO-based cathode achieved a discharge capacity of more than 200 mAh g^{-1} and an energy density of more than 500 Wh kg^{-1} when used in both Li- and Na-ion batteries [115].

5.2.1.2.2 Anode materials for metal-ion batteries

(a) Conjugated organic scaffolds

Various organic molecules are mainly classified as conjugated carbonyls, thioethers, organodisulfides, nitroxyl radical polymers, etc. Such molecules can be given a reversible lithiation and de-lithiation mechanism. In general, carbonyl-bearing small molecules are promising candidates for cathode materials. In addition, most of them are suitable for use as anode electrode materials [116]. Conjugated carbonyl-bearing molecules are very favorable electrode materials due to their miscellaneous structural design, high theoretical capacity, and fast reaction kinetics [117–119]. Carbonyl groups need electron-donating groups (EDGs) in order to obtain a large potential window for device operation when used in anodes [120]. For instance, an aromatic molecule, 1,4,5,8-naphthalenetetracarboxylic dianhydride fused with C_6 aromatic rings displayed a charge capacity of 724 $mAhg^{-1}$ in an LIB [121]. Similar structures such as itaconic acid ($C_5O_4H_4$) can achieve reversible lithiation and de-lithiation with six lithium ions per unit and therefore demonstrated a capacity of 1270 mAh g^{-1} [122].

With the increasing energy demand in various areas, such as in warmer climates and in batteries, the greatest challenge of working at high temperatures is still ongoing. LIBs have often been studied for development in this respect; however, the scarcity of Li resources has posed a major obstacle to this development. Potassium-ion batteries (KIBs) are being studied to develop high-temperature batteries to overcome this problem. K is nearly 1200 times as abundant as Li in nature, and the redox potential of K^+/K is lower than that of Na^+/Na [123]. The volume change of electrodes in KIBs is greater than that of LIBs and Na-ion batteries (NIBs) due to the bigger ionic size of K. However, the surface morphology is the main point in work on high temperatures, particularly in LIBs. It is also more challenging for KIBs to resist temperature because of the uneven interface between the electrode and the electrolyte. However, some organic anode materials can overcome this issue due to their effective redox centers and their ability to store the K ion in their structures. For instance, azobenzene-4,4′-dicarboxylic acid potassium salts (ADAPTS) were found to be suitable for use as anode materials, exhibiting an extended cycle life and increased capacity [124]. At 50 °C for 400 cycles, such an anode material had a reversible capacity of 109 mAh g^{-1} at a current density of 1C and a capacity of 77 mAh g^{-1} at a current density of 2 C for 1000 cycles. When this KIB was tested at 60 ° C, it exhibited a capacity of 113 mAh g^{-1} at 2 C and little capacity fading (81%) over 80 cycles.

Potassium terephthalate (K_2TP) and potassium 2,5-pyridinedicarboxylate (K_2PC) (small organic dicarboxylates) were utilized as extremely beneficial anode materials in the structure of KIBs [125]. Their specific capacities were found to be 181 and 190 mAh g^{-1} for around 100 cycles. In addition to this material, disodium terephthalate (Na_2TP or $Na_2C_8H_4O_4$) and its derivatives were introduced as organic anode materials for NiBs [126]. Two Na cations can reversibly de-intercalate to a terephthalate structure in this design. In NiBs, it is important to prevent the Na metal from forming dendrites because of safety issues and to maintain high energy densities. Na_2TP has a redox potential of 0.4 V, which is more suitable for preserving the Na metal in anode materials; such a material was shown to have a reversible capacity of 295 mAh g^{-1}.

(b) Polymer-based organic structures

Poly(1,6-dihydropyrazino[1, 3g]quinoxaline-2,3,8-triyl-7-(2*H*)-ylidene-7,8-dimethylidene) (PQL) is an polyazaacene analog that has conjugated ladder polymer nanostructures and a large number of nitrogen heteroatoms [127]. Therefore, PQL is a good candidate for use as an LIB anode material due to its high capacitance even at high temperatures. At 50 °C and a low rate of 0.05 C, PQL attains a high capacity of 1770 mAh g^{-1}, and at a high rate of 5 C, PQL shows a reversible capacity of 303 mAh g^{-1}. At a capacity of 500 mAh g^{-1}, it can remain stable even after 1000 charges and discharges at a rate of 2.5 C.

Polymer-based organic electrode materials have expanded the application areas of electronics due to their ability to be flexible. Poly(chalcogenoviologen)s (polyBnV) is known to be a typical electron acceptor molecule which also possesses electrochromic behavior. The addition of heavy metals (S, Se, Te) into viologen moieties significantly increases the cycle life, stability, and capacity of LIBs. In particular, poly(Te-BnV) has demonstrated a capacity of 502 $mAhg^{-1}$ at 100 mA g^{-1} following the removal of 20 Li ions from its structure [128].

Two polyterthiophene derivatives, poly(3′-styryl-4,4″-didecyloxyterthiophene) (poly(OC_{10}DASTT)) and poly(4,4″-didecyloxyterthiophene) (poly(OC_{10}STT)) were electrochemically deposited on a Ni/Cu-coated fabric and used as an anode alongside a polypyrrole cathode [129]. Poly(OC_{10}DASTT) and poly(OC_{10}STT) exhibited capacities of 45.2 mAh g^{-1} and 94.7 mAh g^{-1}, respectively.

A 1,4,5,8-naphthalenetetracarboxylic dianhydride (NTCDA)-derived polyimide has been introduced as an anode material in aqueous rechargeable LIBs and NIBs [130]. Polyimide is a well-known material in the battery field due to its high capacity, great structural stability, and favorable operating voltage characteristics. A polyimide/$LiCoO_2$ cell structure demonstrated a high specific energy and a specific capacity of 80 Wh kg^{-1} and 71 mAh g^{-1}, respectively, at a current rate of 100 mA g^{-1}.

5.2.1.2.3 Bipolar materials

Bipolar electrode materials are exceptionally attractive and possess clear benefits in terms of enhanced power performance, volumetric/gravimetric energy density, and, above all, a simplified construction process and reduced production expense. Another advantage is their application-oriented and easily targeted designable battery configuration.

Polyisothianaphthene has been used as bipolar material, which was reversibly able to accept of both Li^+ (730 mAh g^{-1}) and PF_6^- ions (106 mAh g^{-1}). This low-bandgap polymer also displayed a capacity of 92 mAh g^{-1} in a bipolar battery structure [131].

Symmetric batteries based on bipolar electrodes have attracted great attention due to their simple fabrication process. The alkaline salt of 4,4′-(phenazine-5,10-diyl)dibenzoate (PZDB) transports both anions and cations and can remain in both the oxidized and reduced states. The large potential difference between phenazine and benzoate leads to a high energy density of 127 Wh kg h^{-1} at a high voltage of 2.5 V [132].

5.2.1.2.4 Fully organic batteries

There are several examples of fully organic batteries based on different organic molecules (figure 5.8). Organic lithium salts, such as 2,5-dihydroxyterephthalic acid ($Li_4C_8H_2O_6$), with different nanomorphologies have been used as the active material of positive and negative electrodes in rechargeable batteries [133]. Nanosheet

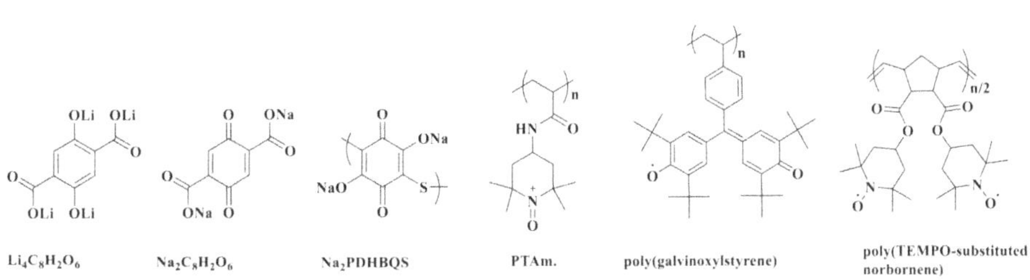

Figure 5.8. Several examples of organic molecules used in battery electrodes.

structures of $Li_4C_8H_2O_6$ exhibited higher capacity than nanoparticle structures of $Li_4C_8H_2O_6$. In addition, nanosheet structures of this compound exhibited a high energy density (130 W h kg^{-1}) in an all-organic battery (symmetric, both electrodes prepared using $Li_4C_8H_2O_6$ nanosheets) at an operational voltage of 1.8 V. A similar material, $Li_2C_6O_6$, was used as an electroactive material at the same operational voltage and exhibited low potential. However, these studies focused on multifunctional (i.e. both positive and negative electrode) structures [134]. Organometallic salts can be used in Na-ion batteries as an electroactive material for both the positive and negative electrodes. $Na_4C_8H_2O_6$, was prepared by a one-pot reaction and exhibited a stable capacity of 180 mA h g^{-1} over wide ranges of operating voltages of 1.6–2.8 V and 0.1–1.8 V when used as the positive and negative electrode, respectively [135]. Another example of a Na-ion battery fabricated with Na salts of poly(2,5-dihydroxy-p-benzoquinonyl sulfide)/rGO (Na_2PDHBQS/rGO) as a composite cathode and a disodium terephthalate anode displayed a reversible capacity of 210 mA h g^{-1} at 0.1 C [133, 135–139].

A class of polymers bearing stable radical groups on their polymer chains is used in organic radical batteries (ORBs) and is suitable for use in dual-ion cells. For this purpose, poly(galvinoxylstyrene) and poly(TEMPO-substituted norbornene) were used as an n-type redox-active polymer and a p-type polymer, respectively, in the preparation of an all-organic battery. The all-organic radical battery remained stable and retained 60% of its initial capacity after 250 cycles [137]. Another radical polymer, poly-(tripyridiniomesitylene) (PTPM), was used as an anode material with a poly(2,2,6,6-tetramethylpiperidinyloxy-4-yl acrylamide) (PTAm)-based cathode in a full cell which was rechargeable for more than 2000 cycles at a discharge voltage of 1.5 V [138]. Another TEMPO-substituted polyacrylamide, PTAm, was used to fabricate a totally organic rechargeable cell which displayed excellent stability for more than 2000 cycles and a high coulombic efficiency of 95% [139].

In the case of fully organic batteries, a solid electrolyte is another important consideration for sustainability and long-term performance. A poly(aniline/o-nitroaniline) (P(AN-NA)) cathode and a poly(anthraquinonyl sulfide) (PAQS) anode were assembled to fabricate a cell with a solid-state Na^+ electrolyte prepared in the organic crystal succinonitrile. This solid-state battery exhibited a good operational voltage of 3.5 V and a high-rate cyclic stability [140].

P-type polytriphenylamine and n-type polynaphthalene diimide polymers were used as the cathode and anode, respectively, with LiTFSI as the electrolyte in a dual-ion battery. The device displayed a specific capacity of 105 mAh g^{-1} at a current density of 0.5 A g^{-1} based on the cathode material's weight and a retention of 85% after 700 cycles [141]. A similar device was constructed using the same electrodes but a different electrolyte concentration. This device configuration exhibited a discharge capacity of 90 mAh g^{-1} at an output voltage of 1.2 V [142].

A dual-ion battery was constructed using poly(2-vinylthianthrene) as the cathode and poly(2-methacrylamide-tetracyanoanthraquinodimethane) as the anode. This device offered 105 mAh g^{-1} at a current density of 1 C, based on the cathode material's weight. Moreover, this dual-ion battery exhibited a retention of 67% at 250 cycles [143].

A polyimide cathode and a pre-reduced perylenetetracarboxylate anode were explored for use in an LIB; the device displayed a specific capacity of 81 mAh g^{-1} at a current density of 11.3 A g^{-1} [144]. A NIB was assembled using *N*,*N*-diamino-3,4,9,10-perylenetetracarboxylic polyimide as the cathode and pre-reduced terephthalate (Na_2TP) as the anode. This cell exhibited a specific capacity of 73 mAh g^{-1} at a current density of 50 A g^{-1} [145].

Poly(vinylcarbazole) was used as cathode with a potassium terephthalate anode and an ether electrolyte in a long-life KIB. The results revealed that the constructed full cell could be operated in a voltage range of 2–4.1 V; it achieved a discharge capacity of 67 mAh g^{-1} (cathode) with an energy density of 214 Wh kg^{-1} (3.2 V × 67 mA g^{-1}). The constructed full cell also outperformed most other K-ion full cells in terms of long lifespan by achieving an average discharge capacity of 46 mAh g^{-1} over 3000 cycles at a current density of 2 A g^{-1} [146]. Poly(*N*-vinylcarbazole) (PVK) was used as the positive electrode material as a counter for poly(1,1′-pentyl-4,4′-bipyridinium dihexafluorophosphate) (PBPy) in a viologen skeleton [147]. In a full-cell design, the negative electrode capacity was shown to be twice the positive electrode capacity (100 mAh g^{-1}) due to operation in the low potential region of 1.8 V.

Many organic battery materials need metal-ion electrolytes, such as Li^+ or Na^+. However, anthraquinone- and benzonquinone-functionalized poly(3,4-ethylenedioxythiophene) (PEDOT-AQ) and (PEDOT-BQ) were used as the negative and positive electrodes to assemble batteries without metal ions or any other conductive additive [148]. For this purpose, a substituted pyridinium triflate and a slurry containing the corresponding pyridine base were used to obtain $2e^-/2H^+$ quinone/hydroquinone redox reactions. As a result, PEDOT-AQ and PEDOT-BQ demonstrated specific capacities of 103 and 120 mAh g^{-1}, respectively, and cyclic stability over 1000 cycles.

Polytriphenylamine (PTPAn), which leads to PF_6^- doping–dedoping reactions, and PAQS, which leads to Na^+ doping–dedoping reactions, were used as the cathode and the anode, respectively, in an all-organic NIB [149]. The full device offered the very high capacity of 3200 mA g^{-1} at a high rate of 16 C.

Poly(3,4-dihexylthiophene) (PDHT) was polymerized on carbon nanofibers (C), and this composite displayed an exceedingly high capacity of 300 mAh g^{-1} [150]. A fully organic Li-ion cell was then assembled using the PDHT/C as the anode and pTPA as the cathode. The usage of this redox-active component led to a discharge capacity of 250 mAh g^{-1}.

5.2.2 Self-healing batteries

5.2.2.1 Self-healing chemistry and mechanisms

The self-healing ability of materials is one of their extraordinary properties. Self-healing behavior can be seen in a single molecule (such as DNA), as well as in the fusion of broken bones, the healing of wounds, etc. In some cases, it can be observed macroscopically. The self-healing behavior of a wound on the skin is triggered in an area by the chemical signals produced by broken or injured tissue; thus, self-healing behavior is observed in biological tissues. Synthetic materials, on the other hand, lack the ability to repair themselves, as they do not have a similar mechanism.

It has been realized that the accumulation of undesirable material damage due to volume changes, abrasion, cutting, breakage, or operational fatigue often occurs during practical use and that, over time, degradation results in the deterioration of a device's properties and a significant shorting of its life. For example, the large volume changes in the electrodes used for high-energy LIBs due to the repeated insertion and extraction of lithium ions can lead to cracking or even pulverization of the electrodes and therefore short cycle lifetimes of full LIBs. Fractures and other damage that occur in materials start microscopically, and since the fracture energy cannot be effectively distributed throughout the structure, the fracture grows and spreads throughout the material. The synthesis of polymers in which damage can be easily controlled or repaired has become very important due to the difficulty of application, which leads to long delays, high cost, and reduced polymer life due to bonding, welding, etc. Instead of the classical methods that require external application, it has become more important to include repairability due to the internal structure of the polymer. For this reason, since the 1980s, researchers have tried to develop polymers that have the ability to repair themselves by modeling the mechanisms found in biological systems [151–155].

Materials that have the ability to heal themselves can partially or completely return to their original form after damage to their structures. Materials that can heal themselves to their ideal state, on the other hand, can suddenly and completely return to their original condition as soon as the factor causing the damage is removed. A self-healing mechanism can sometimes spontaneously occur in polymers, or it can be added by an external effect. Self-healing in polymers can be achieved in many ways, such as microencapsulation, hydrogen bond interactions, and disulfide bonds [156–158]. In self-healing materials, the size of the reaction determines the efficiency of the healing. Many chemical reactions have been investigated during the rapid development of self-healing polymeric materials. The most suitable healing chemistry can be selected according to the type of material requiring repair, environmental conditions, and user needs. For example, room-temperature reactions that are not affected by air are often used for autonomous self-healing. Therefore, chemical reactions that cannot be carried out at room temperature cannot be used in such a self-healing process. Also, unlike chemical reactions carried out under conventional reaction conditions, self-healing chemical processes that operate in narrow gaps (e.g. the internal cracks of polymers) are limited by certain factors. For example, mixing to achieve the optimum mixture of reactants is not possible, so stoichiometric ratios cannot be maintained accurately. As a result, reactions with operational solutions cannot provide satisfactory recovery efficiencies. Self-healing properties are not seen in common polymers due to limited chain mobility and a lack of re-linking ability. The last decade has witnessed the development of several reaction classes aimed at self-healing through the division and rearrangement of specific bonds. These units are included in polymer networks and are typically classified as reversible covalent and non-covalent bonds [155, 159].

Self-healing in biological systems can be defined as the healing process of an unbalanced, diseased, or damaged organism. The recovery process of an organism exposed to physical damage or disease consists of the recovery of living tissue, organs, and the biological system as a whole and a return to their normal functioning. In mammals, the repair of cells in a necrotic or damaged area can take place in two different ways, depending on the extent of the damage: cells in the decayed area can regenerate and form tissue similar to the old tissue, or tissues in the damaged area can form scar tissue. Healing by the formation of scar tissue has an organized progression through four stages: hemostasis, inflammation, proliferation, and maturation. The hemostatic phase, which occurs when a blood vessel is cut, begins with blood leaving the body. Platelets in the blood then stick together to close the rupture in the vessel wall. Finally, the coagulated platelets are reinforced with fibrin threads. In the inflammatory phase, bleeding is controlled, scar tissue is formed and the protection of the area against bacteria and viruses is restored. In the proliferative phase, new tissues similar to the old ones are formed and the wound shrinks. In the maturation phase, which is also called the remodeling phase of wound healing, the wound is completely closed and the cells used for healing but no longer needed are removed by programmed cell death. In the case of self-healing polymers, the main approaches involve either physical or chemical events at the molecular level, although there is an overlap between the two approaches. Examples of physical self-healing processes are interchain diffusion, phase-separated morphologies, shape-memory effects, and the introduction of superparamagnetic nanoparticles [152, 159–162].

Self-healing polymers are usually cross-linked via reversible non-covalent interactions or dynamic covalent interactions, and the resulting macroscale properties (such as the self-healing temperature and tensile modulus) are determined by the overall strength and density of such crosslinking interactions. These interactions are, in turn, tightly coupled to the spatial structure of the polymers at the molecular level. Most previous studies took the logical approach of altering macroscale properties through the reengineering of molecular structure, including the modification of polymer backbones and side chains, copolymerization, and the hybridization of polymers and inorganic materials. When this approach is used, each new target application requires an entirely new polymer design, which makes most individual materials only suitable for a small group of applications, due to the narrow range of properties achievable for a particular structure [153, 157, 163, 164].

Interchain diffusion: the curing of thermoplastic polymers by intermolecular diffusion was extensively investigated in the 1980s. The polymers examined were amorphous, semicrystalline, block copolymers, and reinforced fiber composites. If two parts of the same polymer are combined at temperatures greater than the glass transition (T_g), the crack heals and the interface gradually disappears due to molecular diffusion at the interface, thus increasing the mechanical strength between the polymer surfaces. This healing process has been investigated at atmospheric pressure and under vacuum; the healing time requires intervals from minutes to years, and the healing temperature is greater than the glass transition temperature, ranging from −50 °C to +100 °C [155].

Van der Waals interactions: self-healing systems are systems in which the repair of the structure takes place due to the performance of the matrix without any external intervention after the formation of cracks in polymers or polymer composites. The self-healing mechanism of the composite structure can be triggered by a stimulus such as heat, light, or a solvent. Internal self-healing materials have the advantages of allowing multiple repair cycles and not requiring an external agent for the self-healing process. However, the healing mechanism is usually limited to small damage zones. The first studies in this area used thermoplastic polymers. Cracks that occurred in the thermoplastic structure were healed by rejoining the broken chain end groups by heating the matrix. In further studies, the improvement mechanism was also applied to thermoset structures. Intrinsic self-healing mechanisms depend on both covalent bonds and physical and chemical interactions. Van der Waals interactions have been of interest for over two centuries, and the presence of van der Waals interchain forces in polymers was established decades ago [161, 164, 165].

Reversible covalence-based self-healing systems. In the design of self-healing materials, the dissociation and union rates of dynamic bonds are important factors, in addition to chain mobility. Reversible covalent bonds are good candidates for the development of self-healing polymers due to their high bond strength, which increases the mechanical strength of materials. Four types of self-healing covalent bond have been evaluated: (1) reversible cycloading reactions, (2) exchange reactions, (3) stable free-radical-mediated exchange reactions, and (4) heterocyclic compounds/carbohydrates that facilitate bond re-formation. One of the disadvantages of covalent self-healing bonds is their high activation energies [154, 166, 167].

Designs based on H-bonding. H-bonding is typically among the strongest of non-covalent interactions and its directionality and high per-volume concentration confer acceptable mechanical strength. As a result of these attributes and its reversibility, it can be used in the self-healing of thermoplastic polymers. In studies, researchers have confirmed that crack repair occurs very effectively in thermoplastics at the T_g of some materials or at temperatures above T_g. In polystyrene-b-poly(*n*-butylacrylate)-b-polystyrene triblock copolymer systems, the T_g of poly(*n*-butylacrylate) (PBA) as a soft block is −40 °C, and that of polystyrene (PS) as a hard block is 100 °C. In previous studies of PS-*b*-PBA-*b*-PS it was revealed that it has a microphase separation morphology and thermoplastic elastomer properties. However, the mechanical deformation of this covalent system causes irreversible covalent bond rupture and permanent loss of its properties [168–171].

In thermally triggered healing, there are different types of dynamic bond that have been incorporated into materials to make them intrinsically self-healing. With only a few exceptions, materials based on non-covalent reversibly formed bonds tend to heal without any addition of energy following mechanical damage/stress, whereas ruptured covalent dynamic bonds require the input of some form of energy in order for healing to occur. Although self-healing systems based on covalent bonds are not capable of independent healing, these materials benefit from improved mechanical properties, which may be necessary for certain applications. For this reason, many material chemists have invested time into designing self-healing materials using thermally controllable covalent dynamic bonds, such as the bonds formed by Diels–Alder (DA) cycloaddition, Friedel–Crafts reactions, alkoxyamine bonds, and imine bonds. In this system, thermally cured benzoxazine reacts with the polysulphone group's meta position via a Friedel–Crafts reaction [172].

Another system of thermally triggered self-healing materials is that comprised of materials which are based on π–π stacking. Although such materials are composed of non-covalent dynamic bonds, it has been demonstrated that a robust free-standing film that heals within seconds at 80 °C can be formed from a mixture of a diimide-containing polymer and a pyrene end-capped polymer [166, 172].

5.2.2.2 Self-healing systems for energy storage devices

The purpose of self-healing is to restore function that has been lost or weakened due to damage to a material system. Filling the damaged volume completely and regenerating bonds along damaged surfaces can restore/repair fracture properties and improve other functions. The damaged volume depends on the loading conditions, the material's geometry and the properties of the undamaged material. Despite the advances made, only limited kinds of self-healing material have been successfully used in energy storage applications (ESA). This is because more requirements need to be considered when self-healing materials are applied in ESAs, including their mechanical, electrical, and self-healing properties, their electrochemical and thermal stability, and their processability [173, 174].

Self-healing materials including hydrogen-bond-based materials (e.g. supramolecular rubber, urea pyrimidinone-based polymers, carboxylated polyurethane, polyacrylic acid, etc.), borate ester bond-based materials, electrostatic interaction-based materials, and so on, have demonstrated great promise for battery applications. For example, hydrogen-bond-based self-healing materials have been widely applied in ESAs, which include styrene-rubber (SR), ureidopyrimidone (UPy)-containing polymers, and carboxylated polyurethane (CPU). SR has excellent chemical and electrochemical stabilities and high self-healing efficiency. Consequently, it has been used as a self-healing binder for the Si anodes of LIBs, as the electrolyte or artificial solid electrolyte interphase (SEI) layer for lithium-metal batteries (LMBs), and as an electrode substrate for SCs. UPy-containing polymers are facile to design, fabricate, and functionalize, enabling them to be good candidates for self-healing electrolytes. Due to its high mechanical strength and chemical stability, CPU is suitable for use in self-healing encapsulation shells and substrates. The applications of other self-healing materials are still in their infancy [155, 159, 168, 175].

Self-healing materials can be introduced into different components of ESAs to fulfill different functions. Material types that are susceptible to damage, such as plastics/polymers, paints/coatings, metals/alloys, and ceramics/concrete can be produced with self-healing mechanisms. The main purpose here is to enable materials with self-healing capabilities to quickly regain their physical and mechanical integrity after damage. To discover how to equip composite materials with self-healing properties and how to improve self-healing efficiency, many experimental studies and discoveries have been carried out over several years and many innovative strategies have been developed. Self-healing materials can be designed using various technology strategies (e.g. electrohydrodynamics, conductivity, the shape-memory effect, nanoparticle migration, co-deposition) [155, 176].

Depending on the healing agent release strategy adopted, fibers, channels, or microcapsules containing liquid active ingredients such as monomers, dyes, catalysts, or hardeners can be embedded in polymeric systems during the manufacturing phase. In the case of any damage, the fibers, channels, or microcapsules break, leaving the healing agent on the damaged surface. The curing agent is released and polymerizes by reacting with the catalyst in the matrix. Thus, repair is carried out autonomously in the area of damage.

A self-healing polymer binder with balanced rigidity and softness should be able to effectively withstand the high stress caused by volume change with minimum deformation, and it should self-heal rapidly. Inspiring by the fast self-healing concept, researchers have employed a poly(ether-thiourea) (PETU) polymer as the self-healing PETU (SHPETU) binder for a silicon anode. As in the reported self-healing polymers, the hydrogen bond pairs endow the SHPETU polymer with a self-repair function which effectively eliminates cracks and pulverization in the electrode. More importantly, the cross-linked thiourea units in the SHPETU polymer add robustness that balances the softness of the hydrogen-bond-containing self-healing polymer. In addition, the thiourea units provide a strong binding force between the SHPETU and the silicon surface [177].

Considering the low cost and high safety of water as a solvent, another approach has been presented for the design and easy preparation of a water-soluble and rigid–soft modulated polymer binder, poly(acrylic acid)-poly(2-hydroxyethyl acrylate-co-dopamine methacrylate) (PAA-P (HEA-co-DMA)), which *in situ* forms multiple self-healing network structures during electrode preparation. It can accommodate dramatic volume changes, prevent the disintegration of Si microparticles (MPs) and electrodes, and improve the electrode wettability by the electrolyte solution, resulting in excellent cycling and rate performances even under high areal capacities [157].

Ideally, a self-healing system designed according to the healing agent release strategy should be readily available and cost effective. A suitable healing agent should be easily exposed and fast enough to migrate to areas where damage has occurred and should not damage the surface on which it is coated (by corrosion, dissolution, etc). In addition, the healed areas should ideally have mechanical properties equal to or greater than those of the coating matrix. In order for the material's long-term self-healing ability to be stable, the matrix and the repair ability must be compatible. In recent years, a series of studies focused on self-healing materials with different mechanisms for ESAs has been completed, as presented in table 5.11.

5.3 Conclusions

In this chapter, energy storage systems were summarized and next-generation storage systems were elaborated to introduce the reader to new ideas.

Table 5.11. Recent studies of hydrogen-bond-based and ionic-interaction-based self-healing materials.

Self-healing material	Healing condition	Self-healing mechanism	Usage	Performance	Reference
PVA-*g*-PAA/borax	RT, 20 min	Borate ester bonds	Electrolyte	84.8 F g^{-1} after first healing ≈ 100% after tenth healing	[178]
Cellulose–PDA–PAM	RT, 10 min	π–π stacking/ hydrogen bonds	Electrolyte	27.92 mF cm^{-2} at 200 mV s^{-1} 95% after 1000 cycles	[179]
Alginate/Ca^{2+}	RT	Ionic interactions	Anode	1500 mA h g^{-1} at 1 A g^{-1}, 83% after 300 cycles	[180]
Poly (lithium 2-methyl-2-(4-vinylbenzyl) malonate)	RT	Ion–dipole interactions	Anode	51% after 500 cycles at 1 C	[181]
PAA/Fe^{3+}	RT, 24 h	Ionic interactions/ hydrogen bonds	Electrolyte	87.4 F g^{-1} at 0.5 C 89% after 5000 cycles at 1 C	[182]

RT: room temperature, Cellulose–PDA–PAM: cellulose–polydopamine–polyacrylamide, PVA: polyvinyl alcohol, PAA: polyacrylic acid.

Supercapacitors are widely used devices for applications which need high specific power capacities. Here, new approaches to materials for supercapacitors have been discussed. We have seen that organic and inorganic carbon-based materials are very promising for supercapacitors. Organic rechargeable batteries have also been examined. The development of organic materials for many battery components will offer essential alternatives in the field of batteries and solve the many problems encountered.

The recent progress in self-healing polymers for energy storage applications has been summarized as well. In the first part, the self-healing chemistry and mechanisms were presented. The application of self-healing materials for various self-healing-based energy storage applications were then systematically discussed according to the self-healing mechanisms used.

Despite the rapid development of self-healing materials, the availability of self-healing materials for energy storage applications has been limited. In the field of energy storage applications, only a small fraction of the self-healing materials and healing mechanisms have been successfully implemented, because the addition of self-healing components can subtly affect the electrochemical stability window, capacity, cycle stability, and safety in energy applications.

The integration of self-healing encapsulation shells, electrodes, and electrolytes into a single device based on the use of curable materials is highly desirable in practical applications. A high curing efficiency, achievable curing conditions, and high mechanical strength are the essential requirements for self-healing protective shells and substrates. For self-healing electrodes, the binders developed must have good chemical, thermal, and electrochemical stability. Most self-healing materials are electrically and ionically insulating and do not provide the capacity to perform as single-function components in energy storage applications.

For a deeper understanding of materials that can be used for energy storage, the energy storage applications of organic and self-healing materials should be studied using theoretical simulation, modeling, and detailed characterization techniques.

References and further reading

[1] Chen H, Cong T N, Yang W, Tan C, Li Y and Ding Y 2009 Progress in electrical energy storage system: a critical review *Prog. Nat. Sci.* **19** 291–312

[2] Lin F C, Dai X, Xu Z A, Li J and Zhao Z G 2003 High density capacitors *High Power Laser Part. Beams* **15** 94–6 http://www.hplpb.com.cn/en/article/id/354

[3] Arepalli S, Fireman H, Huffman C, Moloney P, Nikolaev P and Yowell L *et al* 2005 Carbon-nanotube-based electrochemical double-layer capacitor technologies for spaceflight applications *JOM* **57** 26–31

[4] https://energy.gov/eere/fuelcells/fuel-cells (accessed 20 April 2021)

[5] Dincer I and Bicer Y 2020 System integration for multigeneration *Integrated Energy Systems for Multigeneration* (Amsterdam: Elsevier Science)

[6] Largeot C, Portet C, Chmiola J, Taberna P L, Gogotsi Y and Simon P 2008 Relation between the ion size and pore size for an electric double-layer capacitor *J. Am. Chem. Soc.* **130** 2730–1

[7] Salunkhe R R, Tang J, Kamachi Y, Nakato T, Kim J H and Yamauchi Y 2015 Asymmetric supercapacitors using 3D nanoporous carbon and cobalt oxide electrodes synthesized from a single metal–organic framework *ACS Nano* **9** 6288–96

[8] Fan W, Zhang C, Tjiu W W, He C B and Liu T X 2013 Graphene-wrapped polyaniline hollow spheres as novel hybrid electrode materials for supercapacitor applications *ACS Appl. Mater. Interfaces* **5** 3382–91

[9] Raghu M S, Yogesh Kumar K, Rao S, Aravinda T, Prasanna B P and Prashanth M K 2018 Fabrication of polyanilinee—few-layer MoS_2 nanocomposite for high energy density supercapacitors *Polym. Bull.* **75** 4359–75

[10] Zhao J, Wang G, Cheng K, Ye K, Zhu K, Yan J, Cao D and Wang H 2020 Growing NiS_2 nanosheets on porous carbon microtubes for hybrid sodium-ion capacitors *J. Power Sources* **451** 227737

[11] Snook G A, Kao P and Best A S 2011 Conducting-polymer-based supercapacitor devices and electrodes *J. Power Sources* **196** 1–12

[12] Dreyer D R, Park S, Bielawski C W and Ruoff R S 2010 The chemistry of graphene oxide *Chem. Soc. Rev.* **39** 228–40

[13] Zheng J, Zhang H, Dong S, Liu Y, Nai C T and Shin H S *et al* 2014 High yield exfoliation of two-dimensional chalcogenides using sodium naphthalenide *Nat. Commun.* **5** 2995–3002

[14] Ioannidou O and Zabaniotou A 2007 Agricultural residues as precursors for activated carbon production—a review *Renew. Sustain. Eng. Rev.* **11** 1966–2005

[15] Wang X, Zhi L J and Mullen K 2008 Transparent, conductive graphene electrodes for dye-sensitized solar cells *Nano Lett.* **8** 323–7

[16] Tung V C, Allen M J, Yang Y and Kaner R B 2009 High-throughput solution processing of large-scale graphene *Nat. Nanotechnol.* **4** 25–9
[17] Park S H, Kim H K, Yoon S B, Lee C W, Ahn D and Lee S I *et al* 2015 Spray-assisted deep-frying process for the *in situ* spherical assembly of graphene for energy-storage devices *Chem. Mater.* **27** 457–65
[18] Meng Y, Zhao Y, Hu C, Cheng H, Hu Y and Zhang Z *et al* 2013 Graphene coresheath microfibres for all-solid-state, stretchable fibriform supercapacitors and wearable electronic textiles *Adv. Mater.* **25** 2326–31
[19] Lee J A, Shin M K, Kim S H, Cho H U, Spinks G M and Wallace G G *et al* 2013 Ultrafast charge and discharge biscrolled yarn supercapacitors for textiles and microdevices *Nat. Commun.* **4** 1970
[20] Bonaccorso F, Colombo L, Yu G, Stoller M, Tozzini V and Ferrari A C *et al* 2015 2D materials. Graphene, related two-dimensional crystals, and hybrid systems for energy conversion and storage *Science* **347** 1246501
[21] Weinstein L and Dash R 2013 Supercapacitor carbons *Mater. Today* **16** 356–7
[22] Li X, Cai W, Colombo L and Ruoff R S 2009 Evolution of graphene growth on Ni and Cu by carbon isotope labeling *Nano Lett.* **9** 4268–72
[23] Yang X, Shi K, Zhitomirsky I and Cranston E D 2015 Cellulose nanocrystal aerogels as universal 3D lightweight substrates for supercapacitor materials *Adv. Mater.* **27** 6104–9
[24] Yu M, Li J and Wang L 2016 KOH-activated carbon aerogels derived from sodiumcarboxymethyl cellulose for high-performance supercapacitors and dye adsorption *Chem. Eng. J.* **310** 300–6
[25] Deng J, Xiong T, Xu F, Li M, Han C and Gong Y *et al* 2015 Inspired by bread leavening: one-pot synthesis of hierarchically porous carbon for supercapacitors *Green Chem.* **17** 4053–60
[26] Li Z, Ahadi K, Jiang K, Ahvazi B, Li P and Anyia A O *et al* 2017 Freestanding hierarchical porous carbon film derived from hybrid nanocellulose for high-power supercapacitors *Nano Res.* **10** 1847–60
[27] Li Y, Wang G, Wei T, Fan Z and Yan P 2016 Nitrogen and sulfur co-doped porous carbon nanosheets derived from willow catkin for supercapacitors *Nano Energy* **19** 165–75
[28] Zhuo H, Hu Y, Tong X, Zhong L, Peng X and Sun R 2016 Sustainable hierarchical porous carbon aerogel from cellulose for highperformance supercapacitor and CO_2 capture *Ind. Crops Prod.* **87** 229–35
[29] Vijayakumar M, Sankar A B, Rohita D S, Rao T N and Karthik M 2019 Conversion of biomass waste into high performance supercapacitor electrodes for real-time supercapacitor applications *ACS Sustain. Chem. Eng* **7** 17175–85
[30] Davis T A, Volesky B and Mucci A 2003 A review of the biochemistry of heavy metal biosorption by brown algae *Water Res.* **37** 4311–30
[31] Wang B, Li D, Tang M, Ma H, Gui Y and Tian X *et al* 2018 Alginate-based hierarchical porous carbon aerogel for high-performance supercapacitors *J. Alloys Compd.* **749** 517–22
[32] Geng Z, Wang H, Wang R, Zhang P, Lang J and Wang C 2016 Facile synthesis of hierarchical porous carbon for supercapacitor with enhanced electrochemical performance *Mater. Lett.* **182** 1–5
[33] Wang K, Cao Y, Wang X, Castro M A, Luo B and Gu Z *et al* 2016 Rod-shape porous carbon derived from aniline modified lignin for symmetric supercapacitors *J. Power Sources* **307** 462–7

[34] Ryu K S, Kim K M, Park N G, Park Y J and Chang S H 2002 Symmetric redox supercapacitor with conducting polyaniline electrodes *J. Power Sources* **103** 305–9
[35] Nyholm L, Nyström G, Mihranyan A and Strømme M 2011 Toward flexible polymer and paper-based energy storage devices *Adv. Mater.* **23** 3751–69
[36] Zhang L, Du W, Nautiyal A, Liu Z and Zhang X 2018 Recent progress on nanostructured conducting polymers and composites: synthesis, application and future aspects *Sci. China Mater.* **61** 1–50
[37] Peng H, Ma G, Sun K, Mu J, Wang H and Lei Z 2014 High-performance supercapacitor based on multi-structural CuS@ polypyrrole composites prepared by *in situ* oxidative polymerization *J. Mater. Chem.* A **2** 3303–7
[38] Nguyen D N and Yoon H 2016 Recent advances in nanostructured conducting polymers: from synthesis to practical applications *Polymers* **8** 118
[39] Babu K F, Senthilkumar R, Noel M and Kulandainathan M A 2009 Polypyrrole microstructure deposited by chemical and electrochemical methods on cotton fabrics *Synth. Met.* **159** 1353–8
[40] Li X and Zhitomirsky I 2013 Electrodeposition of polypyrrole–carbon nanotube composites for electrochemical supercapacitors *J. Power Sources* **221** 49–56
[41] Snook G A and Best A S 2009 Co-deposition of conducting polymers in a room temperature ionic liquid *J. Mater. Chem.* **19** 4248–54
[42] Ghosh S and Inganas O 2000 Electrochemical characterization of poly(3,4-ethylene dioxythiophene) based conducting hydrogel networks *J. Electrochem. Soc.* **147** 1872
[43] Rudge A, Raistrick I, Gottesfeld S and Ferraris J P 1994 A study of the electrochemical properties of conducting polymers for application in electrochemical capacitors *Electrochim. Acta* **39** 273–87
[44] Lu X, Yu M, Wang G, Tong Y and Li Y 2014 Flexible solid-state supercapacitors: design, fabrication and applications *Energy Environ. Sci.* **7** 2160–81
[45] Mu S 2003 The electrocatalytic oxidation of gallic acid on polyaniline film synthesized in the presence of ferrocene phosphonic acid *Synth. Met.* **139** 287–94
[46] Dominis A J, Spinks G M, Kane-Maguire L A and Wallace G G 2002 A de-doping/re-doping study of organic soluble polyaniline *Synth. Met.* **129** 165–72
[47] Guimard N K, Gomez N and Schmidt C E 2007 Conducting polymers in biomedical engineering *Prog. Polym. Sci.* **32** 876–921
[48] Ryu K S, Kim K M, Park Y J, Park N G, Kang M G and Chang S H 2002 Redox supercapacitor using polyaniline doped with Li salt as electrode *Solid State Ion.* **152** 861–66
[49] Ryu K S, Lee Y G, Kim K M, Park Y J, Hong Y S, Wu X L, Kang M G, Park N G, Song R Y and Ko J M 2005 Electrochemical capacitor with chemically polymerized conducting polymer based on activated carbon as hybrid electrodes *Synth. Met.* **153** 89–92
[50] Gurunathan K, Amalnerkar D P and Trivedi D C 2003 Synthesis and characterization of conducting polymer composite (PAn/TiO_2) for cathode material in rechargeable battery *Mater. Lett.* **57** 1642–8
[51] Ryu K S, Lee Y G, Han K S, Park Y J, Kang M G, Park N G and Chang S H 2004 Electrochemical supercapacitor based on polyaniline doped with lithium salt and active carbon electrodes *Solid State Ion.* **175** 765–8
[52] Li J, Xie H, Li Y, Liu J and Li Z 2011 Electrochemical properties of graphene nanosheets/polyaniline nanofibers composites as electrode for supercapacitors *J. Power Sources* **196** 10775–81

[53] Yang W, Lu J, Weng J, Jia W, Ji L, Xiao J and Zhu D 2010 Prevalence of diabetes among men and women in China *New Engl. J. Med.* **362** 1090–101
[54] Hughes M, Chen G Z, Shaffer M S P, Fray D J and Windle A H 2004 Controlling the nanostructure of electrochemically grown nanoporous composites of carbon nanotubes and conducting polymers *Compos. Sci. Technol.* **64** 2325–31
[55] Chen G Z, Shaffer M S P, Coleby D, Dixon G, Zhou W Z, Fray D J and Windle A H 2000 Carbon nanotube and polypyrrole composites: coating and doping *Adv. Mater.* **12** 522–6
[56] Kim B C, Ko J M and Wallace G G 2008 A novel capacitor material based on Nafion-doped polypyrrole *J. Power Sources* **177** 665–68
[57] Valles C, Jimenez P, Munoz E, Benito A M and Maser W K 2011 Simultaneous reduction of graphene oxide and polyaniline: doping-assisted formation of a solid-state charge-transfer complex *J. Phys. Chem.* C **115** 10468–74
[58] Wang K, Wu H, Meng Y and Wei Z 2014 Conducting polymer nanowire arrays for high performance supercapacitors *Small* **10** 14–31
[59] Song Y, Li-Xu J and Liu X 2014 Electrochemical anchoring of dual doping polypyrrole on graphene sheets partially exfoliated from graphite foil for high-performance supercapacitor electrode *J. Power Sources* **249** 48–58
[60] Liu T, Finn L, Yu M, Wang H, Zhai T, Lu X, Tong Y and Li Y 2014 Polyaniline and polypyrrole pseudocapacitor electrodes with excellent cycling stability *Nano Lett.* **14** 2522–7
[61] Shao M, Li Z, Zhang R, Ning F, Wei M, Evans D G and Duan X 2015 Hierarchical conducting polymer@clay core–shell arrays for flexible all-solid-state supercapacitor devices *Small* **11** 3530–8
[62] Huang Y, Zhong M, Huang Y, Zhu M, Pei Z, Wang Z, Xue Q, Xie X and Zhi C 2015 A self-healable and highly stretchable supercapacitor based on a dual crosslinked polyelectrolyte *Nat. Commun.* **6** 10310
[63] Wang J, Xu Y L, Chen X and Du X F 2007 Electrochemical supercapacitor electrode material based on poly(3,4-ethylenedioxythiophene)/polypyrrole composite *J. Power Sources* **163** 1120–5
[64] Li W K, Chen J, Zhao J J, Zhang J R and Zhu J J 2005 Application of ultrasonic irradiation in preparing conducting polymer as active materials for supercapacitor *Mater. Lett.* **59** 800–3
[65] Aouada F A, Moura M R, Girotto E M, Rubira A F and Muniz E C 2007 The synthesis and capacitive properties of poly(3,4-ethylene dioxythiophene)/poly(styrene–sulfonate) and poly (acrylamide) conducting hydrogels *e-Polymers* **7** 6
[66] Berggren M and Malliaras G G 2019 How conducting polymer electrodes operate *Science* **364** 233–4
[67] Wang X *et al* 2018 High electrical conductivity and carrier mobility in oCVD PEDOT thin films by engineered crystallization and acid treatment *Sci. Adv.* **4** 1–9
[68] Wang H, Diao Y, Lu Y, Yang H, Zhou Q, Chrulski K and D'Arcy J M 2020 Energy storing bricks for stationary PEDOT supercapacitors *Nat. Commun.* **11** 3882
[69] Ryu K S, Lee Y G, Hong Y S, Park Y J, Wu X L, Kim K M, Kang M G, Park N G and Chang S H 2004 Poly(ethylenedioxythiophene) (PEDOT) as polymer electrode in redox supercapacitor *Electrochim. Acta* **50** 843–47
[70] Estrada L A, Liu D Y, Salazar D H, Dyer A L and Reynolds J R 2012 Poly[bis-EDOT-isoindigo]: an electroactive polymer applied to electrochemical supercapacitors *Macromolecules* **45** 8211–20

[71] DiCarmine P M, Schon T B, McCormick T M, Klein P P and Seferos D S 2014 Donor–acceptor polymers for electrochemical supercapacitors: synthesis, testing, and theory *J. Phys. Chem.* C **118** 8295–307
[72] Jin E *et al* 2017 Two-dimensional sp^2 carbon-conjugated covalent organic frameworks *Science* **357** 673–76
[73] Cheng Z, Pan H, Zhong H, Xiao Z, Li X and Wang R 2018 Porous organic polymers for polysulfide trapping in lithium–sulfur batteries *Adv. Funct. Mater.* **28** 1707597
[74] Xu Y, Jin S, Xu H, Nagai A and Jiang D 2013 Conjugated microporous polymers: design, synthesis and application *Chem. Soc. Rev.* **42** 8012–31
[75] Dogru M, Sonnauer A, Gavryushin A, Knochel P and Bein T 2011 A covalent organic framework with 4 nm open pores *Chem. Commun.* **47** 1707–9
[76] Liu W *et al* 2018 Two-dimensional polymer synthesized via solid-state polymerization for high-performance supercapacitors *ACS Nano* **12** 852–60
[77] DeBlase C R, Silberstein K E, Truong T T, Abruña H D and Dichtel W R 2013 β-Ketoenamine-linked covalent organic frameworks capable of pseudocapacitive energy storage *J. Am. Chem. Soc.* **135** 16821–4
[78] DeBlase C R, Burgos K H, Silberstein K E, Calero G G R, Bisbey R P, Abruña H D and Dichtel W R 2015 Rapid and efficient redox processes within 2D covalent organic framework thin films *ACS Nano* **9** 3178–83
[79] Mulzer C R, Shen L, Bisbey R P, McKone J R, Zhang N, Abruña H D and Dichtel W R 2016 Superior charge storage and power density of a conducting polymer-modified covalent organic framework *ACS Cent. Sci.* **2** 667–73
[80] Wu Y, Yan D, Zhang Z, Matsushita M M and Awaga K 2019 Electron highways into nanochannels of covalent organic frameworks for high electrical conductivity and energy storage *ACS Appl. Mater. Interfaces* **11** 7661–5
[81] Yang H *et al* 2016 High conductive two-dimensional covalent organic framework for lithium storage with large capacity *ACS Appl. Mater. Interfaces* **8** 5366–75
[82] Khayum M A, Vijayakumar V, Karak S, Kandambeth S, Bhadra M, Suresh K, Acharambath N, Kurungot S and Banerjee R 2018 Convergent covalent organic framework thin sheets as flexible supercapacitor electrodes *ACS Appl. Mater. Interfaces* **10** 28139–46
[83] Halder A, Ghosh M, Khayum M A, Bera S, Addicoat M, Sasmal H S, Karak S, Kurungot S and Banerjee R 2018 Interlayer hydrogen-bonded covalent organic frameworks as high-performance supercapacitors *J. Am. Chem. Soc.* **140** 10941–5
[84] Sun B, Liu J, Cao A, Song W and Wang D 2017 Interfacial synthesis of ordered and stable covalent organic frameworks on amino-functionalized carbon nanotubes with enhanced electrochemical performance *Chem. Commun.* **53** 6303–6
[85] Han Y, Hu N, Liu S, Hou Z, Liu J, Hua X, Yang Z, Wei L, Wang L and Wei H 2017 Nanocoating covalent organic frameworks on nickel nanowires for greatly enhanced-performance supercapacitors *Nanotechnology* **28** 1–8
[86] Liu S, Yao L, Lu Y, Hua X, Liu J, Yang Z, Wei H and Mai Y 2019 All-organic covalent organic framework/polyaniline composites as stable electrode for high-performance supercapacitors *Mater. Lett.* **236** 354–7
[87] Zha Z, Xu L, Wang Z, Li X, Pan Q, Hu P and Lei S 2015 3D graphene functionalized by covalent organic framework thin film as capacitive electrode in alkaline media *ACS Appl. Mater. Interfaces* **7** 17837–43

[88] Feng X, Liang Y, Zhi L, Thomas A, Wu D, Lieberwirth I, Kolb U and Müllen K 2009 Synthesis of microporous carbon nanofibers and nanotubes from conjugated polymer network and evaluation in electrochemical capacitor *Adv. Funct. Mater.* **19** 2125–29
[89] Kou Y, Xu Y, Guo Z and Jiang D 2011 Supercapacitive energy storage and electric power supply using an aza-fused π-conjugated microporous framework *Angew. Chem.* **50** 8753–7
[90] Li X-C, Zhang Y, Wang C-Y, Wan Y, Lai W-Y, Pang H and Huang W 2017 Redox-active triazatruxene-based conjugated microporous polymers for high-performance supercapacitors *Chem. Sci.* **8** 2959–65
[91] Topal S, Sezer E, Eroglu M S and Ozturk T 2020 Surface characterization, electrochromic and capacitive properties of poly(*N*-(4-(4-thieno[3,2-b]thiophen-3-yl-phenyl)phenyl)-*N*-phenylbenzenamine) film *Polymer* **209** 122954
[92] Wang H, Cheng Z, Liao Y, Li J, Weber J, Thomas A and Faul C F J 2017 Conjugated microporous polycarbazole networks as precursors for nitrogen-enriched microporous carbons for CO_2 storage and electrochemical capacitors *Chem. Mater.* **29** 4885–93
[93] Zhuang X, Zhang F, Wu D and Feng X 2014 Graphene coupled Schiff-base porous polymers: towards nitrogen-enriched porous carbon nanosheets with ultrahigh electrochemical capacity *Adv. Mater.* **26** 3081
[94] Peurifoy S R, Russell J C, Sisto T J, Yang Y, Roy X and Nuckolls C 2018 Designing three-dimensional architectures for high-performance electron accepting pseudocapacitors *J. Am. Chem. Soc.* **140** 10960–4
[95] Qiu H, Cheng H, Meng J, Wu G and Chen S 2020 Magnetothermal microfluidic-assisted hierarchical microfibers for ultrahigh-energy-density supercapacitors *Angew. Chem.* **59** 7934–43
[96] Muench S, Wild A, Friebe C, Häupler B, Janoschka T and Schubert U S 2016 Polymer-based organic batteries *Chem. Rev.* **116** 9438–84
[97] Song Z and Zhou H 2013 Towards sustainable and versatile energy storage devices: an overview of organic electrode materials *Energy Environ. Sci.* **6** 2280
[98] Dai G, Wang X, Qian Y, Niu Z, Zhu X, Ye J, Zhao Y and Zhang X 2019 Manipulation of conjugation to stabilize N redox-active centers for the design of high-voltage organic battery cathode *Energy Storage Mater.* **16** 236–42
[99] Park S Y and Kwon J E *et al* 2020 Phenoxazine as a high-voltage p-type redox center for organic battery cathode materials: small structural reorganization for faster charging and narrow operating voltage *Energy Environ. Sci.* **13** 4142
[100] Acker P, Speer M E, Woessner J S and Esser B 2020 Azine-based polymers with a two-electron redox process as cathode materials for organic batteries *J. Mater. Chem.* A **8** 11195–201
[101] Chen Z, Li W, Dai Y, Xu N, Su C, Liu J and Zhang C 2018 Conjugated microporous polymer based on star-shaped triphenylamine-benzene structure with improved electrochemical performances as the organic cathode material of Li-ion battery *Electrochim. Acta* **286** 187–94
[102] Yamamoto K, Suemasa D, Masuda K, Aita K and Endo T 2018 Hyperbranched triphenylamine polymer for ultrafast battery cathode *ACS Appl. Mater. Interfaces* **10** 6346–53
[103] Zhan L, Song Z, Zhang J, Tang J, Zhan H, Zhou Y and Zhan C 2008 PEDOT: cathode active material with high specific capacity in novel electrolyte system *Electrochim. Acta* **53** 8319–23

[104] Ozerova V V, Stenina I A, Kuzmina A A, Kulova T L and Yaroslavtsev A B 2020 Cathode materials based on lithium iron phosphate/PEDOT composites for lithium-ion batteries *Inorg. Mater.* **56** 648–56
[105] Schwartz P O, Pejic M, Wachtler M and Bäuerle P 2018 Synthesis and characterization of electroactive PEDOT–TEMPO polymers as potential cathode materials in rechargeable batteries *Synth. Met.* **243** 51–7
[106] Zhang S, Huang W, Hu P, Huang C, Shang C, Zhang C, Yang R and Cui G 2015 Conjugated microporous polymers with excellent electrochemical performance for lithium and sodium storage *J. Mater. Chem.* A **3** 1896
[107] Nokami T *et al* 2012 Polymer-bound pyrene-4,5,9,10-tetraone for fast-charge and -discharge lithium-ion batteries with high capacity *Am. Chem. Soc* **134** 19694–700
[108] Akihiro S, Hiroki K, Yutaka T, Toshiki N, Yuu I, Nobuhiko H, Hirotetsu S and Jun-ichi Y 2014 Introduction of two lithiooxycarbonyl groups enhances cyclability of lithium batteries with organic cathode materials *J. Power Sources* **260** 211–17
[109] Song Z, Qian Y, Zhang T, Otani M and Zhou H 2015 Poly(benzoquinonyl sulfide) as a High-energy organic cathode for rechargeable Li and Na batteries *Adv. Sci.* **2** 1500124
[110] Li Q L D, Wang H, Wang H, Li Y, Si Z and Duan Q 2019 Conjugated carbonyl polymer-based flexible cathode for superior lithium-organic batteries *ACS Appl. Mater. Interfaces* **11** 28801–8
[111] Tang M, Wu Y, Chen Y, Jiang C, Zhu S, Zhuo S and Wang C 2019 An organic cathode with high capacities for fast charge potassium-ion batteries *J. Mater. Chem.* A **7** 486
[112] Xie J, Chen W, Wang Z, Choo K, Jie W, Liu M and Zhang Q 2017 Synthesis and exploration of ladder-structured large aromatic dianhydrides as organic cathodes for rechargeable lithium-ion batteries *Chem. Asian J.* **12** 868–76
[113] Zhu L M, Lei A W, Cao Y L, Ai X P and Yang H X 2013 An all-organic rechargeable battery using bipolar polyparaphenylene as a redox-active cathode and anode *Chem. Commun.* **49** 567–69
[114] Su C, Han B, Ma J and Xu L 2020 A novel anthraquinone-contained poly(triphenylamine) derivative: its preparation and electrochemical performances as the cathode of lithium ion battery *Chem. Electro. Chem.* **7** 4101–7
[115] Deng W, Shi W, Liu Q, Jiang J, Wang Q and Guo C 2020 Conductive nonconjugated radical polymer as high capacity organic cathode material for high-energy Li/Na ion batteries *J. Power Sources* **479** 228796
[116] Bhosale M E, Chae S, Kim J M and Choi J Y 2018 Organic small molecules and polymers as an electrode material for rechargeable lithium ion batteries *J. Mater. Chem.* A **6** 19885–911
[117] Häupler B, Wild A and Schubert U S 2015 Carbonyls: powerful organic materials for secondary batteries *Adv. Energy Mater.* **5** 1402034
[118] Armand M, Grugeon S, Vezin H, Laruelle S, Ribière P, Poizot P and Tarascon J-M 2009 Conjugated dicarboxylate anodes for Li-ion batteries *Nat. Mater.* **8** 120–5
[119] Lee S, Kwon G, Ku K, Yoon K, Jung S K, Lim H D and Kang K 2018 Recent progress in organic electrodes for Li and Na rechargeable batteries *Adv. Mater.* **30** 1704682
[120] Schon T B, McAllister B T, Li P-F and Seferos D S 2016 The rise of organic electrode materials for energy storage *Chem. Soc. Rev.* **45** 6345–404
[121] Han X, Qing G, Sun J and Sun T 2012 How many lithium ions can be inserted onto fused C_6 aromatic ring systems *Angew. Chem.* **124** 5237–41

[122] Wang Y, Liu W, Guo R, Qu Q, Zheng H, Zhang J and Huang Y 2019 A high-capacity organic anode with self-assembled morphological transformation for green lithium-ion battery *J. Mater. Chem.* A **7** 22621–30

[123] Lei K, Li F, Mu C, Wang J, Zhao Q, Chen C and Chen J 2017 High K-storage performance based on the synergy of dipotassium terephthalate and ether-based electrolyte *Energy Environ. Sci.* **10** 552–57

[124] Liang Y, Luo C, Wang F, Hou S, Liou S C, Qing T, Li Q, Zheng J, Cui C and Wang C 2019 An organic anode for high temperature potassium-ion batteries *Adv. Energy Mater.* **9** 1802986

[125] Deng Q, Pei J, Fan C, Ma J, Cao B, Li C, Jin Y, Wang L and Li J 2017 Potassium salts of para-aromatic dicarboxylates as the highly efficient organic anodes for low-cost K-ion batteries *Nano Energy* **33** 350–5

[126] Park Y, Shin D S, Woo S H, Choi N S, Shin K H, Oh S M, Lee K T and Hong S Y 2012 Sodium terephthalate as an organic anode material for sodium ion batteries *Adv. Mater.* **24** 3562–7

[127] Wu J, Rui X, Long G, Chen W, Yan Q and Zhang Q 2015 Pushing up lithium storage through nanostructured polyazaacene analogues as anode *Angew. Chem.* **127** 7462–6

[128] Li G, Zhang B, Wang J, Zhao H, Ma W, Xu L, Zhang W, Zhou K, Du Y and He G 2019 Electrochromic poly(chalcogenoviologen)s as anode materials for high-performance organic radical lithium-ion batteries *Angew. Chem. Int. Ed.* **58** 8468–73

[129] Wang C Y, Tsekouras G, Wagner P, Gambhir S, Too C O, Officer D and Wallace G G 2010 Functionalised polyterthiophenes as anode materials in polymer/polymer batteries *Synth. Met.* **160** 76–82

[130] Qin H, Song Z P, Zhan H and Zhou Y H 2014 Aqueous rechargeable alkali-ion batteries with polyimide anode *J. Power Sources* **249** 367–72

[131] Kahsay B A, Ramar A, Wang F M, Yeh N H, Lin P L, Luo Z J, Chan T S and Su C H 2019 Investigating an all-organic battery using polyisothianaphthene as a redox-active bipolar electrode material *J. Power Sources* **428** 115–23

[132] Dai G, He Y, Niu Z, He P, Zhang C, Zhao Y, Zhang X and Zhou H 2019 A dual-ion organic symmetric battery constructed from phenazine-based artificial bipolar molecules *Angew. Chem.* **131** 10007–11

[133] Wang S, Wang L, Zhang K, Zhu Z, Tao Z and Chen J 2013 Organic $Li_4C_8H_2O_6$ nanosheets for lithium-ion batteries *Nano Lett.* **13** 4404–9

[134] Chen H, Armand M, Courty M, Jiang M, Grey C P, Dolhem F, Tarascon J M and Poizot P 2009 Lithium salt of tetrahydroxybenzoquinone: toward the development of a sustainable Li-ion battery *J. Am. Chem. Soc.* **131** 8984–8

[135] Wang S, Wang L, Zhu Z, Hu Z, Zhao Q and Chen J 2014 All organic sodium-ion batteries with $Na_4C_8H_2O_6$ *Angew. Chem. Int. Ed.* **53** 5892–6

[136] Li A, Feng Z, Sun Y, Shang L and Xu L 2017 Porous organic polymer/RGO composite as high-performance cathode for half and full sodium ion batteries *J. Power Sour* **343** 424–30

[137] Suga T, Ohshiro H, Sugita S, Oyaizu K and Nishide H 2009 Emerging n-type redox-active radical polymer for a totally organic polymer-based rechargeable battery *Adv. Mater.* **21** 1627–30

[138] Sano N, Tomita W, Hara S, Min C M, Lee J S, Oyaizu K and Nishide H 2013 Polyviologen hydrogel with high-rate capability for anodes toward an aqueous electrolyte-type and organic-based rechargeable device *ACS Appl. Mater. Interfaces* **5** 1355–61

[139] Chikushi N, Yamada H, Oyaizu K and Nishide H 2012 TEMPO-substituted polyacrylamide for an aqueous electrolyte-typed and organic-based rechargeable device *Sci. China Chem.* **55** 822–9
[140] Zhu X, Zhao R, Deng W, Ai X, Yang H and Cao Y 2015 An all-solid-state and all-organic sodium-ion battery based on redox-active polymers and plastic crystal electrolyte *Electrochim. Acta* **178** 55–9
[141] Dong X, Yu H, Ma Y, Bao J L, Truhlar D G, Wang Y and Xia Y 2017 All organic rechargeable battery with reversibility supported by 'water-in-salt' electrolyte *Chem. Eur. J.* **23** 2560–5
[142] Dong X, Guo Z, Guo Z, Wang Y and Xia Y 2018 Organic batteries operatedat −70 °C *Joule* **2** 1–12
[143] Wild A, Strumpf M, Häupler B, Hager M D and Schubert U S 2017 All-organic battery composed of thianthrene- and TCAQ-based polymers *Adv. Energy Mater.* **7** 1601415
[144] Iordache A, Delhorbe V, Bardet M, Dubois L, Gutel T and Picard L 2016 Perylene-based all-organic redox battery with excellent cycling stability *ACS Appl. Mater. Interfaces* **8** 22762–7
[145] Banda H, Damien D, Nagarajan K, Hariharan M and Shaijumon M M 2015 A polyimide based all-organic sodium ion battery *J. Mater. Chem.* A **3** 10453–8
[146] Wang C, Tang W, Wang X L, Zhou Y H, Li D, Jia S, Cao B and Fan C 2021 Long lifespan organic K-ion batteries with working voltage above 2 V in ether electrolytes *Electrochim. Acta* **365** 137365
[147] Yao M, Sano H, Ando H and Kiyobayashi T 2015 Molecular ion battery: a rechargeable system without using any elemental ions as a charge carrier *Sci. Rep.* **5** 10962
[148] Emanuelsson R, Sterby M, Strømme M and Sjödin M 2017 An all-organic proton battery *J. Am. Chem. Soc.* **139** 4828–34
[149] Deng W, Liang X, Wu X, Qian J, Cao Y, Ai X, Feng J and Yang H 2013 A low cost, all-organic Na-ion battery based on polymeric cathode and anode *Sci. Rep.* **3** 2671
[150] Zhu L M, Shi W, Zhao R R, Cao Y L, Ai X P, Lei A W and Yang H X 2013 n-Dopable polythiophenes as high-capacity anode materials for all-organic Li-ion batteries *J. Electroanal. Chem.* **688** 118–22
[151] Bhattacharya S and Alpas A T 2016 Self-healing of cracks formed in silicon–aluminum anodes electrochemically cycled at high lithiation rates *J. Power Sources* **328** 300–10
[152] Chen D *et al* 2017 Self-healing materials for next-generation energy harvesting and storage devices *Adv. Energy Mater.* **7** 1700890
[153] Li F *et al* 2020 silicon anodes for high-performance storage devices: structural design, material compounding, advances in electrolytes and binders *ChemNanoMat* **6** 720–38
[154] Luo C *et al* 2017 Self-healing chemistry between organic material and binder for stable sodium-ion batteries *Chem.* **3** 1050–62
[155] Mai W *et al* 2020 self-healing materials for energy-storage devices *Adv. Funct. Mater.* **30** 1909912
[156] Wu Y *et al* 2017 A room-temperature liquid metal-based self-healing anode for lithium-ion batteries with an ultra-long cycle life *Energy Environ. Sci.* **10** 1854–61
[157] Xu Z *et al* 2018 Silicon microparticle anodes with self-healing multiple network binder *Joule* **2** 950–61
[158] Zhuang J *et al* 2019 A self-healing interface on lithium metal with lithium difluoro (bisoxalato) phosphate for enhanced lithium electrochemistry *J. Mater. Chem.* A **7** 26002–10

[159] Wang S and Urban M W 2020 Self-healing polymers *Nat. Rev. Mater.* **5** 562–83
[160] Wool R P 2008 Self-healing materials: a review *Soft Matter* **4** 400–18
[161] Jin Y *et al* 2017 Self-healing SEI enables full-cell cycling of a silicon-majority anode with a coulombic efficiency exceeding 99.9% *Energy Environ. Sci.* **10** 580–92
[162] Qin J *et al* 2019 Tuning self-healing properties of stiff, ion-conductive polymers *J. Mater. Chem.* A **7** 6773–83
[163] Tasdelen M A, Taskin O S and Celik C 2016 Orthogonal synthesis of block copolymer via photoinduced CuAAC and ketene chemistries *Macromol. Rapid Commun.* **37** 521–26
[164] Huang S *et al* 2018 Low addition amount of self-healing ionomer binder for Si/graphite electrodes with enhanced cycling *New J. Chem.* **42** 6742–49
[165] Taskin O S *et al* 2016 Copper(II) removal from the aqueous solution using microporous benzidine-based adsorbent material *J. Environ. Chem. Eng.* **4** 899–907
[166] Taskin O S *et al* 2015 Synthesis of block copolymers by selective H-abstraction and radical coupling reactions using benzophenone/benzhydrol photoinitiating system *Eur. Polym. J.* **62** 304–11
[167] Bulut E *et al* 2020 Novel approach with polyfluorene/polydisulfide copolymer binder for high-capacity silicon anode in lithium-ion batteries *J. Appl. Polym. Sci.* **137** 48303
[168] Yuca N, Taskin O S and Arici E n.a. An overview on efforts to enhance the Si electrode stability for lithium ion batteries *Energy Storage* **2** e94 pp
[169] Taskin O S *et al* 2015 One-pot, one-step strategy for the preparation of clickable melamine based microporous organic polymer network *Macromol. Mater. Eng.* **300** 1116–22
[170] Chen Z *et al* 2015 High-areal-capacity silicon electrodes with low-cost silicon particles based on spatial control of self-healing binder *Adv. Energy Mater.* **5** 1401826
[171] Chen S 2015 Development of zwitterionic polyurethanes with multi-shape memory effects and self-healing properties *J. Mater. Chem.* A **3** 2924–33
[172] Taskin O S, Kiskan B and Yagci Y 2013 Polybenzoxazine precursors as self-healing agents for polysulfones *Macromolecules* **46** 8773–78
[173] Cao K *et al* 2014 Strategy and mechanism for controlling the direction of defect evolution in graphene: preparation of high quality defect healed and hierarchically porous graphene *Nanoscale* **6** 13518–26
[174] Tao F *et al* 2017 Self-healable and cold-resistant supercapacitor based on a multifunctional hydrogel electrolyte *ACS Appl. Mater. Interfaces* **9** 15541–8
[175] Yuca N *et al* 2018 Highly efficient poly(fluorene phenylene) copolymer as a new class of binder for high-capacity silicon anode in lithium-ion batteries *Int. J. Energy Res.* **42** 1148–57
[176] Wang C *et al* 2013 Self-healing chemistry enables the stable operation of silicon microparticle anodes for high-energy lithium-ion batteries *Nat. Chem.* **5** 1042–8
[177] Chen H *et al* 2021 A mechanically robust self-healing binder for silicon anode in lithium ion batteries *Nano Energy* **81** 105654
[178] Wang Z, Tao F and Pan Q 2016 A self-healable polyvinyl alcohol-based hydrogel electrolyte for smart electrochemical capacitors *J. Mater. Chem.* A **4** 17732–9
[179] Li L *et al* 2018 Flexible double-cross-linked cellulose-based hydrogel and aerogel membrane for supercapacitor separator *J. Mater. Chem.* A **6** 24468–78
[180] Yoon J *et al* 2014 Improvement of desolvation and resilience of alginate binders for Si-based anodes in a lithium ion battery by calcium-mediated cross-linking *Phys. Chem. Chem. Phys.* **16** 25628–35

[181] Kwon T-w *et al* 2014 Systematic molecular-level design of binders incorporating Meldrum's acid for silicon anodes in lithium rechargeable batteries *Adv. Mater.* **26** 7979–85
[182] Guo Y *et al* 2016 A self-healable and easily recyclable supramolecular hydrogel electrolyte for flexible supercapacitors *J. Mater. Chem.* A **4** 8769–76
[183] Luo X, Wang J, Dooner M and Clarke J 2015 Overview of current development in electrical energy storage technologies and the application potential in power system operation *Appl. Energy* **137** 511–36
[184] https://new.abb.com/news/detail/60295/abb-scales-up-fuel-cells-for-shippings-greener-future (accessed 18 April 2021)
[185] http://batteryfaq.org/ (accessed 24 April 2021)
[186] Wang J, Lu K, Ma L, Wang J, Dooner M, Miao S, Li J and Wang D 2017 Overview of Compressed Air Energy Storage and Technology Development *Energies* **10** 991

Chapter 6

Metal oxide semiconductor nanomaterials for heterogeneous photocatalysis

Kasper Wenderich and Marco Altomare

The development of renewable technologies for energy conversion and storage has recently become a major scientific, industrial, and societal challenge, with the aim to mitigate climate change and global warming.

Hydrogen is envisioned as a clean energy vector for our future. In fact, it is an energetic fuel, with a mass-normalized energy density higher than that of fossil feedstock. Moreover, its direct combustion or utilization in fuel cells simply leads to water as the reaction product. However, its production still relies on fossil fuels, as it is carried out on an industrial scale via the steam reforming of methane, thereby contributing a significant carbon footprint.

Water electrolysis for hydrogen production is a promising technology, while from a practical perspective, electrolyzers are still not cost-competitive enough to displace fossil fuels. Ideally, efficient and economical hydrogen generation by water splitting would be a technological breakthrough if it could be attained using solar light as the sole energy input. For this reason, semiconductor materials are at the core of light-to-energy conversion technologies, and photocatalytic processes have received a great deal of research attention aiming at harvesting light to enable the desired chemical reactions.

In this context, metal oxide semiconductors are at the heart of pioneering and ongoing research into photocatalysis and photoelectrochemistry, particularly for hydrogen generation. Among many different photocatalytic materials developed to date, semiconductor metal oxides offer a series of remarkable advantages due to their strong oxidizing ability (e.g. for the decomposition of pollutants), hydrophilicity, (photo-)chemical stability, nontoxicity, environmental friendliness, relatively large abundance, and low cost.

In this chapter, we first cover general aspects of heterogeneous photocatalysis and the materials science of semiconductors, addressing photoinduced redox processes at

doi:10.1088/978-0-7503-3531-7ch6

© IOP Publishing Ltd 2023

semiconductor/liquid interfaces. We then outline key physicochemical and photocatalytic features of metal oxides. We discuss the advantages associated with nanostructuring and provide a selection of relevant techniques for the synthesis of photocatalytic materials. We continue further with an overview of the principles and methods for the modification of oxide photocatalysts with the objective of enhancing their photoactivity. Finally, we critically discuss some practices and trends in current and future research on photocatalysis.

6.1 Introduction

6.1.1 Solar energy conversion

Nowadays, with the increased awareness of climate change and global warming, the development of cost-effective technologies for water splitting to support H_2 generation has become a major industrial and societal challenge. Electrochemical processes such as water electrolysis powered by *green electrons* (i.e. electrical energy obtained from renewable sources such as wind and solar) are pivotal for the energy transition, which aims to develop sustainable alternatives to fossil fuels.

In this context, photocatalytic technologies have also been widely investigated for use in renewable H_2 generation through water splitting. Considering the water-splitting process (6.1)–(6.3):

Overall reaction

$$2H_2O \rightarrow 2H_2 + O_2 \tag{6.1}$$

Water oxidation (O_2 evolution)

$$2H_2O \rightarrow 4H^+ + 4e^- + O_2 \tag{6.2}$$

Proton reduction (H_2 evolution)

$$2H^+ + 2e^- \rightarrow H_2 \tag{6.3}$$

The anodic reaction of water oxidation to O_2 (the oxygen evolution reaction (OER), reaction (6.2)) has essentially no economic value. However, the cathodic reaction of water reduction to H_2 (the hydrogen evolution reaction (HER), reaction (6.3)) can provide an environmentally friendly path for the conversion of an abundant resource, such as water, into an energy vector and widely used chemical, such as H_2. Ideally, efficient and economical water splitting would be a technological breakthrough if it could be attained by a photocatalytic process, that is, using solar light as the sole energy input. Averaged over an entire year, the solar energy received at the Earth's surface is in the order of 10^5 TW, which exceeds by far the current human energy demand (~14 TW) and is hence a huge source of energy that can be exploited.

Today, the production of H_2 relies on fossil fuels, as it is carried out on an industrial scale via the steam reforming of methane. This also yields CO_2 as a by-product, thereby contributing a significant carbon footprint. If it could instead be produced with a net zero-carbon footprint, an H_2-based economy would be remarkably advantageous. H_2 has a higher mass-normalized energy density than fossil fuels (~120–142 MJ kg^{-1}, three to four times higher than that of gasoline at ~45 MJ kg^{-1}) [1–3] and its direct combustion (to a first approximation) simply leads to water as the reaction product. The chemical energy stored in H_2 molecules can be converted into 'clean' electricity in fuel cells. The possibility of implementing sustainable H_2 production technologies, for example in decentralized infrastructures, could also be advantageous for the chemical industry. In fact, H_2 is a common reactant in various chemical processes, such as the refinement of petroleum and the synthesis of ammonia (NH_3) via the Haber–Bosch process, as well as for other sectors, e.g. metallurgy, glass production, and the electronics industry [1, 2].

6.1.2 Metal oxide semiconductors for photocatalysis

Earth-abundant metal oxides are at the core of pioneering work on photocatalysis and photoelectrochemistry for solar energy conversion. In a system in which metal oxide semiconductor particles are suspended in a liquid solution or gas phase, the irradiation of the semiconductor with light (photons) of adequate energy, can enable *photocatalytic reactions*. Photocatalytic reactions are light-induced redox processes that take place in the interfacial region around a semiconductor particle or surface. The field of photocatalysis can be traced back more than a century to early observations of the chalking of oxide-based paints (e.g. titania, TiO_2) and to studies of the darkening (reduction) of metal oxides under illumination and in contact with organic compounds [4, 5]. For an historical overview, the reader is referred to reference [6].

Research into photocatalysis has intensified particularly over the last 50 years, i.e. since the pioneering work by Fujishima and Honda in 1972 [7, 8], who demonstrated the possibility of solar photoelectrochemical (PEC) splitting of water into H_2 and O_2. The discovery of water *photoelectrolysis* at illuminated oxide electrodes established the use of PEC cells for *artificial photosynthesis* or, in a broader sense, for *solar-energy-to-fuel conversion* [9]. In follow-up research, Bard demonstrated the photocatalytic effects of suspended semiconductor particles [10], and Nozik formulated the concept of *photochemical diodes* [11], postulating that the basic functions of a photoelectrosynthetic cell, namely, light absorption, charge separation, and interfacial redox processes (discussed later), could take place in suspended micro- or nanoscale particles that would split water under illumination. Following these seminal contributions, the photocatalytic and PEC properties of semiconductor metal oxides were explored extensively, envisioning future green technologies based on large-scale conversion of solar energy into *clean* fuels such as H_2. A whole spectrum of photocatalytic reactions and conversion processes have been considered in addition to water electrolysis; these have aimed to utilize solar light to oxidize (i.e. decompose) pollutants or bacteria in gas streams or wastewater [12], or to convert abundant molecules in the atmosphere, such as CO_2 or N_2, into useful, high-added-value chemicals such as hydrocarbons or ammonia, respectively [13, 14]. Research efforts continue to the present day in similar directions.

Among many different photocatalytic materials developed to date, semiconductor metal oxides offer a series of remarkable advantages due to their strong oxidizing ability (e.g. for the decomposition of pollutants), hydrophilicity, (photo-)chemical stability, nontoxicity, environmental friendliness, large abundance, and relatively low cost. Various metal oxide semiconductors feature an electronic structure (i.e. band edge positions) that allows for water reduction and can therefore enable photocatalytic H_2 generation. Examples include TiO_2, $SrTiO_3$, $CaTiO_3$, $KTaO_3$, Ta_2O_5, and ZrO_2, among others [15].

As space constraints preclude an all-inclusive treatment of the subject, in this book chapter we focus on the photocatalytic properties of nanostructured n-type metal oxide semiconductor materials. We first cover general aspects of heterogeneous photocatalysis and the materials science of semiconductors. We then outline key physicochemical and photocatalytic features of metal oxides. We discuss a selection of relevant techniques for the synthesis of photocatalytic materials and advantages associated with nanostructuring. We continue further with an overview of the principles and methods used for the modification of oxide photocatalysts with the aim to enhance their photoactivity. Finally, we critically discuss some practices and trends in current and future research on photocatalysis [16].

6.2 Fundamentals of heterogeneous photocatalysis

We discuss below the underlying principles of light energy conversion via heterogenous photocatalysis and address photoinduced redox processes at semiconductor/liquid interfaces, with a particular focus on solar photocatalytic H_2 evolution.

6.2.1 Key steps in heterogeneous photocatalysis

In general terms, *photocatalysis* can be defined as the light-induced (*photo*) acceleration of a redox reaction in the presence of a catalyst (*catalysis*). In *heterogeneous photocatalysis,* the redox reactions of interest occur at the surface of a solid semiconductor, and more precisely at the interface between a semiconductor (a solid phase) and the environment (a liquid or gas phase). The semiconductor is thus referred to as the photocatalyst. In contrast, in a homogeneous photocatalytic process, the reactants exist in the same phase. This is, for instance, the case for redox processes mediated by light-activated molecules [6, 17]. We refer to the latter as *homogeneous photocatalysis*, since both the reactants and the catalyst are molecular species and interact homogeneously within the same liquid or gas phase; hence, there is no solid/liquid interface at which redox reactions occur. Herein, we focus on heterogeneous processes.

As sketched in figure 6.1, a generic heterogeneous photocatalytic process typically implies the following steps: (i) excitation, i.e. light absorption and the generation of electron/hole pairs; (ii) the separation of electron/hole pairs and their diffusion to the surface of the photocatalyst; and (iii) photooxidation and photoreduction reactions at active sites on the surface of the photocatalyst.

6.2.1.1 (i) Excitation: light absorption and charge carrier generation

Figure 6.2 illustrates generic energy level diagrams and relative light absorption spectra for an isolated atom, an isolated molecule, and a solid semiconductor.

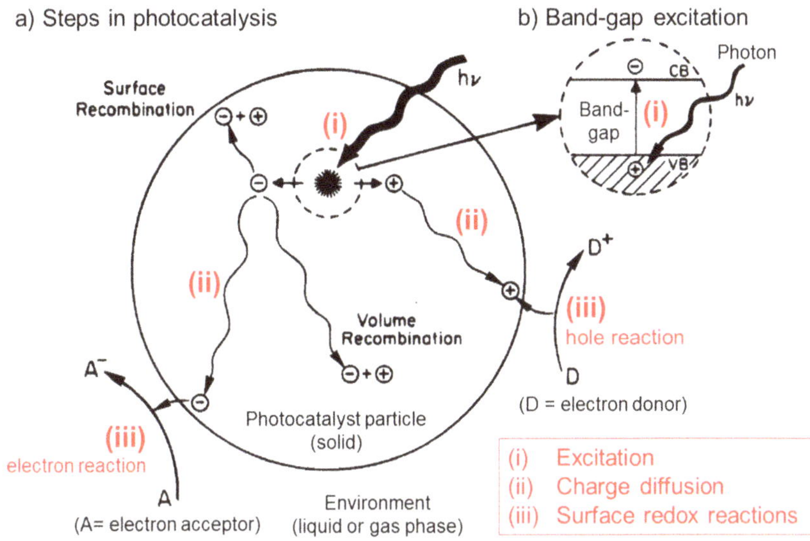

Figure 6.1. Excitation of a semiconductor photocatalyst particle, the diffusion of electron/hole pairs, and surface redox reactions with electron donor and electron acceptor species. Adapted with permission from [18]. Copyright (1995) American Chemical Society.

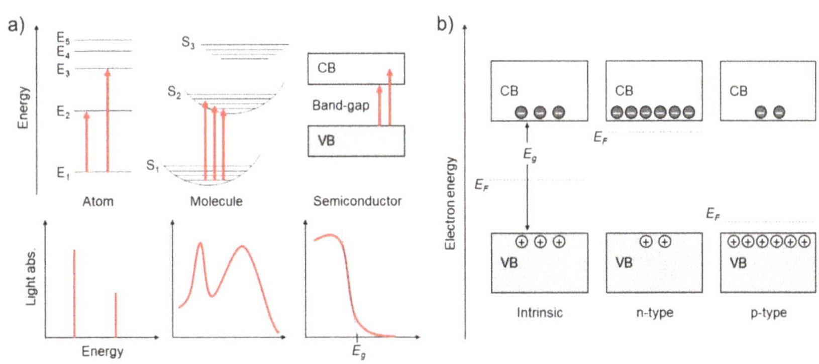

Figure 6.2. (a) Diagrams showing the evolution of energy levels and relative light absorption spectra for an isolated atom, an isolated molecule, and a solid semiconductor. From left to right, there is an increasing interaction between atoms. (b) Energy band diagrams for intrinsic, n-type, and p-type semiconductors.

The interaction between light and isolated atoms gives rise to sharp characteristic absorption lines due to the atoms' quantized electronic energy states. The interaction between light and molecular light absorbers yields defined absorption bands, due to a convolution of electronic transitions with the vibrational or rotational excitation of bonds. The electronic transition in molecules implies the excitation of electrons from the *highest occupied molecular orbital* (*HOMO*) to the *lowest unoccupied molecular orbital* (*LUMO*). In contrast, light absorption by solid materials with an electronic band structure, such as semiconductors, results in a continuous interband absorption spectrum with a cutoff energy corresponding to the bandgap (E_g).

Hence, for semiconductors, light absorption takes place when the energy (E) of an incident photon is higher than the bandgap of the semiconductor (E_g). The bandgap is defined as the energy gap between the valence band (VB) maximum and the conduction band (CB) minimum (figure 6.1(b)). Such an electronic structure is a peculiar property of semiconductors, which provides them with an electrical conductivity that is higher than that of insulators, such as dielectric compounds (e.g. ceramics), which have even larger gaps between the bands, and conductors such as metals that feature a continuum of states in which electrons are highly delocalized. At a temperature of $T = 0$ K, the VB is the highest energy level filled with electrons and the CB is the lowest energy level filled with no electrons. In a simplified view, one can think of the VB and CB for semiconductors as the equivalent of the HOMO and LUMO, respectively, for molecular light absorbers. Hence, when the energy of incident photons $E = h\nu \geqslant E_g$, light can be absorbed, inducing interband absorption, and for each photon absorbed, an electron from the VB is excited to the CB, leaving behind a hole (i.e. an electron vacancy) in the VB (figure 6.1(b)).

The process of semiconductor excitation due to light absorption is also known as the generation of *electron–hole pairs* ($e^-–h^+$), or *charge carrier generation.* One can think of this process as if the energy of the absorbed photons is converted into the internal energy of such excited states, i.e. into the electrochemical potentials of the CB e^- and the VB h^+. The electrochemical potential of the CB e^- corresponds to the energy level of the conduction band minimum (CBM, the bottom of the CB), and that of the *VB* h^+ corresponds to the energy level of the valence band maximum (VBM, the top of the VB).

Pure semiconductors, also called *intrinsic semiconductors,* typically have a relatively low concentration (density) of charge carriers, and the VB *holes* and CB *electrons* are present in equal amounts. In order to increase the charge carrier density, extraneous atoms are introduced into the lattice of the semiconductors in controlled small amounts. This approach is called doping, and the resulting semiconductors are called *extrinsic semiconductors.* Here, we differentiate between *n-* and *p-type* semiconductors. The n-type character implies that a semiconductor has an excess of negative carriers, i.e. electrons in the valence band. This can be caused by the presence of 'donor' atom impurities, which are atoms with more electrons in their outer shell than the constituent atoms of the semiconductor. For elemental semiconductors such as Si (group IV in the periodic table), donors are nitrogen or phosphorous atoms (both group V). Thus, both nitrogen- and phosphorous-doped Si have an n-type character. Considering TiO_2 as an example of a metal oxide semiconductor, the donors can be Nb or Ta, among others. However, oxygen vacancies (OVs), i.e. point defects in the oxide lattice where oxygen atoms are missing, also contribute to the n-type nature of TiO_2 and of other metal oxide semiconductors. On this basis, the majority carriers in n-type semiconductors are electrons, which, in simple words, means that the conductivity of such materials is due to the movement of electrons, since they are present in larger concentrations than holes. Likewise, a semiconductor with a p-type character has an excess of positive carriers, i.e. holes. P-type behavior in Si is obtained by doping it with boron (group III). Accordingly, the majority carriers in p-type semiconductors are holes. For semiconductors, the Fermi level, E_F, can be defined (in a simplified manner) as the average energy level of the electrons in the solid. We come back to doping and bandgap engineering in section 6.1.

When light strikes the surface of a semiconductor, it can essentially be absorbed, reflected, or transmitted. This depends primarily on the photon energy with respect to the semiconductor bandgap, the angle of incidence of the light, and the semiconductor's intrinsic properties, including structure and size.

The bandgap can be modified by bandgap engineering, that is, by introducing adequate amounts of extraneous atoms into the semiconductor lattice. While doping, as discussed above, aims to increase the charge carrier density without altering the band structure, the aim of bandgap engineering is typically to reduce the bandgap size (i.e. to obtain a narrower E_g) by shifting the VBM 'upward' to higher energy levels, or the CBM 'downward' to lower levels. This can improve the ability of semiconductors to harvest light, or, in simple terms, it enables the absorption of a larger fraction of the incoming photons. Note, however, that shifting the band positions may also cause an undesired change in the energy or electrochemical potential of CB e^- or VB h^+.

If we consider a photocatalyst slurry (i.e. semiconductor particles dispersed in a liquid phase), photon propagation, absorption, and scattering need to be considered, as light absorption is affected by the size of the nanoparticles and by their concentration in the suspension. In a more ideal case, e.g. when light illuminates the flat surface of a bulk semiconductor, incident light can penetrate the material to a certain depth as a function of the semiconductor's light absorption coefficient at a given wavelength $\alpha^{-1}(\lambda)$ [19]. According to the Lambert–Beer law, the light intensity is exponentially attenuated as it propagates through the material [9, 20], and the absorption coefficient depends on the nature of the electron transition [3].

In fact, a semiconductor can exhibit two types of bandgap, i.e. direct or indirect (figure 6.3). The minimum energy state in the CB and the maximum energy state in the VB are characterized by a certain crystal momentum (k-vector, in the Brillouin zone) [3]. The bandgap is 'direct' if the crystal momentum of electrons and holes is the same in both the CB and the VB. In this case, photon absorption and the associated electron transition imply only the exchange of energy between the photon and the electron. However, if the k-vectors are instead different, the material has an 'indirect bandgap.' The process of absorption of photons in an indirect bandgap material involves not only the exchange of energy between the photon and the electron but also the absorption or emission of a phonon, i.e. a quantum of energy, through lattice vibration. The phonon momentum equals the difference between the electron momentum and the hole momentum. Thus, light absorption by indirect bandgap semiconductors is much less likely to occur in a given span of time. In simple words, photon absorption is 'less probable,' which is in line with experimental observations that light absorption and carrier generation/diffusion are far less efficient in indirect bandgap materials than in direct bandgap counterparts. As a result, light with a photon energy close to or higher than the

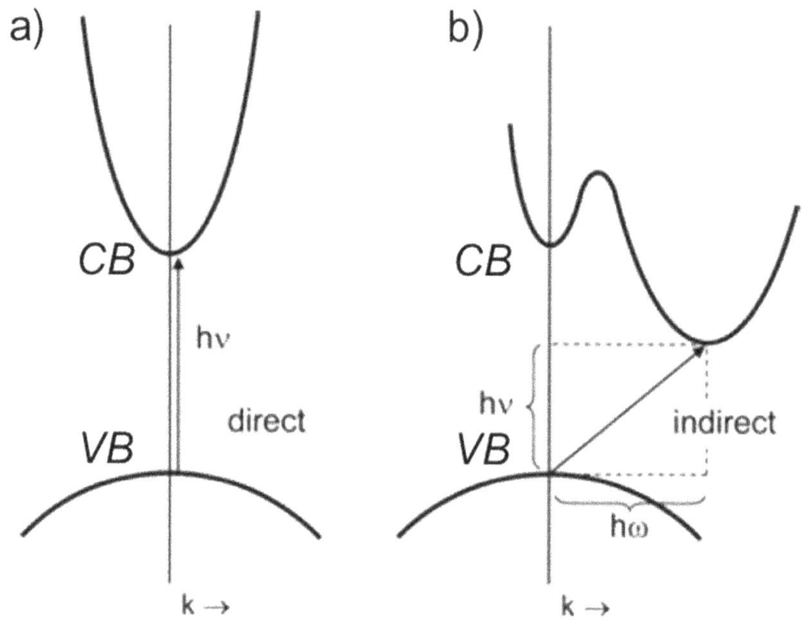

Figure 6.3. Optical transitions, or excitation, of semiconductors with a direct and an indirect bandgap. The indirect transition requires the assistance of a phonon [3] (2012) (Springer Science+Business Media, LLC 2012). With permission of Springer.

bandgap can penetrate much farther before being fully absorbed in an indirect bandgap material. This is relevant since most common metal oxide photocatalysts are indirect bandgap semiconductors. Hence, they must be designed (in terms of structure and physicochemical properties) to provide maximized photon harvesting and optimized charge carrier management.

6.2.1.2 (ii) The separation of electron/hole pairs

As introduced above, the filled electronic levels in the semiconductor (that form the VB) are separated from the empty electronic levels (that form the CB) by the forbidden gap (E_g, figure 6.1(b)). Upon bandgap excitation e^-–h^+ pairs are formed. Once formed, charge carriers may have different fates [21].

To enable surface redox reactions, the desired fate for charge carriers is their diffusion towards the semiconductor surface, where they can react with species in the environment (at the semiconductor/liquid or semiconductor/gas interface). The process by which charge carriers reach the semiconductor surface and react with an electron donor or acceptor species is called *charge collection*. The charge collection efficiency can be defined as the number of charge carriers that react at the photocatalyst surface with respect to the total number of carriers generated by excitation (light absorption). Hence, this parameter describes the quantum yield of the photocatalyst.

A major drawback in photocatalysis is that charge collection is low for many pristine materials. Charge carriers in the semiconductor have finite mobility and lifetime, which depend on factors such as the properties and morphology of the semiconductor material and the carrier type (e^- or h^+). Considering the semiconductor bulk, and in the absence of an external electric field (i.e. with no electric bias applied), charge carriers move within the lattice of the semiconductor by diffusion. The range of transport for charge carriers is defined by the mean free diffusion length L, i.e. the average length a charge carrier travels between generation/separation and recombination (see the next paragraph). L depends on the carrier diffusion constant D and the carrier lifetime τ according to $L = \sqrt{D \times \tau}$. D and τ are material-specific parameters. We mentioned earlier that the introduction of dopants into the lattice of the semiconductor can alter the density and mobility of charge carriers. Doping, as discussed in section 6.1, is thus an effective strategy for improving the concentration and mobility of carriers, so as to enhance the performance of photocatalysts.

Charge carriers can also recombine, i.e. a CB electron fills a VB hole, with the consequent emission of a photon or heat. These processes, also known as radiative emission (associated with photoluminescence) and thermalization, respectively, are undesired in photocatalysis, as they prevent carriers from reacting at the surface of the photocatalyst. Recombination (figure 6.1(a)), which can occur both in the bulk or at the photocatalyst surface, represents a major cause of energy loss and low quantum yield in photocatalytic processes.

Another possible fate for charge carriers is trapping, which typically takes place at crystallographic (lattice) defects, such as oxygen vacancies in metal oxide semiconductors, or, more generally, at bulk or surface defect states. Like recombination, trapping is also undesired, as it hinders the mobility of electrons and holes and therefore reduces their collection efficiency and the rate of surface redox reactions.

6.2.1.3 (iii) Surface (or interface) redox reactions

The photocatalyst surface represents the solid/liquid (or solid/gas) interface where charge carriers are collected by electron donors or acceptors in the environment. This is therefore the region of interest when describing redox reactions in photocatalysis.

Figure 6.4 illustrates the interfacial energetics involved in a generic photocatalytic process, in which the energy levels of the charge carriers in the semiconductor and of the reactants in the contacting solution (the reaction environment) are shown on a common diagram.

The vertical arrow (bottom to top) indicates the increase of the electron energy; the energy level of charge carriers and the electrochemical potential of species in the environment increase in the opposite direction (top to bottom).

The reaction environment can be a liquid medium, e.g. an aqueous solution, or a gas phase. To exemplify the process, we consider the redox couples D/D^+ and A/A^-, where D is a generic electron donor (a reductant), and A is a generic electron acceptor (an oxidant). As mentioned earlier, while the bandgap of a semiconductor determines its optical (light absorption) features, the positions of

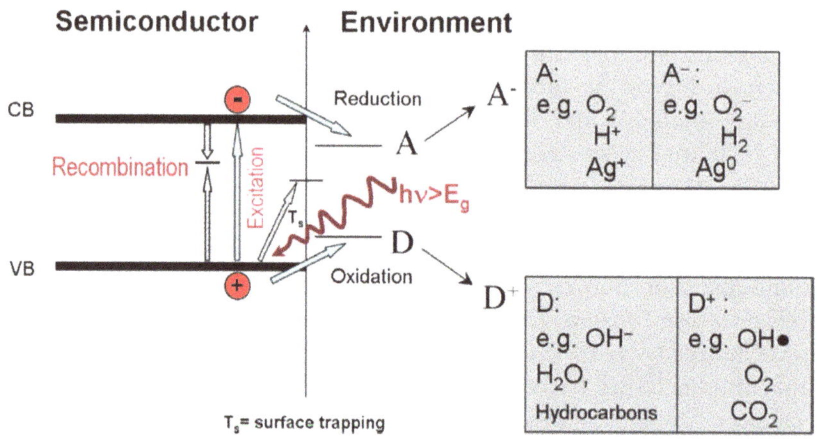

Figure 6.4. A scheme of photoinduced processes at the solid/liquid (or solid/gas) interface of a generic semiconductor [22] John Wiley & Sons. [Copyright © 2012 WILEY-VCH Verlag GmbH & Co. KGaA, Weinheim].

the CB and the VB govern its electrochemical (redox) properties, that is, the 'exit energy' of electrons and holes determine their ability to respectively reduce or oxidize species in the contacting solution. This means that, as indicated in figure 6.4, for A and D to be reduced and oxidized, respectively, the semiconductor CB and VB edges must 'bracket' the two redox levels corresponding to the two generic reactions, i.e. $A + e^{-} \rightarrow A^{-}$ (reduction or the electron reaction) and $D + h^{+} \rightarrow D^{+}$ (oxidation or the hole reaction). In other words, the CBM should be above (at a more negative potential than) the electrochemical potential of the A/A^{-} couple ($E^{0}_{A/A^{-}}$), and the VBM should be below (at a more negative potential than) the electrochemical potential of the D/D couple ($E^{0}_{D/D^{+}}$). Another equivalent way to describe this situation is that CB electrons can only reduce A species whose electrochemical potential is more positive than the energy level of such CB electrons, and VB holes can only oxidize D species whose electrochemical potential is less positive than the energy level of such VB holes. For the oxidation of D, note that the hole transfer from the semiconductor to the donor species can be seen as an electron transfer reaction in the opposite direction, i.e. from the donor to the photocatalyst.

The considerations above only indicate whether a certain reaction is thermodynamically spontaneous ($\Delta G < 0$, i.e. an energetically 'downhill' process), while no kinetic information is given. A given redox process may in fact be thermodynamically spontaneous but kinetically hampered. Moreover, to establish whether electron or hole transfer is possible, it is more accurate to consider not only the redox potentials under standard conditions but also the energy levels and electrochemical potentials of interest under the experimental conditions adopted (the pH, for example, is a crucial parameter) [23]. Furthermore, note that photocatalysis is, in general, a nonselective process, and thus charge carriers can react indiscriminately with any redox species available, if the thermodynamic requirements are met.

The process is thus governed by the relative concentrations of species and reaction kinetics. Multiple redox process can therefore proceed in parallel when their kinetics are comparably fast. However, selectivity can be achieved, to some extent, by adjusting the reaction conditions or the properties of the photocatalyst, or by the use of composite materials [24]. Given its complexity, we do not cover the selectivity of photocatalytic reactions in this book.

Having established the groundwork, we outline in the following a few final generic implications of photocatalytic processes, after which we describe the photocatalytic properties of metal oxides.

6.2.2 The stability of semiconductors

Stability is a key prerequisite of a catalysts; thus, the semiconductors used for photocatalysis (or photoelectrochemistry) must be stable against corrosion. The corrosion of semiconductor materials can occur in three different forms: chemical, electrochemical, or photocorrosion [25].

Chemical corrosion is the degradation of the semiconductor due to chemical dissolution and is ascribed to the intrinsic thermodynamic instability of a given semiconductor in a specific chemical environment. Electrochemical corrosion involves electron transfer at the semiconductor/solution interface that triggers the decomposition of the semiconductor; hence, the semiconductor degradation is due to red-ox reactions. The chemical and electrochemical corrosion reactions are classic corrosion reactions that can be triggered by pH, redox species in the electrolyte (e.g. oxidizing or reducing agents) or by an applied electrochemical potential. In this sense, the (electro)chemical stability of a semiconductor can be assessed through its relative Pourbaix diagram, i.e. as a function of the electrolyte pH and the applied electrochemical potential. Note that both chemical corrosion and electrochemical corrosion can take place in the dark.

Photocorrosion, on the other hand, is a specific (typically undesired) feature of some semiconductors. Essentially, photocorrosion is the process of semiconductor degradation caused by photogenerated charge carriers produced in the semiconductor under illumination. As described earlier, a photocatalytic process must include at least two reactions that occur simultaneously, one involving oxidation via photogenerated holes, and another involving reduction via photopromoted electrons. Both processes must be balanced, so that the photocatalyst itself does not accumulate VB holes or CB electrons. This can be compared to the two half-cell reactions in an electrochemical cell: for charge neutrality, electrons that are given up at the anode must be fully consumed at the cathode. Charge accumulation in the semiconductor can cause changes in the oxidation state of the constituent atoms, thereby compromising the stability of the material. For example, p-type Cu (I) oxides tend to undergo disproportionation reactions under illumination, i.e. oxidation to CuO (Cu(II)) followed by dissolution, and reduction to Cu metal (Cu^0) [26]. Photooxidation would instead take place upon the accumulation of holes, as in the case of *n*-type Si, III–V compounds (GaAs or InP), but also metal oxides such as $BiVO_4$. Photooxidation represents the most severe limitations to the stability of Si or III–V compounds under illumination in aqueous solutions [10, 27]. Stability, along with considerations regarding the supply chain, the cost of materials, and the preference towards environmentally friendly atomic constituents, make metal oxides the most suitable candidates for use in the design of photocatalytic materials [28–30].

6.2.3 Sacrificial agents

In a photocatalytic process, one of the two photoinduced redox reactions may be kinetically hampered; this means that the rate of such reaction is slow with respect to the other. This can be ascribed to a sluggish (i.e. slow) transfer of charge carriers to the donor or acceptor species across the solid/liquid interface. Moreover, from a mechanistic standpoint, a certain reaction may imply the transfer of multiple charge carriers. For example, in water splitting (reaction (6.1), section 6.1.1), the oxidation of water to O_2 (reaction (6.2)) is a four-hole process [31]. This represents a kinetic hurdle (compared to the hydrogen evolution reaction). If we assume that the hole (anodic) reaction proceeds slowly in a generic photocatalytic process, hole accumulation will occur in the semiconductor, which can in turn increase the probability of hole recombination with electrons, consequently limiting the rate of the electron (cathodic) reaction as well. This can, for example, slow down the kinetics of the reduction of protons to H_2 (reaction (6.3)). Moreover, as discussed above, charge accumulation in the semiconductor can also undermine the stability of the material. Thus, charge accumulation should be avoided, or, in other words, photocatalytic processes must be designed in such a way that charge collection at the semiconductor surface is optimized. This can be attained by boosting the kinetics of the slow reaction by modifying the photocatalyst's surface with a co-catalyst. In section 6.2, we discuss the use of co-catalysts and their specificity, i.e. how co-catalysts can be used to accelerate certain reactions, with emphasis on the hydrogen evolution reaction (HER).

Alternatively, a fast electron acceptor or a fast electron donor can be added to the reaction environment to rapidly react with the charge carriers that would otherwise tend to accumulate. Such species are also named *sacrificial agents* or *scavengers*, as their fast consumption limits charge accumulation in the photocatalyst and hence favors the kinetics of the photocatalytic reaction of interest. For instance, a hole scavenger is an electron donor species that can rapidly consume VB holes. Typical examples of hole scavengers are organics (alcohols such as methanol, ethanol, or glycerol) or sulfite ions. Alcohol molecules, if present in the reaction environment, rapidly capture the VB holes in the semiconductor and eventually undergo oxidation to CO_2. This in turn reduces the recombination rate for CB electrons, since holes are rapidly depleted from the photocatalyst. Thus, a hole scavenger in an aqueous environment can aid in accelerating the kinetics of H_2 evolution.

Sacrificial species can also be used to scavenge electrons; that is, an electron scavenger is an electron acceptor species that can rapidly consume CB electrons.

Ag^+ ions can efficiently do so, thereby undergoing reduction and forming Ag^0 (metal) deposits at the surface of the semiconductor. We will come back to the use of sacrificial agents later (in section 6.2.5), and provide more details and examples in the context of photocatalytic H_2 evolution, hence underpinning the difference between photocatalytic overall water splitting and the photocatalytic reforming of organics.

6.2.4 The semiconductor/electrolyte interface in photoelectrochemistry

In the pioneering work of Fujishima and Honda in 1972 [7], the possibility of solar PEC water splitting was demonstrated using a rutile TiO_2 single-crystal photoelectrode. Such a TiO_2 photoelectrode, electrically connected to a platinum counter electrode, was irradiated with photons of wavelength shorter than ~415 nm (see the schematic in figure 6.5(a)). Under these conditions, photocurrent, i.e. a current generated upon illuminating a semiconductor, was found to flow through the PEC system, and an oxidation reaction (the oxygen evolution reaction (OER)) occurred at the TiO_2 surface, while a reduction reaction (the HER) took place at the Pt electrode. These findings proved that water could be split (converted) into hydrogen and oxygen on an illuminated metal oxide semiconductor surface.

In the following, we briefly discuss the basics of PEC water splitting. We aim at clarifying similarities and differences between photocatalytic and photoelectrochemical reactions. We hope this is useful to better understand the criteria used to design efficient photocatalysts. A detailed treatment of photoelectrochemistry is out of the scope of this chapter. Interested readers are referred to references [3, 9, 32, 33] for a thorough overview.

The system in figure 6.5 can be described as a Schottky-type PEC cell, in which an n-type semiconductor is used as photoanode, and a metal (Pt) is used as cathode.

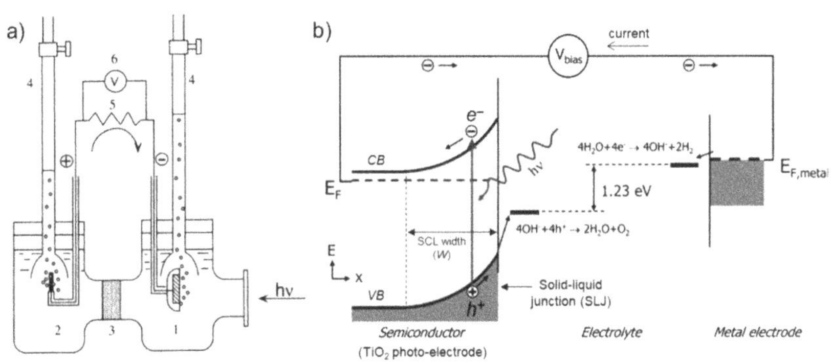

Figure 6.5. (a) Schematic diagram of a PEC cell: (1) n-type TiO_2 photoelectrode; (2) platinum electrode; (3) proton exchange membrane; (4) gas burettes (for gas collection and volume determination); (5) load resistance; and (6) voltmeter. Reprinted from [15], Copyright (2000), with permission from Elsevier. (b) Schematic representation of PEC water splitting as shown in (a). Illuminated conditions are shown for a TiO_2 photoelectrode with an E_{CB} of −0.65 V (vs. the reversible hydrogen electrode (RHE)) and an E_{VB} of 2.35 V (vs. RHE) at the semiconductor/electrolyte interface [3] (2012) (Springer Science+Business Media, LLC 2012). With permission of Springer.

The two electrodes are wired up through an external circuit, through which an external electric bias can be applied to flow current (electrons).

When the metal oxide semiconductor is placed in contact with an electrolyte solution, thermodynamic equilibration takes place at the interface. This originates from the difference in the Fermi levels between the semiconductor and the electrochemical potential of (redox species in) the electrolyte. In this example, before equilibrium is reached, the Fermi level (E_F) of the semiconductor is higher than the electrochemical potential of the redox species in the electrolyte. While the junction is being established, electrons flow from the n-type semiconductor into the electrolyte. This leaves a net positive charge in the semiconductor, which generates an electric field, i.e. a potential gradient or a potential barrier (Schottky barrier) at the solid/liquid junction (SLJ)—see figure 6.5(b).

The potential gradient leads to the formation of a space charge layer (SCL) within a thin surface region of the semiconductor, in which the energy bands are generally bent upwards or downwards, respectively, in the cases of n- and p-type semiconductors. The thickness of the SCL is usually of the order of 1–10^3 nm, depending on the carrier density, the dielectric constant of the semiconductor, and the doping level. Upon illumination, electron–hole pairs are generated and separated in the SCL due to the electric field. In the case of an n-type semiconductor, the electric field across the SCL drives photogenerated holes (minority carriers) toward the interfacial region (i.e. the SLJ) and electrons (majority carriers) toward the interior (bulk) of the semiconductor and from there via the electrical connection to the external circuit (note that the reverse process takes place at a p-type semiconductor photoelectrode). The effectiveness of the spatial separation of e^-–h^+ pairs can be improved by applying an external electric bias (potential), by which it is possible to tune the width of the SCL and extent of band bending. Increased band bending and widening of the depletion layer can be obtained when an anodic bias is applied to the photoanode. This leads to a larger field and more efficient e^-/h^+ separation. Conversely, when a p-type semiconductor is used (a photocathode), a cathodic bias must be applied to increase band bending and widen the depletion layer. In general, the external bias applied to separate charge carriers makes it possible to drive the anodic and cathodic processes at two distinct, physically separated surfaces, i.e. the (photo)anode and (photo)cathode, respectively. Therefore, the criteria for materials design in photoelectrochemistry imply that each (photo) electrode must be 'competent' for one specific reaction, i.e. either an oxidation or a reduction reaction.

Compared to a PEC configuration as described above, a generic photocatalytic system features crucial differences. First, photocatalytic systems are typically based on suspensions containing photocatalytic particles (slurries) in a desired solution under illumination. The suspended photocatalyst particles cannot be connected to an external circuit; thus, an external electric bias cannot be applied in this case. One could consider that photocatalysis occurs under 'wireless' (or 'open-circuit') conditions, as no external electric bias is applied. If we consider a single photocatalyst particle, the cathodic and anodic reactions take place at different sites on the surface of the same particle (see figure 6.1(a) and cf with figure 6.5(b)). Charge separation in the photocatalyst particle is driven by the built-in potential ascribed to the space charge region formed upon contacting the solution, and to charge carrier generated upon illumination. Hence, photocatalyst particles must be designed to feature catalytic sites that are specific both for the reduction and the oxidation processes desired (we provide examples in sections 6.2.5 and 6.6.2). In addition to their use in suspensions, photocatalysts can be in the form of supported semiconductor nanoparticles or a nanocrystalline layer grown on a proper substrate material. Also in such cases, the electron and hole reactions take place at different sites on the surface of the same photocatalyst particle or layer. Therefore, also photocatalysts used in the form of supported layers must be designed to feature catalytic sites specific to both the desired reduction and oxidation reactions.

6.2.5 Photocatalytic water splitting and H_2 generation

From a thermodynamic standpoint, the overall water-splitting reaction is endergonic ($2H_2O \rightarrow 2H_2 + O_2$, reaction (6.1) in section 6.1.1), with a Gibbs free energy of $\Delta_r G^\circ = 237$ kJ per mol of H_2 produced; according to (6.4):

$$\Delta_r G^\circ = -nFE^\circ \tag{6.4}$$

where n = moles of electrons; F = Faraday constant (96 485 C mol^{-1}); E°, i.e. the standard electrochemical potential of the reaction of interest (water splitting) is 1.23 V. Hence, when a single semiconductor light absorber is used to split water into oxygen and hydrogen, such photocatalyst must feature a bandgap larger than 1.23 eV. That is, the semiconductor should exhibit a light absorption edge at $\lambda \leqslant 1000$ nm. However, taking into account the overpotentials required for the hydrogen and oxygen reactions, a potential $E \sim 2.0$ V is a reasonable estimate of the minimum energy required to enable water splitting at reasonable rates [34, 35]. This corresponds to a light absorption edge $\leqslant$ 610 nm. Furthermore, the semiconductor can drive water splitting if its *CBM* lies at more negative potentials than the water reduction potential ($E^\circ_{H_2O/H_2} = 0$ V) and a *VBM* that lies at more positive potentials than the water oxidation potential ($E^\circ_{O_2/H_2O} = +1.23$ V vs. RHE)—we introduced these aspects earlier in section 6.2.1. It should be noted that the position of the band edges of the semiconductor, and therefore the energy levels of charge carriers, may vary depending on the pH of the reaction environment [36]. The pH affects also the adsorption on the photocatalyst surface of donor and acceptor species, as well as the electrochemical potential of redox couples involved in the reactions, according to the Nernst equation (6.5):

$$E = E^\circ - \frac{RT}{nF} \ln \frac{[\text{red}]}{[\text{ox}]} \tag{6.5}$$

where E° = standard electrochemical potential of the redox couple of interest; R = universal gas constant (8.314 J K^{-1} mol^{-1}); T = temperature in kelvin; n = moles of electrons; F = the Faraday constant (96485 C mol^{-1}); [red] = the concentration of the reduced species; and [ox] = the concentration of the oxidized species. For the hydrogen and oxygen reactions, the electrochemical potential varies by −0.059 V per pH unit from their standard potential, i.e. 0 and 1.229 V_{RHE}, respectively.

Multiple semiconductors can be combined to form heterojunction-based photocatalysts. As we will discuss later in section 6.2, this allows the spectral range of activity of the photocatalyst to be widened or the charge carrier separation and collection to be improved. In both cases, the aim is to achieve enhanced photocatalytic performance.

When the thermodynamic prerequisites we have just discussed are met, an illuminated photocatalyst can produce excited carriers that are sufficiently energetic to drive the HER and the OER. If both the hydrogen and oxygen reactions occur at a reasonable rate, e.g. due to the presence of efficient co-catalysts, the overall reaction is referred to as *overall water splitting*, i.e. H_2 and O_2 are evolved from *pure* water. This situation is shown in figure 6.6(a), where the splitting of water is driven by an illuminated composite metal oxide photocatalyst, i.e. Pt, TiO_2, or RuO_2 suspended in pure water. Various oxide-based photocatalytic systems can drive overall water splitting. Interested readers are referred to the review articles [37, 38].

Conversely, in the absence of co-catalysts, i.e. when pristine TiO_2 is used, the hydrogen reaction and particularly the oxygen reaction are sluggish. As introduced in section 6.2.3, a way to accelerate the kinetics of hydrogen evolution is to add an electron donor as a *sacrificial agent*. For example, if the reaction environment is an aqueous solution containing methanol, illuminated TiO_2 can drive the photocatalytic evolution of hydrogen from water (an electron reaction), while at the same time methanol is oxidized to CO_2 (hole reaction). Due to the less positive oxidation potential of methanol ($CH_3OH + H_2O \rightarrow CO_2 + 6H^+ + 6e^-$; $E^\circ = 0.016$ V vs. RHE) compared to water ($2H_2O \rightarrow 4H^+ + 4e^- + O_2$; $E^\circ = 1.229$ V vs. the RHE), the oxidation of ethanol via VB holes is more favorable than the oxidation of water. In general, not only is the oxidation of organic molecules (to CO_2) thermodynamically 'downhill' but it is also kinetically fast, meaning that it occurs relatively fast at the

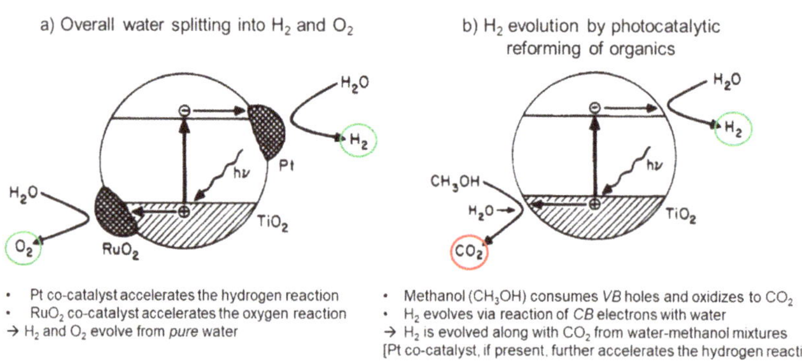

Figure 6.6. A comparison between overall water splitting into H_2 and O_2 and the evolution of H_2 by photocatalytic reforming of organics. Adapted with permission from [18]. Copyright (1995) American Chemical Society.

surface of pristine TiO_2 (i.e. in the absence of any specific co-catalyst). This situation is shown in figure 6.6(b)—cf with figure 6.6(a).

The trend of the photocatalytic H_2 evolution rate in the presence of different sacrificial agents has been widely studied, particularly for metal oxide photocatalysts [39, 40]. The ability of the scavenger to capture valence band holes depends primarily on factors such as the electrochemical potential of the scavenger redox couple (e.g. CH_3OH/CO_2) and the polarity of the scavenger molecule. The less positive the electrochemical potential and the higher the polarity, the better the efficacy of hole capture. A less positive electrochemical potential provides a larger thermodynamic driving force for the oxidation of the scavenger via VB holes. For example, the electrochemical potential of methanol (CH_3OH/CO_2 couple, $E^\circ = 0.016\ V_{RHE}$) is less positive than that of ethanol (CH_3CH_2OH/CO_2 couple, $E^\circ = 0.084\ V_{RHE}$), which explains the higher photocatalytic rate typically measured in the presence of methanol. The polarity affects the scavenger oxidation rate by influencing its adsorption kinetics on the photocatalyst surface.

The use of water–organic mixtures is of particular interest in the context of the *photocatalytic reforming* of organics [41, 42], in which aqueous–organic mixtures (for example, derived from biomasses) are used as feedstocks for H_2 generation. Indeed, the downside of this approach is that H_2 evolution occurs along with mineralization of the organic phase to CO_2 [43], hence this approach has a significant net CO_2 footprint.

6.2.6 Photocatalysis as an advanced oxidation process

In addition to the development of clean energy technologies, another global challenge is to provide clean water by recycling and purifying wastewater streams. Industrial and agricultural wastewater contains a variety of noxious aromatic compounds, while the presence of xenobiotic substances (i.e. antibiotics, drugs, toxins, and pesticides) in surface and ground water is becoming an increasingly growing issue, as such substances are often resistant to degradation by current purification techniques. In this context, photocatalysis has been widely explored as an advanced oxidation process (AOP). Here, the goal is not the synthesis of a specific product but rather the degradation of a target molecule or a class of compounds. A large body of literature dealing with metal oxide photocatalysts, and with TiO_2 in particular, documents the suitability of photocatalysis for oxidizing toxic molecules of pollutant species of different types, including volatile organic compounds in the gas phase, aromatic molecules, pharmaceutical compounds, and microorganisms in wastewater.

The energy band diagram for TiO_2 is shown in figure 6.7. The redox potential for photogenerated holes is +2.53 V (vs. the standard hydrogen electrode (SHE)). Hence, VB holes in TiO_2 can directly oxidize adsorbed organic molecules to CO_2, as mentioned above. TiO_2 VB holes can also react with water to form hydroxyl radicals (OH*), whose redox potential is only slightly less positive than that of VB holes, hence OH* are also strong oxidizing species. The redox potential of CB electrons is $-0.52\ V_{SHE}$, which is negative enough to reduce dissolved O_2 to O_2^{-*} superoxide

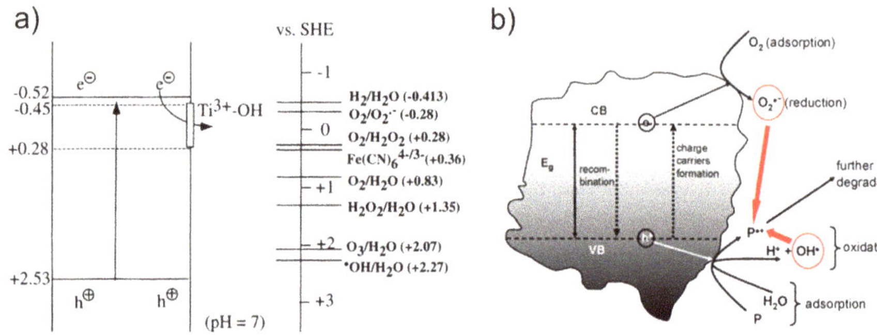

Figure 6.7. (a) Schematic diagram showing the potentials of various redox processes occurring on the surface of TiO_2 at pH 7. Reprinted from [15], Copyright (2000), with permission from Elsevier. (b) The fates of electrons and holes in a semiconductor particle in the presence of water containing a pollutant (P). Adapted from [44], Copyright (2009), with permission from Elsevier.

radicals or to hydrogen peroxide (H_2O_2). Such species, namely OH*, O_2^{-}*, and H_2O_2, also referred to as 'reactive oxygen species' (ROS), can play a dominant role in photocatalytic oxidation processes. Such species are, in fact, highly oxidizing (and at the same time unselective) and are able to break a variety of noxious recalcitrant molecules. For an overview of photocatalytic oxidation processes, the reader is referred to [44, 45].

6.2.7 Materials choice

To summarize the discussion above, an ideal photocatalyst must meet several requirements. It should feature adequate positions of the CBM and the VBM, a broad spectral range of light absorption, and efficient charge separation and transport properties. Furthermore, a photocatalyst must be composed of earth-abundant and environmentally friendly elements, and it must be cheap, relatively easy to synthesize, and stable. Note that given the intermittent nature of sunlight, photocatalysts should remain stable not only under illumination but also in dark periods [46, 47].

No semiconductor material has yet been found or developed that comes close to meeting these highly demanding prerequisites, and hence a trade-off must be made. This still prevents photocatalysis (and photoelectrochemistry) from becoming competitive with fossil-based technologies and from finding applications at the industrial level. Research in this field, however, has progressed significantly in recent decades, opening promising application perspectives for the future. In fact, nanostructuring, doping, bandgap engineering, and photosensitization methods, along with the development of composite photocatalysts, have led to remarkable leaps ahead in terms of stability and efficiency of photocatalytic materials [48].

If we consider single materials, the best properties in terms of visible-light absorption and charge carrier transport are exhibited by single crystalline elemental semiconductors such as Si (and Ge), or II–VI (CdSe, CdS, CdTe, ZnSe, ZnS, etc) and III–V (GaP, GaAs, InP, etc.) semiconductor compounds. Such semiconductors typically feature a narrow bandgap, e.g. 2 eV or less. Hence, they absorb the solar spectrum optimally and are therefore widely applied in the design of high-efficiency photovoltaic (PV) and optoelectronic devices. However, they require energy-intensive processes for their preparation. Moreover, some of these materials are composed of toxic elements. In addition, they easily oxidize or photocorrode when illuminated in aqueous environments. To prevent degradation and performance decay they must be *passivated*. This means that their surface needs to be coated by protective layers, typically produced via expensive processes. Atomic layer deposition (ALD) is a most suitable method with which to coat ultrathin oxide layers (e.g. Al_2O_3 or TiO_2) onto Si-, GaAs-, and GaP-based photoelectrodes [49].

The requirement for stability in aqueous environments and under illumination makes metal oxide semiconductors a most promising choice for the design of photocatalysts and photoelectrodes [32]. Metal oxides are, in fact, generally stable under the conditions adopted in photocatalytic or PEC reactions. In the following section, we describe the most relevant metal oxides for photocatalysis and focus on advanced properties that originate from their nanostructuring.

6.3 Metal oxide photocatalysts

6.3.1 n- and p-type oxide semiconductors

Metal oxide semiconductors are valence compounds with a high degree of ionic bonding. Their conduction band minimum (CBM) and valence band maximum (VBM) mainly consist of the metal (M) *ns* or *nd* empty orbitals and the oxygen (O) 2*p* filled orbitals, respectively [50]. Consequently, while the CB position varies according to the natures of different metals and their electronic structures, the energy level of the VB edge typically lies at 3.0 ± 0.5 V for most metal oxides [51, 52], as illustrated in figure 6.8.

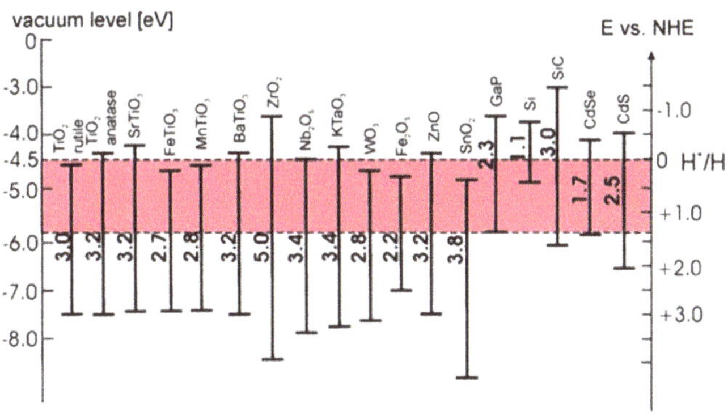

Figure 6.8. Band positions and bandgap values for various metal oxide semiconductors. As reference, a few non-oxide semiconductors are also included. The energy scale is indicated with reference to the vacuum level (left side) or the normal hydrogen electrode (NHE, right side). The electrochemical potentials for hydrogen and oxygen evolution from water are defined by the upper and lower edges of the red area, respectively. Reproduced from [53] with permission of The Royal Society of Chemistry.

The interaction between the metal and oxygen orbitals determines the charge carrier transport properties of the oxide. In general, metal *ns* and *nd* orbitals are highly dispersive (delocalized), whereas the O 2*p* orbitals are localized, which yields a smaller effective mass for electrons than for holes. Since the carrier mobility μ is inversely proportional to the carrier effective mass m^* ($\mu = e\tau/m^*$, where τ is the free carrier scattering time, and e the electron charge), the smaller electron effective mass of metal oxides is the origin of their better electron transport in comparison to hole transport. Thus, most of the relevant metal oxides, such as TiO_2, WO_3, Fe_2O_3, ZnO, Ta_2O_5, Nb_2O_5, and SnO_2, among others, as well as their solid solutions and multinary oxides (e.g. ternary oxides such as perovskite-like titanates or tantalates) are n-type semiconductors with an indirect bandgap.

In n-type metal oxide semiconductors, intrinsic point defects, most commonly oxygen vacancies (VOs), serve as donor species, thus yielding an n-type semiconductor character with reasonable electron conductivity. Intrinsic point defects such as oxygen vacancies form rather easily in metal oxides, and their formation is typically referred to as *unintentional doping*. In other words, oxygen vacancies are typically present even if not purposely introduced. However, while the electronic band structure of metal oxides can facilitate electron transport to some extent, it should be noted that good n-type conductivity cannot be achieved without *intentional doping* (i.e. the deliberate introduction of dopants) because the bandgap of these metal oxides is relatively wide, and the energetic levels associated with intrinsic defects such as VOs may be too deep to provide negative carriers at a significant concentration. Thus, intentional doping is required to achieve enhanced charge transport properties and conductivity.

Some metal oxides exhibit p-type conductivity. For this to occur, the hole effective mass should be significantly smaller, which requires a more dispersive VBM. The key to realizing a dispersive VBM in metal oxides is to design materials with metal cations that introduce occupied *d* or *s* states near the *VBM*. The resulting *p-d* and *p-s* coupling between the metal cation and oxygen orbitals enhances the dispersion of the VBM, hence reducing the hole effective mass. Among p-type metal oxide semiconductors, nickel (II) oxide (NiO) is a well-known example that has been widely employed as p-type transparent conductive oxide (TCO) [54]. Cu-based oxides, e.g. Cu_2O and CuO, have also attracted much attention.

6.3.2 Relevant oxide photocatalysts

The variety of semiconductor oxides used in photocatalysis is immense, spanning from binary to ternary and multinary compounds; they may be pure or doped materials, used as single light absorbers or in the form of junctions. For a complete overview of the field, the reader is referred to the following reviews [9, 37, 48, 55, 56]. In the following, we describe a selection of most popular and widely investigated metal oxide photocatalysts. In selecting these compounds, our intention is to highlight oxides that offer a reasonable trade-off between high stability, great abundance and environmentally friendliness, suitable light absorbance, and adequate band edge energies.

6.3.2.1 – TiO_2

Titanium dioxide or titania (TiO_2), a d^0 metal oxide, is the most investigated transition-metal-oxide material in photocatalysis. In addition to photocatalysis and photoelectrochemistry, it has been explored for a wide range of applications, including Grätzel-type (dye-sensitized) solar cells, perovskite solar cells, gas sensors, pigments, batteries, and electrochromic devices, among others [15, 18, 57, 58].

The most common polymorphs of TiO_2 are tetragonal anatase, tetragonal rutile, and orthorhombic brookite. In contrast to the anatase and rutile phases, which are relatively easy to synthesize as pure polymorphs, the synthesis of pure brookite free from rutile or anatase contaminations is laborious. Reliable protocols for the preparation of pure brookite nanomaterials via synthetic routes at mild temperatures and pressures have only become available recently. Hence, brookite has been investigated in a comparably limited number of studies [59–62]. In general, the most stable form of TiO_2 is rutile. In contrast, the anatase and brookite phases are metastable and convert into rutile when treated at temperatures above ca. 600 °C–700 °C [63]. The anatase and brookite phases are typically found to outperform the rutile phase in terms of photocatalytic performance. In all three forms, the titanium (Ti^{4+}) atoms are coordinated with six oxygen (O^{2-}) atoms, forming TiO_6 octahedra.

TiO_2 is an n-type semiconductor (due to the presence of oxygen vacancies) and has bandgaps of 3.2 eV for anatase, 3.0 eV for rutile, and 3.2 eV for brookite. The VB of TiO_2 is composed of $2p$ orbitals of oxygen hybridized with the $3d$ orbitals of titanium, while the CB is composed only of the $3d$ orbitals of titanium. The CB energy level of the anatase phase lies at −0.51 V (at pH 7), and hence CB electrons have enough energy to reduce water to hydrogen as well as to reduce oxygen gas dissolved in water. Its VB lies at a potential of +2.69 V (at pH 7), which is far more positive than the oxidation potentials of most organic and inorganic compounds (electron donors). Thus, holes in TiO_2 are highly oxidative and can be transferred to surface-adsorbed water molecules or hydroxide groups (i.e. titanol groups) to form hydroxyl radicals. Therefore, as also discussed in previous sections, the band positions of anatase (and brookite) TiO_2 make this oxide semiconductor particularly suitable for water splitting, for generating reactive oxygen species (ROSs), and for the degradation of organic pollutants. Note that rutile, in contrast, is not able to evolve hydrogen, since its CB edge is more positive than (lies below) the water reduction potential (figure 6.8).

In general, despite its stability and ease of availability, the major limitation of TiO_2 is its poor light-harvesting ability. Depending on the crystallographic phase and therefore on the bandgap, the light absorption threshold for TiO_2 is at ~390–410 nm, which makes it active only under UV light illumination. Note that UV light constitutes only ca. 5% of the solar spectrum, and thus a large portion of the solar irradiation (i.e. the visible spectral range, ~400–700 nm) cannot be used to photoexcite TiO_2-based materials. The light absorption properties of TiO_2, and of metal oxides generally, can be modified by various methods including bandgap engineering or photosensitization, the latter of which is achieved by the attachment of molecular light absorbers or plasmonic nanoparticles (we cover these aspects in section 6.2). For TiO_2, bandgap engineering by nitrogen incorporation is a most successful approach [64], which allows the optical bandgap to be extended to ~500 nm. On the other hand, nanostructuring of TiO_2 and its surface modification with co-catalyst nanoparticles (also described in section 6.2) are commonly carried out to improve the efficiency of charge carrier separation and collection.

6.3.2.2 – WO_3

Tungsten trioxide, WO_3, is a widely investigated d^0 metal oxide that features functional properties in many applications including photocatalysis, photoelectrochemistry, and phototherapy due to its light absorption properties (12% of the solar spectrum, with E_g ~ 2.5–2.8 eV) and better electron transport (12 $cm^2\ V^{-1}\ s^{-1}$) than TiO_2 (0.3 $cm^2\ V^{-1}\ s^{-1}$) [65].

Various stable oxygen-deficient tungsten oxides ($WO_{x<3}$) exist that can feature intense light absorption up to the near-infrared (NIR) region at >700 nm. WO_3 crystals are generally formed by corner- and edge-sharing of WO_6 octahedra. Several crystalline modifications exist, such as monoclinic I (γ-WO_3), monoclinic II (ε-O_3), triclinic (δ-WO_3), orthorhombic (β-WO_3), tetragonal (α-WO_3), hexagonal (h-WO_3), and cubic (c-WO_3) [66]. Since the triclinic, monoclinic, orthorhombic, and tetragonal WO_3 phases basically have the same chessboard-like arrangement of WO_6 octahedra, they can reversibly transform into each other by adequate annealing and cooling steps [67]. They differ only in the extent by which the W atoms are displaced from the center of each WO_6 octahedron.

At room temperature, γ-WO_3 is the most stable form; its E_g of 2.6 eV corresponds to the difference between the energy levels of the VB (which consists of filled O 2*p* orbitals) and the CB (formed by empty W 5*d* orbitals). For nanostructured WO_3 with a reduced grain (crystallite) size, the bandgap generally increases due to quantum size effects.

WO_3 is a n-type semiconductor with a conduction band edge that is slightly more positive than the H_2/H_2O reduction potential and a valence band edge that lies at a much more positive level than the H_2O/O_2 oxidation potential. Thus, while WO_3 can efficiently photooxidize a wide range of organic compounds, dyes, and bacterial pollutants, it is not able to evolve hydrogen. Moreover, WO_3 has remarkable stability in acidic and neutral environments, making it a promising candidate for treating water contaminated by organic acids. It is, however, unstable under alkaline conditions due to dissolution.

Pure WO_3-based materials are usually inefficient photocatalysts because their low charge mobility results in a high electron–hole recombination rate, which significantly reduces the efficiency of the photocatalytic reaction. Thus, multiple approaches for improving the photocatalytic performance of WO_3 have been explored, among which, doping and surface co-catalyst modification can help to improve the charge carrier mobility and the collection efficiency, respectively.

6.3.2.3 – Fe_2O_3

Due to the multiple oxidation states of iron, its oxides can have different stoichiometry and crystalline structure, including wüstite (FeO), hematite (α-Fe_2O_3), maghemite (ν-Fe_2O_3), and magnetite (Fe_3O_4). Iron oxides are n-type semiconductors. α-Fe_2O_3 is the thermodynamically most stable polymorph among iron oxides. α-Fe_2O_3 has a hexagonal crystal structure in which each iron atom is surrounded by six oxygen atoms. In hematite, the iron and oxygen atoms are organized in a corundum structure, i.e. a trigonal–hexagonal scalenohedral structure with six formula units per cell. Due to its relatively narrow bandgap (E_g ~ 2.0–2.2 eV), α-Fe_2O_3 has been widely studied not only for solar photocatalysis [68] but also as photoanode for PEC water splitting [69–71]. Hematite, in fact, absorbs light up to 600 nm, and hence it can harvest up to 40% of solar radiation. Moreover, it is stable in most aqueous solutions (pH > 3) and is one of the cheapest semiconductor materials available. The main application of hematite in photocatalysis is for the degradation of organic pollutants. Note, in fact, that its CB edge is more positive than (lies below) the water reduction potential (figure 6.8).

As it features high chemical stability, α-Fe_2O_3 has also been studied as an anode material in lithium-ion batteries, for water treatment applications, in chemical sensors, and for biomedical applications (due to its biocompatibility).

6.3.2.4 – Multinary oxides

In addition to the simple binary oxides such as those described above, various multinary metal oxide semiconductors have also been widely studied in view of their photocatalytic properties. In general, multinary oxides, which are composed of two or more metals, have attracted a great deal of research attention. Among ternary oxides, the family of perovskite-type oxides has the general formula ABO_3, where A is a rare or alkaline earth metal and B is a first-row transition metal. In perovskite-type oxides, cations with a large ionic radius coordinate with 12 oxygen atoms and occupy A sites, while cations with a smaller ionic radius coordinate with six oxygen atoms and occupy B sites [72]. The absence of lattice distortion in perovskite crystals, i.e. their perfect octahedral structure, results in a cubic lattice. However, ABO_3 perovskites can exhibit lattice distortion to different extents, in which the octahedra are tilted around their centers. An increasing rotation of the octahedral structure results in a gradual transformation of the crystal phase in the following sequence: orthogonal, rhombohedral, tetragonal, monoclinic, and triclinic phases. Distortion of the crystal structure, such as that caused by the introduction of extraneous atoms, makes it possible to tune the bandgap of such perovskite-type oxides, attain visible-light absorption, and manipulate the band edge levels to enable specific photocatalytic reactions. Moreover, lattice distortion in perovskite oxides strongly influences the separation of charge carriers. Perovskite materials such as titanate, tantalate, vanadium- and niobium-based perovskites, and ferrite perovskites can exhibit visible-light activity [73].

Most perovskite photocatalysts are prepared by conventional solid-state reactions, in which the produced particles have a typical size in the range of several hundred nm or some μm. To some extent, this limits their performance due to suboptimum light harvesting and charge collection. Alternative routes for the synthesis of perovskite oxides have also been developed, such as the sol–gel method, combustion synthesis, thermal decomposition of bimetallic compounds, sonochemistry, microemulsion methods, electrospinning, co-precipitation, and microwave-assisted methods. Hydrothermal approaches are most popular for the synthesis of anisotropic nanocrystals [73].

Among perovskites of titanium, i.e. titanates, strontium titanate, $SrTiO_3$, is a most popular and widely explored photocatalyst. It is a cubic perovskite n-type semiconductor with an indirect bandgap of ~3.3 eV. It has high (photo)chemical stability, relatively good availability, and a suitable band edge position to drive overall water splitting (figure 6.8) [74]. $SrTiO_3$ co-doped with Rh and Sb can exhibit a bandgap as narrow as 2.5 eV. When loaded with IrO_2 (as an HER/OER co-catalyst), a suspension of Rh- and Sb-co-doped $SrTiO_3$ particles ($SrTiO_3$:Rh,Sb) can stoichiometrically evolve H_2 and O_2 under visible light illumination ($\lambda < 500$ nm) [75]. $SrTiO_3$ particle photocatalysts used in combination with other co-catalysts, i.e. RuO_2 or Ru, or even more abundant co-catalytic materials such as Ni and Ni oxides, can also enable H_2 and O_2 evolution, though with lower efficiencies. Other titanates such as $BaTiO_3$ and $CaTiO_3$ (E_g ~ 3.3 and 3.5 eV, respectively) have also been well studied as photocatalyst and photoelectrode materials, and when used as photocatalysts can drive overall water splitting upon surface decoration with suitable co-catalysts.

Tantalates, niobates, and ferrites are also interesting perovskite compounds for photocatalytic applications. A thorough description of perovskite photocatalysts is outside the scope of this book chapter. We refer interested readers to the following references [37, 38, 55].

6.3.3 Beyond metal oxides

Finally, it is worth mentioning that in addition to metal oxides, a number of alternative semiconductor materials have also been studied for photocatalytic applications, particularly in recent years. Examples are carbon-based compounds (e.g. two-dimensional materials such as graphene or graphitic carbon nitride, g-C_3N_4), metal–organic frameworks (MOFs), covalent organic frameworks (COFs), transition-metal nitrides (e.g. Ta_3N_5), oxynitrides (e.g. TaON), and metal oxyhalides (e.g. BiOCl), among others.

6.4 Nanostructuring

A powerful approach used to develop advanced photocatalytic systems is to exploit some specific effects at the nanoscale to overcome the main limitations of bulk photocatalysts. This aims to overcome the poor photocatalytic efficiencies associated with limited light absorption and the short lifetime and low mobility of electron–hole pairs. Nanostructuring strategies have attracted significant research interest in the last 30 years. Looking at nature, the earliest examples of nanostructured solar energy conversion systems date back to the beginning of life on this planet: these are the photosynthetic systems of bacteria. Artificial nanostructures for photocatalysis and solar-energy-to-fuel-conversion have emerged only in the last four decades, and research in this field is still in progress. The effects of the nanostructuring of semiconductors on their light absorption, charge carrier collection, surface area, and quantum confinement are discussed in the following sections.

6.4.1 Charge carrier collection

To drive redox reactions, photogenerated charge carriers need to reach the surface of the photocatalyst, i.e. the interface between the semiconductor and the contacting solution (the solid/liquid interface). For optimum collection of both carrier types (holes and electrons) at the solid/liquid interface, the key is to shorten the carrier collection pathways by nanostructuring the photocatalyst. In simple words, the size of a photocatalyst's features can be reduced by nanostructuring until it reaches dimensions in the same range as the electron and hole diffusion lengths, L_e and L_h, respectively [76, 77]. In the case of supported photocatalyst layers, this can be achieved by increasing the surface roughness or by surface nanostructuring (figure 6.10(a)(*ii*)). For suspended photocatalyst nanoparticles, ideal electron–hole collection is possible if their particle size is smaller than the charge carrier diffusion length, i.e. when $d < L_e$, L_h (figure 6.10(a)(*iii*)). This was experimentally verified with nano bulk particles of $KCa_2Nb_3O_{10}$ for hydrogen evolution and by nanoscaling Fe_2O_3 into colloidal dispersions for oxygen evolution [78]. This approach is particularly useful for first-row transition-metal oxides, which suffer from low hole mobility and short hole lifetimes.

6.4.2 Light absorption

As mentioned earlier in this chapter (section 6.2.1), the ability of a material to absorb light is determined by the Lambert–Beer law and the wavelength-dependent absorption coefficient α. When light penetrates a semiconductor, the penetration depth α^{-1} defines the distance beyond which the photon intensity is reduced to $1/e$ of the incident irradiation power [76, 77]. This principle is shown in figure 6.9(b). If we consider an ideal model, such as a bulk semiconductor layer of thickness d, to ensure the absorption of more than 90% of the incident light, the film thickness d must be >2.3 times the value of α^{-1}. The typical dimensions of nanostructured photocatalysts are usually smaller than α^{-1}, so each nanoparticle only absorbs a small fraction of the incident light. However, full absorption of the incoming light by a photocatalyst suspension can be achieved by optimizing its concentration in the suspension and by designing the optical path length of the photocatalytic reactor along with a proper illumination geometry. A photocatalyst particle suspension can in fact increase the light distribution in three dimensions. This occurs due to refraction at the surfaces of the particles and by Mie scattering. As a result, the light in a suspension impinges on the particles from all directions, which can maximize the overall absorption (cf figure 6.9(b)(*iii*) vs. (*i*)).

Nanostructuring is also desirable when supported photocatalyst layers are used (figure 6.9(b)(*ii*)), since nanostructured semiconductor surfaces reduce losses due to light reflection and increase internal light scattering. Moreover, nanostructured

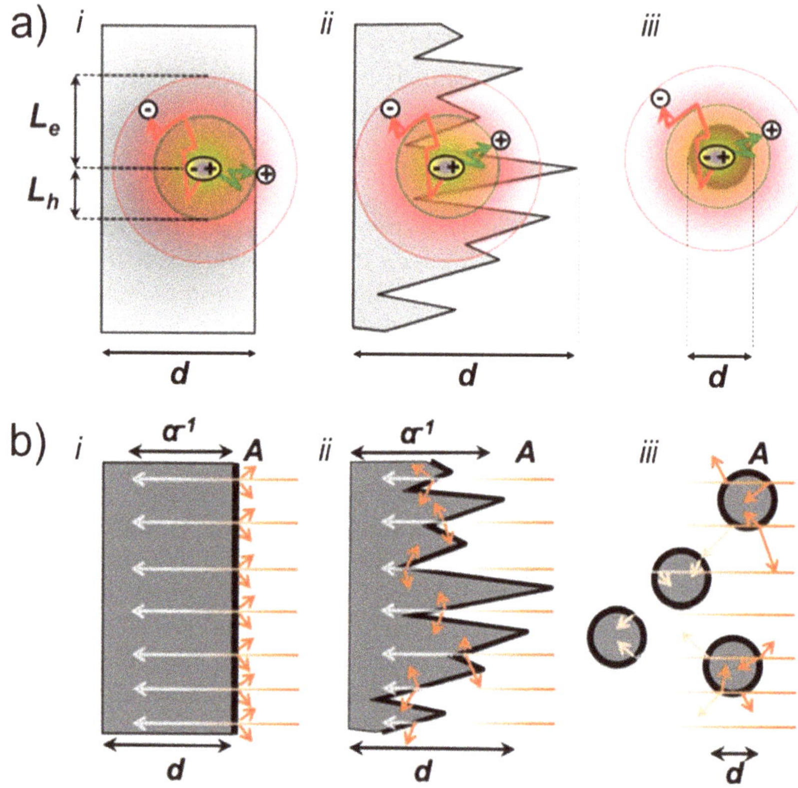

Figure 6.9. (a) Charge collection in (*i*) a flat film, (*ii*) a nanostructured film, and (*iii*) a particle suspension. d = the film or particle thickness; L_e = the electron diffusion length; L_h = the hole diffusion length. (b) Light distribution in (*i*) a flat film, (*ii*) a nanostructured film, and (*iii*) a particle suspension. A = surface area; d = film or particle thickness; α^{-1} = optical penetration depth. Short arrows denote scattered or reflected light [78] (2015) (© 2015 Springer International Publishing Switzerland). With permission of Springer.

photocatalyst surfaces increase the horizontal light distribution, which can maximize photon harvesting.

6.4.3 Surface area

Nanostructuring typically leads to an increase in the surface area of materials. For metal oxides, this effect is considered beneficial in enhancing the photocatalytic activity (in general, this holds true in heterogeneous catalysis). Strategies for increasing the surface area of semiconductor metal oxides are based on ball milling of bulk materials, direct synthesis of nanoparticles (e.g. via wet chemical methods), and nano-/mesostructuring of bulk materials to produce porous catalysts, typically using soft or hard templates. We describe a selection of nanostructuring techniques for photocatalysts in section 6.5.

The engineering of mesopores in photocatalysts to facilitate access to a large interior surface is gaining more and more popularity. However, the main question is whether such surface is available and accessible to charge carriers for collection at the SLJ as well as to reactants in the contacting solution for diffusion toward active sites. Hence, models are required to interpret the increase in the activity of nanostructured porous semiconductors in relation to the increase in their surface area. In a recent study, Cherevan and coworkers [79] proposed a model based on the extraction of two unit-free terms: the *relative activity increase* and the *relative surface area increase*. If we compare a *high-surface-area* semiconductor to a *low-surface-area reference material* (e.g. a bulk, non-porous material), these parameters describe, respectively, the factor of the activity increase (in terms of the increment in the gas evolution rate, photonic efficiency, incident photon-to-current efficiency (IPCE) performance, or photocurrent values), and the factor of the specific surface area increase. The specific surface area is typically measured by inert gas physisorption measurements (e.g. using N_2) and calculated by the Brunauer–Emmet–Teller (BET) method using specific isotherms. A key assumption is that an increase in surface area should lead to a linear increase in the number of catalytically active surface sites, in the absence of mass transfer limitations, and assuming a homogeneous density of surface active sites. Therefore, the photocatalytic activity should increase proportionally. Contrarily to this assumption, the majority of the analyzed data from a few hundred scientific articles indicated that the expected linear correlation was not seen. Instead, the data set was widely scattered, showing deviations as large as >50% from the expected correlation. This analysis suggests that, in addition to the measured surface area, other factors must be considered, such as limited mass transport of reactants through the semiconductor structure when small pores are present. Moreover, surface recombination centers (the density of which can increase with increasing surface area) may become a dominant factor that limits the concentration of available charge carriers. This can indeed hamper carrier collection and the kinetics of surface redox processes. In other words, if surface trapping dominates, only small performance enhancements or even a deterioration in activity can be observed as a consequence of the increase in surface area. The efficiency losses ascribed to surface recombination in nanostructured semiconductors can, however, be limited, e.g. by reducing the density of surface defects and traps through surface passivation. Other aspects to consider are the semiconductor crystallinity and solid-state charge transport properties. Charge carrier mobility in the semiconductor may change upon an increase in surface area, because the methods adopted for nanostructuring often also induce changes in the physicochemical properties of the semiconductor (e.g. the formation of surface or bulk defects, doping, surface adsorption, etc.), thus introducing another contribution to the nonlinear correlation between activity and surface area increase. Moreover, pore geometry (ordered vs. disordered, broad vs. narrow size distribution) and pore wall thickness are rarely evaluated, while these can be crucial structural features affecting the photo (electro)catalytic performance. The *optimum* pore wall thickness in view of efficient charge transport to the solid/liquid interface may vary for differently structured materials along with the diffusion length of charge carriers.

6.4.4 Quantum confinement

Downscaling size and morphology of semiconductor features can enable a range of optical and electronic *quantum* properties. An example is optical gap widening due to quantum confinement. When quantum sizes are reached, there is a gradual transition from semiconductor to molecular properties. Hence, nanomaterials represent a bridge between the bulk and molecular domains, and can provide both molecular and bulk features simultaneously [80]. The size threshold below which a material becomes quantum confined can be quite different from material to material and is determined by the quantum-mechanical nature of the electrons and holes in the given materials. Optical bandgap confinement can be described for zero-dimensional (0D) semiconductors, e.g. *quantum dots*, using the classic Brus model [81], as shown in equation (6.6):

$$\Delta E_g = E_g^q - E_g^{\text{bulk}} = \frac{h^2}{8m_0R^2}\left(\frac{1}{m_e^*} + \frac{1}{m_h^*}\right) - \frac{1.8e^2}{4\pi\varepsilon_0\varepsilon_r R}, \tag{6.6}$$

where ΔE_g is the bandgap variation associated with the size effect, E_g^q is the (wider) bandgap of the quantum-sized semiconductor, E_g^{bulk} is the (narrower) bandgap of the bulk semiconductor, h is Planck's constant, R is the radius of the semiconductor nanoparticles, e is the charge of the electron, m_0 is the free electron mass, m_e^* is the effective mass of the electron, m_h^* is the effective mass of the hole, ε_0 is the vacuum relative permittivity, and ε_r is the relative permittivity of the semiconductor (which is material specific). The limited availability of data for the effective masses of the electron m_e^* and the hole m_h^* as well as for the semiconductor relative permittivity ε_r allow only for rough predictions of the expected size for the onset of quantum confinement effects. In general, quantization effects become apparent when electrons and holes are confined to a dimension that approaches a critical size called the exciton Bohr radius in the bulk material. In other terms, quantum effects are enabled when charge carriers are spatially confined to a volume that is less than their De Broglie wavelength. From an experimental point of view, although this critical size depends on the particular semiconductor, it typically lies in the 1–10 nm range [82].

Bandgap tunability and the resulting shifts in absorption/emission energies have been extensively investigated for quantum dots [83, 84]. The dependence of the semiconductor bandgap on the quantum dot size follows the $1/d^n$ law, where d is the quantum dot diameter and n is an exponential factor ($1 \leqslant n \leqslant 2$) [85]. The increase in E_g observed for quantum-confined nanomaterials translates into a blueshift in their optical absorption edge with decreasing particle size. Another effect of size quantization is the appearance of fine structure in the absorption spectrum, which is associated with the increasing possibility of discrete transitions with decreasing particle size, i.e. due to a transition from semiconductor to molecular properties. However, for most semiconductor materials, this effect is typically attenuated by a broad size distribution or surface-state effects. In addition, with increasing E_g, the CB edge shifts to more reducing (more negative) potentials and the VB to more oxidizing (more positive) potentials. This is a consequence of the blueshift (a shift to shorter wavelengths) in the absorption edge with decreasing size. One could expect that these effects would increase the thermodynamic driving force and consequently the rate of interfacial charge transfer. In other words, quantum semiconductors should be more photoactive than their bulk counterparts to an extent that exceeds the plain contribution of the larger surface area. However, studies of semiconductor quantum dots made of TiO_2 [86, 87], ZnO [88, 89], WO_3 [90, 91], CdS [92], CdSe [93, 94] for photocatalytic or PEC reactions have reported controversial results. It has been suggested that the positive effects of quantization (more reducing or oxidizing charge carriers) can be undermined by the larger bandgap of quantum-sized semiconductors, which translates into the absorption of a smaller portion of the solar spectrum. Another interesting aspect is that an indirect semiconductor can, when it reaches a quantum size, become a direct semiconductor and hence feature stronger light absorption [95]. Furthermore, quantization of the density of states and bandgap widening may enable selective photocatalytic processes. In other words, adjusting the size of the semiconductor makes it possible to tune the exit energy of electrons and holes with respect to the electrochemical potentials of specific reactants [80].

6.5 Metal oxide nanostructuring techniques

The nanostructuring of metal oxides is important in achieving high photocatalytic activity. In this section, we briefly describe examples of different synthesis techniques typically used to produce nanostructured photocatalytic oxide materials. Note that there are many types of synthesis, and only a selection is described here. We make a distinction between the synthesis of powders and of films. The synthesis of powders is relatively easy, and slurry systems for photocatalysis are simple to realize. The downside of such particulate systems is that the separation of a photocatalyst powder from a liquid phase is generally more complicated. Thin photocatalyst films, which are well attached to a substrate, can instead be easily recovered from liquids. However, with an increase in the layer thickness, the photocatalyst may not be fully illuminated, which means that the incident radiation may not be fully absorbed. Whether this phenomenon happens depends, apart from the photocatalyst thickness, on parameters such as the semiconductor light absorption coefficient, the illumination wavelength, and the metal oxide morphology (e.g. compactness, mesoporous properties, 1D, etc). Moreover, reactants in the liquid (or gas) phase may experience limited access to active sites deep in the photocatalyst layer, which means that not all the available surface area can be used. As discussed earlier (see section 6.4), nanostructures can overcome these issues to some extent by enhancing light management and mass transfer. If the substrate is non-transparent, illumination can only be provided from the semiconductor side. To allow for back-side illumination (in which light first passes through the substrate and then reaches the photocatalyst layer), an alternative is the use of optically transparent substrates.

These can be made of glass, quartz, or polymers, while in photoelectrochemical (PEC) setups, fluorine-doped tin oxide (FTO) or indium-doped tin oxide (ITO).

Another issue to consider is the separation of reaction products. This is a particular problem in the case of photocatalytic water splitting from photocatalyst slurries, in which formed H_2 and O_2 might react backward on the photocatalyst surface to form H_2O again. This one-pot approach is simple in terms of design and construction, but the evolution—in the same solution and in the same headspace—of products from both anodic (hole-mediated) and cathodic (electron-mediated) reactions makes it necessary to employ separation methods, i.e. the separation of downstream products. Films, in contrast to powders, may allow for the separation of such products when used in photocatalysis if carefully structured. In such a case, the photocatalytic metal oxide film needs to be supported and the cathodic and anodic sites separated. An example of such a configuration is photocatalytic water splitting using noble metal nanostructured films (e.g. Pt) as a cathode interfaced with a suitable photocatalyst (e.g. TiO_2) as an anode. The metal–photocatalyst junction can be established through a metal support (e.g. a thin Ti plate) [96]. This configuration, although operating with no external electric bias, implies that the photocatalytic cell features two compartments, and thus products can evolve separately. Although it is outside the scope of this chapter, films can also be used for photoelectrochemistry. For more information on PEC water splitting, the reader is referred to some books published on this topic [3, 97].

Different strategies can be employed to grow crystalline metal oxides [98, 99]. Crystallization can take place through the nucleation of particles and subsequent growth from a precursor solution (e.g. in sol–gel synthesis or in hydrothermal synthesis). Such an approach gives opportunities for the growth of well-defined nanostructures with proper control of the reaction environment. After the formation of nuclei, crystal growth takes place in such a way that the total surface energy of the crystal decreases. Thus, there is a preferred growth direction. This preferred orientation is dependent on the interaction of the crystal surface with the chemical environment. When the solvent is changed, or when impurities and/or additives are added to the solvent in which crystal growth takes place, the relative surface energy and the corresponding facet stability change. As a result, the preferred growth orientation also changes. If the reaction conditions are carefully tweaked, crystal facet engineering of well-defined particles can take place. A schematic of this approach is included in figure 6.10. For instance, through the addition of fluorine to induce a fluorine-terminated surface during hydrothermal synthesis (see the corresponding section for an explanation of this technique), anatase TiO_2 particles with a relatively high percentage of {001} facets can be grown [100]. Alternatively, $BiVO_4$ with dominant (040) facet exposure can be hydrothermally synthesized when $TiCl_3$ is used as a directing agent [101]. Another good example of how additives can have an influence on the final crystal shape has been demonstrated by Xie *et al* [102]. Here, hydrothermal treatment of a WB precursor in an HF solution containing water and ethanol resulted in well-defined $H_{0.23}WO_3$ nanocubes with {002}, {020}, and {200} in similar exposure ratios, which could be converted to quasi-cubic-like WO_3 through calcination in air (see figure 6.10). When the WB precursor was treated

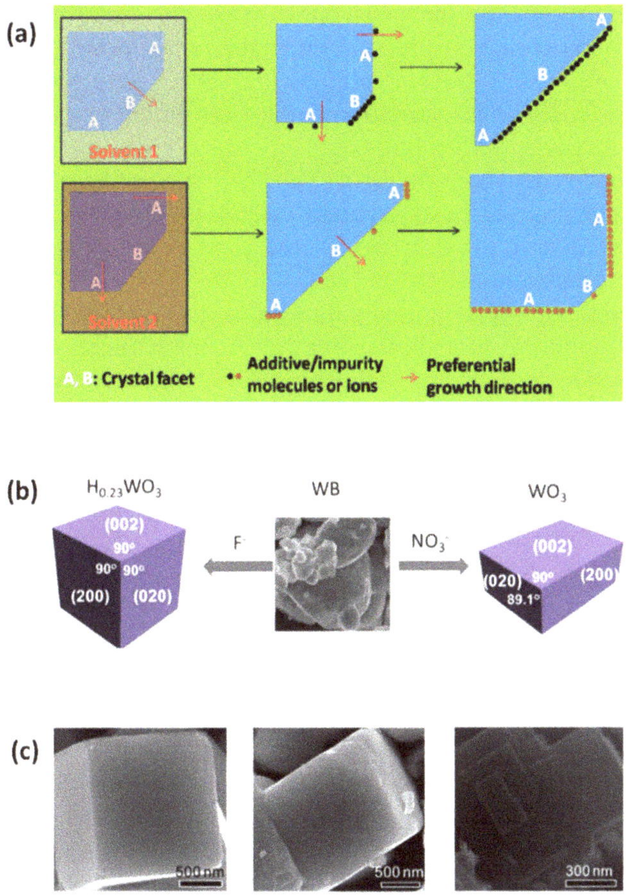

Figure 6.10. (a) The controlled growth of nanostructures through wet chemistry techniques. In the case of solvent 1, particles tend to grow in the direction perpendicular to facet B. However, when an additive/impurity is added, the favored growth site is blocked, forcing the particle to grow in the directions perpendicular to facets A. Changing the solvent (solvent 2) can also have an effect on the preferred growth orientation. Similar effects to those caused by solvent 1 can be seen when additives or impurities are added. Reproduced from [98] with permission of The Royal Society of Chemistry. (b) An example of crystal facet engineering: starting from a WB precursor, either well-defined cubic $H_{0.23}WO_3$ particles were grown using F^- as an additive, or well-defined rectangular WO_3 was formed when NO_3^- was used as an additive. (c) SEM images of $H_{0.23}WO_3$ nanocubes (left), quasi-cubic WO_3 nanocubes (middle) and WO_3 nanosheets (right). (b) and (c) were adapted from [102] with permission of The Royal Society of Chemistry.

in a HNO_3 solution, rectangular WO_3 nanosheets were formed with the dominant exposure of {002} facets. Interestingly, the authors reported a shift of the conduction band and valence band to higher (i.e. more negative) potentials for the WO_3 nanosheets compared to those of the quasi-cubic crystals, as well as a slightly wider bandgap. The quasi-cubic WO_3 nanocubes had a lower (i.e. more positive) VB and showed higher normalized activity for oxygen evolution from water oxidation. In contrast, due to the favorable positioning of the CB, the WO_3 nanosheets were able to photoreduce CO_2, while the quasi-cubic WO_3 particles were not capable of driving such a photoreduction process due to the unfavorable position of the CB. Proper control of crystal growth conditions may hence have consequences for the activity of metal oxides. Thus, careful optimization should be performed to obtain enhanced photocatalytic performances.

Other approaches used to engineer well-defined nanostructures can be based on altering the morphology of existing parent crystals [98, 99]. Specifically, through a chemical process, dissolution of (parts of) the parent crystals takes place, possibly followed by recrystallization.

Another interesting approach used to produce nanostructured metal oxide films is electrochemical anodization. It allows, for example, the synthesis of ordered arrays of oriented metal oxide nanotubes or nanopores directly on metal substrates, as in the case of TiO_2 nanotubes produced by anodization of the metal substrate in a fluoride-containing solution; we will touch upon this later in this chapter [103, 104]. The chemical environment plays a critical role in the growth of nanostructures (including possible faceting), and thus proper etching and capping agents should be chosen.

6.5.1 Wet chemistry methods

Methods in which metal oxides are grown in a liquid environment are called wet chemistry methods. Wet chemistry methods typically allow for the production of homogeneous samples, the possibility of synthesizing composite materials (e.g. multinary metal oxides), control over stoichiometry (crucial in the case of doping), and control over the formation of complex shapes [105].

6.5.1.1 Sol–gel methods

One of the easiest methods for the synthesis of nanocrystalline metal oxides is the sol–gel synthesis method. It is a wet chemistry technique that requires a solvent (either aqueous or organic) and a metal precursor (or multiple precursors) [67, 99, 105–107]. Typically, these metal precursors are either metal alkoxides or inorganic metal salts. In the case of metal oxides, oxygen (either as a gas or present within a molecule) should be present as well. First, a homogeneous mixture of these chemicals is prepared under magnetic stirring, which is followed by hydrolysis to convert the mixture into a sol. The addition of an acid or a base might be necessary to induce the hydrolysis. Subsequently, a gel is formed due to the condensation of adjacent molecules. This is the step in which M–O–M (metal–oxide–metal) bonds are formed, which form the skeleton of the metal oxide structure. The size and cross-linking of the colloidal particles formed can be changed by alternating the precursor or the pH of the solution. Initially, a porous structure that maintains the liquid phase is formed. Over time, aging of the gel takes place. During this stage, polycondensation along with reprecipitation of the gel network occurs. The decrease in porosity and increase in the size of the colloidal particles ultimately results in the separation (from the liquid phase) of a solid nanocrystalline metal oxide phase. Afterwards, drying of the solid precipitate takes place (either through drying under atmospheric conditions, thermal drying, supercritical drying, or freeze drying). As a last step, calcination of the dried gel takes place, which is necessary for crystallization and the removal of residues and water molecules from the metal oxide. Here, the calcination temperature has an important role in controlling the material pore size and density. The described procedure aims to form powdered nanostructures. However, the sol–gel method can also be easily utilized to grow metal oxide films. In such a case, the sol–gel can be drop-cast, spin-coated, or dip-coated on a substrate during the gelation process. Alternatively, the metal oxide grown can also be deposited in its final powder form on the substrate through similar methods.

The advantages of the sol–gel method include its simplicity (both in terms of the procedure and the lab accessories required), the homogeneity of the samples formed, the option of controlling the stoichiometry, and the easy introduction of dopants. The disadvantages can include the high price of some metal alkoxide precursors and the long-lasting and sometimes laborious synthesis procedure.

6.5.1.2 Hydrothermal synthesis and solvothermal synthesis

Another popular technique used to grow nanostructured oxide powders or films is the hydrothermal synthesis technique [67, 105, 107, 108]. In the case of the hydrothermal synthesis of a powder, either an aqueous solution containing a precursor or a slurry solution containing 'seeds' (e.g. formed through a co-precipitation method) is transferred to an autoclave, which is sealed and heated up to relatively mild temperatures (usually roughly in the range of 100 °C–250 °C). Due to the build-up of pressure within the autoclave, the solution remains in the liquid phase, and thus crystal growth can take place at temperatures and pressures above those of ambient conditions. Crystal phase transitions during hydrothermal crystallization can be studied using an *in situ* Raman probe. An example is the synthesis of WO_3 powder, which uses sodium tungstate dihydrate ($Na_2WO_4{\cdot}2H_2O$) as a precursor in an aqueous solution containing citric acid as a capping agent [109]. Upon the addition of sufficient HCl (i.e. lowering the pH to a value less than one), hydrolysis of the solution takes place and tungsten oxide dihydrate ($WO_3{\cdot}2H_2O$) is formed. Upon heating to more than 70 °C, a transformation to tungsten oxide monohydrate ($WO_3{\cdot}H_2O$) takes place, which at more than 150 °C is then converted to mainly monoclinic WO_3 (and to a minor extent to $WO_3{\cdot}0.33H_2O$). Over time, the $WO_3{\cdot}0.33H_2O$ is converted to monoclinic WO_3 during the isothermal stage.

Hydrothermal synthesis can also be used to grow metal oxides on substrates (e.g. ITO, FTO, or Si wafers) [108]. In this case, a seed layer is formed at the substrate (either prior to or during the hydrothermal synthesis procedure), followed by crystallization at this surface. For example, single-crystal TiO_2 rutile nanorod arrays can be grown on FTO by exposing the FTO to a solution containing titanium butoxide in a mixture of deionized water and concentrated hydrochloric acid under hydrothermal conditions [110]. Through alteration of the growth parameters (such as the reaction time, temperature, reactant concentration, and pH and the possible inclusion of additives), the material's properties such as the diameter, length, and the density of the nanorods can be controlled. Metals can also be used as substrates. In this case, the metal itself acts as a precursor for the metal oxide formation through its initial dissolution [108].

Solvothermal synthesis is similar to hydrothermal synthesis, with one major difference: the aqueous solution is replaced by organic media such as alcohols, toluene, ketones, or carboxylic acids [57, 105, 111, 112]. The advantages of the solvothermal method over the hydrothermal method are that the temperature and pressure can be elevated to higher values when a solvent with a boiling temperature higher than that of water is used. Furthermore, the solubility, reactivity, and diffusion behavior of the reactants are influenced by the physical and chemical properties of the solvents. Specifically, the morphology and crystallization behavior of the final products are influenced by the polarity and coordinating behavior of the solvent. Finally, the solvothermal method can yield narrow distributions of size and dispersity and minimal agglomeration.

6.5.1.3 Combustion

Metal oxide formation through combustion is based on the induction of a fast and self-sustaining redox reaction [105, 111, 113, 114]. In the case of solution combustion synthesis, a precursor solution is prepared containing a fuel and an oxidant with metal cations. The oxidant with metal cations can easily be realized using metal precursors such as metal nitrates. The fuel, for instance citric acid, urea, or glycine, can form complexes with the metal ions. After a solution containing a combustion mixture is formed, the solution is dehydrated to form a gel. The gel is then ignited by exposing it very briefly to a thermal or electric source, preferably as a single step in a narrow temperature range to prevent the formation of unwanted side-products. The advantages of using the combustion method include the formation of highly crystalline nanoparticles, and, due to the brief reaction time, the prevention of unwanted crystal phase transitions (such as a phase transition that yields rutile formation in the case of TiO_2). Furthermore, it is a simple, time-efficient, and relatively cheap method, and it allows for easy modification. For instance, the fuel acts as reducer, complexing agent, and microstructural template. By changing the fuel, the reaction's exothermicity can be changed, and thus the phase and properties of the powder can be controlled. Dopants can also be introduced. Due to the self-sustaining nature of the process, it can also be considered to be energy-efficient. A disadvantage of combustion synthesis could be that carbonaceous residue might still be present. If this occurs, it will need to be burnt off in an oven afterwards, e.g. by thermal treatment in an oxygen-rich environment.

6.5.1.4 Anodization

As introduced above, electrochemical anodization is an effective way to create defined 1D nanostructured oxides of various transition metals, among which, anodic alumina 'nanopores' [115] and titania 'nanotubes' [22, 103, 104, 116, 117] are the most explored oxides. In contrast to alumina (which is a dielectric), TiO_2 nanotubes as well as many other semiconductor oxide nanotube or nanopore arrays such as Fe_2O_3 [118–124], WO_3 [125], Ta_2O_5 [126], Nb_2O_5 [127], and SnO_2 [128], have found wide applications in photocatalysis (and photoelectrochemistry). Electrochemical anodization (i) is scalable, (ii) offers precise control over the oxide nanoscale geometry, and (iii) allows for such defined oxide nanostructures to be produced on virtually any shape of metal substrate (these oxides are therefore also back-contacted for use as photoelectrodes, if desired) [103, 104, 116]. Depending on the anodizing conditions and on the metal oxide being formed, such layers of 1D oxide nanostructures can be as thick as a few hundred micrometers, and the nanopores or nanotubes can have diameters in the range from a few nanometers to a few hundred nanometers.

Classically, electrochemical anodization is carried out in a two-electrode electrochemical arrangement in which a metal substrate (M) is immersed in a suitable aqueous electrolyte and subjected to a positive (anodic) electric bias (see figure 6.11). In this configuration, with the metal substrate M as the working electrode (positively polarized, anode) and a platinum mesh or graphite rod as the counter electrode (negatively polarized, cathode), oxidation of the metal substrate occurs, yielding a metal oxide coating ($MO_{x/2}$) that covers the metal substrate evenly. The oxide growth, ascribed to field-assisted ion migration, occurs as metal cations (M^{x+}) are subject to outward migration from the metal substrate toward

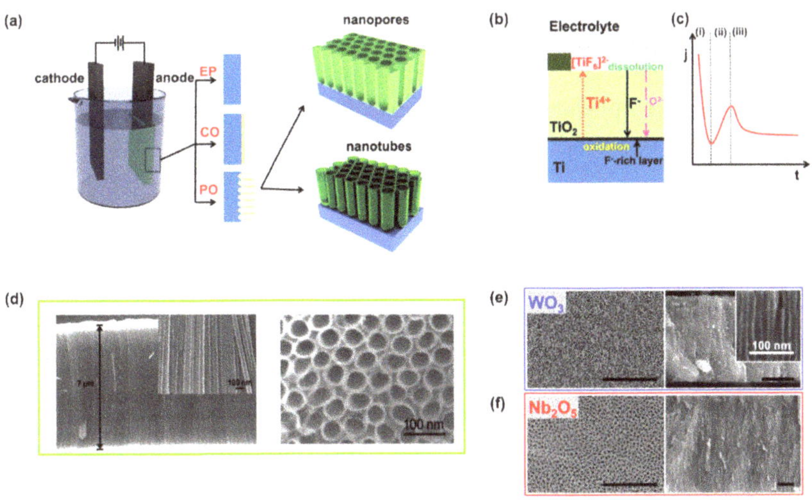

Figure 6.11. (a) Typical setup used for electrochemical anodization using two electrodes. Possible outcomes of the anodization, depending on the electrochemical conditions: electropolishing (EP, when oxide dissolution proceeds faster than oxide formation), the formation of a compact anodic oxide (CO), and the formation of an oxide layer composed of self-ordered nanostructures (nanopores or nanotubes) (PO). (b) TiO_2 formation during electrochemical anodization in the presence of F^- ions. At the inner interface, oxide formation is taking place, while at the outer interface, dissolution and complexation of Ti^{4+} as TiF_6^{2-} (soluble in the electrolyte) takes place. A fluoride-rich layer is formed between TiO_2 and Ti due to rapid fluoride migration. (c) Typical current density (j) vs. time (t) curve during nanotube formation by electrochemical anodization. In timeframe (i), a compact oxide layer is formed. In (ii), irregular nanoscale pores are formed. Finally, in (iii), a regular nanoporous/nanotubular layer is formed. (a)–(c) adapted from [116] with permission of The Royal Society of Chemistry. (d) Side (left) and top view (right) of smooth TiO_2 nanotubes prepared in a glycerol electrolyte with 0.5 wt% NH_4F [133] John Wiley & Sons. [Copyright © 2005 WILEY-VCH Verlag GmbH & Co. KGaA, Weinheim]. (e) WO_3 and (f) Nb_2O_5 nanochannels grown in molten pure o-H_3PO_4 on tungsten and niobium foil, respectively. (e) and (f) adapted with permission from [125]. Copyright (2015) American Chemical Society.

the electrolyte, while at the same time, O^{2-} ions migrate (from the electrolyte) toward the anode. Thus, a metal oxide layer forms at the interface between the metal substrate and the anodizing electrolyte. When the oxide formed is stable in the anodizing electrolyte, as in the case of anodically grown titania layers in aqueous H_2SO_4- or H_3PO_4-based electrolytes, the growth process is self-limiting: with increasing the anodic oxide thickness, the electric field rapidly drops and so does the inward/outward ion migration; consequently, the oxide growth ceases. This results in the formation of an anodic oxide layer with a finite thickness. Such an oxide film is compact or, more generally, does not exhibit specific morphological features.

Instead, adequate conditions for the formation of defined high aspect ratio (1D) oxide nanostructures can be established when, as a result of adding specific ions to the electrolyte, a steady-state equilibrium is achieved between the electrochemical oxide formation and its dissolution (in the electrolyte). Typically, the use of electrolytes containing F^- is key for the anodic growth in the form of arrays of nanopores or nanotubes of many of the semiconductor oxides mentioned above [103, 104, 116, 129–131]. A key advantage of anodization is that the oxide nanostructure and physicochemical properties can be adjusted by tuning simple electrochemical parameters (electrolyte composition and temperature, applied voltage, anodization time, etc). The effects of such parameters on the growth of anodic oxide nanostructures have been documented in recent decades [103, 104, 116, 117, 125, 127, 128, 132–137].

6.5.1.5 Other wet chemistry techniques

As stated, only some of the wet chemistry synthesis techniques are described here. However, there are many more types of synthesis. For instance, (co-)precipitation methods can also be used to synthesize metal oxides. Examples include the alkalinization of titanium chlorides in solution to form titanium hydroxide as a precipitate, which can be annealed to form TiO_2 [105], or the acidification of tungstate-based solutions, in which crystalline $WO_3{\cdot}nH_2O$ particles are produced which form monoclinic WO_3 upon calcination [67, 109]. Another technique which can be used is spray pyrolysis [67, 105, 138, 139]. Here, a precursor solution is brought into an aerosol phase, which is sprayed onto a heated substrate, forming a metal oxide layer through thermal decomposition. The morphology and size of the particles can be controlled by tweaking the composition and concentration of the precursor (from which the aerosol is formed), the gas flow, the distance between the substrate and the nozzle, and the substrate temperature. The advantages of spray pyrolysis (compared to other thin-film deposition methods) are that it is simple, financially attractive, and reproducible, it allows for the deposition of large areas, and the deposited films demonstrate good optical and electrical properties. Nevertheless, spray pyrolysis does not yield uniform and smooth films in general.

Other examples include microemulsion methods (in which, for example, water micelles in oil are utilized for nanoparticle synthesis) [105, 107, 139], sonochemistry (in which ultrasonic waves are involved in the synthesis to form, grow, and burst bubbles) [107, 139], and electrodeposition (in which, in contrast to anodization, electrochemistry is used to grow metal oxide particles from solution on a substrate) [105, 139].

6.5.2 Vacuum-based deposition techniques

In addition to wet chemistry methods, vacuum-based deposition techniques can also be used for the synthesis of photocatalytic oxide materials. A concrete advantage of these methods is that they allow for the direct deposition of thin films on substrates. A distinction can be made between physical and chemical deposition techniques [105, 107, 139–141]. In the case of physical vapor deposition (PVD), a solid target of a metal oxide is used as a precursor and vaporized. It is then deposited on a substrate. In PVD, no chemical transformation occurs, and thus it is essential that the precursor is stable in the gas phase. In practical terms, the material is transferred from the target to the substrate, typically in the form of polycrystalline metal oxide layers. It should be noted that large quantities of energy are required to vaporize a metal oxide. In contrast, chemical vapor deposition (CVD) makes use of chemical reactions and gas phase precursors, e.g., an organometallic compound. The precursor is typically volatile, and reacts in the gas phase at the surface of the substrate to form the desired semiconductor layer. Such a chemical transformation takes place at controlled temperatures and pressures through a reaction with, for example, oxygenated terminations (i.e. –OH) at the surface of the substrate.

A popular PVD technique for the growth of metal oxide layers is direct current (DC) or radio frequency (RF) sputtering [67, 105, 142]. Here, high-energy particles (such as argon ions in a plasma) are used in a vacuum chamber to bombard a metal (or metal oxide) target, from which clusters are ejected and deposited on a substrate. The plasma is ignited by applying a high voltage across two electrodes (one of which is the target itself). This causes ionization of the argon gas, which is present in the chamber at a low partial pressure (e.g. in the range of few mbar when striking the plasma). In the case of oxide sputtering, RF power supplies are often used (i.e. RF sputtering), due to the relatively poor conductivity of the oxide targets. On the other hand, when a metal target is used, radio frequencies are not strictly necessary but sputtering should be carried out under reactive conditions, i.e. in an oxygen–argon plasma: clusters of the metal ejected from the target react in the gas phase with oxygen and are hence deposited onto the substrate in the form of a metal oxide. ALD, a CVD approach, is also frequently used for the growth of metal oxide layers [143–145]. ALD uses pulses of chemical precursors in the gas phase that react with or are chemisorbed on a substrate. Between each pulse, the reactor is purged with an inert gas. In this way, a metal oxide film is grown layer by layer on a substrate. The advantage of ALD lies in the fact that very thin and homogeneous layers of metal oxide can be grown, as well as multilayers (suitable for heterojunctions, see section 6.2), which can also be grown on porous or high-aspect-ratio nanostructured substrates. However, a disadvantage is the slow synthesis procedure, which makes industrial scale-up challenging, and the use of toxic or environmentally unfriendly precursors.

6.5.3 Photolithography

The last synthesis technique which we address here is photolithography [107, 146–148]. Photolithography is used to design micro- or nano-patterned films. Typically, it is based on the concept of coating a thin film of photoresist on a layer of interest, which can be a semiconductor used for photocatalytic purposes. The photoresist can either be positive or negative. When a positive photoresist is exposed to illumination through a patterned mask, the parts of the photoresist that were illuminated become 'weaker' and are, in a following step, easily dissolved by a liquid called photoresist developer. The parts which were instead shadowed by the mask remain. In contrast, illuminating a negative photoresist results in the insolubility of such a resist in the developer. Hence, when a mask containing a pattern of opaque and transparent areas is placed on top of either type of photoresist, illumination results in a patterned structure in which part of the photoresist becomes soluble upon development, whereas the other part does not. After etching the soluble part away, a pattern of insoluble photoresist is left behind. This remaining photoresist can act as a protective mask for the semiconductor underneath. Thus, through the introduction of, for example, a chemical etching agent, only the unprotected semiconductor material is removed, whereas the protected semiconductor material remains untouched. After the photoresist pattern is stripped off, a substrate with a semiconductor pattern is revealed [107].

6.6 Photocatalyst modification

In addition to nanostructuring, an alternative, often complementary, approach to increasing the photocatalytic efficiency of a metal oxide is through chemical modification of the bulk composition. This can provide a means for bandgap engineering, i.e. altering the band positions, with the aim of absorbing a broader wavelength range in the case of bandgap narrowing. Moreover, dopants can be introduced to increase the conductivity and improve the charge transport in the metal oxide. The surface of the metal oxide can also be chemically modified to enhance its photocatalytic activity, for example through the deposition of co-catalysts (which lead to enhanced charge separation and more active surface sites for the desired chemical reactions). Although we focus on metal oxides, these strategies can also be applied to other types of semiconductor material.

6.6.1 Bulk modification

A serious limitation of many metal oxides for their use in efficient photocatalytic conversion is their relatively large bandgap (see figure 6.8). For instance, titanium dioxide (TiO_2) is a stable, non-toxic, and (relatively) cheap photocatalyst [149]. However, a low quantum efficiency due to large bandgaps in the range of 3.0–3.2 eV for rutile and anatase limits its efficiency for solar light harvesting [150, 151]. In practical terms, TiO_2's bandgap only allows for very limited absorption of the solar

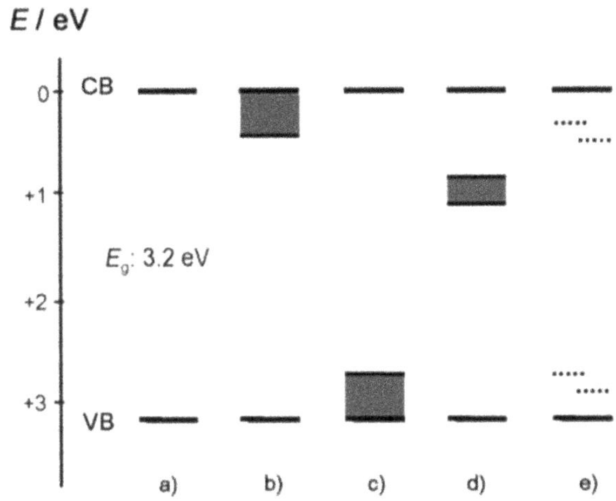

Figure 6.12. Possible ways to alter the band structure of TiO_2. (a) Pristine TiO_2 with a bandgap of $E_g = 3.2$ eV, (b) CB lowering, (c) VB elevation, (d) the introduction of a discrete band within the bandgap, and (e) the introduction of doping levels within the bandgap [150] John Wiley & Sons. [Copyright © 2010 WILEY-VCH Verlag GmbH & Co. KGaA, Weinheim].

spectrum (up to ca. 390 nm for an E_g of 3.2 eV, corresponding to ca. 5% of the solar spectrum) [152, 153]. Meanwhile, it is important to keep in mind that the edges of the CB and the VB should be at favorable positions for the reduction and oxidation reactions, respectively. In the case of TiO_2, the CB mainly consists of Ti 3*d* orbitals, whereas the VB is mainly composed of O 2*p* orbitals [3, 150, 154]. A strategy used to increase the efficiency of a photocatalyst is the incorporation of 'extraneous' atoms [150, 152, 155]. The introduction of such atoms can result in: (i) the introduction of new doping levels within the bandgap, (ii) the lowering of the CB (i.e. inducing a shift to a more positive potential) or (iii) elevating the VB (i.e. inducing a shift to a more negative potential). As a result, the absorption of lower-energy photons becomes possible, due to additional electronic transitions in scenario (i) and a redshift of the absorption edge in scenarios (ii) and (iii). The conductivity of the metal oxide can also be increased through doping. The different scenarios resulting from the incorporation of extraneous atoms are depicted in figure 6.12.

6.6.1.1 Doping

Typically, doping of a material can take place through (i) self-doping, or through the introduction of extraneous (ii) metal or (iii) non-metal atoms. In the case of self-doping, it is critical that reduction of the metal species occurs within the metal oxide (for instance, the reduction of Ti^{4+} to Ti^{3+} in TiO_2) [150, 154–157]. This can be realized through the introduction of oxygen vacancies in the crystal lattice. For the self-doping of TiO_2, it has been theorized that intragap bands form just below the CB minimum, resulting in a virtual decrease in the bandgap. The result is that the TiO_2 is able to absorb photons with wavelengths higher than 400 nm. The color change of the titania reflects this, and the type of color change depends on the reduction conditions. Known examples in the literature are 'blue' titania [158–161] and 'black' titania [61, 162–168]. The methods used to introduce oxygen vacancies include heat treating the metal oxide in vacuum or inert gas conditions (i.e. in the absence of oxidizing species) and exposing the metal oxide to reducing conditions in a gas environment (e.g. annealing in H_2 or Ar/H_2) or in aqueous electrolytes. It should be noted that the self-doping of TiO_2 does not always result in a response of the material to visible light. Indeed, when the electronic states introduced are at a lower position than, for example, the HER redox potential, electrons excited to those states do not have sufficient energy to actually reduce protons to form H_2. As doping levels can also act as recombination centers, an over excess can also result in a large number of photogenerated charge carriers that must be recombined before they can be utilized for photocatalytic reactions.

Doping can also occur due to the introduction of extraneous metal ions (i.e. cation doping) [150, 154, 155, 169–173]. The metal types used for doping include alkaline earth metals, rare-earth metals, transition metals, and post-transition metals. Phase transitions and structural distortions can be induced or prevented. Doping with transition metals is the most common approach, particularly with first-row transition metals and d^8 and d^9 transition metals. Due to the unique electronic $3d$ and spectral configuration of the transition metals, doping results in intragap electronic states or a shift in the CB minimum or the VB maximum [155, 171]. Therefore, the absorption edge is redshifted, allowing for the absorption of photons with lower energy within the solar spectrum. In addition, the electrical performance of the metal oxide can be enhanced due to better charge transport and electron mobilities, which are the results of the presence of defects and changes in diffusion behavior. In some cases, codoping or even doping with multiple metals, or with a mixture of metals and non-metals, can be more effective than monodoping in enhancing visible-light absorption [155].

Despite these advantages of metal doping, there are also a couple of disadvantages to take into account. For instance, metal-doped TiO_2 is thermally instable, and doping results in the formation of charge carrier recombination centers [64, 154, 174]. Some transition-metal dopants are also relatively scarce. A vast body of literature describes the doping of metal oxides with non-metals as an interesting alternative. Typically, non-metal doping (i.e. anion doping) in metal oxides can use nitrogen, carbon, fluorine, chlorine, iodine, phosphorous, sulfur or selenium [151, 152, 155, 173, 175, 176]. Here, a distinction can be made between substitutional and interstitial doping. In the case of substitutional doping, oxygen atoms within the crystal lattice are typically replaced by non-metal atoms, resulting in depopulation of the VB by one to three electrons, depending on the electronic configuration of the dopant [176]. Here, it is possible that not only one, but even two anions can replace one O^{2-} anion [173]. When interstitial doping occurs, a non-metal is additionally incorporated into the empty spaces within the crystal lattice [152, 173, 175]. One of the proposed theories states that doping with non-metals typically creates mixed or localized states near the valence band edges due to the $2p$ orbital contributions of those non-metals [152, 155]. As a result, the visual bandgap is narrowed and thus visible-light activity (caused by the absorption of lower-energy photons) is increased. A concrete, attractive example of non-metal doping is nitrogen doping of a metal oxide such as TiO_2 [64, 173–175, 177–179]. Although the research literature disagrees about how N-doping affects the electronic structure of TiO_2, it is altered in such a way that the maximum wavelength that can be absorbed is redshifted and thus a larger part of the solar spectrum can be utilized. It should be noted that reports of doping are not always precise in stating whether the redshift in the absorption edge of metal oxides is actually due to the introduction of intragap localized states or to bandgap narrowing [150, 171, 180]. However, regardless of which mechanism causes the redshift, doping is a promising tool for increasing absorption of the solar spectrum.

In practical terms, doping can be introduced into a crystal lattice through different synthesis techniques. These techniques include, but are not limited to, hydrothermal synthesis, sol–gel synthesis, co-precipitation, co-sputtering, thermal treatment (or sputtering) in an atmosphere containing the doping species, solvothermal synthesis, solid-state reactions, electrochemical anodization, etc [150, 155]. In principle, doping can be achieved in material synthesis when a precursor of a parent metal or non-metal is introduced into the synthesis procedure. The amount of doping can be controlled by adjusting the ratio between the precursors used for the synthesis of the metal oxide and for the introduction of the dopant. An overview of the different synthesis methods, as well as other aspects described in this paragraph, is given in figure 6.13.

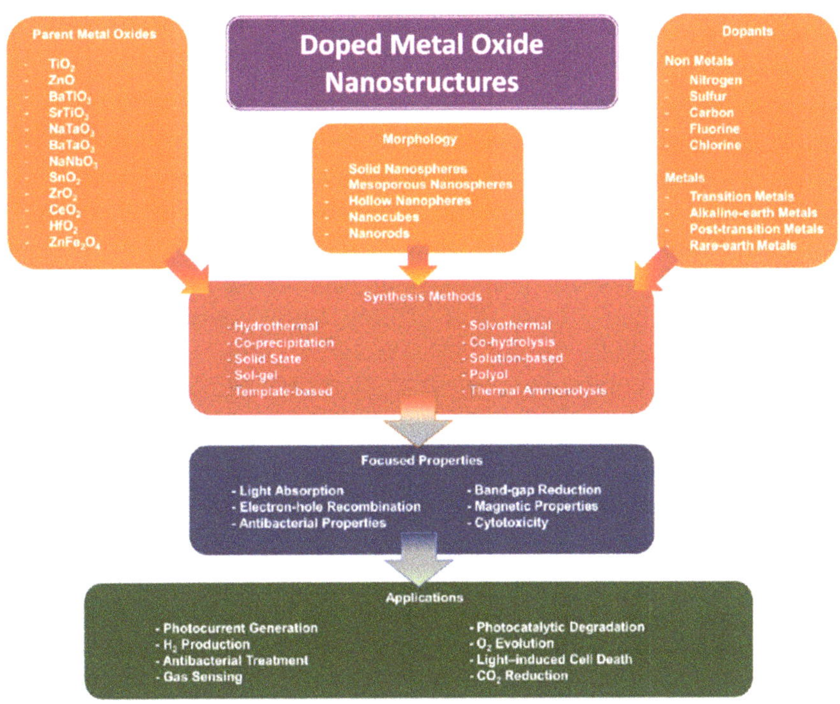

Figure 6.13. Overview of the types of doping used in different metal oxides, with their synthesis procedures, induced properties, and applications. Adapted with permission from [155]. Copyright (2020) American Chemical Society.

Finally, we have discussed doping as a means of changing the bulk properties of a material. However, it should be noted that doping can also result in modifications to the surface, which may enhance the photocatalytic activity of the material [155]. In a theoretical study, García-Mota *et al* compared the surface reactivities of undoped vs. doped TiO_2 surfaces (specifically, those doped with transition metals) using (Hubbard-U) density functional theory (DFT) calculations [181]. They demonstrated that the oxygen adsorption bond is stronger on the doped TiO_2 surface due to an increase in localized surface resonances and surface states near the Fermi level. For non-metal doping, the catalytic performance in, for example, the oxidative coupling of methane or the oxidative hydrogenation of ethane is improved due to an increase in surface basicity [173].

6.6.1.2 Bandgap engineering in oxide solid solution photocatalysts

In addition to doping, the formation of solid solutions of binary and/or ternary metal oxides, i.e. metal oxides with ABO_2, ABO_3, ABO_4, or AB_2O_4 compositions (where A and B are metals), offers interesting opportunities for bandgap engineering [182]. An important difference between doping and solid solutions is the concentration of the extraneous element incorporated in the crystal lattice of the 'original' metal oxide: in the case of doping, this amount is substantially smaller than the specific stoichiometric amounts typical of solid solutions. In the latter case, a wide bandgap semiconductor can be mixed with a narrow bandgap semiconductor, resulting in continuous modulation of the conduction band, the valence band, or both. In addition, hybridization of the CB and the VB results in faster charge transfer. A well-known example of such a solid solution is bismuth vanadate ($BiVO_4$) [182–184]. This material is not only stable and cheap, due to the abundancy of its constituent elements, but its narrow bandgap of 2.4 eV also allows for visible-light absorption. $BiVO_4$ is a promising photocatalyst for water oxidation due to its suitable VB position. Nevertheless, some drawbacks of $BiVO_4$ are the poor transport of charge carriers (especially electrons), the slow kinetics of O_2 evolution, and the unsuitable position of the CB for proton reduction. The latter means that overall photocatalytic water splitting cannot be achieved by $BiVO_4$ without the addition of a second semiconductor in a Z-scheme or heterojunction configuration (see section 6.2), or without an external bias (thus using $BiVO_4$ as a photoanode).

Another example of a solid solution suited for photocatalysis is that of β-$AgAl_{1-x}Ga_xO_2$ in solution (where $0 \leqslant x \leqslant 1$), in which the wide-gap β-$AgAlO_2$ is mixed with the narrow-gap β-$AgGaO_2$ through a sol–gel method [185]. With increasing x, the conduction band minimum of the β-$AgAl_{1-x}Ga_xO_2$ is shifted to more positive potentials, i.e. closer to the valence band edge maximum (see figure 6.14). Because the electronic position of the valance band edge maximum remains similar, the bandgap narrows. By tweaking the ratio of Al to Ga, it is possible to significantly increase the production of acetone from the photocatalytic degradation of isopropanol (IPA), and optimum activity is obtained through the use of the β-$AgAl_{0.6}Ga_{0.4}O_2$ composition. Here, the IPA is oxidized to acetone either through $^{\bullet}O_2^-$ radicals (formed through the reduction of O_2) or directly by holes. It has been

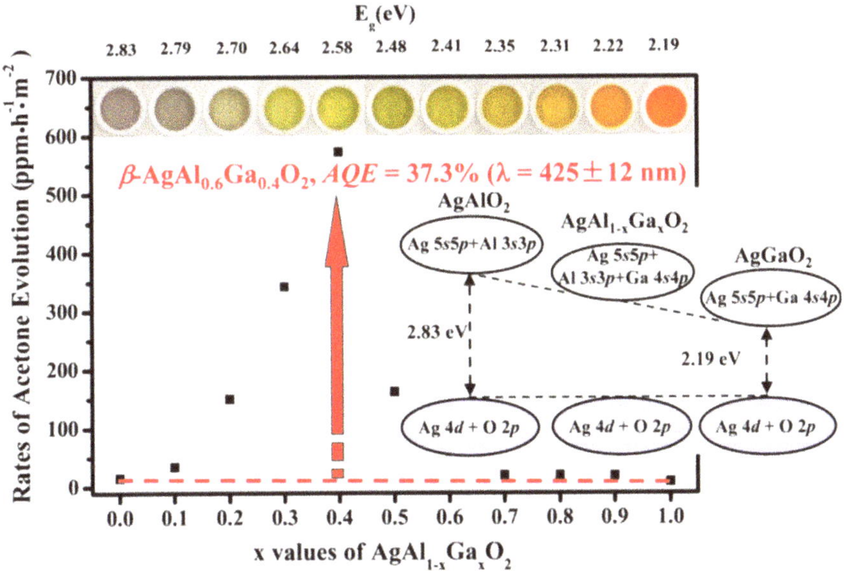

Figure 6.14. Rates of acetone formation from the photocatalytic degradation of isopropanol using different configurations of β-$AgAl_{1-x}Ga_xO_2$, where $0 \leqslant x \leqslant 1$. The corresponding bandgaps and colors of the samples are also demonstrated. When the Al is replaced by Ga, the conduction band shifts to lower (i.e. less negative) potentials, narrowing the bandgap. At $x = 0.4$, an apparent quantum efficiency (AQE) of 37.3% is found at an absorption wavelength of 425 nm. Reprinted with permission from [185]. Copyright (2011) American Chemical Society.

theorized that this stoichiometry offers the optimal balance between achieving as narrow a bandgap as possible (to allow for the absorption of low-energy photons) and retaining the correct positioning of the band positions with respect to the redox potentials involved in this IPA degradation. Clearly, solid-state solutions have benefits in terms of enhancing the photocatalytic performance of metal oxides.

6.6.2 Surface modification

The surfaces of metal oxides can be modified in such a way that their activity is enhanced, for example by decoration with other nanostructured compounds, or by the anchoring of molecular moieties. In the following, we discuss the surface modification of oxide photocatalysts with co-catalytic nanoparticles, light photosensitization through the incorporation of dyes or plasmonic nanoparticles, and we finally provide a brief description of the advantages of heterojunctions and Z-schemes. Care should be taken that the deposition ('loading') of such nanoparticles is not confused with doping (as explained in section 6.1) [186].

6.6.2.1 Co-catalytic nanoparticles

The activity of a photocatalyst, for example, in the water-splitting half-reactions (i.e. the hydrogen and oxygen evolution reactions), can be further enhanced by the deposition of co-catalytic nanoparticles on its surface [187–196]. Note, in fact, that even if the band positions of a given metal oxide semiconductor are suitable for H_2 or O_2 evolution, these reactions tend to occur at extremely slow rates in the absence of a suitable co-catalyst decorated on the surface of the semiconductor. In other words, such co-catalyst nanoparticles (NPs) substantially contribute to enhancing the kinetics of either the reductive or the oxidative reactions. The increased performance attained with co-catalysts can be explained through two phenomena. In some cases, the co-catalytic nanoparticles can lower the activation energy barrier for a reaction to take place; i.e. from a catalytic point of view, the surface of the co-catalytic material is better suited for the reaction than the surface of the semiconductor generating charge carriers. Moreover, due to a difference in work function between the semiconductor and the co-catalytic nanoparticle, a Schottky junction can form when the work function of the co-catalytic nanoparticle is smaller (relevant for reductive co-catalysts) or larger (relevant for oxidative co-catalysts) than that of the semiconductor. When such a junction is formed, a built-in electric field is created. As a result, the co-catalytic nanoparticles act as charge carrier beacons, attracting either electrons or holes, enhancing electron/hole separation and thus suppressing charge carrier recombination within the semiconductor (figure 6.15(left)). In other words, co-catalyst nanoparticles can trap charge carriers from the semiconductor and mediate their transfer to redox species in the environment.

Noble metal particles (e.g. Pt, Au, Ag, Pd, Ru, and Rh) are frequently used as reductive co-catalysts for H_2 evolution; Pt in particular is known to be effective due to its high work function, low overpotential, and intrinsic catalytic activity for proton reduction to hydrogen [190, 191, 193, 196, 197]. As noble metals are costly, other materials have also been investigated as co-catalytic particles for H_2 evolution: earth-abundant metals not only in their metallic form (Co, Ni, and Cu) but also in the form of oxides (NiO, NiO_x, CuO, Cu_2O), hydroxides ($Co(OH)_2$, $Ni(OH)_2$, and $Cu(OH)_2$), sulfides (FeS, CoS, NiS, NiS_2, CuS,

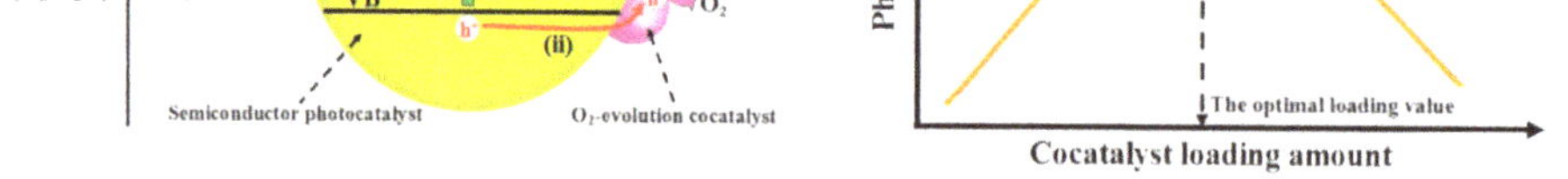

Figure 6.15. (Left) Conceptual view of co-catalytic nanoparticles in photocatalytic water splitting. After light excitation (i), reductive and oxidative co-catalysts attract electrons and holes, respectively (ii). These surfaces are suited for either proton reduction or water oxidation (iii). (Right) The relationship between photocatalytic activity and the amount of co-catalyst loading. Adapted from [190] with permission of The Royal Society of Chemistry.

MoS_2, MoS_3, and WS_2), phosphides (Co_2P, NiP, and MoP), and carbides (WC) [187, 190]. One should keep in mind that the chemical inertness of earth-abundant metals and metal oxides also plays a role during photocatalysis. For instance, cuprous oxides can be reduced to metallic copper [198], and conversely, metallic copper can be oxidized to cuprous oxides [199]. Similar mechanisms have been reported for nickel [47, 200]. *In situ* techniques (such as operando x-ray absorption spectroscopy) can shed light on the relationship between the products formed and the oxidation state of the co-catalytic nanoparticles [46]. Carbon (such as carbon nanotubes and graphene) can be utilized as a co-catalyst for H_2 evolution as well. This is interesting, as carbon is cost-efficient, abundant, stable, non-toxic, highly conductive, can have a large surface area, and can expose novel morphologies. Molecular complexes such as hydrogenases can also be employed. Similarly, for CO_2 reduction, noble metals, metal oxides, and cobaltates ($ZnCo_2O_4$ and $MnCo_2O_4$) are efficient co-catalysts [187].

Oxidative co-catalytic nanoparticles, i.e. NPs that co-catalyze hole-mediated processes, are typically metal oxides. For O_2 evolution, examples include noble metal oxides such as RuO_2 and IrO_2, earth-abundant compounds such as CoO_x, MnO_x, NiO_x, and FeO_x, cobalt–phosphate (Co–Pi), and $B_2O_{3-x}N_x$ [187, 190]. Among those compounds, while noble metal oxides (RuO_2 and IrO_2) are the most active O_2 evolution co-catalysts, their downside is their expense, making other oxidation co-catalysts interesting substitutions.

Although a co-catalyst contributes to enhancing the photocatalytic activity of the metal oxide, it should be kept in mind that there is an optimal loading (figure 6.15(right)) [190, 195, 201]. When the loading amount becomes too high, effects that limit the photocatalytic activity come into play. The metal oxide ('free') surface required for the counter reaction becomes blocked and the co-catalytic particles might shield the metal oxide from light (thus causing inefficient light absorption). The size of the co-catalyst particle also matters: with an identical loading, smaller NPs are generally more active due to the exposure of a higher surface area and more active sites. In addition to size, the morphology of the co-catalyst can also play a role in photocatalytic activity, as facets suitable for the photocatalytic reaction can be exposed [188].

In the literature, the engineering of co-catalysts can result in even higher activities. Examples include alloys [46, 192] and core–shell particles such as Rh–Cr_2O_3, in which the shell can suppress undesired reactions (e.g. in the water-splitting reverse reaction of H_2 and O_2), protect the metal from corrosion, and improve the robustness of the co-catalyst [190, 202].

For the loading of co-catalytic nanoparticles, several chemical synthesis routes can be utilized, including, but not limited to, sol–gel methods, impregnation, chemical precipitation, chemical reduction, and hydrothermal methods [190]. To decorate a photocatalyst surface, charge carriers generated upon illumination of the photocatalyst can be utilized. This technique, unique to semiconductors, allows for the deposition of metal (or metal oxide) nanoparticles on the catalyst surface by photodeposition [186, 203, 204]. In other words, in photodeposition, the photocatalytic properties of a material are used to deposit either metal particles (M)

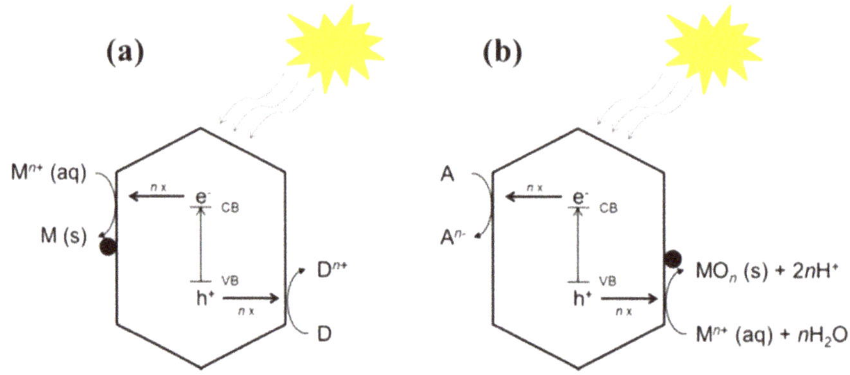

Figure 6.16. The mechanisms responsible for (a) reductive and (b) oxidative photodeposition. M denotes metals; D and A denote electron donors and acceptors, respectively; *n* denotes the numbers of electrons and holes involved. Reprinted with permission from [186]. Copyright (2016) American Chemical Society.

through reduction via reaction (6.7) or metal oxide particles (MO_n) through oxidation on the surface via reaction (6.8):

$$M^{n+}(aq) + ne^- \rightarrow M(s) \tag{6.7}$$

$$M^{n+}(aq) + nh^+ + nH_2O \rightarrow MO_n(s) + 2nH^+. \tag{6.8}$$

The concept of photodeposition is demonstrated in figure 6.16. As with other photocatalytic reactions, the reduction and oxidation potentials of these reactions need to be at favorable positions compared to the CBM and the VBM of the oxide semiconductor 'support,' respectively. During the process of photodeposition, a sacrificial reagent for the opposing oxidation/reduction reaction can be used. Such sacrificial reagents (also discussed earlier in this chapter, see section 6.2.3) have an influence on the properties of the materials deposited. For instance, the use of different organic compounds results in different sizes, shapes, loadings, and dispersions of Pt when photodeposited on TiO_2 [205].

Similarly, for the photodeposition of Pt on WO_3, the presence or absence of methanol has a huge influence on the oxidation state, loading, size, and dispersion of the co-catalytic material [206]. Finally, photodeposition is interesting, as metals and metal oxides are deposited on the electron and hole hotspots of the material, respectively, i.e. surface spots where electrons or holes are most efficiently transferred to the 'environment.' This allows for site-selective structure-directed photodeposition [186, 207, 208]. By engineering metal oxide particles through crystal facet engineering (described earlier in section 6.5), the selective photodeposition of metals and metal oxides (i.e. reductive and oxidative co-catalysts) can take place on facets suited for reduction and oxidation, respectively. Because these co-catalysts can even further promote charge carrier separation, high photocatalytic activities can be achieved. An example of such successful structure-directed photodeposition has been provided by Li *et al*, in which the impregnation of Pt and MnO_x on facetted $BiVO_4$ did not yield increases in photocatalytic water oxidation performance, whereas photodeposition resulted in a more than 60-fold increase. This can be explained by considering that photodeposition allows for site-selective co-catalyst deposition on very active spots (facets) of the oxide photocatalyst [208].

Another approach used to produce metal co-catalyst nanoparticles is based on solid-state dewetting of a thin film of a metal of choice. Continuous films of a chosen metal (some nanometers to tens of nanometers thick) are first deposited onto the oxide photocatalyst surface by a PVD technique such as evaporation or sputtering. Such metal films can then break up into individual metal nanoparticles when heated to a sufficiently high temperature [209, 210]. The heat treatment can be carried out in air or in an inert or reducing atmosphere, depending on the metal of interest. For example, Au films on titania surfaces easily dewet in air [211–213]. An inert (Ar) or reducing (H_2/Ar) atmosphere can be better suited to metals that are susceptible to oxidation, such as Ni or Cu [214]. Obviously, this process relies on the thermal stability of the metal oxide photocatalyst under the adopted dewetting conditions. The overall driving force for dewetting is the minimization of the free surface energy of the metal film and substrate and of the metal–substrate interface energy. The thinner the metal film, the higher the system surface energy, and hence the lower the activation energy required to enable metal atom surface mobility (toward agglomeration). Thus, metal films can dewet even at temperatures well below the metal's melting point, i.e. while the metal remains in the solid state, and even at temperatures between half and two thirds of the metal's melting point [215, 216].

Overall, this approach may offer advantages such as: (i) the accuracy of the co-catalyst's size and loading on the photocatalyst and precise tuning of the co-catalyst's coverage vs. the free ('exposed') oxide surface, which help to optimize light harvesting and charge carrier management; [201] (ii) facile manipulation of the co-catalyst's composition—for example, bimetallic nanoparticles can be produced by dewetting metal films of a desired composition (by sequential or co-sputtering); [46, 217] and (iii) in combination with the use of vacuum-deposition techniques to deposit metal films, the direct solid-state formation of the metal co-catalyst NPs at the oxide photocatalyst surface offers the advantage of a 'neat' co-catalyst/oxide interface, in contrast to those produced using conventional wet chemical methods.

6.6.2.2 Visible-light photosensitization

Inorganic particles or organic compounds (molecules or molecular complexes) can also be coated onto the surface of a metal oxide to allow the absorption of photons with wavelengths in the visible range. This phenomenon is called light photosensitization [218]. Here, we make a distinction between *dye photosensitization* and *localized surface plasmon resonance (LSPR)*.

In the case of dye photosensitization, an organic dye is adsorbed on the surface of a metal oxide [219–223]. Such a dye is able to absorb photons, possibly resulting in the excitation of electrons from the highest occupied molecular orbital (HOMO) to the lowest unoccupied molecular orbital (LUMO) (figure 6.17(a)). When the LUMO is more negative than the CB of the metal oxide, the excited electrons can be

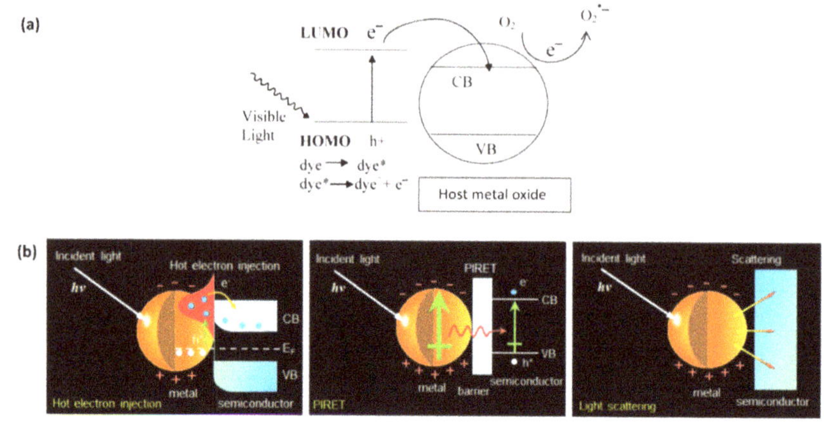

Figure 6.17. (a) A conceptual view of dye photosensitization. Visible light is used to excite an electron from the HOMO to the LUMO of a dye, after which, it is injected into the host metal oxide. There, it is used for a photocatalytic reduction reaction (in this example, O_2 to ${}^{\bullet}O_2^-$). Adapted from [222], Copyright (2009), with permission from Elsevier. (b) Conceptual views of three different mechanisms that can occur in dye photosensitization: hot electron injection, plasmon-induced resonance energy transfer (PIRET) and light scattering. PIRET can still take place when a thin barrier is present between the metal and the semiconductor. Adapted from [233] with permission of The Royal Society of Chemistry.

transferred from the dye into the metal oxide. Such electrons can be used for catalytic reactions. Meanwhile, the dye becomes oxidized, but it can revert to its original structure by accepting electrons from an electron donor (i.e. an oxidant in the reaction environment), thus regenerating the adsorbed dye. Examples of dyes used include phthalocyanine, porphyrin, anthocyanin, and squaraine [220]. The added value lies in the fact that such dyes can have smaller bandgaps than the metal oxide, and therefore the harvesting of visible light (and thus a broader part of the solar spectrum) becomes possible in addition to the photons absorbed by the metal oxide itself (i.e. photons with a higher energy than the bandgap of the metal oxide). In fact, for longer wavelengths, the metal oxide only acts as a carrier and provides active surface sites; the light harvesting is done by the dye itself. Typically, dye sensitization can form radicals such as ${}^{\bullet}O_2^-$ (through oxygen reduction), ${}^{\bullet}OOH$, and ${}^{\bullet}OH$ (through a series of protonation and reduction steps). ${}^{\bullet}OH$ can also simply form through the oxidation of water. Although such radicals are useful in the decomposition of pollutants (as discussed earlier in this chapter, see section 6.2.6), they can also attack the dye molecule, thus decomposing it. In the absence of reagents, it is often observed that the dye itself can be degraded. Therefore, care should be taken that dye-sensitized metal oxides are always used in the vicinity of such reagents. Still, dye adsorption could be a powerful technique for making metal oxides sensitive to visible light. Note that this approach is used in Grätzel-type solar cells [224] (i.e. dye-sensitized solar cells, DSSCs), in which the stability of the dye (a Ru complex) is achieved by (i) using acetonitrile as a solvent (i.e. no ROS are produced that can degrade the dye); and (ii) the use of a redox couple (I^-/I_3^-) to regenerate the dye.

In the case of plasmon-induced photocatalysis, small metallic nanoparticles are deposited on the surface of, for example, a metal oxide semiconductor, using a similar approach to that of co-catalyst loading [225–235]. A major difference from co-catalysis, however, is that such particles allow the photocatalytic complex to absorb visible light. When a plasmonic nanoparticle is illuminated by light, the electromagnetic nature of the photons can induce coherent oscillation of free charges within the metallic particles, provided that such particles are considerably smaller than the photon wavelength. This is called LSPR. The maximum oscillation amplitude is reached at a specific wavelength. Due to the plasmon resonance, the extinction cross-section (the sum of the absorption and scattering cross-sections) increases significantly, and a confined local electric field is formed, 50–10 000 times larger than the incident field. The LSPR wavelength, the extinction cross-section, and the magnitude of the local electric field enhancement can be tuned by controlling the size, shape, and crystalline structure of the material. In this way, it has been demonstrated that Au and Ag nanoparticles, for example, can have a high extinction cross-section in the visible range, and can therefore act very well as visible-light photosensitizers for metal oxides. More earth-abundant elements such as Cu and Al can also be used for LSPR.

Due to the LSPR, energy transitions can take place from the plasmonic nanoparticles to the metal oxide. The mechanisms responsible for the energy transfer from the plasmonic nanoparticles to the metal oxide are complex and still the subject of much debate in the research literature. Here, we highlight three important mechanisms which can occur (figure 6.17(b)). The first theory to explain plasmon-induced photocatalysis states that due to LSPR, 'hot' electrons are generated and transferred to the conduction band of the adjacent metal oxide. The electrons in the CB and the holes in the plasmonic nanoparticles can then be used for catalytic reactions. Alternatively, plasmon-induced resonance energy transfer (PIRET) could take place. Here, plasmonic energy is transferred from the plasmonic particles to the metal oxide through dipole–dipole interactions in the near field. Because PIRET is a blueshift energy transfer process, the energy can be used to create electron/hole pairs in the metal oxide. Finally, plasmonic nanoparticles can also give rise to 'light trapping.' Here, the plasmonic nanoparticles do not absorb light, but scatter it so that it penetrates the metal oxide. This effect leads to enhanced light absorption and charge separation.

It should be noted that as in the case of molecular photosensitization, the resulting nanocomposite photocatalyst can become active under excitation of the metal oxide's bandgap as well as under plasmonic excitation.

6.6.2.3 Heterojunctions and Z-schemes

Metal oxide photocatalysts can also be modified by 'attaching' other semiconductors to them; i.e. other semiconductors can be loaded onto metal oxides with the aim of improving their overall performance for photocatalytic applications. As in the case of decoration with co-catalyst or plasmonic particles, we are dealing with heterojunctions [48, 105, 182, 195, 219, 221, 222, 236], except that in this situation, the junction allows two semiconductors to interface, which yields intriguing effects for photocatalytic applications.

Three different configurations are possible (see also figure 6.18(a)): (i) a straddling gap configuration, in which the first semiconductor has a higher ('more negative') CB edge and a lower ('more positive') VB edge than the second semiconductor; (ii) a staggered gap configuration, in which both the CB and VB edges of the first semiconductor are at higher positions compared to the band edges of the second semiconductor, while the VB of the first semiconductor is still at a lower position than the CB of the second semiconductor; and (iii) a broken gap configuration, in which the VB and CB edges of the first semiconductor are both at a higher position than the CB edge of the other semiconductor, and the VB of the first semiconductor is at a higher position than the CB of the second semiconductor. Of those configurations, the staggered gap, also known as a type-II heterojunction, is

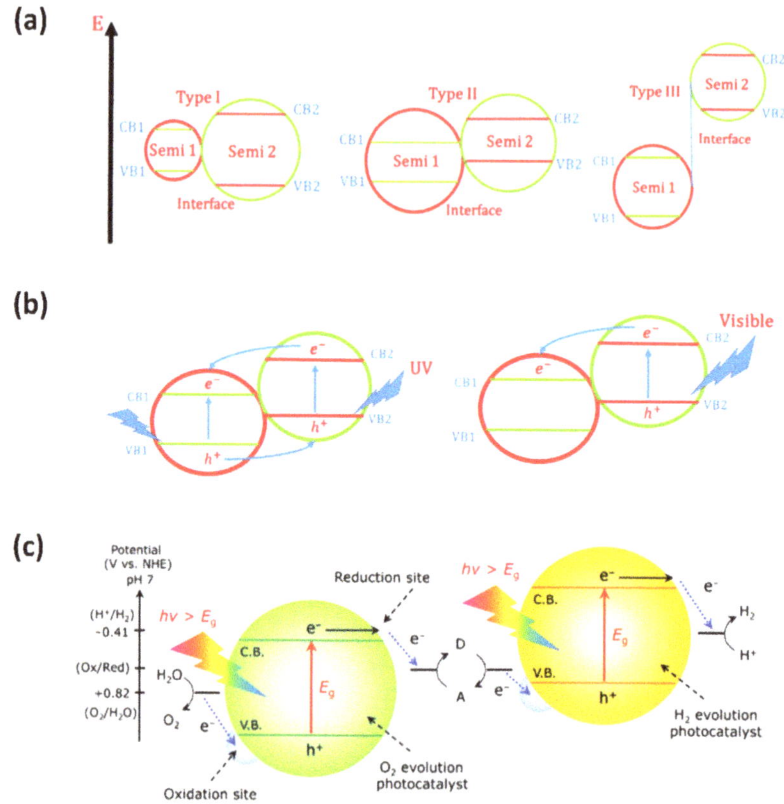

Figure 6.18. (a) Different types of semiconductor–semiconductor heterojunction: type-I (straddling gap), type-II (staggered cap), and type-III (broken gap). (b) Left: charge carrier transfer when both semiconductors are excited in a type-II SC–SC heterojunction. Right: sensitization behavior if one semiconductor is able to absorb visible light and the other is not. (c) A Z-scheme consisting of two semiconductors with reduction and oxidation co-catalysts and a redox mediator. This Z-scheme can drive the photocatalytic splitting of water. Adapted from [182] with permission of The Royal Society of Chemistry.

preferred for photocatalysis. When the two semiconductors come into contact with one another, band bending occurs at the interface, invoking the presence of an electric field. Consequently, upon excitation of both semiconductors, electrons can migrate from the semiconductor with the highest CB minimum to the semiconductor with the lowest CB minimum; alternatively, holes can migrate from the semiconductor with the lowest VB maximum to the semiconductor with the highest VB maximum. Thus, spatial charge separation takes place. As a consequence, properly designed metal oxide heterojunctions can aid in suppressing recombination, and consequently the lifetime of the charge carriers can be increased and photocatalytic activity can be enhanced. Examples of heterojunctions include TiO_2/ZnO [237] (two n-type semiconductors) and Cu_2O/TiO_2 [238] (a p-type and an n-type semiconductor forming a p–n junction).

An alternative situation can occur when one of the two semiconductors has a sufficiently small bandgap for visible-light absorption (e.g. quantum dots), and the other is instead a wide bandgap material which thus only absorbs in the UV spectral range (figure 6.18(b)). When visible-light illumination is present and the CBs (or VBs) of the semiconductors are properly aligned, electrons (or holes) can be injected from the narrow bandgap semiconductor into the wide bandgap semiconductor; thus, the narrow bandgap semiconductor acts as a light photosensitizer in the system (this effect can resemble hot electron injection induced by plasmonic nanoparticles, in which the plasmonic nanoparticles act as a visible-light sensitizer). Efficient charge carrier separation also takes place in this way, as the holes (or electrons) remain in the narrow bandgap semiconductor. Thus, charge recombination can also be suppressed in this scenario, and photocatalytic activity can be enhanced. Examples of such type-II heterojunctions include the abovementioned Cu_2O/TiO_2, which consists of narrow-gap Cu_2O (2.0 eV) in contact with wide-gap TiO_2 (3.2 eV) [238] and CdS/TiO_2, in which the CdS has a narrow bandgap of 2.4 eV and the TiO_2 has a large bandgap of 3.2 eV [182, 239].

Junctions can also consist of semiconductors with identical chemical composition, but with different crystal phases and/or band structures (for instance p–n junctions made from the same material) [48, 236, 240, 241]. In the literature, these junctions are also known as homojunctions, although this term has been criticized when the bandgaps are unequal [48]. Here, to avoid confusion, we call heterojunctions consisting of the same material with different crystal phases 'multiphase,' whereas junctions consisting of p–n configurations are referred to as homojunctions. An example of a material which can exist in different crystalline phases is TiO_2, for which three crystalline forms are known: anatase, rutile, and brookite [242]. A combination of anatase and rutile would act as a multiphase heterojunction capable of good photocatalytic performance [48, 243, 244]. Evonik Aeroxide TiO_2 P25 (often simply abbreviated to 'P25') is a nanocatalyst powder composed of multiphase heterojunctions consisting of 80% anatase and 20% rutile. P25 is known for high photocatalytic activity due to the synergy between the two crystal phases. An example of a homojunction of p-type and n-type TiO_2 has been reported by Pan *et al* [241]. As with heterogeneous p–n junctions, efficient charge carrier separation takes place, thus increasing the lifetime of such carriers. Indeed, Pan *et al* showed much higher photoactivity in H_2 generation for p–n junctions than for pure p-type or pure n-type TiO_2 (using Pt as a co-catalyst and methanol as a sacrificial reagent).

In addition to heterojunctions, the use of Z-schemes is another promising strategy for the enhancement of photocatalytic activities [55, 182, 195]. As with type-II heterojunctions, Z-schemes consist of two semiconductor materials, in which the CB and VB of one semiconductor are shifted to a higher position (i.e. more negative) than those of the other semiconductor. However, these semiconductors do not need to be in contact with each other. Instead, a redox mediator in an aqueous solution such as IO_3^-/I^- can be used for charge transfer. Let us consider semiconductors SC 1 and SC 2, among which, SC 2 has the higher CB and VB (see also figure 6.18(c)). When both semiconductors are illuminated, electron/hole pairs are formed in both. The electrons in SC 1 are used to reduce the oxidant of the redox couple, whereas the holes in SC 2 are used to oxidize the reductant. Meanwhile, the high-energy electrons of SC 2 and the high-energy holes of SC 1 can be used for reduction and oxidation reactions, respectively. The main advantage of the Z-scheme is that favorable positions for the VB (SC 1) and the CB (SC 2) can be achieved in relation to the oxidation and reduction potentials of interest, respectively, while keeping the bandgap of both semiconductors narrow. For example, tungsten oxide (WO_3) has a relatively narrow bandgap of 2.7 eV (in comparison to other metal oxides). Therefore, it can absorb at least part of the visible spectrum. However, although the VB of WO_3 is at a favorable position for water oxidation, the CB of WO_3 is at an unfavorable position for proton reduction, making it unusable as a sole photocatalyst for hydrogen evolution or overall water splitting. This problem can be circumvented by combining WO_3 in a Z-scheme with another semiconductor that has a narrow bandgap but a favorable CB edge position. A successful example of overall water splitting under visible light utilizing a Z-scheme has been reported by Abe *et al*, in which TaON (which has a bandgap of 2.5 eV) was used as a second semiconductor next to WO_3, and IO_3^-/I^- was used as a redox mediator [245, 246]. In these studies, both TaON and WO_3 were coated with co-catalytic Pt particles.

Z-schemes do not always require redox mediators. Examples of semiconductor–metal–semiconductor configurations are also known (in which the metal acts as an electron mediator) and even Z-schemes without mediators (i.e. 'direct' Z-schemes) are known. The latter do resemble type-II heterojunctions, but it should be kept in mind that the electron/hole transfer mechanism is very different.

6.7 A brief outlook on practices in photocatalytic research

In this chapter, we have covered some key fundamental aspects of metal oxide photocatalysis and described a variety of metal oxide nanostructured semiconductors, particularly their properties and synthesis methods, as well as the promise they hold for photocatalytic applications. With proper execution, a vast variety of test reactions and experimental protocols can be used to assess the performance of metal oxides for photocatalytic applications.

A very popular method in academic research on photocatalysis is to demonstrate the activity of a photocatalytic material by monitoring its dye degradation performance under visible light illumination. In such experiments, the photocatalyst, either a slurry (an oxide suspension in a liquid phase) or a film, is exposed under suitable illumination to a solution containing a dye (e.g. methylene blue, methyl orange, or acid orange 7). For metal oxide photocatalysts, the dye can generally be directly degraded by holes (this implies adsorption of dye molecules at the oxide surface and direct reaction with charge carriers) or indirectly through reactive oxygen species, i.e. ROS, such as $^{\bullet}OH$ radicals formed either through water oxidation or through the protonation of $^{\bullet}O_2^-$ radicals (the latter are formed through O_2 reduction by CB electrons) [247]. The dye solution generally has a very distinct color prior to the photocatalytic degradation and displays an absorption spectrum with clear absorption band(s) when measured by UV–vis spectrophotometry. When photocatalytic degradation takes place, the intensity of such absorption band(s) decreases. According to the Beer–Lambert law, the dependency between absorbance and concentration is linear [248]. Thus, by monitoring the change in absorbance intensity over time, the change in concentration and thus the conversion and reaction rates can be calculated. However, care should be taken when dye degradation is used in photocatalysis, particularly when dye degradation tests are used to prove the visible-light activity of a given metal oxide photocatalyst [249–251]. Dyes typically absorb visible light, due to excitation of electrons from the HOMO to the LUMO, which can then be injected into the CB of the metal oxide. Subsequently, the electrons can be used to reduce O_2 to $^{\bullet}O_2^-$ radicals, which can form $^{\bullet}OH$ radicals through protonation. The latter can then degrade the dye, i.e. via formation of ROS. Thus, in this mechanism, the dye undesirably acts as a light photosensitizer for its own degradation. Light sources with a broad spectrum, e.g. those that also emit visible-light photons, are used in some cases in order to test the visible-light activity of the photocatalyst. When the photocatalytic activity is reported, it can be (falsely) assumed that the dye is purely degraded by the metal oxide. If dye degradation is used, it is crucial that the light used for the photocatalyst excitation is not absorbed by the dye. In other words, the use of a dye degradation test is not an adequate experimental approach to prove the visible-light activity of the photocatalyst if the spectral range of the light used can also excite the dye. Note two additional aspects: (i) the dye can simultaneously degrade by 'photolysis,' i.e. by excitation due to photon absorption leading to bond cleavage in the dye molecule (this typically occurs with high photon energies and thus upon deep UV light illumination), and (ii) the absorption decrease is due to the cleavage of the diazo (–N = N–) bond (typical of some dye molecules), but the molecular fragments formed thereby become new 'substrates' that can undergo further degradation. This makes the study of dye degradation kinetics quite complicated.

Despite its simplicity, dye degradation should be used consciously, bearing in mind that 'photosensitization' can take place, which would affect the photocatalytic behavior or lead to an overestimation of the photocatalytic activity of the metal oxide. A more appropriate approach is to study the photocatalytic decomposition of molecules that do not absorb photons in the spectral range used to excite the photocatalyst. Examples include the decomposition of organic substances (either in the liquid or gas phase) that are not sensitive to visible light. An example is formic acid, which offers advantages such as: it does not absorb light in the visible spectral range, and it oxidizes to CO_2, providing a simple degradation mechanism compared to more complex molecules such as dyes [252, 253].

Another aspect to consider is the proper engineering of nanostructured metal oxides. To achieve high photocatalytic activities, proper and careful nanostructuring is essential. However, caution must be taken when engineering composite nanostructures based on heterojunctions. For example, we have outlined that type-II heterojunctions (or direct Z-schemes) can be attractive for efficient charge carrier separation. This concept is often used for nanopowdered photocatalysts used as slurries for photocatalytic reactions. Such powders, however, are often also coated onto a conductive film to enable their further characterization as photoelectrodes. While various characterization techniques (including photoluminescence, transient absorption spectroscopy, and impedance spectroscopy) can provide solid evidence of enhanced charge separation and photocatalytic performance ascribed to the heterojunction formed, it may not be straightforward to support these properties using results of PEC experiments.

This issue arises because in a nanostructured heterojunction intended for photocatalysis, for example a composite metal oxide nanostructure, the sites for anodic (hole-mediated) and cathodic (electron-mediated) reactions are located on the same solid entity (the nanoparticle). Upon illumination, with the proper band alignment, charge separation occurs, as discussed in section 6.6.2 (see also figure 6.19(a)). In a growing number of research articles, such composite nanoparticulate materials, besides being tested for photocatalysis, are also deposited on a conductive substrate (carbon, conductive glass, or a metal foil) with the aim to test their PEC properties. The enhanced photocatalytic activity is often substantiated by an improved PEC performance—a direct connection between the photocatalytic and PEC performance is then often made to corroborate the improved charge separation properties provided by the heterojunction. However, it should be pointed out that for PEC investigations, these composite oxide nanoparticles are often anchored to a conductive scaffold (the back contact) with a random orientation (see figure 6.19(b)).

This is clearly not ideal considering that a precise orientation of the junction with respect to the charge collector would be required to make the junction effective in improving charge separation. This is so because in a PEC configuration, the anodic and cathodic processes occur at the surfaces of two different electrodes and are hence spatially separated; in other words, when oxide heterojunctions are used as

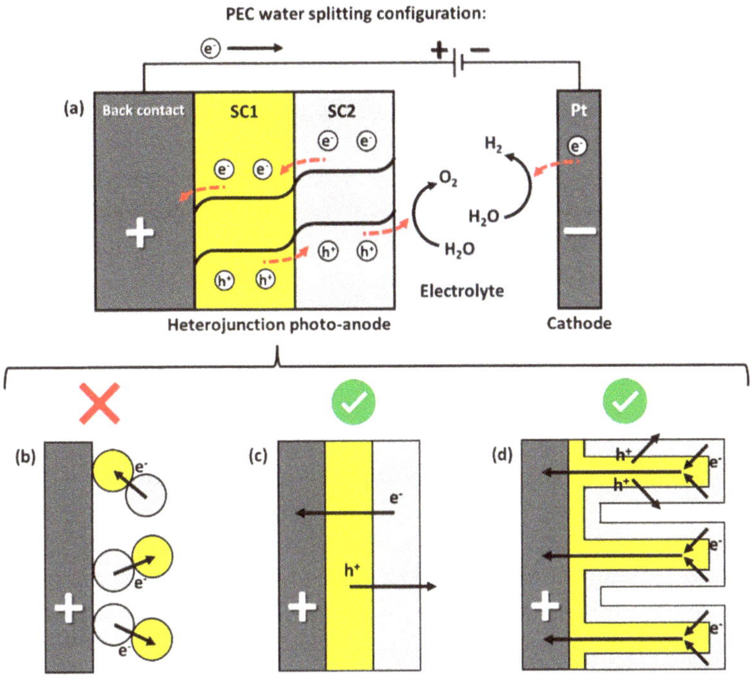

Figure 6.19. (a) A photoanode based on a heterojunction of two semiconductors on a conductive substrate (the back contact). Band bending drives electron and hole separation: holes move toward the solid/liquid junction, electrons move toward the back contact (and to the cathode through the external circuit). (b) A heterojunction of two semiconductors coated onto a substrate in powder form with random orientations. In this configuration, a significant part of the electrons may not be guided towards the conductive substrate (see the band alignment in (a)). (c) Stacked nanostructured layers of two semiconductors. In this case, charge separation may be enhanced. (d) A semiconductor coated onto another semiconductor in a 1D shape. In this case, charge separation may also be enhanced. The photoelectrode's performance may also be further enhanced by the increased surface area and more efficient light harvesting provided by the 1D nanoarchitecture (see section 6.4).

photoanodes in a PEC configuration, their design should be such that holes can be transferred towards the solid–liquid junction where they oxidize water to O_2, while electrons can be driven towards the photoanode's back contact to reach the counter electrode (a cathode or a photocathode), where they reduce water to H_2. If the junction's orientation is inadequate, the PEC results for metal oxide composites may not be consistent with the results of photocatalysis or may even lead to inaccurate conclusions on the charge transfer dynamics. Thus, oxide heterojunctions in the form of composite nanoparticles are not optimal for study in a PEC configuration, unless suitable methods are used for their deposition in a coherent orientation within the photoelectrode layer and with respect to the desired direction of charge collection. On the contrary, other composite materials, prepared by methods other than conventional powder technologies or wet chemistry (e.g. the sol–gel method), may be more suitable for PEC studies. This is the case of stacked nanostructured layers (see figure 6.19(c)): here, the layers can be deposited sequentially (by any suitable technique) to provide directional hole and electron transfer. An alternative approach is to use layers of 1D or 2D semiconductor materials which can subsequently be coated with a secondary semiconductor, thus also forming a junction that allows for directional charge separation (see figure 6.19(d)).

For a further careful evaluation of how to engineer proper tests to determine the photocatalytic activity of metal oxides, we refer the interested reader to some critical papers in the literature [249–251].

An example of how proper engineering of a photocatalytic system (with a metal oxide as the photocatalyst) can result in high quantum efficiencies was recently demonstrated by Domen and coworkers [254]. Here, water splitting was achieved at record external quantum efficiencies of up to 96% (at illumination wavelengths from 350 to 360 nm). To accomplish this, site-selective photo-deposition was used to load HER and OER co-catalysts onto $SrTiO_3$ doped with Al (at an Al:Ti ratio of less than 0.01). Rh/Cr_2O_3 and CoOOH were used as HER and OER co-catalysts, respectively. In addition to the systematic engineering of photocatalytic systems, studies are also emerging that describe the possibility of combining photocatalysis with other fields of research. Specifically, photothermal chemistry (i.e. the (localized) heating of photocatalytic particles) [255, 256], or the combination of photocatalysis with magnetism [257, 258] could be promising.

Progress in the utilization of metal oxides for photocatalytic purposes has been made not only on a fundamental level. Promising developments are also taking place at the industrial scale. For instance, pilot plants could contribute to investigations of whether photocatalysis can take place at larger scales than the laboratory scale. An example of a location where experiments take place at the pilot-plant scale is the Plataforma Solar de Almería (PSA) [259]. Here, TiO_2 is used for the removal of organic pollutants while simultaneously producing hydrogen under direct solar simulation. Another example of how metal oxides could contribute to the development of a greener environment can be found in the city of Hengelo, The Netherlands. In the period of 2008–11, a study was performed to investigate whether TiO_2 could practically be applied to reduce NO_x emissions in a residential area [260]. To do so, a street was paved for a length of 150 meters with a concrete pavement containing TiO_2 (this was called the 'DeNO_x street'), whereas another part of the street was paved for a length of 100 m with normal paving blocks (called the 'Control street'). The researchers saw a clear difference in NO_x concentrations between the DeNO_x street and the Control street. On average, the concentration was 19% lower in the DeNO_x street, while under ideal weather conditions a decrease of 45% could be achieved. Similar photocatalytically driven self-cleaning concepts were previously also used to design paints. For instance, Maggos *et al* demonstrated that applying a TiO_2-containing paint in an indoor car park could contribute to the photocatalytic degradation of NO_x gases (from car exhausts) [261].

Questions

1. Two oxide photocatalysts based on pristine WO_3 and TiO_2 nanopowders are tested under solar light illumination in view of their water-splitting activity in plain water (in the absence of any sacrificial species). Qualitatively discuss the activity trend for such oxides, the reaction products expected based on their bandgaps and band positions, and the experimental conditions adopted. Assume then that both oxides are decorated with comparable loadings of Pt nanoparticles, and ethanol is added to the reaction solution. Discuss how their activity would change, why, and what products would be formed.
2. Discuss the use of an oxide photocatalytic slurry compared to an oxide photocatalytic film, including advantages/disadvantages. Describe at least two synthesis techniques you would use for each case, including their underlying principles and experimental conditions.
3. What is hydrothermal synthesis? Describe how it can be used to synthesize oxides and the physicochemical feature such oxides can exhibit.
4. Describe the growth mechanism for anodic nanoporous (e.g. nanotube or nanopore) oxide layers; sketch the setup you would use and a typical current–time profile recorded during the growth. Describe the motivation for this cell design and discuss the current–time profile.
5. How can bandgap engineering be achieved for a metal oxide semiconductor? Describe the principle of bandgap engineering and discuss whether it is beneficial or detrimental in photocatalysis, i.e. what the pros and cons could be. Elaborate your answer by considering aspects such as light absorption and charge carrier generation and collection.
6. Discuss the role of co-catalysts in photocatalysis. Describe a typical co-catalyst for H_2 evolution and a typical co-catalyst for O_2 evolution. Elaborate on the activity trend with respect to the co-catalyst loading.
7. What is ‘photodeposition’? Describe its underlying principle. How, why, and for what would you use this method?
8. What are plasmonic nanoparticles, what materials can they be composed of, and why can they be relevant in photocatalysis? Elaborate your answer by describing the underlying physics along with an example.
9. Describe the use of junctions between two metal oxide semiconductors in photocatalysis. Discuss and illustrate (with simple sketches) the charge carrier dynamics in relation to the oxides’ bandgaps and band energy positions.
10. What is a ‘Z-scheme’ in photocatalysis and what are its advantages? Elaborate your answer by considering aspects such as light harvesting and the redox potential of charge carriers. When would you need a redox couple shuttle mediator in the reaction environment?
11. Assume you intend to design an oxide photocatalyst for ‘overall water splitting.’ Which oxide(s) and co-catalyst(s) would you use? Describe the motivation for your choice and sketch the photocatalyst structure,

including a qualitative description of the ideal band positions of the oxide(s) with respect to the redox potential for H_2 and O_2 evolution from water.
12. Dye degradation tests are often used to evaluate the activity of photocatalysts. Describe the underlying principle of such tests and elaborate on their suitability for testing a photocatalyst under visible-light illumination.

References

[1] Ramachandran R 1998 An overview of industrial uses of hydrogen *Int. J. Hydrog. Energy* **23** 593–8
[2] Abdalla A M, Hossain S, Nisfindy O B, Azad A T, Dawood M and Azad A K 2018 Hydrogen production, storage, transportation and key challenges with applications: a review *Energy Convers. Manage.* **165** 602–27
[3] Van de Krol R and Grätzel M 2012 *Photoelectrochemical Hydrogen Production* (New York: Springer)
[4] Renz C 1921 Lichtreaktionen der Oxyde des Titans, Cers und der Erdsäuren *Helv. Chim. Acta* **4** 961–8
[5] Baur E and Perret A 1924 Über die Einwirkung von Licht auf gelöste Silbersalze in Gegenwart von Zinkoxyd *Helv. Chim. Acta* **7** 910–5
[6] Fujishima A, Zhang X and Tryk D A 2008 TiO_2 photocatalysis and related surface phenomena *Surf. Sci. Rep.* **63** 515–82
[7] Fujishima A and Honda K 1972 Electrochemical photolysis of water at a semiconductor electrode *Nature* **238** 37–8
[8] Fujishima A and Honda K 1971 Electrochemical evidence for the mechanism of the primary stage of photosynthesis *Bull. Chem. Soc. Jpn.* **44** 1148–50
[9] Osterloh F E 2013 Inorganic nanostructures for photoelectrochemical and photocatalytic water splitting *Chem. Soc. Rev.* **42** 2294–320
[10] Bard A J 1979 Photoelectrochemistry and heterogeneous photo-catalysis at semiconductors *J. Photochem.* **10** 59–75
[11] Nozik A J 1978 Photoelectrochemistry: applications to solar energy conversion *Annu. Rev. Phys. Chem.* **29** 189–222
[12] Chong M N, Jin B, Chow C W K and Saint C 2010 Recent developments in photocatalytic water treatment technology: a review *Water Res.* **44** 2997–3027
[13] White J L *et al* 2015 Light-driven heterogeneous reduction of carbon dioxide: photocatalysts and photoelectrodes *Chem. Rev.* **115** 12888–935
[14] Chen X, Li N, Kong Z, Ong W-J and Zhao X 2018 Photocatalytic fixation of nitrogen to ammonia: state-of-the-art advancements and future prospects *Mater. Horiz.* **5** 9–27
[15] Fujishima A, Rao T N and Tryk D A 2000 Titanium dioxide photocatalysis *J. Photochem. Photobiol.* C **1** 1–21
[16] Melchionna M and Fornasiero P 2020 Updates on the roadmap for photocatalysis *ACS Catal.* **10** 5493–501
[17] Ravelli D, Dondi D, Fagnoni M and Albini A 2009 Photocatalysis. A multi-faceted concept for green chemistry *Chem. Soc. Rev.* **38** 1999–2011
[18] Linsebigler A L, Lu G and Yates J T 1995 Photocatalysis on TiO_2 surfaces: principles, mechanisms, and selected results *Chem. Rev.* **95** 735–58
[19] Tang S, Qiu W, Xiao S, Tong Y and Yang S 2020 Harnessing hierarchical architectures to trap light for efficient photoelectrochemical cells *Energy Environ. Sci.* **13** 660–84

[20] Razeghi M 2009 Optical properties of semiconductors *Fundamentals of Solid State Engineering* (Boston, MA: Springer), 1–53
[21] Kafizas A, Godin R and Durrant J R 2017 Charge carrier dynamics in metal oxide photoelectrodes for water oxidation *Semiconductors for Photocatalysis Semiconductors and Semimetals* vol 97 1st edn (Amsterdam: Elsevier), 1 3–46
[22] Paramasivam I, Jha H, Liu N and Schmuki P 2012 A review of photocatalysis using self-organized TiO_2 nanotubes and other ordered oxide nanostructures *Small* **8** 3073–103
[23] Cheng H and Selloni A 2010 Hydroxide ions at the water/anatase TiO_2 (101) interface: structure and electronic states from first principles molecular dynamics *Langmuir* **26** 11518–25
[24] Lazar M A and Daoud W A 2013 Achieving selectivity in TiO_2-based photocatalysis *RSC Adv.* **3** 4130
[25] Nandjou F and Haussener S 2017 Degradation in photoelectrochemical devices: review with an illustrative case study *J. Phys. D: Appl. Phys.* **50** 124002
[26] Kwon Y, Soon A, Han H and Lee H 2015 Shape effects of cuprous oxide particles on stability in water and photocatalytic water splitting *J. Mater. Chem.* A **3** 156–62
[27] Weng B, Qi M-Y, Han C, Tang Z-R and Xu Y-J 2019 Photocorrosion inhibition of semiconductor-based photocatalysts: basic principle, current development, and future perspective *ACS Catal.* **9** 4642–87
[28] Bae D, Seger B, Vesborg P C K, Hansen O and Chorkendorff I 2017 Strategies for stable water splitting: via protected photoelectrodes *Chem. Soc. Rev.* **46** 1933–54
[29] Ros C, Andreu T and Morante J R 2020 Photoelectrochemical water splitting: a road from stable metal oxides to protected thin film solar cells *J. Mater. Chem.* A **8** 10625–69
[30] Chen S and Wang L W 2012 Thermodynamic oxidation and reduction potentials of photocatalytic semiconductors in aqueous solution *Chem. Mater.* **24** 3659–66
[31] Mesa C A *et al* 2020 Multihole water oxidation catalysis on haematite photoanodes revealed by operando spectroelectrochemistry and DFT *Nat. Chem.* **12** 82–9
[32] Sivula K and van de Krol R 2016 Semiconducting materials for photoelectrochemical energy conversion *Nat. Rev. Mater.* **1** 15010
[33] Walter M G, Warren E L, McKone J R, Boettcher S W, Mi Q, Santori E A and Lewis N S 2010 Solar water splitting cells *Chem. Rev.* **110** 6446–73
[34] Bolton J R, Strickler S J and Connolly J S 1985 Limiting and realizable efficiencies of solar photolysis of water *Nature* **316** 495–500
[35] Murphy A, Barnes P, Randeniya L, Plumb I, Grey I, Horne M and Glasscock J 2006 Efficiency of solar water splitting using semiconductor electrodes *Int. J. Hydrog. Energy* **31** 1999–2017
[36] Tschurl M 2018 Semiconductor/metal (oxide) hybrid materials for applications in photocatalysis *Encyclopedia of Interfacial Chemistry* (Amsterdam: Elsevier), 573–80
[37] Fabian D M, Hu S, Singh N, Houle F A, Hisatomi T, Domen K, Osterloh F E and Ardo S 2015 Particle suspension reactors and materials for solar-driven water splitting *Energy Environ. Sci.* **8** 2825–50
[38] Chen S, Takata T and Domen K 2017 Particulate photocatalysts for overall water splitting *Nat. Rev. Mater.* **2** 17050
[39] Al-Azri Z H N, Chen W-T, Chan A, Jovic V, Ina T, Idriss H and Waterhouse G I N 2015 The roles of metal co-catalysts and reaction media in photocatalytic hydrogen production:

performance evaluation of M/TiO_2 photocatalysts (M=Pd, Pt, Au) in different alcohol–water mixtures *J. Catal.* **329** 355–67

[40] Galińska A and Walendziewski J 2005 Photocatalytic water splitting over Pt–TiO_2 in the presence of sacrificial reagents *Energy Fuels* **19** 1143–7

[41] Chiarello G L, Ferri D and Selli E 2011 Effect of the CH_3OH/H_2O ratio on the mechanism of the gas-phase photocatalytic reforming of methanol on noble metal-modified TiO_2 *J. Catal.* **280** 168–77

[42] Naldoni A, D'Arienzo M, Altomare M, Marelli M, Scotti R, Morazzoni F, Selli E and Dal Santo V 2013 Pt and Au/TiO_2 photocatalysts for methanol reforming: role of metal nanoparticles in tuning charge trapping properties and photoefficiency *Appl. Catal.* **B 130-31** 239–48

[43] Muggli D S, McCue J T and Falconer J L 1998 Mechanism of the photocatalytic oxidation of ethanol on TiO_2 *J. Catal.* **173** 470–83

[44] Malato S, Fernández-Ibáñez P, Maldonado M I, Blanco J and Gernjak W 2009 Decontamination and disinfection of water by solar photocatalysis: recent overview and trends *Catal. Today* **147** 1–59

[45] Rueda-Marquez J J, Levchuk I, Fernández Ibañez P and Sillanpää M 2020 A critical review on application of photocatalysis for toxicity reduction of real wastewaters *J. Clean. Prod.* **258** 120694

[46] Spanu D, Minguzzi A, Recchia S, Shahvardanfard F, Tomanec O, Zboril R, Schmuki P, Ghigna P and Altomare M 2020 An operando X-ray absorption spectroscopy study of a NiCu–TiO_2 photocatalyst for H_2 evolution *ACS Catal.* **10** 8293–302

[47] Han K, Kreuger T, Mei B and Mul G 2017 Transient behavior of Ni@NiO_x functionalized $SrTiO_3$ in overall water splitting *ACS Catal.* **7** 1610–4

[48] Marschall R 2014 Semiconductor composites: strategies for enhancing charge carrier separation to improve photocatalytic activity *Adv. Funct. Mater.* **24** 2421–40

[49] Hu S, Shaner M R, Beardslee J A, Lichterman M, Brunschwig B S and Lewis N S 2014 Amorphous TiO_2 coatings stabilize Si, GaAs, and GaP photoanodes for efficient water oxidation *Science* **344** 1005–9

[50] He H 2020 Metal oxide semiconductors and conductors *Solution Processed Metal Oxide Thin Films for Electronic Applications* (Amsterdam: Elsevier), 7–30

[51] Scaife D E 1980 Oxide semiconductors in photoelectrochemical conversion of solar energy *Sol. Energy* **25** 41–54

[52] Matsumoto Y 1996 Energy positions of oxide semiconductors and photocatalysis with iron complex oxides *J. Solid State Chem.* **126** 227–34

[53] Batzill M 2011 Fundamental aspects of surface engineering of transition metal oxide photocatalysts *Energy Environ. Sci.* **4** 3275

[54] Sato H, Minami T, Takata S and Yamada T 1993 Transparent conducting p-type NiO thin films prepared by magnetron sputtering *Thin Solid Films* **236** 27–31

[55] Kudo A and Miseki Y 2009 Heterogeneous photocatalyst materials for water splitting *Chem. Soc. Rev.* **38** 253–78

[56] Marschall R and Wang L 2014 Non-metal doping of transition metal oxides for visible-light photocatalysis *Catal. Today* **225** 111–35

[57] Chen X and Mao S S 2007 Titanium dioxide nanomaterials: synthesis, properties, modifications, and applications *Chem. Rev.* **107** 2891–959

[58] Hashimoto K, Irie H and Fujishima A 2005 A historical overview and future prospects *Jpn. J. Appl. Phys.* **44** 8269
[59] Beltram A, Romero-Ocaña I, Josè Delgado Jaen J, Montini T and Fornasiero P 2016 Photocatalytic valorization of ethanol and glycerol over TiO_2 polymorphs for sustainable hydrogen production *Appl. Catal.,* A **518** 167–75
[60] Di Paola A, Cufalo G, Addamo M, Bellardita M, Campostrini R, Ischia M, Ceccato R and Palmisano L 2008 Photocatalytic activity of nanocrystalline TiO_2 (brookite, rutile and brookite-based) powders prepared by thermohydrolysis of $TiCl_4$ in aqueous chloride solutions *Colloids Surf.,* A **317** 366–76
[61] Wierzbicka E *et al* 2021 Reduced grey brookite for noble metal free photocatalytic H_2 evolution *J. Mater. Chem.* A **9** 1168–79
[62] Di Paola A, Bellardita M and Palmisano L 2013 Brookite, the least known TiO_2 photocatalyst *Catalysts* **3** 36–73
[63] Hanaor D A H and Sorrell C C 2011 Review of the anatase to rutile phase transformation *J. Mater. Sci.* **46** 855–74
[64] Asahi R, Morikawa T, Ohwaki T, Aoki K and Taga Y 2001 Visible-light photocatalysis in nitrogen-doped titanium oxides *Science* **293** 269–71
[65] Janáky C, Rajeshwar K, de Tacconi N R, Chanmanee W and Huda M N 2013 Tungsten-based oxide semiconductors for solar hydrogen generation *Catal. Today* **199** 53–64
[66] Huang Z-F, Song J, Pan L, Zhang X, Wang L and Zou J-J 2015 Tungsten oxides for photocatalysis, electrochemistry, and phototherapy *Adv. Mater.* **27** 5309–27
[67] Zheng H, Ou J Z, Strano M S, Kaner R B, Mitchell A and Kalantar-zadeh K 2011 Nanostructured tungsten oxide—properties, synthesis, and applications *Adv. Funct. Mater.* **21** 2175–96
[68] AlSalka Y, Granone L I, Ramadan W, Hakki A, Dillert R and Bahnemann D W 2019 Iron-based photocatalytic and photoelectrocatalytic nano-structures: facts, perspectives, and expectations *Appl. Catal.* B **244** 1065–95
[69] Kment Š, Sivula K, Naldoni A, Sarmah S P, Kmentová H, Kulkarni M, Rambabu Y, Schmuki P and Zbořil R 2020 FeO-based nanostructures and nanohybrids for photoelectrochemical water splitting *Prog. Mater Sci.* **110** 100632
[70] Bassi P S, Gurudayal, Wong L H and Barber J 2014 Iron based photoanodes for solar fuel production *Phys. Chem. Chem. Phys.* **16** 11834
[71] Sivula K, Le Formal F and Grätzel M 2011 Solar water splitting: progress using hematite (α-Fe_2O_3) photoelectrodes *Chem. Sus. Chem.* **4** 432–49
[72] Tanaka H and Misono M 2001 Advances in designing perovskite catalysts *Curr. Opin. Solid State Mater. Sci.* **5** 381–7
[73] Grabowska E 2016 Selected perovskite oxides: characterization, preparation and photocatalytic properties—a review *Appl. Catal.* B **186** 97–126
[74] van Benthem K, Elsässer C and French R H 2001 Bulk electronic structure of $SrTiO_3$: experiment and theory *J. Appl. Phys.* **90** 6156–64
[75] Asai R, Nemoto H, Jia Q, Saito K, Iwase A and Kudo A 2014 A visible light responsive rhodium and antimony-codoped $SrTiO_3$ powdered photocatalyst loaded with an IrO_2 cocatalyst for solar water splitting *Chem. Commun.* **50** 2543–6
[76] Yang P, Brittman S and Liu C 2015 Nanowires for photovoltaics and artificial photosynthesis *Semiconductor Nanowires: From Next-Generation Electronics to Sustainable EnergySmart Materials Series* (London: Royal Society of Chemistry), 6 277–311

[77] Hochbaum A I and Yang P 2010 Semiconductor nanowires for energy conversion *Chem. Rev.* **110** 527–46
[78] Osterloh F E 2015 Nanoscale effects in water splitting photocatalysis *Solar energy for fuelsTopics in Current Chemistry* 371 ed H Tüysüz *et al* (Cham: Springer), 105–42
[79] Cherevan A S, Deilmann L, Weller T, Eder D and Marschall R 2018 Mesoporous semiconductors: a new model to assess accessible surface area and increased photocatalytic activity? *ACS Appl. Energy Mater.* **1** 5787–99
[80] Edvinsson T 2018 Optical quantum confinement and photocatalytic properties in two-, one- and zero-dimensional nanostructures *R. Soc. Open Sci.* **5** 180387
[81] Brus L E 1983 A simple model for the ionization potential, electron affinity, and aqueous redox potentials of small semiconductor crystallites *J. Chem. Phys.* **79** 5566–71
[82] Mills A and Le Hunte S 1997 An overview of semiconductor photocatalysis *J. Photochem. Photobiol.* A **108** 1–35
[83] Murray C B, Kagan C R and Bawendi M G 2000 Synthesis and characterization of monodisperse nanocrystals and close-packed nanocrystal assemblies *Annu. Rev. Mater. Sci.* **30** 545–610
[84] Alivisatos A P 1996 Semiconductor clusters, nanocrystals, and quantum dots *Science* **271** 933–7
[85] Law M, Goldberger J and Yang P 2004 Semiconductor nanowires and nanotubes *Annu. Rev. Mater. Res.* **34** 83–122
[86] Anpo M, Shima T, Kodama S and Kubokawa Y 1987 Photocatalytic hydrogenation of propyne with water on small-particle titania: size quantization effects and reaction intermediates *J. Phys. Chem.* **91** 4305–10
[87] Kormann C, Bahnemann D W and Hoffmann M R 1988 Preparation and characterization of quantum-size titanium dioxide *J. Phys. Chem.* **92** 5196–201
[88] Hoffman A J, Yee H, Mills G and Hoffmann M R 1992 Photoinitiated polymerization of methyl methacrylate using Q-sized ZnO colloids *J. Phys. Chem.* **96** 5540–6
[89] Jacobsson T J and Edvinsson T 2012 Antireflective coatings of ZnO quantum dots and their photocatalytic activity *RSC Adv.* **2** 10298
[90] Tanaka D, Oaki Y and Imai H 2010 Enhanced photocatalytic activity of quantum-confined tungsten trioxide nanoparticles in mesoporous silica *Chem. Commun.* **46** 5286
[91] Johansson M B, Zietz B, Niklasson G A and Österlund L 2014 Optical properties of nanocrystalline WO_3 and WO_{3-x} thin films prepared by DC magnetron sputtering *J. Appl. Phys.* **115** 213510
[92] Hoffman A J, Mills G, Yee H and Hoffmann M R 1992 Q-sized CdS: synthesis, characterization, and efficiency of photoinitiation of polymerization of several vinylic monomers *J. Phys. Chem.* **96** 5546–52
[93] Holmes M A, Townsend T K and Osterloh F E 2012 Quantum confinement controlled photocatalytic water splitting by suspended CdSe nanocrystals *Chem. Commun.* **48** 371–3
[94] Zhao J, Holmes M A and Osterloh F E 2013 Quantum confinement controls photocatalysis: a free energy analysis for photocatalytic proton reduction at CdSe nanocrystals *ACS Nano* **7** 4316–25
[95] Fondell M, Jacobsson T J, Boman M and Edvinsson T 2014 Optical quantum confinement in low dimensional hematite *J. Mater. Chem.* A **2** 3352–63

[96] Altomare M, Pozzi M, Allieta M, Bettini L G and Selli E 2013 H_2 and O_2 photocatalytic production on TiO_2 nanotube arrays: effect of the anodization time on structural features and photoactivity *Appl. Catal.* B **136-37** 81–8

[97] Chen Z, Dinh H N and Miller E 2013 *Photoelectrochemical Water Splitting: Standards, Experimental Methods, and ProtocolsSpringerBriefs in Energy* (Berlin: Springer)

[98] Liu G, Yu J C, Lu G Q and Cheng H-M 2011 Crystal facet engineering of semiconductor photocatalysts: motivations, advances and unique properties *Chem. Commun.* **47** 6763–83

[99] Parashar M, Shukla V K and Singh R 2020 Metal oxides nanoparticles via sol–gel method: a review on synthesis, characterization and applications *J. Mater. Sci., Mater. Electron.* **31** 3729–49

[100] Yang H G, Sun C H, Qiao S Z, Zou J, Liu G, Smith S C, Cheng H M and Lu G Q 2008 Anatase TiO_2 single crystals with a large percentage of reactive facets *Nature* **453** 638–41

[101] Wang D, Jiang H, Zong X, Xu Q, Ma Y, Li G and Li C 2011 Crystal facet dependence of water oxidation on $BiVO_4$ sheets under visible light irradiation *Chem. Eur. J.* **17** 1275–82

[102] Xie Y P, Liu G, Yin L and Cheng H-M 2012 Crystal facet-dependent photocatalytic oxidation and reduction reactivity of monoclinic WO_3 for solar energy conversion *J. Mater. Chem.* **22** 6746–51

[103] Roy P, Berger S and Schmuki P 2011 TiO_2 nanotubes: synthesis and applications *Angew. Chem. Int. Ed.* **50** 2904–39

[104] Lee K, Mazare A and Schmuki P 2014 One-dimensional titanium dioxide nanomaterials: nanotubes *Chem. Rev.* **114** 9385–454

[105] Carp O, Huisman C L and Reller A 2004 Photoinduced reactivity of titanium dioxide *Prog. Solid State Chem.* **32** 33–177

[106] Macwan D P, Dave P N and Chaturvedi S 2011 A review on nano-TiO_2 sol–gel type syntheses and its applications *J. Mater. Sci.* **46** 3669–86

[107] Sohrabi S, Moraveji M K and Iranshahi D 2020 A review on the design and development of photocatalyst synthesis and application in microfluidic reactors: challenges and opportunities *Rev. Chem. Eng.* **36** 687–722

[108] Shi W, Song S and Zhang H 2013 Hydrothermal synthetic strategies of inorganic semiconducting nanostructures *Chem. Soc. Rev.* **42** 5714–43

[109] Wenderich K, Noack J, Kärgel A, Trunschke A and Mul G 2018 Effect of temperature and pH on phase transformations in citric acid mediated hydrothermal growth of tungsten oxide *Eur. J. Inorg. Chem.* **2018** 917–23

[110] Liu B and Aydil E S 2009 Growth of oriented single-crystalline rutile TiO_2 nanorods on transparent conducting substrates for dye-sensitized solar cells *J. Am. Chem. Soc.* **131** 3985–90

[111] Phoon B L, Lai C W, Juan J C, Show P-L and Chen W-H 2019 A review of synthesis and morphology of $SrTiO_3$ for energy and other applications *Int. J. Energy Res.* **43** 5151–74

[112] Wang Y, He Y, Lai Q and Fan M 2014 Review of the progress in preparing nano TiO_2: an important environmental engineering material *J. Environ. Sci.* **26** 2139–77

[113] Rajeshwar K and de Tacconi N R 2009 Solution combustion synthesis of oxide semiconductors for solar energy conversion and environmental remediation *Chem. Soc. Rev.* **38** 1984–98

[114] Deganello F and Tyagi A K 2018 Solution combustion synthesis, energy and environment: best parameters for better materials *Prog. Cryst. Growth Charact. Mater.* **64** 23–61

[115] Masuda H, Yamada H, Satoh M, Asoh H, Nakao M and Tamamura T 1997 Highly ordered nanochannel-array architecture in anodic alumina *Appl. Phys. Lett.* **71** 2770–2
[116] Riboni F, Nguyen N T, So S and Schmuki P 2016 Aligned metal oxide nanotube arrays: key-aspects of anodic TiO_2 nanotube formation and properties *Nanoscale Horiz.* **1** 445–66
[117] Kowalski D, Kim D and Schmuki P 2013 TiO_2 nanotubes, nanochannels and mesosponge: self-organized formation and applications *Nano Today* **8** 235–64
[118] Prakasam H E, Varghese O K, Paulose M, Mor G K and Grimes C A 2006 Synthesis and photoelectrochemical properties of nanoporous iron (III) oxide by potentiostatic anodization *Nanotechnology* **17** 4285–91
[119] Rangaraju R R, Panday A, Raja K S and Misra M 2009 Nanostructured anodic iron oxide film as photoanode for water oxidation *J. Phys. D: Appl. Phy.* **42** 10
[120] Mohapatra S K, John S E, Banerjee S and Misra M 2009 Water photooxidation by smooth and ultrathin α-Fe_2O_3 nanotube arrays *Chem. Mater.* **21** 3048–55
[121] LaTempa T J, Feng X, Paulose M and Grimes C A 2009 Temperature-dependent growth of self-assembled hematite (α-Fe_2O_3) nanotube arrays: rapid electrochemical synthesis and photoelectrochemical properties *J. Phys. Chem.* C **113** 16293–8
[122] Albu S P, Ghicov A and Schmuki P 2009 High aspect ratio, self-ordered iron oxide nanopores formed by anodization of Fe in ethylene glycol/NH_4F electrolytes *Phys. Status Solidi* **3** 64–6
[123] Lee C-Y, Wang L, Kado Y, Killian M S and Schmuki P 2014 Anodic nanotubular/porous hematite photoanode for solar water splitting: substantial effect of iron substrate purity *Chem. Sus. Chem.* **7** 934–40
[124] Wang L *et al* 2015 Anodic self-organized transparent nanotubular/porous hematite films from Fe thin-films sputtered on FTO and photoelectrochemical water splitting *Res. Chem. Intermed.* **41** 9333–41
[125] Altomare M, Pfoch O, Tighineanu A, Kirchgeorg R, Lee K, Selli E and Schmuki P 2015 Molten o-H_3PO_4: a new electrolyte for the anodic synthesis of self-organized oxide structures – WO_3 nanochannel layers and others *J. Am. Chem. Soc.* **137** 5646–9
[126] Wei W, Macak J M, Shrestha N K and Schmuki P 2009 Thick self-ordered nanoporous Ta_2O_5 films with long-range lateral order *J. Electrochem. Soc.* **156** K104
[127] Altomare M, Cha G and Schmuki P 2020 Anodic nanoporous niobium oxide layers grown in pure molten ortho-phosphoric acid *Electrochim. Acta* **344** 136158
[128] Palacios-Padrós A, Altomare M, Tighineanu A, Kirchgeorg R, Shrestha N K, Díez-Pérez I, Caballero-Briones F, Sanz F and Schmuki P 2014 Growth of ordered anodic SnO_2 nanochannel layers and their use for H_2 gas sensing *J. Mater. Chem.* A **2** 915–20
[129] Assefpour-Dezfuly M, Vlachos C and Andrews E H 1984 Oxide morphology and adhesive bonding on titanium surfaces *J. Mater. Sci.* **19** 3626–39
[130] Zwilling V, Aucouturier M and Darque-Ceretti E 1999 Anodic oxidation of titanium and TA6V alloy in chromic media. An electrochemical approach *Electrochim. Acta* **45** 921–9
[131] Beranek R, Hildebrand H and Schmuki P 2003 Self-organized porous titanium oxide prepared in H_2SO_4/HF electrolytes *Electrochem. Solid-State Lett.* **6** B12
[132] Albu S P, Ghicov A, Macak J M and Schmuki P 2007 250 μm long anodic TiO_2 nanotubes with hexagonal self-ordering *Phys. Status Solidi* **1** R65–7
[133] Macak J M, Tsuchiya H, Taveira L, Aldabergerova S and Schmuki P 2005 Smooth anodic TiO_2 nanotubes *Angew. Chem. Int. Ed.* **44** 7463–5

[134] Macak J M, Tsuchiya H, Ghicov A, Yasuda K, Hahn R, Bauer S and Schmuki P 2007 TiO_2 nanotubes: self-organized electrochemical formation, properties and applications *Curr. Opin. Solid State Mater. Sci.* **11** 3–18
[135] Ruff T, Hahn R and Schmuki P 2011 From anodic TiO_2 nanotubes to hexagonally ordered TiO_2 nanocolumns *Appl. Surf. Sci.* **257** 8177–81
[136] So S, Lee K and Schmuki P 2012 Ultrafast growth of highly ordered anodic TiO_2 nanotubes in lactic acid electrolytes *J. Am. Chem. Soc.* **134** 11316–8
[137] Kment S, Riboni F, Pausova S, Wang L, Wang L, Han H, Hubicka Z, Krysa J, Schmuki P and Zboril R 2017 Photoanodes based on TiO_2 and α-Fe_2O_3 for solar water splitting—superior role of 1D nanoarchitectures and of combined heterostructures *Chem. Soc. Rev.* **46** 3716–69
[138] Perednis D and Gauckler L J 2005 Thin film deposition using spray pyrolysis *J. Electroceram.* **14** 103–11
[139] Rani A, Reddy R, Sharma U, Mukherjee P, Mishra P, Kuila A, Sim L C and Saravanan P 2018 A review on the progress of nanostructure materials for energy harnessing and environmental remediation *J. Nanostructure Chem.* **8** 255–91
[140] Yang X-Y, Chen L-H, Li Y, Rooke J C, Sanchez C and Su B-L 2017 Hierarchically porous materials: synthesis strategies and structure design *Chem. Soc. Rev.* **46** 481–558
[141] Li Puma G, Bono A, Krishnaiah D and Collin J G 2008 Preparation of titanium dioxide photocatalyst loaded onto activated carbon support using chemical vapor deposition: a review paper *J. Hazard. Mater.* **157** 209–19
[142] Takeda S, Suzuki S, Odaka H and Hosono H 2001 Photocatalytic TiO_2 thin film deposited onto glass by DC magnetron sputtering *Thin Solid Films* **392** 338–44
[143] Johnson R W, Hultqvist A and Bent S F 2014 A brief review of atomic layer deposition: from fundamentals to applications *Mater. Today* **17** 236–46
[144] Leskelä M and Ritala M 2002 Atomic layer deposition (ALD): from precursors to thin film structures *Thin Solid Films* **409** 138–46
[145] Knisley T J, Kalutarage L C and Winter C H 2013 Precursors and chemistry for the atomic layer deposition of metallic first row transition metal films *Coord. Chem. Rev.* **257** 3222–31
[146] Das S and Srivastava V C 2016 Microfluidic-based photocatalytic microreactor for environmental application: a review of fabrication substrates and techniques, and operating parameters *Photochem. Photobiol. Sci.* **15** 714–30
[147] Luo C, Xu C, Lv L, Li H, Huang X and Liu W 2020 Review of recent advances in inorganic photoresists *RSC Adv.* **10** 8385–95
[148] Wallraff G M and Hinsberg W D 1999 Lithographic imaging techniques for the formation of nanoscopic features *Chem. Rev.* **99** 1801–22
[149] Nakata K and Fujishima A 2012 TiO_2 photocatalysis: design and applications *J. Photochem. Photobiol.* C **13** 169–89
[150] Nah Y-C, Paramasivam I and Schmuki P 2010 Doped TiO_2 and TiO_2 nanotubes: synthesis and applications *Chem. Phys. Chem.* **11** 2698–713
[151] Dozzi M V and Selli E 2013 Doping TiO_2 with p-block elements: effects on photocatalytic activity *J. Photochem. Photobiol.* C **14** 13–28
[152] Nasirian M, Lin Y P, Bustillo-Lecompte C F and Mehrvar M 2018 Enhancement of photocatalytic activity of titanium dioxide using non-metal doping methods under visible light: a review *Int. J. Environ. Sci. Technol.* **15** 2009–32

[153] Magalhães P, Andrade L, Nunes O C and Mendes A 2017 Titanium dioxide photocatalysis: fundamentals and application on photoinactivation *Rev. Adv. Mater. Sci.* **51** 91–129 https://www.ipme.ru/e-journals/RAMS/no_25117/01_25117_magalhaes.html
[154] Kang X, Liu S, Dai Z, He Y, Song X and Tan Z 2019 Titanium dioxide: from engineering to applications *Catalysts* **9** 191
[155] Medhi R, Marquez M D and Lee T R 2020 Visible-light-active doped metal oxide nanoparticles: review of their synthesis, properties, and applications *ACS Appl. Nano Mater.* **3** 6156–85
[156] Macak J M, Gong B G, Hueppe M and Schmuki P 2007 Filling of TiO_2 nanotubes by self-doping and electrodeposition *Adv. Mater.* **19** 3027–31
[157] Wang L, Wang Y, Cheng Y, Liu Z, Guo Q, Ha M N and Zhao Z 2016 Hydrogen-treated mesoporous WO_3 as a reducing agent of CO_2 to fuels (CH_4 and CH_3OH) with enhanced photothermal catalytic performance *J. Mater. Chem.* A **4** 5314–22
[158] Hamdy M S, Amrollahi R and Mul G 2012 Surface Ti^{3+}-containing (blue) titania: a unique photocatalyst with high activity and selectivity in visible light-stimulated selective oxidation *ACS Catal.* **2** 2641–7
[159] Zhang Y, Xing Z, Liu X, Li Z, Wu X, Jiang J, Li M, Zhu Q and Zhou W 2016 Ti^{3+} self-doped blue TiO_2(B) single-crystalline nanorods for efficient solar-driven photocatalytic performance *ACS Appl. Mater. Interfaces* **8** 26851–9
[160] Zhou Y, Chen C, Wang N, Li Y and Ding H 2016 Stable Ti^{3+} self-doped anatase-rutile mixed TiO_2 with enhanced visible light utilization and durability *J. Phys. Chem.* C **120** 6116–24
[161] Zuo F, Wang L, Wu T, Zhang Z, Borchardt D and Feng P 2010 Self-doped Ti^{3+} enhanced photocatalyst for hydrogen production under visible light *J. Am. Chem. Soc.* **132** 11856–7
[162] Chen X, Liu L, Yu P Y and Mao S S 2011 Increasing solar absorption for photocatalysis with black hydrogenated titanium dioxide nanocrystals *Science* **331** 746–50
[163] Liu N, Schneider C, Freitag D, Hartmann M, Venkatesan U, Müller J, Spiecker E and Schmuki P 2014 Black TiO_2 nanotubes: cocatalyst-free open-circuit hydrogen generation *Nano Lett.* **14** 3309–13
[164] Liu N *et al* 2014 Hydrogenated anatase: strong photocatalytic dihydrogen evolution without the use of a co-catalyst *Angew. Chem. Int. Ed.* **53** 14201–5
[165] Liu N, Schneider C, Freitag D, Zolnhofer E M, Meyer K and Schmuki P 2016 Noble-metal-free photocatalytic H_2 generation: active and inactive 'black' TiO_2 nanotubes and synergistic effects *Chem. Eur. J.* **22** 13810–4
[166] Naldoni A, Allieta M, Santangelo S, Marelli M, Fabbri F, Cappelli S, Bianchi C L, Psaro R and Dal Santo V 2012 Effect of nature and location of defects on bandgap narrowing in black TiO_2 nanoparticles *J. Am. Chem. Soc.* **134** 7600–3
[167] Naldoni A, Altomare M, Zoppellaro G, Liu N, Kment Š, Zbořil R and Schmuki P 2019 Photocatalysis with reduced TiO_2: from black TiO_2 to cocatalyst-free hydrogen production *ACS Catal.* **9** 345–64
[168] Naldoni A, Fabbri F, Altomare M, Marelli M, Psaro R, Selli E, Salviati G and Dal Santo V 2015 The critical role of intragap states in the energy transfer from gold nanoparticles to TiO_2 *Phys. Chem. Chem. Phys.* **17** 4864–9
[169] Peng B, Meng X, Tang F, Ren X, Chen D and Ren J 2009 General synthesis and optical properties of monodisperse multifunctional metal-ion-doped TiO_2 hollow particles. *The J. Phys. Chem.* C **113** 20240–5

[170] Guo M, Lu J, Wu Y, Wang Y and Luo M 2011 UV and visible Raman studies of oxygen vacancies in rare-earth-doped ceria *Langmuir* **27** 3872–7

[171] Wang Y, Zhang R, Li J, Li L and Lin S 2014 First-principles study on transition metal-doped anatase TiO_2 *Nanoscale Res. Lett.* **9** 46

[172] Zhang Z *et al* 2019 Generalized synthetic strategy for transition-metal-doped brookite-phase TiO_2 nanorods *J. Am. Chem. Soc.* **141** 16548–52

[173] Liu Y, Wang W, Xu X, Marcel Veder J-P and Shao Z 2019 Recent advances in anion-doped metal oxides for catalytic applications *J. Mater. Chem.* A **7** 7280–300

[174] Qiu X and Burda C 2007 Chemically synthesized nitrogen-doped metal oxide nanoparticles *Chem. Phys.* **339** 1–10

[175] Dunnill C W and Parkin I P 2011 Nitrogen-doped TiO_2 thin films: photocatalytic applications for healthcare environments *Dalton Trans.* **40** 1635–40

[176] Di Valentin C and Pacchioni G 2013 Trends in non-metal doping of anatase TiO_2: B, C, N and F *Catal. Today* **206** 12–8

[177] Chen H and Dawson J A 2015 Nature of nitrogen-doped anatase TiO_2 and the origin of Its visible-light activity *J. Phys. Chem.* C **119** 15890–5

[178] Huang L, Fu W, Fu X, Zong B, Liu H, Bala H, Wang X, Sun G, Cao J and Zhang Z 2017 Facile and large-scale preparation of N doped TiO_2 photocatalyst with high visible light photocatalytic activity *Mater. Lett.* **209** 585–8

[179] Piątkowska A, Janus M, Szymański K and Mozia S 2021 C-, N- and S-doped TiO_2 photocatalysts: a review *Catalysts* **11** 144

[180] Serpone N 2006 Is the band gap of pristine TiO_2 narrowed by anion- and cation-doping of titanium dioxide in second-generation photocatalysts? *J. Phys. Chem.* B **110** 24287–93

[181] García-Mota M, Vojvodic A, Abild-Pedersen F and Nørskov J K 2013 Electronic origin of the surface reactivity of transition-metal-doped TiO_2(110) *J. Phys. Chem.* C **117** 460–5

[182] Xu C, Ravi Anusuyadevi P, Aymonier C, Luque R and Marre S 2019 Nanostructured materials for photocatalysis *Chem. Soc. Rev.* **48** 3868–902

[183] Huang Z-F, Pan L, Zou J-J, Zhang X and Wang L 2014 Nanostructured bismuth vanadate-based materials for solar-energy-driven water oxidation: a review on recent progress *Nanoscale* **6** 14044–63

[184] A M, J M, Ashokkumar M and Arunachalam P 2018 A review on $BiVO_4$ photocatalyst: activity enhancement methods for solar photocatalytic applications *Appl. Catal.,* A **555** 47–74

[185] Ouyang S and Ye J 2011 β-$AgAl_{1-x}Ga_xO_2$ solid-solution photocatalysts: continuous modulation of electronic structure toward high-performance visible-light photoactivity *J. Am. Chem. Soc.* **133** 7757–63

[186] Wenderich K and Mul G 2016 Methods, mechanism, and applications of photodeposition in photocatalysis: a review *Chem. Rev.* **116** 14587–619

[187] Bai S, Yin W, Wang L, Li Z and Xiong Y 2016 Surface and interface design in cocatalysts for photocatalytic water splitting and CO_2 reduction *RSC Adv.* **6** 57446–63

[188] Chen Y, Feng X, Guo X and Zheng W 2019 Toward a fundamental understanding of factors affecting the function of cocatalysts in photocatalytic water splitting *Curr. Opin. Green Sustain. Chem.* **17** 21–8

[189] Li X, Yu J, Jaroniec M and Chen X 2019 Cocatalysts for selective photoreduction of CO_2 into solar fuels *Chem. Rev.* **119** 3962–4179

[190] Ran J, Zhang J, Yu J, Jaroniec M and Qiao S Z 2014 Earth-abundant cocatalysts for semiconductor-based photocatalytic water splitting *Chem. Soc. Rev.* **43** 7787–812

[191] Yang J, Wang D, Han H and Li C 2013 Roles of cocatalysts in photocatalysis and photoelectrocatalysis *Acc. Chem. Res.* **46** 1900–9

[192] Zhong S, Xi Y, Wu S, Liu Q, Zhao L and Bai S 2020 Hybrid cocatalysts in semiconductor-based photocatalysis and photoelectrocatalysis *J. Mater. Chem.* A **8** 14863–94

[193] Lettieri S, Pavone M, Fioravanti A, Santamaria Amato L and Maddalena P 2021 Charge carrier processes and optical properties in TiO_2 and TiO_2-based heterojunction photocatalysts: a review *Materials* **14** 1645

[194] Lee I, Kim W-J and Lee D C 2020 Design of metallic cocatalysts in heterostructured nanoparticles for photocatalytic CO_2-to-hydrocarbon conversion *J. Phys. D: Appl. Phys.* **53** 123001

[195] Maeda K 2011 Photocatalytic water splitting using semiconductor particles: history and recent developments *J. Photochem. Photobiol.* C **12** 237–68

[196] Meng A, Zhang L, Cheng B and Yu J 2019 Dual cocatalysts in TiO_2 photocatalysis *Adv. Mater.* **31** 1807660

[197] Liu J, Li Y, Zhou X, Jiang H, Yang H G and Li C 2020 Positively charged Pt-based cocatalysts: an orientation for achieving efficient photocatalytic water splitting *J. Mater. Chem.* A **8** 17–26

[198] Polliotto V, Livraghi S, Krukowska A, Dozzi M V, Zaleska-Medynska A, Selli E and Giamello E 2018 Copper-modified TiO_2 and $ZrTiO_4$: Cu oxidation state evolution during photocatalytic hydrogen production *ACS Appl. Mater. Interfaces* **10** 27745–56

[199] Muñoz-Batista M J, Motta Meira D, Colón G, Kubacka A and Fernández-García M 2018 Phase-contact engineering in mono- and bimetallic Cu–Ni co-catalysts for hydrogen photocatalytic materials *Angew. Chem. Int. Ed.* **57** 1199–203

[200] Schubert J S, Popovic J, Haselmann G M, Nandan S P, Wang J, Giesriegl A, Cherevan A S and Eder D 2019 Immobilization of Co, Mn, Ni and Fe oxide co-catalysts on TiO_2 for photocatalytic water splitting reactions *J. Mater. Chem.* A **7** 18568–79

[201] Altomare M, Nguyen N T and Schmuki P 2016 Templated dewetting: designing entirely self-organized platforms for photocatalysis *Chem. Sci.* **7** 6865–86

[202] Maeda K, Teramura K, Lu D, Saito N, Inoue Y and Domen K 2006 Noble-metal/Cr_2O_3 core/shell nanoparticles as a cocatalyst for photocatalytic overall water splitting *Angew. Chem. Int. Ed.* **45** 7806–9

[203] Tossi C, Hällström L, Selin J, Vaelma M, See E, Lahtinen J and Tittonen I 2019 Size- and density-controlled photodeposition of metallic platinum nanoparticles on titanium dioxide for photocatalytic applications *J. Mater. Chem.* A **7** 14519–25

[204] Lee Y, Kim E, Park Y, Kim J, Ryu W, Rho J and Kim K 2018 Photodeposited metal-semiconductor nanocomposites and their applications *J. Materiomics* **4** 83–94

[205] Nakamatsu H, Kawai T, Koreeda A and Kawai S 1986 Electron-microscopic observation of photodeposited Pt on TiO_2 particles in relation to photocatalytic activity *J. Chem. Soc., Faraday Trans. 1* F **82** 527–31

[206] Wenderich K, Han K and Mul G 2018 The effect of methanol on the photodeposition of Pt nanoparticles on tungsten oxide *Part. Part. Syst. Charact.* **35** 1700250

[207] Ohno T, Sarukawa K and Matsumura M 2002 Crystal faces of rutile and anatase TiO_2 particles and their roles in photocatalytic reactions *New J. Chem.* **26** 1167–70

[208] Li R, Zhang F, Wang D, Yang J, Li M, Zhu J, Zhou X, Han H and Li C 2013 Spatial separation of photogenerated electrons and holes among {010} and {110} crystal facets of $BiVO_4$ *Nat. Commun.* **4** 1432

[209] Thompson C V 2012 Solid-state dewetting of thin films *Annu. Rev. Mater. Res.* **42** 399–434

[210] Leroy F, Borowik Ł, Cheynis F, Almadori Y, Curiotto S, Trautmann M, Barbé J C and Müller P 2016 How to control solid state dewetting: a short review *Surf. Sci. Rep.* **71** 391–409

[211] Yoo J E, Lee K, Altomare M, Selli E and Schmuki P 2013 Self-organized arrays of single-metal catalyst particles in TiO_2 cavities: a highly efficient photocatalytic system *Angew. Chem. Int. Ed.* **52** 7514–7

[212] Licklederer M, Mohammadi R, Nguyen N T, Park H, Hejazi S, Halik M, Vogel N, Altomare M and Schmuki P 2019 Dewetted Au nanoparticles on TiO_2 surfaces: evidence of a size-independent plasmonic photoelectrochemical response *J. Phys. Chem.* C **123** 16934–42

[213] Tesler A B, Sannomiya T, Hejazi S, Mohammadi R, Vogel N, Altomare M and Schmuki P 2021 Metallic nanoparticle-on-mirror: multiple-band light harvesting and efficient photocurrent generation under visible light irradiation *Nano Energy* **90** 106609

[214] Yoo J, Altomare M, Mokhtar M, Alshehri A, Al-Thabaiti S A, Mazare A and Schmuki P 2016 Photocatalytic H_2 generation using dewetted Pt-decorated TiO_2 nanotubes: optimized dewetting and oxide crystallization by a multiple annealing process *J. Phys. Chem.* C **120** 15884–92

[215] Mullins W W 1957 Theory of thermal grooving *J. Appl. Phys.* **28** 333–9

[216] Jiran E and Thompson C V 1990 Capillary instabilities in thin films *J. Electron. Mater.* **19** 1153–60

[217] Spanu D, Recchia S, Mohajernia S, Tomanec O, Kment Š, Zboril R, Schmuki P and Altomare M 2018 Templated dewetting–alloying of NiCu bilayers on TiO_2 nanotubes enables efficient noble-metal-free photocatalytic H_2 evolution *ACS Catal.* **8** 5298–305

[218] Chen H and Wang L 2014 Nanostructure sensitization of transition metal oxides for visible-light photocatalysis *Beilstein J. Nanotechnol.* **5** 696–710

[219] Dong S, Feng J, Fan M, Pi Y, Hu L, Han X, Liu M, Sun J and Sun J 2015 Recent developments in heterogeneous photocatalytic water treatment using visible light-responsive photocatalysts: a review *RSC Adv.* **5** 14610–30

[220] Liu B, Fang Y, Li Z and Xu S 2015 Visible-light nanostructured photocatalysts—a review *J. Nanosci. Nanotechnol.* **15** 889–920

[221] Pelaez M *et al* 2012 A review on the visible light active titanium dioxide photocatalysts for environmental applications *Appl. Catal.* B **125** 331–49

[222] Rehman S, Ullah R, Butt A M and Gohar N D 2009 Strategies of making TiO_2 and ZnO visible light active *J. Hazard. Mater.* **170** 560–9

[223] Hagfeldt A, Boschloo G, Sun L, Kloo L and Pettersson H 2010 Dye-sensitized solar cells *Chem. Rev.* **110** 6595–663

[224] O'Regan B and Grätzel M 1991 A low-cost, high-efficiency solar cell based on dye-sensitized colloidal TiO_2 films *Nature* **353** 737–40

[225] Hou W and Cronin S B 2013 A review of surface plasmon resonance-enhanced photocatalysis *Adv. Funct. Mater.* **23** 1612–19

[226] Ueno K and Misawa H 2013 Surface plasmon-enhanced photochemical reactions *J. Photochem. Photobiol.* C **15** 31–52

[227] Brongersma M L, Halas N J and Nordlander P 2015 Plasmon-induced hot carrier science and technology *Nat. Nanotechnol.* **10** 25–34
[228] Zheng Z, Xie W, Huang B and Dai Y 2018 Plasmon-enhanced solar water splitting on metal–semiconductor photocatalysts *Chem. Eur. J.* **24** 18322–33
[229] Zhou X, Liu G, Yu J and Fan W 2012 Surface plasmon resonance-mediated photocatalysis by noble metal-based composites under visible light *J. Mater. Chem.* **22** 21337–54
[230] Jiang R, Li B, Fang C and Wang J 2014 Metal/semiconductor hybrid nanostructures for plasmon-enhanced applications *Adv. Mater.* **26** 5274–309
[231] Valenti M, Jonsson M P, Biskos G, Schmidt-Ott A and Smith W A 2016 Plasmonic nanoparticle–semiconductor composites for efficient solar water splitting *J. Mater. Chem. A* **4** 17891–912
[232] Linic S, Christopher P and Ingram D B 2011 Plasmonic–metal nanostructures for efficient conversion of solar to chemical energy *Nat. Mater.* **10** 911–21
[233] Wu N 2018 Plasmonic metal–semiconductor photocatalysts and photoelectrochemical cells: a review *Nanoscale* **10** 2679–96
[234] Cushing S K and Wu N 2016 Progress and perspectives of plasmon-enhanced solar energy conversion *J. Phys. Chem. Lett.* **7** 666–75
[235] O'Neill D B, Prezgot D, Ianoul A, Otto C, Mul G and Huijser A 2020 Silver nanocubes coated in ceria: core/shell size effects on light-induced charge transfer *ACS Appl. Mater. Interfaces* **12** 1905–12
[236] Li H, Zhou Y, Tu W, Ye J and Zou Z 2015 State-of-the-art progress in diverse heterostructured photocatalysts toward promoting photocatalytic performance *Adv. Funct. Mater.* **25** 998–1013
[237] Zhang M, An T, Liu X, Hu X, Sheng G and Fu J 2010 Preparation of a high-activity ZnO/TiO_2 photocatalyst via homogeneous hydrolysis method with low temperature crystallization *Mater. Lett.* **64** 1883–6
[238] Huang L, Peng F, Wang H, Yu H and Li Z 2009 Preparation and characterization of Cu_2O/TiO_2 nano–nano heterostructure photocatalysts *Catal. Commun.* **10** 1839–43
[239] Gopidas K R, Bohorquez M and Kamat P V 1990 Photophysical and photochemical aspects of coupled semiconductors: charge-transfer processes in colloidal cadmium sulfide–titania and cadmium sulfide–silver(I) iodide systems *J. Phys. Chem.* **94** 6435–40
[240] Phang S J, Wong V-L, Tan L-L and Chai S-P 2020 Recent advances in homojunction-based photocatalysis for sustainable environmental remediation and clean energy generation *Appl. Mater. Today* **20** 100741
[241] Pan L, Wang S, Xie J, Wang L, Zhang X and Zou J-J 2016 Constructing TiO_2 p–n homojunction for photoelectrochemical and photocatalytic hydrogen generation *Nano Energy* **28** 296–303
[242] Allen N S, Mahdjoub N, Vishnyakov V, Kelly P J and Kriek R J 2018 The effect of crystalline phase (anatase, brookite and rutile) and size on the photocatalytic activity of calcined polymorphic titanium dioxide (TiO_2) *Polym. Degrad. Stab.* **150** 31–6
[243] Scanlon D O *et al* 2013 Band alignment of rutile and anatase TiO_2 *Nat. Mater.* **12** 798–801
[244] Li H, Sun B, Yang F, Wang Z, Xu Y, Tian G, Pan K, Jiang B and Zhou W 2019 Homojunction and defect synergy-mediated electron–hole separation for solar-driven mesoporous rutile/anatase TiO_2 microsphere photocatalysts *RSC Adv.* **9** 7870–7

[245] Abe R, Takata T, Sugihara H and Domen K 2005 Photocatalytic overall water splitting under visible light by TaON and WO_3 with an IO_3^-/I^- shuttle redox mediator *Chem. Commun.* 3829–31
[246] Abe R, Higashi M and Domen K 2011 Overall water splitting under visible light through a two-step photoexcitation between TaON and WO_3 in the presence of an iodate–iodide shuttle redox mediator *Chem. Sus. Chem.* **4** 228–37
[247] Ajmal A, Majeed I, Malik R N, Idriss H and Nadeem M A 2014 Principles and mechanisms of photocatalytic dye degradation on TiO_2 based photocatalysts: a comparative overview *RSC Adv.* **4** 37003–26
[248] Swinehart D F 1962 The Beer–Lambert law *J. Chem. Educ.* **39** 333
[249] Herrmann J-M 2010 Fundamentals and misconceptions in photocatalysis *J. Photochem. Photobiol.* A **216** 85–93
[250] Ohtani B 2008 Preparing articles on photocatalysis—beyond the illusions, misconceptions, and speculation *Chem. Lett.* **37** 216–29
[251] Ohtani B 2010 Photocatalysis A to Z—what we know and what we do not know in a scientific sense *J. Photochem. Photobiol.* C **11** 157–78
[252] Dozzi M V, Livraghi S, Giamello E and Selli E 2011 Photocatalytic activity of S- and F-doped TiO_2 in formic acid mineralization *Photochem. Photobiol. Sci.* **10** 343–9
[253] Dozzi M V, D'Andrea C, Ohtani B, Valentini G and Selli E 2013 Fluorine-doped TiO_2 materials: photocatalytic activity vs time-resolved photoluminescence *J. Phys. Chem.* C **117** 25586–95
[254] Takata T, Jiang J, Sakata Y, Nakabayashi M, Shibata N, Nandal V, Seki K, Hisatomi T and Domen K 2020 Photocatalytic water splitting with a quantum efficiency of almost unity *Nature* **581** 411–4
[255] Ghoussoub M, Xia M, Duchesne P N, Segal D and Ozin G 2019 Principles of photothermal gas-phase heterogeneous CO_2 catalysis *Energy Environ. Sci.* **12** 1122–42
[256] Keller N, Ivanez J, Highfield J and Ruppert A M 2021 Photo-/thermal synergies in heterogeneous catalysis: towards low-temperature (solar-driven) processing for sustainable energy and chemicals *Appl. Catal.* B **296** 120320
[257] Jacinto M J, Ferreira L F and Silva V C 2020 Magnetic materials for photocatalytic applications—a review *J. Sol–Gel Sci. Technol.* **96** 1–14
[258] Singh P, Sharma K, Hasija V, Sharma V, Sharma S, Raizada P, Singh M, Saini A K, Hosseini-Bandegharaei A and Thakur V K 2019 Systematic review on applicability of magnetic iron oxides–integrated photocatalysts for degradation of organic pollutants in water *Mater. Today Chem.* **14** 100186
[259] Malato S, Maldonado M I, Fernández-Ibáñez P, Oller I, Polo I and Sánchez-Moreno R 2016 Decontamination and disinfection of water by solar photocatalysis: the pilot plants of the Plataforma Solar de Almeria *Mater. Sci. Semicond. Process.* **42** 15–23
[260] Ballari M M and Brouwers H J H 2013 Full scale demonstration of air-purifying pavement *J. Hazard. Mater.* **254–55** 406–14
[261] Maggos T, Plassais A, Bartzis J G, Vasilakos C, Moussiopoulos N and Bonafous L 2008 Photocatalytic degradation of NO_x in a pilot street canyon configuration using TiO_2-mortar panels *Environ. Monit. Assess.* **136** 35–44

IOP Publishing

Sustainable Nanomaterials for Energy Applications

Ana Cremades and David Maestre

Chapter 7

Thermoelectric energy harvesting with nanomaterials

Mario Amado, Francisco Domínguez-Adame and David Sánchez

The current challenges of coping with ever-increasing energy consumption worldwide necessitate efforts to find sustainable resources. This paradigm shift has to take place alongside the quest for more efficient ways of saving the energy wasted in industrial facilities and domestic applications. In this context, research into thermoelectricity explores the possibility of benefiting from dissipated energy and making devices less energy demanding. Achieving the goal of efficient heat-to-electricity conversion requires materials with strongly suppressed thermal conductivity but high electronic conduction. In this chapter, we present an overview of the key experimental and theoretical results related to the basics of thermoelectric nanomaterials. The focus is placed on the physical mechanisms by which the efficiency of nanomaterials can be improved.

7.1 Introduction

According to the BP Statistical Review of World Energy [1], primary energy consumption grew at a rate of 2.9% worldwide in 2019, the fastest growth since 2010. The ever-increasing world energy consumption relies strongly on non-renewable resources (oil, coal, natural gas) and nuclear power, while environmentally friendly resources (traditional biofuels, hydropower, wind, solar energy) are still used to a much lesser extent (see figure 7.1). Limited reserves, environmental risks and deterioration, acid precipitation, pollution, and global warming are major drawbacks of an energy system dominated by fossil fuels. Thus, progress toward decarbonization is urgent and indefeasible. Shifting industrial and residential energy needs toward renewable energy sources will ultimately make non-renewable energy demands fade away.

doi:10.1088/978-0-7503-3531-7ch7 © IOP Publishing Ltd 2023

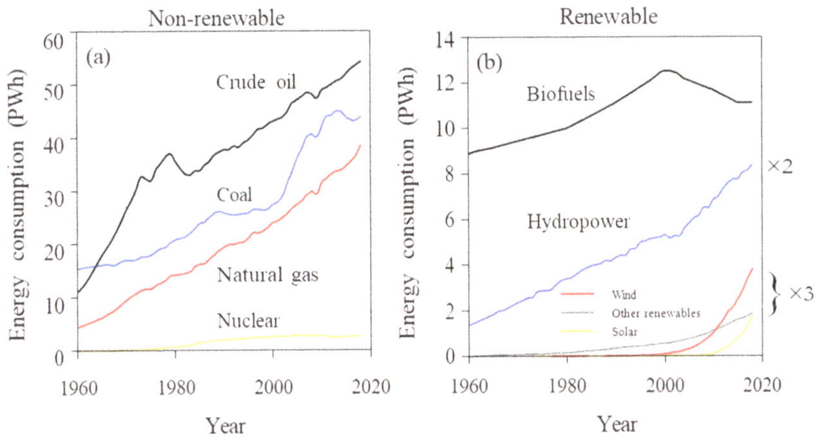

Figure 7.1. (a) Non-renewable and (b) renewable global primary energy consumption in the last six decades, measured in petawatt-hours per year. Note the different vertical scales of both panels and the scale factors in the right-hand panel. Data retrieved from [1, 2].

Investment in sustainable resources must be accompanied by a quest for more efficient ways to save the energy wasted in industrial facilities and domestic applications. Energy harvesting from heat leakages in factories, incinerators, power plants, boilers, automobile and aircraft engines, mobile electronics, and buildings aims for a substantial reduction in overall power consumption. In this context, research into thermoelectricity helps in developing new materials with high performance as well as exploring innovative physical mechanisms for heat-to-electricity conversion, thus paving the way for the efficient harvesting of waste heat. However, heat and electricity are two very different forms of energy [3]. On the one hand, heat is a ubiquitous but low-quality form of energy. On the other hand, electricity is a versatile energy source, but its production is difficult and expensive. Thermoelectricity bridges these two forms of energy by converting the heat emanating from a hot source directly into electricity without the use of moving parts or greenhouse gases. Hence, thermoelectric solid-state devices can work efficiently for a long time with low maintenance, even if temperature gradients are small [4].

Although the phenomenon of thermoelectricity was discovered two centuries ago thanks to the pioneering work of Thomas Johann Seebeck (1770–1831) and Jean Charles Athanase Peltier (1785–1845), significant progress has only been achieved in recent years with the development of tailored inorganic and organic materials (see reference [5] for a recent historical survey of thermoelectricity). Despite their advantages, however, thermoelectric devices are scarce and mainly restricted to niche markets such as power generation in space or Peltier cooling in optoelectronics and small refrigerators. From the materials science perspective, the low efficiency of thermoelectric devices seriously limits their widespread use in energy-harvesting applications. The efficiency of the best known thermoelectric modules is usually lower than 10%, whereas the efficiency of traditional heat engines reaches values of around 40%.

Since the 1950s, the thermoelectric efficiency of devices or materials has been characterized by the so-called dimensionless *figure of merit*, defined as $ZT = \sigma S^2 T/\kappa$ [6, 7]. Here, S stands for the Seebeck coefficient (thermopower), and σ and κ are the electrical and thermal conductivities at temperature T, respectively. In general, electrons (el) and phonons (ph) may contribute to heat transport and consequently the thermal conductivity splits according to $\kappa = \kappa_{el} + \kappa_{ph}$. It is generally accepted that a minimum ZT of the order of three is required for widespread applications but, with a few exceptions, bulk materials fall well below this limit. Approaches to push materials toward efficient thermoelectric performance must address ways to enhance S and σ while keeping κ as low as possible. Unfortunately, the unfavorable interdependence between S, σ, and κ hampers progress toward obtaining efficient thermoelectric materials and devices for real applications.

In this chapter, we focus on strategies for energy harvesting that overcome the limitations of bulk thermoelectric materials. Specifically, we discuss the fundamentals of charge and energy transport when the system size is on the order of a few nanometers. In addition to the physical and chemical properties of the constituent materials, such as their electrical and thermal conductivities, the size of a nanoscale system becomes an additional and controllable parameter that can be used to tune the response of thermoelectric devices, as briefly discussed in section 7.2. In section 7.3, we introduce the scattering approach as a suitable framework in which to study the dependence of the electric and heat currents on the size and geometry of thermoelectric nanodevices. Once the theoretical background has been presented, we proceed to analyze quantum effects and lattice heat conduction in nanowires (NWs) (section 7.4), quantum dots (section 7.5), and quantum rings (section 7.6). In addition, the usefulness and advantages of new two-dimensional (2D) materials in thermoelectricity are thoroughly discussed in section 7.7. Section 7.8 is devoted to applications and devices. Finally, section 7.9 concludes this chapter with a brief summary of the main concepts.

7.2 Why nanoscale materials?

Mildred Dresselhaus (1930–2017)

Recent advances in nanotechnology have made possible new routes for achieving unprecedented thermoelectric efficiency. It is now established on solid grounds that nanometer-sized objects may exhibit thermoelectric efficiencies unachievable with bulk materials [8–17]. Several physical mechanisms can be considered at the nanoscale to overcome the limitations of bulk thermoelectric materials. For instance, phonon mean free paths, ℓ_{ph}, are usually much longer than those of electrons, ℓ_{el}. Dresselhaus and coworkers predicted long ago that NWs may exploit this difference in order to achieve a significant reduction of the lattice thermal conductivity while maintaining good electric properties [8, 9], provided that their diameter is larger than ℓ_{el} but smaller than ℓ_{ph}. Furthermore, quantum effects can be used to overcome the limitations that stem from the classical Wiedemann–Franz law [18]. For instance, nanodevices that display sharp resonances in their electric conductivity (such as Fano line shapes [19]) are considered ideal candidates for thermoelectricity at the nanoscale because the ratio σ/κ_{el} increases well above the Wiedemann–Franz limit [18, 20–27]. This fact can be understood from Mott's famous formula for thermopower [28]:

$$S = -\frac{\pi^2}{3}\frac{k_B^2 T}{e}\frac{\mathrm{d}\ln\sigma(E)}{\mathrm{d}E}\bigg|_{E=E_F}, \tag{7.1}$$

where E_F is the Fermi energy. A large thermopower is available if E_F coincides with a steep slope in the electric conductivity. To this end, nanostructures are particularly useful due to the discrete nature of their energy spectrum. As their size decreases, their energy level spacing increases and nanoconductors can thus work as energy filters, allowing the passage of only one charge type (electrons or holes). In this way, the particle and energy currents can become proportional to each other and the thermoelectric efficiency can be maximized. In nanostructures, therefore, enhanced phonon scattering due to their reduced size, even in the absence of phonon–phonon interactions or Umklapp processes, combined with sharp resonances in their electric conductivity, result in high values of ZT. An additional advantage of nanostructures is their easy integration into modern electronic technology, which permits the recovery of some of the large thermal dissipation of integrated circuits and chips (a major challenge nowadays) and the local cooling of individual elements, thereby increasing their reliability.

7.3 Charge and energy transport at the nanoscale

A convenient theoretical framework for the investigation of electron transport in nanostructures is the Landauer–Büttiker approach, which is based on the quantum scattering formalism [29–32]. Its conceptual simplicity and broad applicability allow us to make straightforward calculations that are valid for a wide variety of systems. This approach relies on a few general assumptions that make its predictions largely independent of the chosen theoretical model. It conceives the nanostructure to be a quantum system coupled to macroscopic electronic reservoirs (metallic leads) characterized by well-defined electrochemical potentials and temperatures. Computations can be done with arbitrary electrostatic profiles that define the confinement potential across the nanostructure. Different disorder realizations and dephasing sources can also be included without too much difficulty. Electron–electron interactions are treated at the mean-field level, a type of approximation that is sufficiently accurate for many practical purposes.

The scattering approach sees the transport problem as a quantum transmission phenomenon. A conduction electron with energy E can be transmitted across the nanostructure with a probability $\mathcal{T}(E)$. If the nanostructure is attached to two reservoirs (the minimal transport configuration with static fields), $\mathcal{T}(E)$ can be calculated using the transmission probability from, say, the left (L) to the right (R) reservoir. Both terminals can differ in their electrochemical potentials (μ_L and μ_R) and temperatures (T_L and T_R), but they are assumed to be reservoirs at equilibrium. This implies that the transmitted electron dissipates its excess energy inside the terminals, i.e. far from the nanostructure. The electric (I) and heat (J) currents carried by spin-degenerate electrons are then given by:

$$I = \frac{2e}{h}\int dE\ \mathcal{T}(E)\left[f_L(E) - f_R(R)\right], \tag{7.2a}$$

$$J = \frac{2}{h}\int dE\ (E - \mu_L)\mathcal{T}(E)\left[f_L(E) - f_R(R)\right], \tag{7.2b}$$

where the Fermi–Dirac function is $f_\alpha(E) = 1/1 + \exp[(E - \mu_\alpha)/k_B T_\alpha]$ for $\alpha = L, R$ with $\mu_\alpha = E_F + eV_\alpha$ and $T_\alpha = T + \theta_\alpha$ (T is the background temperature). Thus, the voltage bias applied to the nanostructure is $V = V_L - V_R$, while the temperature gradient is given by the difference $\theta = T_L - T_R$. The above equations are valid when both currents are measured at the left lead, but this is not essential, since in the steady state the electric current at the right lead is $-I$ due to charge conservation. The heat current is not conserved due to dissipation. Let $J_L = J$ be the heat flowing in the left lead. J_R can then be obtained from equation (7.2b) by making the replacements $L \to R$ and $R \to L$. As a consequence, the total heat produced in the system equals the Joule heating inferred from the entropy produced, namely, $J_L + J_R = -IV$. The minus sign arises because we adopt the convention that fluxes are positive when currents are directed towards the nanostructure.

Importantly, equations (7.2a) and (7.2b) are independent of the reservoirs' density of states. This formalism then reduces the transport properties of a given nanodevice to the calculation of its quantum-mechanical transmission. This can be better analyzed if we expand I and J to the first order in V and θ. This is the linear regime of transport, which is illustrative to discuss not only charge and thermal fluxes at low driving fields, but also the symmetries that the transport coefficients L must satisfy. We then have [33]:

$$I = L_{11}\frac{V}{T} + L_{12}\frac{\theta}{T^2}, \tag{7.3a}$$

$$I = L_{21}\frac{V}{T} + L_{22}\frac{\theta}{T^2}. \tag{7.3b}$$

Here, V/T and θ/T^2 are the thermodynamic affinities. These are suitable quantities for a careful analysis of the thermoelectric transport. The L_{ij} coefficients with $i, j = 1, 2$ can be expressed in terms of the integral

$$\mathcal{J}_n = \frac{2}{h} \int dE (E - E_F)^n \mathcal{T}(E)[-f'(E)], \tag{7.4}$$

as $L_{11} = e^2 T\mathcal{I}_0$, $L_{12} = L_{21} = eT\mathcal{I}_1$, and $L_{22} = T\mathcal{I}_2$. The reciprocal relation $L_{12} = L_{21}$ is dictated by the underlying microreversibility of the Schrödinger equation from which the scattering amplitudes are calculated. Reciprocity states that in response to small external fields, the charge current induced by a temperature difference and the heat flux driven by a voltage bias are described by precisely the same transport coefficient. This is the celebrated Kelvin–Onsager relation that connects the Seebeck and Peltier coefficients, as we will now show. We note in passing that in the presence of a magnetic field $\overrightarrow{B}$ (or any other force that breaks time-reversal symmetry) the reciprocal relation is transformed accordingly: $L_{12}(\overrightarrow{B}) = L_{21}(-\overrightarrow{B})$.

The linear electric conductance G, Seebeck thermopower S, Peltier coefficient Π, and thermal conductance K are determined from the above coefficients:

$$G \equiv \left.\frac{I}{V}\right|_{\theta=0} = \frac{L_{11}}{T} \tag{7.5a}$$

$$S \equiv \left.\frac{V}{\theta}\right|_{I=0} = \frac{L_{12}}{TL_{11}}, \tag{7.5b}$$

$$\Pi \equiv \left.\frac{J}{I}\right|_{\theta=0} = \frac{L_{21}}{TL_{22}}, \tag{7.5c}$$

$$K \equiv \left.\frac{J}{\theta}\right|_{I=0} = \frac{1}{T^2}\left(L_{22} - \frac{L_{12}\,L_{21}}{L_{11}}\right). \tag{7.5d}$$

Clearly, the Seebeck and Peltier coefficients are not independent of each other but satisfy $\Pi = TS$ (Kelvin–Onsager). It follows that thermoelectric generation and refrigeration are essentially the same phenomena. A temperature gradient induces a charge flux in the same way that a voltage bias induces a heat current. Importantly, the Peltier heat is reversible, unlike Joule heating: if one reverses the sign of V, the Peltier component of J also reverses, whereas the dissipated Joule heat IV is always positive since the leading-order term follows V^2. Therefore, up to the first order of the external forces (the linear response regime), a nanodevice cannot dissipate by itself, although far away from the scatterer, thermalization is always present due to the implicit assumption of reservoirs in local equilibrium.

Equation (7.4) tells us that the integral has its largest contribution from energies around the Fermi energy, since the Fermi–Dirac function derivative forms a peak located around E_F with a width of $4k_BT$. Thus, at low temperatures, electrons that fall inside the transport window are those lying near E_F. A Sommerfeld expansion of equation (7.4) sheds further light on this reasoning. At very low temperatures, the system responses become

$$G \simeq \frac{2e^2}{h}\mathcal{T}(E_F), \tag{7.6a}$$

$$S \simeq \frac{\pi^2 k_B^2 T}{3e}\frac{d \ln \mathcal{T}}{dE}\bigg|_{E_F}, \tag{7.6b}$$

$$\Pi \simeq \frac{\pi^2 k_B^2}{3e}\frac{d \ln \mathcal{T}}{dE}\bigg|_{E_F}, \tag{7.6c}$$

$$K \simeq \frac{\pi^2 k_B^2}{3e}\mathcal{T}(E_F). \tag{7.6d}$$

We observe from equation (7.6a) that the electric conductance is given by the transmission at the Fermi energy, a key result from scattering theory. When the transmission is quantized, as is the case in electronic waveguides (quantum point contacts), the measured G shows a nice quantization effect. The maximum conductance supported by a perfectly transmitting quantum mode (full transparency) is then $2e^2/h$, which depends on universal constants only. The thermal conductance [equation (7.6d)] also depends solely on the transmission probability. However, it becomes zero at zero temperature because no heat can flow at $T = 0$. The proportionality factor between G and K is the Lorenz number, $L_0 = \pi^2 k^2\ T/3e$. This is a manifestation of the Wiedemann–Franz law, which states that since both charge and energy are transported by the same particle in conductors (the electron), the electric and thermal conductances must be connected to each other. This principle assumes that electrons can freely move in the conductor. Therefore, deviations from the Wiedemann–Franz law are to be expected when electron–electron interactions are important [34]. At higher temperatures, the law also breaks down due to the excitation of different thermal carriers (phonons, magnons). However, if it holds, it has important consequences for the performance of thermoelectric devices.

The thermopower given by equation (7.6b) [or, equivalently, the Peltier coefficient given by equation (7.6c)] depends on the derivative of the transmission with respect to energy. Hence, S is rather sensitive to changes in $\mathcal{T}$ and provides us with information complementary to that of G. Typical thermopower values are of the order of k_B/e = 86 meV K^{-1} for transmission functions that vary $k_B T$ across the Fermi energy. As a consequence, rapidly changing transmission functions are beneficial for thermoelectric energy harvesting or cooling. When equation (7.6a) is used in equation (7.6b), one finds the Mott relation [cf equation (7.1)]. As in the case of the Wiedemann–Franz law discussed above, the Mott relation is valid only for low temperatures and independent electrons. Violations of the Mott law can then be the signatures of deviations from single-electron transport [35]. In any case, equations (7.6a), (7.6b), (7.6c), and (7.6d) highlight an interconnection between all linear responses, which has crucial implications for the efficiency of thermoelectric devices, as shall be seen next.

7.3.1 A general discussion of efficiency

The thermodynamic efficiency of a heat engine operating between two reservoirs at temperatures $T_L > T_R$ (equivalently, $\theta > 0$) is given by

$$\eta = -\frac{IV}{J} \leqslant \eta_C, \tag{7.7}$$

where the numerator expresses the power extracted from the system by an external load (the negative sign emphasizes the fact that power is produced when current flows against the electric bias), while in the denominator, J represents the heat flux transferred from the hot terminal to the cold one during system operation. This heat-to-work conversion must be due to the second law, bounded from above by the Carnot limit, $\eta_C = 1 - T_R/T_L$.

The output power $P = -IV$ is zero either when $V = 0$ or when $I = 0$. The latter occurs when the voltage is $V = -S\theta$, as indicated in equation (7.5b). This is the stopping voltage V_{stop} that counteracts the flow driven by the temperature difference. In between the values $V = 0$ and V_{stop}, the power attains a maximum. We can find this maximum by substituting equations (7.5a) and (7.5b) into equation (7.3a). We then obtain $P = -GV^2 + GSV\theta$. For a fixed θ, this function becomes maximal when $V = V_{stop}/2$, in which case the maximum output power is $P_{max} = GS^2\theta^2/4$. The combination GS^2 is called the power factor (PF), and it is clear that one should try to find systems in which both the electric conductance and the Seebeck coefficient are high in order to generate large PFs.

An analogous optimization procedure can be carried out for the efficiency [equation (7.7)], where J is given as a linear response by equation (7.3b). The solution reads

$$\eta = \eta_c \frac{\sqrt{ZT + 1} - 1}{\sqrt{ZT + 1} + 1}, \tag{7.8}$$

ZT being the dimensionless figure of merit introduced in section 7.1:

$$ZT = \frac{GS^2\,T}{K}. \tag{7.9}$$

Its intuitive meaning can be derived from the above definitions: ZT is the ratio between the PF (which measures the strength of the heat-to-work conversion) and the thermal conductance (which determines the flow of energy from the hot to the cold reservoir). An ideal engine should then be an excellent electric conductor (with high G) and a poor thermal conductor (with low K), ensuring that most of the extracted energy is transformed into useful (electric) work while minimizing the heat dumped into the cold reservoir.

From equation (7.9), it is obvious that the figure of merit should be positive, since the second law demands that both G and K should be positive and S^2 and T are manifestly positive. On the other hand, ZT should be as large as possible. The limit $ZT \to \infty$ is attained at the Carnot value. This point is interesting because both fluxes (electric and heat) become proportional, $I \propto J$, independently of the driving fields. This is the so-called tight-binding condition, which causes the Onsager matrix constructed from the L_{ij} coefficients in equation (7.3) to be singular.

It can be shown that the efficiency at the maximum power is given by [36]

$$\eta = \frac{\eta_C}{2} \frac{ZT}{2 + ZT}. \tag{7.10}$$

In the tight-binding case, this efficiency becomes $\eta_C/2$, which is the leading-order term of the Curzon–Ahlborn efficiency [37], $\eta_{CA} = 1 - \sqrt{1 - \eta_C}$.

We can offer an appealing interpretation for the figure of merit based on the integrals given by equation (7.4). If we notice that $\mathcal{T}(E)[-f'(E)]$ can be regarded as a probability density, then

$$<\cdots> = \frac{\int dE(\ldots)\, \mathcal{T}(E)[-f'(E)]}{\int dE\ \mathcal{T}(E)[-f'(E)]} \tag{7.11}$$

is a normalized average. Hence, the thermopower can be expressed as $S = \langle(E - \mu)\rangle/eT$, i.e. the average energy transferred during an electron tunneling event divided by temperature.

S is then a measure of the entropy generated through electric transport per unit of elementary charge. Analogously, ZT can be cast as

$$ZT = \frac{< (E - \mu) >^2}{< (E - \mu)^2 > - < (E - \mu) >^2}. \tag{7.12}$$

The denominator represents the energy fluctuations that occur during a transport event. Therefore, ZT quantifies the ratio between the transferred energy and its fluctuations. When $ZT \to \infty$, the fluctuations are minimal. This case corresponds to tunneling through a resonance so narrow that fluctuations due to quantum coherence are not allowed. In other words, maximum efficiency is achieved when the transmission is described by a delta function, $\mathcal{T}(E) \propto \delta(E - \varepsilon_0)$, where ε_0 is the resonance position [20]. This implies that a quasilocalized level, such as one present in a weakly coupled quantum dot, is the ideal system in which to observe a Carnot efficiency. We will further elaborate on this prediction in section 7.5.

7.3.2 Multiterminal setups and strong fields

The scattering approach to quantum thermoelectric transport can be extended to conductors in contact with multiple terminals. It is then possible to define a transverse thermopower, in which the temperature difference is applied between two terminals while the voltage response is measured across an additional pair of unheated terminals. As a consequence, a thermovoltage is observed even when both thermal contacts are heated simultaneously [38]. More importantly, multiterminal setups allow us to test fundamental symmetries of quantum transport. Thus, the microreversibility principle is confirmed in four-terminal ballistic rectifiers when a magnetic field $\overrightarrow{B}$ pierces the cavity [39]. This enables us to distinguish two types of reciprocity. On the one hand, both G and K are symmetric when the direction of B is reversed. This symmetry also holds in the classical regime of transport. On the other hand, both S and Π are reciprocal to each other when $\overrightarrow{B}$ is inverted (the generalized Kelvin–Onsager principle mentioned earlier). We note, however, that neither S nor Π alone is generally symmetric if $\overrightarrow{B} \rightarrow -\overrightarrow{B}$ in macroscopic conductors. Quantum elastic transport introduces additional constraints such that the thermoelectric responses become symmetric. When dephasing or inelastic scattering is present in samples with enough disorder or large sizes, the quantum constraints no longer hold and the Seebeck and Peltier coefficients exhibit magnetic-field asymmetries [40, 41]. These asymmetries affect the thermoelectric efficiencies in a dramatic way [42].

Three-terminal geometries have recently received a lot of attention [43]. Their advantage over two-terminal heat engines is their ability to spatially separate particle and heat currents [44, 45]. The underlying physical mechanism is a rectification of thermal fluctuations emitted by the extra terminal, which is heated to a temperature higher than those of the current-carrying terminals [46]. Energy-harvesting efficiencies are hence large, because there is little heat leakage and the generated electric power can be fully delivered to an external load. In the optimal configuration, these devices can achieve (at least theoretically) the tight-binding condition between I and J.

Nevertheless, these structures strongly rely on electron–electron interactions, whereas the Landauer–Büttiker formalism is valid for free electrons. Despite this, the framework can conveniently be extended to tackle interactions at the mean-field level. This is particularly important for the nonlinear regime of transport, since out-of-equilibrium responses depend on the screening properties of the nanoconductor [47]. A novel phenomenon arises even in the weakly nonlinear regime, namely, the buildup of charge layers that accumulate mainly in interfaces when the driving fields are strong. An analytical treatment can be applied to the simplified situation in which the screening potential is discretized into portions with constant values. In this case, interactions can be phenomenologically described by capacitance coefficients, and the non-equilibrium screening is thus written in terms of injectivities and emissivities that depend on both capacitances and the local densities of states. Charge buildup can occur due to either voltage or temperature biases, and the screening functions are different. In this manner, it is possible to formulate a gauge-invariant self-consistent theory for nonlinear quantum thermoelectric transport [48, 49]. One of its main predictions is the violation of Onsager reciprocity relations, quantifying their departures from the equilibrium condition [50, 51]. Nevertheless, nonlinear transport is also subject to bounds which limit the amount of heat that can be extracted from a thermal bath [52].

The issue of charge buildups and the need for a gauge-invariant description of transport are also crucial when the external fields are time dependent. In the adiabatic regime (low frequencies), the quantum admittance is a function of the oscillatory electric and thermal shifts [53]. Interestingly, a dynamic forcing can be beneficial for the performance of a nanoscale heat engine [54]. This can be achieved by breaking the Onsager reciprocity relation and simultaneously decreasing the entropy production rate [55]. The figure of merit is accordingly modified and even generalized to quantum pumping effects (of both charge and energy) [56].

7.4 Nanowires

As anticipated in section 7.2, the pioneering theoretical work by Hicks and Dresselhauss [8, 9] demonstrated that quasi one-dimensional nanostructures such as semiconductor NWs overcome the trade-off relationship between thermopower and electrical conductivity (see references [57–59] for a comprehensive review of NWs as a route for efficient thermoelectrics). The electron density of states close to the band edges [see figure 7.2(a)] is enhanced in one-dimensional (NW) and

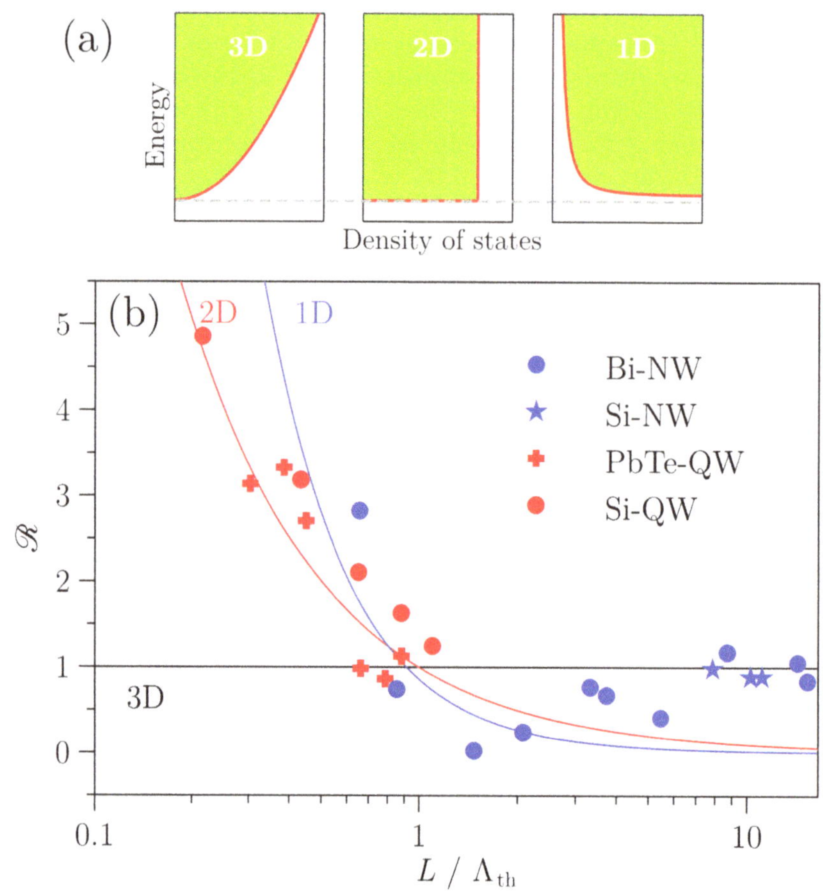

Figure 7.2. (a) The free electron densities of states close to the band edge (dashed line) in bulk semiconductors (3D), QWs (2D), and NWs (1D). In the last two cases, only one sub-band is shown. (b) The ratio $R = PF_{opt} / PF^{3D}_{opt}$ as a function of L/Λ_{th}. Solid lines indicate the theoretical prediction and symbols denote experimental data from quantum wells (QWs) and NWs. The regime of strong quantum confinement ($L < \Lambda_{th}$) enhances the PF compared to the bulk value. Adapted from [60].

two-dimensional (quantum well (QW)) materials compared to those of bulk (three-dimensional) materials, increasing the PF defined earlier, PF = $S\sigma^2$ when the confinement length L is smaller than the thermal de Broglie wavelength of the electrons $\Lambda_{th} = h/\sqrt{2\pi m^* k_B T}$, m^* being the electron effective mass [60]. A theoretical analysis based on the use of a one-band model for nondegenerate semiconductors to describe quantum effects demonstrated that the PF can be maximized by setting the proper value of the chemical potential [60]. The solid lines in figure 7.2(b) show the calculated optimum PF, PF_{opt}, scaled by the same magnitude of a bulk material PF_{opt}^{3D} as a function of L/Λ_{th}. The symbols denote experimental data from QWs and NWs. The theoretical approach predicts a remarkable enhancement of the power factor in QWs and NWs over the bulk value in the regime of strong quantum confinement ($L < \Lambda_{th}$). This condition is difficult to achieve in Si-based nanostructures because the thermal de Broglie wavelength of the electrons is small ($\Lambda_{th,Si}$ = 4.5 nm). In contrast, other materials such as Bi are more favorable for exploiting quantum confinement effects in nanostructures ($\Lambda_{th,Bi}$ = 32 nm) [60].

7.4.1 Quantum confinement and nanostructuring effects

Quantum confinement effects plays a major role in the observed enhancement of the PF in nanostructures. However, bulk semiconductors have low carrier density compared to metallic materials even at room temperature, and their thermal conductivity is typically dominated by the phonon contribution to heat transport. In this sense, current electronic technologies are mainly based on Si, but unfortunately its thermal conductivity at room temperature is too high for thermoelectric applications ($\kappa \sim 150$ W $(\mathrm{mK})^{-1}$ [61]). At lower temperatures, the thermal conductivity of Si decreases ($\kappa \sim 35$ W $(\mathrm{mK})^{-1}$ [62]), but its PF is reduced as well. Therefore, the combination of Si-based thermoelectric modules with electronic devices represents a challenge for engineers and designers.

Remarkably, the nanostructuring of Si is feasible and offers possibilities for overcoming the poor thermoelectric performance of bulk Si. It is known that reduced-geometry effects can lead to stronger phonon scattering, thus reducing the lattice contribution to thermal conductivity. At the same time, the electric conduction is not severely deteriorated and, consequently, a significant increment in ZT is achieved. In this regard, Zhang *et al* used a simple metal-assisted chemical etching method to fabricate high-density and large-area vertically aligned porous Si NW arrays with different lengths, porosities, and diameters [63]. They observed that the figure of merit was $ZT = 0.493$, almost two orders of magnitude larger than that measured in bulk Si ($ZT = 0.0064$). A similar enhancement of the figure of merit was obtained by Calaza *et al* in ordered dense arrays of *p*-type Si NWs produced using a vapor–liquid–solid method; they estimated that the figure of merit ranged from 0.30 to 0.93 [64]. Boukai *et al* independently measured the Seebeck coefficient, the electrical conductivity, and the thermal conductivity of narrow Si NWs 10 and 20 nm wide and reported a 200-fold reduction in thermal conductivity and a 100-fold enhancement in ZT (see figure 7.3). The improvement in the thermoelectric efficiency was attributed to phonon scattering effects in the NWs [15].

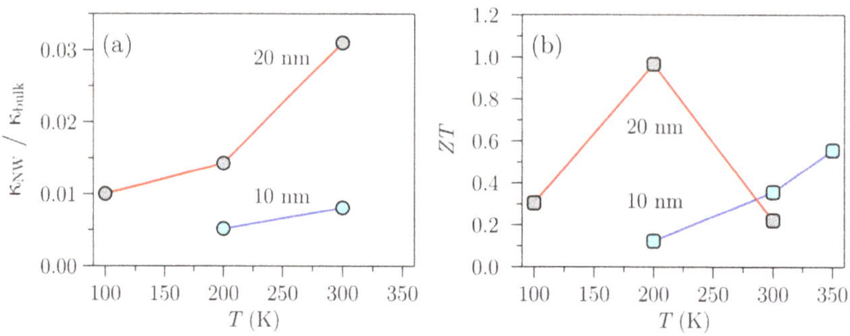

Figure 7.3. Temperature dependence of (a) the thermal conductivity of Si NWs, κ_{NM}, expressed in units of the thermal conductivity of bulk Si κ_{bulk}, and (b) the figure of merit, ZT, of Si NWs of two different widths, as indicated in the plots. Lines are guides for the eye. Data retrieved from [15].

Intentionally induced point defects can even further increase phonon scattering and lead to lower thermal conductivity. Bennet *et al* proved that the introduction of large concentrations of vacancies into Si nanofilms yields a 20-fold reduction in the thermal conductivity [65]. The thermoelectric properties of cubic zinc-blend SiC NWs with nitrogen impurities and vacancies were theoretically addressed by Xu *et al* [66]. They claimed that the figure of merit, ZT, could be significantly enhanced by doping this material with N impurities and creating Si vacancies, reaching $ZT = 1.78$ at 900 K in SiC in n-type NWs with a diameter of 1.1 nm. Chen *et al* observed that cation doping along with the modulation of cation vacancies induced vacancy-based defects that boosted the thermoelectric performance of GeTe-based materials [67]. These authors found that alloying Sb_2Te_3 into GeTe increased the concentration of cation vacancies, determining the thermoelectric response of the alloy by intensifying phonon scattering. A high PF of PF = 32.6 $\mu W\ cm^{-1}\ K^{-2}$ was realized for $Ge_9Sb_2Te_{11.91}$ at 725 K. These alloys presented high values for the figure of merit, $ZT \sim 1.6$–2.1, in the range of 700–800 K (see figure 7.4).

7.4.2 Nanowires based on topological insulators

Topological insulators are generally defined as a class of materials that supports gapless electron states at the surface, whilst the bulk remains an insulator [68–72]. Examples of topological insulators are compound semiconductors such as Bi_2Te_3, Bi_2Se_3, Sb_2Te_3, and SnTe. Most topological insulator materials are based on heavy elements and display relatively narrow bandgaps. The combination of these two features causes topological insulators to perform as highly efficient thermoelectric materials at room temperature [73]. On the one hand, heavy elements enhance relativistic effects and lead to narrow gaps which are favorable for room-temperature thermoelectricity [74]. Furthermore, the electrons in heavy atoms undergo strong relativistic corrections, leading to the band inversion that enables conducting states to occur at the surface. On the other hand, heavy atoms tend to reduce the lattice thermal conductivity of the material. All these features lead to a significant enhancement in the thermoelectric efficiency [75]. As introduced previously in

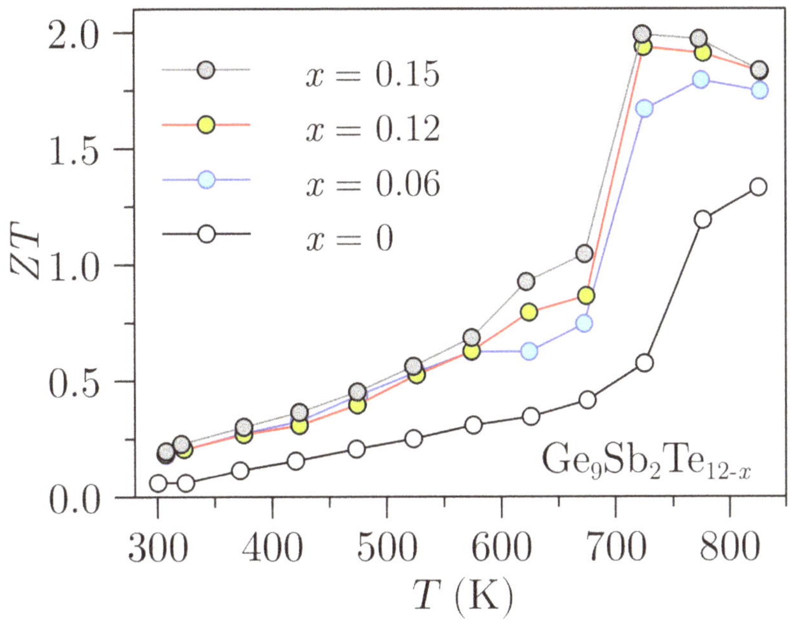

Figure 7.4. The temperature dependence of the figure of merit, ZT, for $Ge_9Sb_2Te_{12-x}$ alloys. Adapted with permission from [67]. Copyright (2020) American Chemical Society.

section 7.2, quantum confinement [8, 9] leads to a further increase in the thermoelectric performance of selected materials that, in the case of topological insulators, have undergone deep scrutiny. Figure 7.5 displays the electrical and thermal conductivities, Seebeck coefficients, and figures of merit of some nanostructured topological insulators normalized to the corresponding values for their respective bulk materials. Quantum confinement leads to a noticeable increment in the thermoelectric performance of low-dimensionally structured topological insulators (thin films, NWs, and nanoribbons), which outperform their bulk counterparts. Moreover, the robustness of the topological surface states prevents the detrimental effects of surface roughness and defects on electrical transport observed in NWs based on topologically trivial materials. Surface-dominated thermoelectric properties have also been observed in Bi NWs by Huber *et al* [76, 77]. They reported nearly pure surface electronic transport for temperatures $T < 100$ K in Bi NW arrays with diameters around the critical value, 50 nm, for the semimetal-to-semiconductor transition. They claimed that surface electronic transport together with phonon scattering at the surface enhanced the thermoelectric performance of Bi NW arrays at low temperatures.

7.5 Quantum dots

Quantum dots are effectively zero-dimensional conductors, since electrons are confined in the three spatial directions. Due to this size quantization, dots show discrete energy levels. This is an appealing feature for quantum thermoelectrics

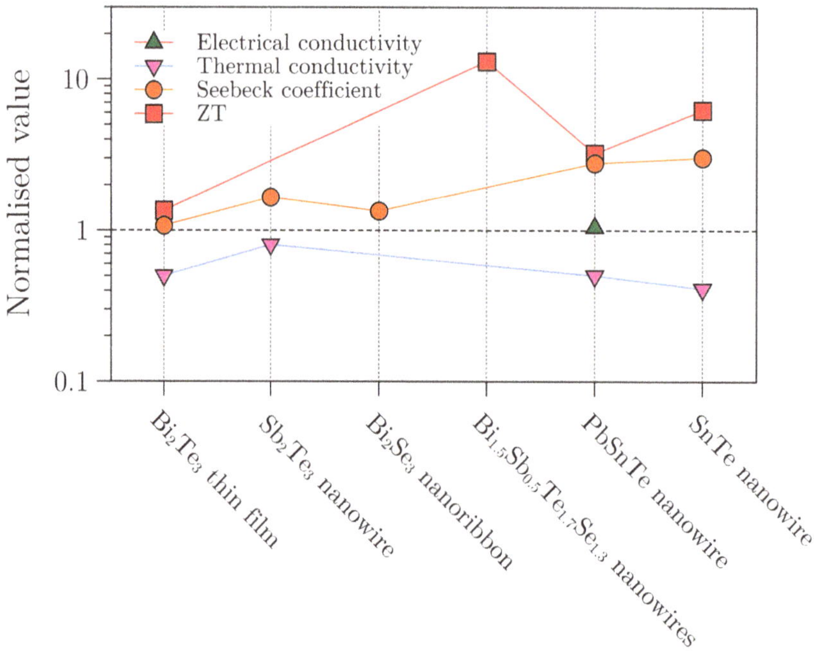

Figure 7.5. Thermoelectric values (electrical conductivity, S, κ, and ZT values) of various topological insulator-based nanostructures normalized to the respective values of their bulk counterparts at room temperature. The lines are guides for the eye. Data retrieved from [73].

because, as discussed in section 7.3, the Carnot limit can be achieved (at least theoretically) in systems with infinitely sharp resonances. When a dot is attached to (possibly multiple) terminals, its energy levels become broadened and acquire a finite lifetime, since electrons have the possibility to hop off to the coupled terminals via the tunnel effect. For simplicity, let us a consider a single-level dot with energy position ε_0 and width Γ. A good approximation that holds when ε_0 is far from the top of the tunnel barriers that separate the dot from the terminals is to consider the Breit–Wigner shape for the transmission function:

$$\mathcal{T}(E) = \frac{\Gamma_L \Gamma_R}{(E - \varepsilon_0)^2 + (\Gamma/2)^2}. \tag{7.13}$$

Here, we have considered that the dot is only coupled to two terminals (L and R), thus $\mathcal{T}(E)$ represents the probability that an electron with energy E traverses the dot. In general, the couplings Γ_L and Γ_R may differ while their sum is Γ. The dot becomes fully transparent when E is aligned with ε_0 and when the couplings are symmetric, $\Gamma_L = \Gamma_R = \Gamma/2$. Otherwise, $\mathcal{T}(E)$ varies between zero and one. In the limit $\Gamma \to 0$, equation (7.13) reduces to a delta function, $\mathcal{T}(E) = 2\pi\Gamma_L\Gamma_R\delta(E - \varepsilon_0)/\Gamma$. The dot then works as a perfect energy filter because only electrons impinging into the dot with energy $E = \varepsilon_0$ are transmitted. The rest are scattered back into the reservoirs. By substituting the delta function above into equations (7.2a) and (7.2b), one can find the electric and heat currents, from which the efficiency readily follows:

$$\eta = \frac{-eV}{\varepsilon_0 - \mu_L}. \tag{7.14}$$

ε_0 can be tuned by an external gate voltage. We would like to find the value of ε_0 that maximizes η. To this end, we compute the entropy production of a such a level coupled to two reservoirs by means of the Clausius expression:

$$-\dot{S} = \frac{(\varepsilon_0 - \mu_L)(f_L - f_R)}{T_L} + \frac{(\varepsilon_0 - \mu_R)(f_R - f_L)}{T_R}, \tag{7.15}$$

where the Fermi functions are evaluated at $E = \varepsilon_0$. The second law of thermodynamics dictates that $\dot{S} \geqslant 0$. The Carnot limit is attained in reversible conditions, namely, $\dot{S} = 0$. These occur when the dot level satisfies the relation

$$\frac{\varepsilon_0 - \mu_L}{T_L} = \frac{\varepsilon_0 - \mu_R}{T_R}, \tag{7.16}$$

which corresponds to the energy when both occupation factors, f_L and f_R, are equal [78]. By substituting the solution of equation (7.16) into equation (7.14) we find the maximum Carnot efficiency $\eta_C = 1 - T_R/T_L$. Recently, a quantum-dot engine has been observed operating close to the Carnot limit in InAs/InP NW quantum dots [79, 80]. Unfortunately, a reversible engine generates zero output power and is not thus useful for practical purposes. A finite power can be produced for nonzero values of the broadening Γ. It has been theoretically shown that the output power is maximal for $\Gamma = 2.25k_BT$ [81] and then decreases if Γ is further increased, since the dot loses its energy-filtering properties. These features appear to be universal [82]. In the experiment, a few femtowatts have been achieved for efficiencies of around 70% of the Carnot efficiency. For comparison, the efficiency of a macroscopic thermoelectric generator is around ten times smaller.

The thermopower line shape provides useful insights into the transport regime of quantum dots. If we apply Mott's law (equation (7.1)) to a bulk metal, we find the thermopower $S \sim k_BT/E_F$, which is much smaller than (in units of k_B/e) at low and intermediate temperatures. For a dot with infinitely narrow resonances, the same law yields S of the order of one (in units of k_B/e) at low. However, this calculation neglects Coulomb repulsion effects, which are particularly strong in quantum dots due to their zero-dimensional character. In fact, most of the dots studied experimentally exhibit the so-called Coulomb blockade effect, which is a measurable manifestation of electron–electron interactions at the nanoscale. Electron transport through a dot is forbidden even at finite bias unless a charging energy U is supplied to overcome the Coulomb repulsion that takes place between two electrons in the dot. When this effect is taken into account, theory predicts [83] that $S \sim U/T$, which is typically much larger than one. Further, because the dot energy levels can be shifted by a nearby gate, the thermopower line shape is sawtooth-like as a function of the gate voltage. From these results, one can infer that S changes sign each time an energy level crosses the Fermi energy. The thermopower sign reveals the nature of the current carriers: electrons (particles traveling above E_F) or holes (particles traveling below E_F). This prediction has been experimentally confirmed [84, 85], although S turns out to be much smaller than U/T. The reason for this discrepancy is that a real dot level always has a finite Γ. At very small Γ, electrons tunnel through the dot one by one (sequential tunneling), but when the coupling to the terminals increases, simultaneous tunneling of two electrons (cotunneling) starts to make a significant contribution to the Seebeck coefficient, thus lowering the maximum value of S [86].

7.5.1 Interaction, nonlinear, and interference effects

For even stronger coupling, the system enters the Kondo regime, characterized by a crossover temperature T_K. When $T < T_K$, the Coulomb blockade is lifted and the dot can transmit electrons between two consecutive conductance peaks whenever the dot spin increases by 1/2. Indeed, the Kondo effect is due to an effective antiferromagnetic interaction between the dot's localized electron and the reservoirs' itinerant electrons. As a consequence, a many-body singlet forms whose most salient feature is a new resonance that sharply emerges at the Fermi level (the Abrikosov–Shul resonance). As a first approximation, one can describe the Abrikosov–Shul resonance with the aid of a Breit–Wigner expression [equation (7.13)] and renormalized parameters, i.e. Γ is replaced by $k_B T_K$ and ε_0 by a value slightly greater than E_F. This approximation is valid if $T \ll T_K$ (the Fermi liquid regime). The conductance G given by equation (7.6a) then increases as T approaches T_K and saturates at the quantized value $2e^2/h$ when $T \to 0$. Meanwhile, S reverses its sign around T_K, since for $T > T_K$ the Abrikosov–Shul resonance disappears and transport changes from electron-like to hole-like [87, 88]. Therefore, the thermopower sign changes are a sensitive signature of the presence of Kondo correlations [89]. Another way to quench the Kondo resonance is to apply an external magnetic field B that breaks spin degeneracy. The Kondo peak then splits and eventually disappears when the ensuing Zeeman energy Δ_Z is much larger than $k_B T_K$. At temperatures higher than the Kondo temperature, S remains unaffected in the presence of Zeeman splitting. However, for $T < T_K$, the absolute value of the thermopower decreases with increasing Δ_Z and finally becomes zero for sufficiently strong values of B. The Seebeck coefficient can also work as a probe that detects Kondo states of higher symmetry. In addition to the SU(2) spin Kondo effect, in which the antiferromagnetic interaction leads to flips of the quantum-dot local spin, if the system possesses another degenerate pseudospin degree of freedom (stemming, e.g. from orbital motion in carbon nanotubes) then entanglement between spin and pseudospin states gives rise to an SU(4) Kondo effect whose Abrikosov–Shul resonance displays an enlarged broadening since $T_K^{SU(4)} > T_K^{SU(2)}$. More importantly, the SU(4) peak becomes asymmetric, causing an enhancement of the thermopower seen as a dip in S as a function of temperature [90]. This shows the power of thermoelectric measurements to obtain spectroscopic information from strongly correlated nanostructures.

An additional advantage of using quantum dots for thermoelectrics is that it is much easier to explore the nonlinear regime of transport as compared with that of a bulk metal. As discussed earlier, the transmission function of a dot changes rapidly with energy, and the validity of a Sommerfeld expansion leading to equations (7.6) is restricted to very small voltage and temperature shifts. Far from equilibrium, the thermovoltage V_{th} is no longer a linear function of the applied thermal gradient θ but shows an intriguing behavior [91]. First, V_{th} decreases with θ and then, with a further increase in θ, the thermovoltage changes its sign. We emphasize that this phenomenon differs from the sign reversal of the linear Seebeck coefficient as a gate voltage is swept across the dot level. It has been demonstrated that the peculiar property of the nonlinear V_{th} is due to the presence of two resonances that arise from the Coulomb blockade splitting of the dot level, i.e. ε_0 and $\varepsilon_0 + U$ [92]. When the temperature bias θ is small, electrons preferentially tunnel through the ε_0 resonance, thereby transmitting charges from the hot terminal to the cold one. As θ grows, the thermovoltage shrinks until $V_{th} = 0$; at this point, the thermally driven fluxes from the hot and cold sides counterbalance each other. If the temperature difference continues to increase, V_{th} changes sign and the electrons are preferentially transmitted from the cold terminal to the hot one, even though the latter becomes hotter and hotter. This is a purely nonlinear effect and is not unique to quantum dots. Both core–shell and topological NWs also show a similar reversal of the thermoelectric current [93]. In the strong coupling limit, large thermal gradients quench the Kondo effect, but an analysis of the thermocurrent and its nontrivial zeros is capable of yielding both the Kondo temperature and the charging energy of the system [94].

Another useful ability of dots is that these systems can be coupled to reservoirs other than normal metals. An interesting avenue of recent research has utilized hybrid structures such as superconductors and ferromagnets [95]. Superconductors would be ideal thermoelectric materials due to their high (low) values of $G(K)$, yet their Seebeck coefficients are zero due to the electron–hole symmetry present in the superconducting density of states, and as a consequence, ZT becomes vanishingly small. One way to break electron–hole symmetry is by means of a magnetic field, such as the stray fields emanating from a nearby ferromagnet [96, 97]. Alternatively, one can insert a dot between a normal terminal and a superconducting contact, which electrically breaks the symmetry by tuning the dot level away from the symmetry point $\varepsilon_0 = U/2$ [98]. Dots can also act as splitters of Cooper pairs exiting a superconducting lead, which in turn originate nonlocal thermoelectric currents and hence heat conversion [99] or cooling [100].

Double quantum dots are the natural extensions of single dots. Two dots can be tunnel coupled either in series or in parallel. In the former case, molecular-type states (bonding and antibonding) form; in the latter, wave interference due to path scattering is a prominent effect. Interference can be destructive, giving rise to asymmetric transmission line shapes termed Fano antiresonances. Around a Fano antiresonance, $\mathcal{T}(E)$ changes sharply and, as we know, thermoelectric effects are strong. This effect has been exploited to enhance the thermopower and the thermoelectric efficiency of quantum-dot systems [21, 101], nanoscale junctions [24], and single-molecule devices [102]. The thermoelectric response L_{12} [equation (7.3a)] shows an oscillatory pattern as a function of the Fermi energy, with an amplitude as high as $0.05ek_B/h \sim 0.1$ nA K^{-1} [26]. A crucial difference from a standard resonance is that near a transmission dip, the thermopower is independent of temperature and can thus be enhanced to significantly high values [103, 104]. Quantum interference also amplifies both the thermodynamic efficiency [105, 106] and the efficiency at maximum power by suppressing the back-flow from the cold terminal into the dots [107]. In this manner, heat extracted from the hot reservoir is transformed into useful work with great effectiveness.

7.6 Quantum rings

Quantum rings can exploit the wave properties of carriers in the coherent regime to tailor the transmission coefficient from the source to the drain, in a complete analogy to the beam splitter of an optical interferometer. A side-gate voltage applied across the ring allows for additional control of the interference pattern and the electron current between the two leads can therefore be accurately modulated [108]. According to the Landauer–Büttiker formalism discussed in section 7.3, the conductance of a nanoscale conductor is calculated as the sum of the transmission coefficients of all propagating channels at the Fermi energy. Therefore, the electrical conductivity and hence the thermopower become highly tunable in quantum rings (see Mott's formula (7.1)) by a careful design of their geometry or by means of a side-gate voltage.

7.6.1 Electron transport in quantum rings

Graphene is a material with a combination of many remarkable properties [109] and has also received attention as a potential thermoelectric material. Although bulk graphene is a poor thermoelectric material, nanostructures based on graphene, such as nanoribbons and quantum rings, have been widely studied for their remarkable enhancement of thermoelectric efficiency [26, 110–113]. In this context, Mazzamuto *et al* studied the thermal and electrical properties of perfect nanoribbons as a function of their width and edge orientation [113]. They found that nanostructures with armchair and zigzag sections of different widths present lower phonon thermal conductance but retain high electron conductivity and thermopower, yielding ZT values that exceed unity even at room temperature. Furthermore, heat and thermoelectric currents in quantum rings coupled to a photon cavity in a resonance regime can be tuned by the polarized photon field [114].

Its large electron mobility and long coherence lengths make graphene an ideal candidate for the use of quantum phenomena. Quantum coherence effects, such as the occurrence of Fano antiresonances in the transmission coefficient, open up new ways to achieve higher thermoelectric efficiencies in graphene nanodevices. As an example, we discuss the thermoelectric properties of the rectangular graphene quantum rings (see figure 7.6) proposed in reference [26]. The rings have both type of edge (zigzag and armchair) in the same device, with the objective of

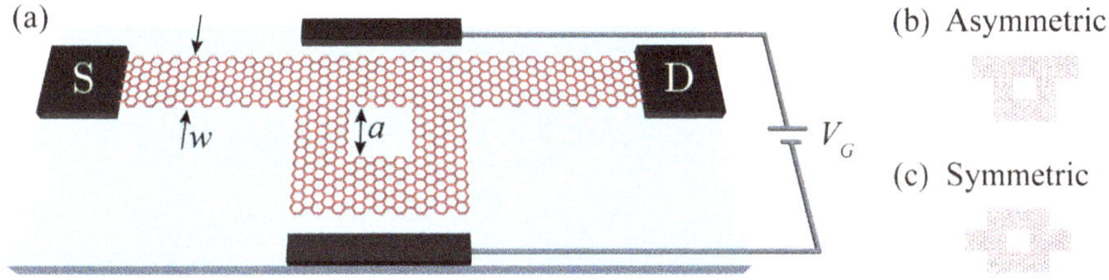

Figure 7.6. (a) Schematic diagram of a graphene nanoring attached to two armchair nanoribbons acting as source (S) and drain (D) contacts. A side-gate voltage V_G is applied across the nanoring. The configuration can be (b) asymmetric or (c) symmetric. Adapted with permission from [115]. Copyright (2020) by the American Physical Society.

enhancing phonon scattering at bends and thus reducing phonon thermal conductivity, as suggested in reference [113]. Two configurations, namely symmetrically and asymmetrically connected square nanorings, are considered. It turns out that the latter display higher thermoelectric efficiency due to the occurrence of Fano antiresonances in the transmission. Furthermore, the thermoelectric coefficients can be controlled by a side-gate voltage between the two arms of the quantum ring. The width of all nanoribbons, including both the leads and the ring, is w and the dimensions of the inner hole are $a \times a$. For simplicity, only the case $w = a$ is discussed. The leads are two semi-infinite armchair graphene nanoribbons with $N \neq 3n - 1$, where N is the number of hexagons across the nanoribbon and n is a positive integer. In this case, the band structure has a width-dependent gap and parabolic dispersion in the vicinity of the gap edges.

The devices are modeled using a nearest neighbor tight-binding Hamiltonian in which the hopping parameter is set to 2.8 eV and the site energy can vary from one atom to another due to the presence of a side-gate voltage drop V_G across the ring. The quantum transmission boundary method combined with the effective transfer matrix method can be used to compute wave functions and the transmission coefficient $\tau(E)$ as a function of energy (see reference [116] for further details). Once $\tau(E)$ is known, the thermopower S, the electric conductance, and the electric thermal conductance can easily be obtained [21] from equations (7.6a) and (7.6b) (here we neglect the phonon contribution that will be discussed later).

Figure 7.7 compares the transmission coefficients and the figures of merit of symmetric (red lines) and asymmetric (blue lines) rings in the absence of a side-gate voltage ($V_G = 0$) in the one-mode regime for $w = 12.0$ nm ($N = 49$) and $w = 12.5$ nm ($N = 51$). The temperature is set to $T = 4$ K. The transmission features remain qualitatively unchanged as the nanoribbon width increases. When $N = 3n - 2$ (panel (a)), the transmission coefficient displays resonant peaks whose shape is Lorentzian close to the resonance energy for both configurations (Breit–Wigner line shapes, see equation (7.13)). The smooth profile of the resonance peaks in the transmission coefficient yields very low thermoelectric efficiency (panel (b)). A similar trend is observed when $N = 3n$ (panels (c) and (d)) for the case of symmetrically connected rings. On the contrary, if the ring is connected asymmetrically, the transmission

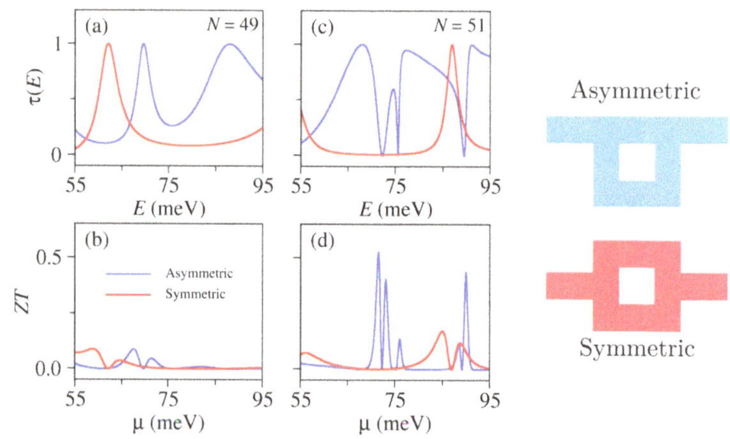

Figure 7.7. (a) The transmission coefficients as a function of the electron energy and (b) the figures of merit as a function of the chemical potential in the absence of a side-gate voltage ($V_G = 0$) in symmetric (red lines) and asymmetric (blue lines) rings for a nanoribbon width of $N = 49$. Panels (c) and (d) show the same magnitudes when $N = 51$. Panels (a)–(d) adapted with permission from reference [115], Copyright (2020) by the American Physical Society.

coefficient shows Fano line shapes (blue line in panel (c)) with an abrupt change over a narrow energy window, leading to higher values of the figure of merit (blue line in panel (d)). The enhancement observed when the chemical potential crosses Fano antiresonances at 72 meV and 90 meV is remarkable compared to the case of Breit–Wigner resonances. In this situation, the thermopower and figure of merit reach maximum values close to $|S| \sim 100\ \mu V\ K^{-1}$ and $ZT \sim 0.5$.

A higher figure of merit is achieved when the quantum ring is subjected to an additional side-gate voltage V_G as shown schematically in figure 7.6(a). Figure 7.8 shows the results for asymmetric rings with $N = 49$ (left-hand panels) and $N = 51$ (right-hand panels) when the side-gate voltage $V_G = 46$ mV. In the case of $N = 49$, the side-gate voltage leads to a maximum value of $ZT \sim 7$ when the chemical potential crosses a sharp Fano antiresonance in the transmission coefficient. Similar results were found for symmetric rings (not shown in the figure). Therefore, the possibility of inducing Fano antiresonances by electrostatic means seems very promising from the point of view of the control and optimization of the thermoelectric properties of quantum rings.

7.6.2 Phonon transport in quantum rings

In the previous section, we theoretically showed that graphene nanorings might be useful as thermoelectric devices due to quantum interference effects that lead to large S and hence high ZT when the heat current is solely due to electrons. Nevertheless, lattice heat conduction is expected to be the most important contribution to heat transport in carbon materials due to the strong covalent sp^2 or sp^3 bonding. We now address this contribution at high and low temperatures. In the former regime, quantum effects are irrelevant and lattice dynamics is well accounted for by

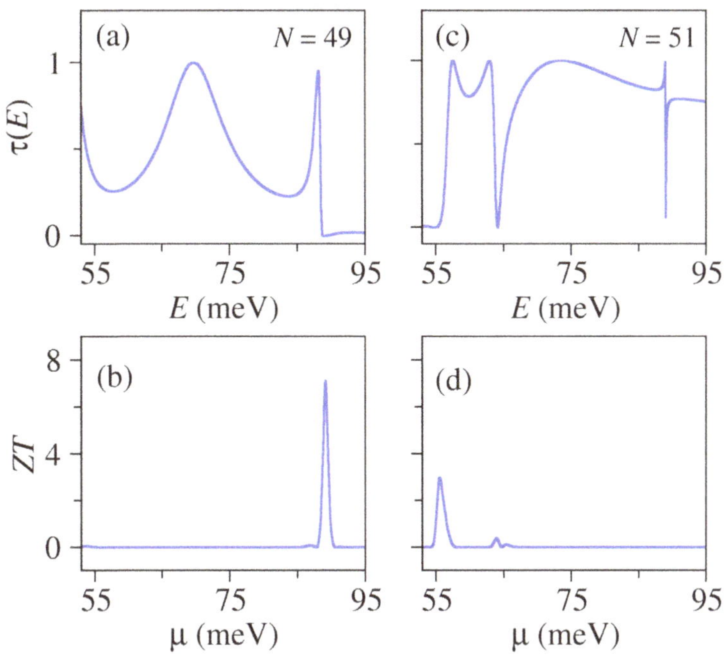

Figure 7.8. (a) The transmission coefficient in an asymmetric quantum ring as a function of the electron energy and (b) the figure of merit as a function of the chemical potential for a nanoribbon width of $N = 49$ and a side-gate voltage $V_G = 46$ mV. Panels (c) and (d) show the same magnitudes when $N = 51$. Adapted with permission from reference [115], Copyright (2020) by the American Physical Society.

Newton's equations of motion. On the other hand, at low temperatures, quantum coherent effects cannot be neglected because phonons travel ballistically through the system. The crossover temperature between the two regimes is estimated to be of the order of 107 K [117].

7.6.3 Lattice heat conduction at high temperatures

The lattice thermal conductivity can be calculated using non-equilibrium molecular dynamics (NEMD) simulations [118, 119], in which the classical trajectories of carbon atoms are calculated by solving Newton's equations of motion. Atoms with initial positions and velocities undergo collisions governed by an empirical interatomic potential, such as the adaptive intermolecular reactive bond order (AIREBO) potential [120]. The classical description holds whenever the average interatomic separation is larger than the thermal de Broglie wavelength $\Lambda_{\text{th}} = h/\sqrt{2\pi M k_B T}$, where M is the atomic mass. This is indeed the scenario in graphene at high temperatures, since the distance between carbon atoms is about 0.142 nm, while $\Lambda_{\text{th}} = 0.030$ nm at room temperature.

Figure 7.9 shows the thermal conductivity for armchair graphene nanoribbons and the two types of quantum ring considered in the previous section (symmetric and asymmetric) for widths W of up to 6 nm. The results show that κ_{ph}

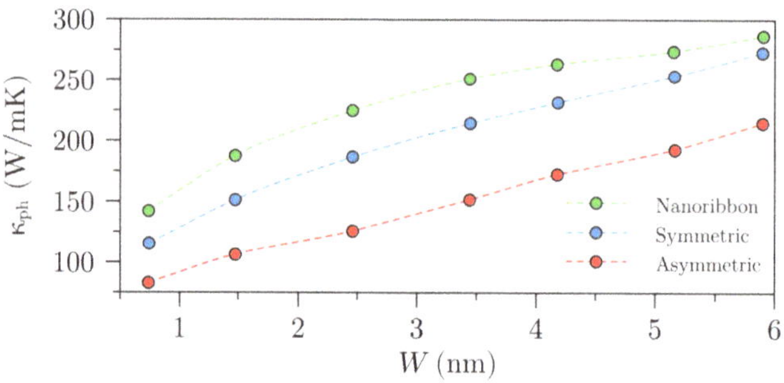

Figure 7.9. The lattice thermal conductivity as a function of width W for armchair nanoribbons and quantum rings (symmetric and asymmetric). The error bars are similar to the symbol size. Adapted from [117], Copyright (2017), with permission from Elsevier.

monotonously increases with the width W for both nanoribbons and quantum rings. This trend can be understood as follows: wider nanoribbons have a larger number of vibrational modes, while the number of edge-localized modes does not change. Thus, the edge effect decreases and κ_{ph} increases with W. Ultimately, for sufficiently wide nanoribbons, κ_{ph} reaches the value observed for bulk graphene (2000–4000 W $(\text{mK})^{-1}$ [121]). Although κ_{ph} also increases with W in quantum rings, it remains always lower than for nanoribbons. This is a consequence of the presence of different edges, both armchair and zigzag, in the quantum rings that leads to a mismatch of the vibrational modes [113] and also of scattering at the bends [122, 123]. The introduction of more asymmetries into such rings, such as different widths of the ring arms or edge disorder, increases the scattering and reduces the thermal conductivity even further [117].

7.6.4 Lattice heat conduction at low temperatures

NEMD simulations exclude quantum effects and consequently cannot be regarded as a valid tool for the study of graphene nanostructures at less than 100 K (see reference [117] and the references therein). The phonon transport properties of graphene nanoribbons and quantum rings in the low-temperature regime can be assessed using a density-functional-based tight-binding (DFTB) method [124] combined with Green's function techniques (technical details can be found in reference [125]). This method is based on density-functional theory, but also uses ideas from the semiempirical tight-binding approach, hence combining accuracy and numerical efficiency (systems can include up to 2000 atoms in a quantum simulation). Interestingly, most of the Γ (also including all the acoustic phonons) and K phonon modes have been shown to be uncoupled from electrons [126]. Therefore, as a reasonable approximation, electron–phonon interactions are not considered in the calculations shown below. Consequently, phonon and electron transports can be treated independently within the Landauer–Büttiker approach.

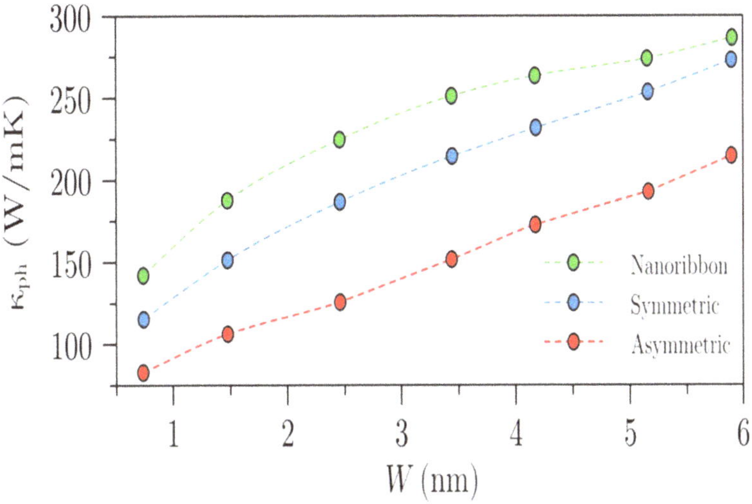

Figure 7.10. Lattice thermal conductance as a function of temperature for armchair nanoribbons and quantum rings. The width of the nanoribbons and the size of the inner hole are $W = a = 1.23$ nm (see figure 7.6). The arrow indicates the critical temperature. Adapted with permission from [125], Copyright (2019) by the American Physical Society.

Figure 7.10 shows the lattice thermal conductance as a function of temperature for armchair nanoribbons and quantum rings (symmetric and asymmetric). We observe that there are two clearly distinguishable regimes in the thermal conductance. The critical temperature separating both regimes is around 30–50 K, marked by an arrow in figure 7.10. Above this temperature, the nanoribbon has the highest thermal conductance, as predicted by classical NEMD simulations. However, below this critical temperature, the behavior is exactly the opposite, and the quantum rings show a larger conductance compared to that of the graphene nanoribbon. This counterintuitive phenomenon has been attributed to phonon interference effects that happen at low temperatures [125]. In other words, the occurrence of resonance peaks in the phonon transmission spectra of the quantum rings yields higher values of the lattice thermal conductance. Fortunately, a careful selection of the ring geometry can tune the interference pattern (e.g. by varying the size of the inner hole) and allow quantum rings to maintain low lattice thermal conductance.

The thermoelectric figure of merit (not shown in the figure) reaches its highest values at energies in the vicinity of Fano resonances and pronounced slopes of the transmission coefficient, as discussed previously. In the case of the symmetric configuration, values of $ZT \approx 1$ are only obtained when there is no hole in the nanostructure, since they always display smooth transmission profiles. On the contrary, for the asymmetric configuration, values of ZT around unity or even higher are obtained in all cases, indicating that this configuration would be more suitable for thermoelectric applications. In conclusion, the addition of the lattice contribution to the thermal transport in the calculation of ZT does not significantly degrade the efficiency in the vicinity of abrupt slopes and Fano resonances, which are the features that yield the largest values of ZT.

7.7 Layered 2D materials

Immediately after the isolation of the first truly 2D material, i.e. monolayer graphene, by Novoselov *et al* in 2004 [127] the field of thermoelectricity in layered 2D materials blossomed. The family of 2D materials encompasses a number of different compounds such as (but not restricted to) graphene, transition-metal dichalcogenides (TMDs), black phosphorous, layered binary IV–VI compounds like GeSe and, ultimately, Xenes and MXene associates, e.g., arsenene or bismuthene for the first family and Ti_3C_2 for the latter. Due to their unique electronic and mechanical properties, layered 2D materials have been recently exploited in high performance thermoelectric devices [128–131] with a variety of unexpected characteristics that we will introduce in this section. 2D materials might present two major drawbacks for a good thermoelectric performance, i.e., if the material behaves like a gapless semimetal it will be difficult to attain the desired imbalance between hot and cold carriers and, moreover, their thermal transport can be enormous compared to bulk 3D materials. Despite the technological complexity to overcome with two problems, it has been demonstrated that with a careful stacking of different layers of 2D materials a bandgap might arise in the vicinity of the Dirac point [132] as well as the thermal conductivity along the z-direction can be dramatically reduced by a controlled interfacial coupling when stacking different 2D layers [133, 134]. Concerning κ it can be adequately reduced nanostructuring the material [135] as well as by choosing the right underlying substrate [121, 136–138], henceforth, 2D layered materials for thermoelectric purposes show an enormous potentiality for their implementation in working thermoelectric devices, still to be explored both experimentally and theoretically.

7.7.1 Graphene

Graphene possesses a sizeable electron mobility [139] which is responsible for its incredibly high electrical and thermal conductivities of 10^6 S cm^{-1} and few thousand W (mK)$^{-1}$, respectively [140, 141], with a maximum reported Seebeck coefficient for pristine single-layer graphene of about 80 μV K^{-1} [142]. This value of the S coefficient was unexpected at its discovery due to the semimetallic behavior of pristine graphene, in which the bands cross each other linearly at the charge neutrality point (CNP). As previously mentioned, a strong asymmetry in the density of states (DOS) favors higher values of S, and graphene does not present such asymmetry. The first in-plane Seebeck and four-probe conductivity measurements [142, 143] showed that anomalous Seebeck ($S_{xx} = \Delta V_{xx}/\Delta T_{xx}$) and Nernst ($S_{xy} = \Delta V_{xy}/\Delta T_{xx}$) signals are present in pristine graphene, violating the established Mott's law [equation (7.1)] for metallic-like structures. This discrepancy with the existing Mott's law was first ascribed to the presence of electron and hole puddles near the CNP [144] resulting from impurities in the sample. More recently, experiments using ultraclean graphene samples on hexagonal BN (hBN) have finally

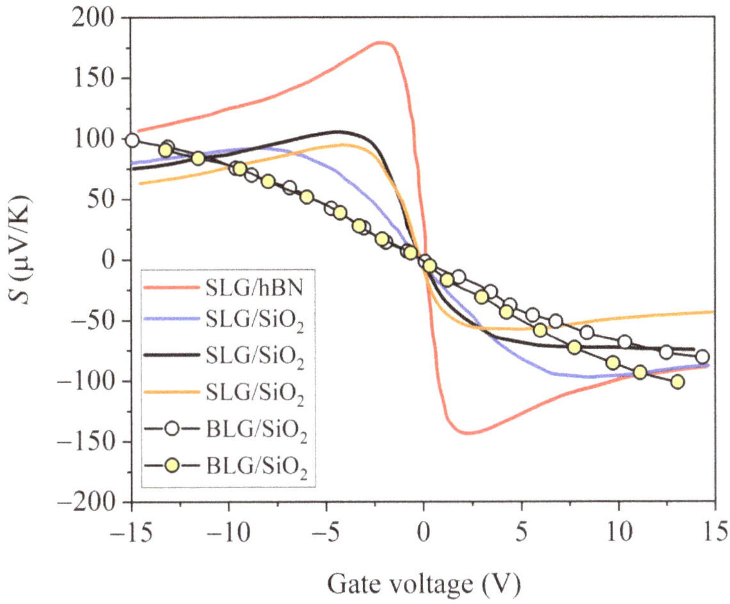

Figure 7.11. The Seebeck coefficient measured as a function of the gate voltage for different graphene layers and substrates. Data for single-layer graphene (SLG) and double layer graphene (DLG) transferred onto SiO_2 or hBN substrates are presented. All the curves were measured at room temperature and the position of the CNPs shifted to zero for clarity. Data retrieved from [130].

shown that this deviation from the standard Mott's law is a consequence of electron–electron interactions [145] in graphene, in which a plasma of electrons and holes is responsible for both this measurable Seebeck coefficient [145, 146] and the violation of the classical Wiedemann–Franz law [18] in this layered 2D material [147].

As previously mentioned, the combination of a high κ and a low S is detrimental for thermoelectric purposes, but there is still plenty of room for further improvements. Nowadays, it is well known that the transport properties of graphene are substantially affected by the phononic contribution of the underlying substrate [121, 136–138, 148] as well as from the number of layers [140, 149–151], as presented in figure 7.11. A way to improve the figure of merit in graphene might rely on its careful nanostructuring [135, 152, 153] and on reducing the phononic transmission and scattering mechanisms arising from the substrate. Silica substrates [154] have been found to be promising candidates to this end, since they can reduce κ to around 1000 W $(mK)^{-1}$. Moreover, as we have already mentioned, suspended graphene [141] or graphene deposited on alkyl-terminated-compatible hexaboron nitride [146, 155] would enhance graphene's figure of merit by reducing the thermal conductivity or by increasing its Seebeck coefficient. In figure 7.12 we show a thermoelectric measurement of graphene at room temperature made by Duan *et al* [146] with an optical micrograph of a graphene-on-hBN device in (a); the difference in S (b) and the PF (c), both measured as a function of the back-gate voltage when the graphene is placed directly onto SiO_2 or hBN; and finally a simulation of the

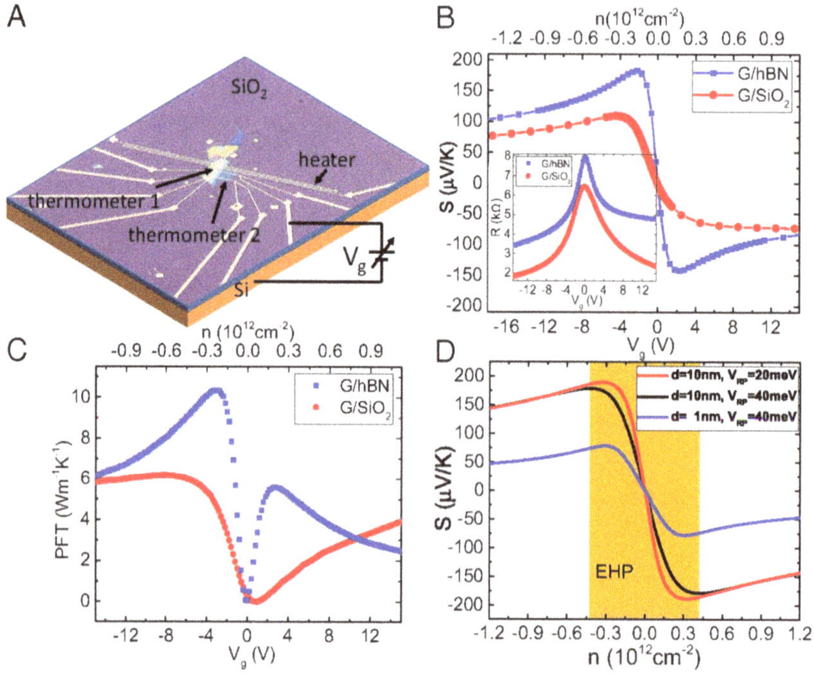

Figure 7.12. Thermoelectric measurements of graphene at room temperature. (a) An optical micrograph of a graphene-on-hBN (G/hBN) device. (b) The measured Seebeck coefficients of G/hBN and G/SiO_2 devices as a function of the back-gate voltage at 290 K. (Inset) Measured resistance vs. V_g in both devices at 290 K. (c) Measured PFT (Power Factor times Temperature) in both samples as a function of the back-gate voltage at 290 K. (d) Simulation of the carrier density dependence of the Seebeck coefficient at 300 K using the screened Coulomb-scattering model for two values of the hBN thickness, *d*, and random potential fluctuations (RPFs) induced by charge impurities. The rectangular shadow corresponds to the EHP region in a sample with $d = 10$ nm and RPF = 40 meV. Reprinted with permission from [146]. © 2016 Proceedings of the National Academy of Science of the United States of America.

carrier density dependence of *S* using the screened Coulomb-scattering model for two values of the hBN thickness (d). Other strategies used to enhance the thermoelectric *ZT* in graphene have included band engineering [156], functionalization [157], defect engineering [158], Rashba interactions [159, 160], and nanostructuring [23, 117, 136, 153]

7.7.2 Transition-metal dichalcogenides

Transition-metal dichalcogenides (TMDs) form a novel class of layered materials. They are composed of a transition-metal atom (such as Mo, W, or Ti) and two chalcogens (S, Se, or Te, for example); they exhibit outstanding physical, chemical, and optical properties. TMDs possess relatively high electrical conductivity and moderately low thermal conductivity, making them ideal candidates for thermoelectric purposes. The most common TMDs (MoS_2, $MoSe_2$, WS_2, and WSe_2) have similar electronic properties. They are semiconducting and show sizeable electronic bandgaps with values in the range of ~1 eV [161–163]. As a result of their quasi-2D structure, TMDs exhibit a much higher Seebeck coefficient [128, 164, 165] than graphene and remain practically unaffected by the number of layers. It is worth mentioning that even though S does not depend on the thickness of these TMDs, their electrical conductivity increases when they are reduced to a few layers, or eventually to the monolayer limit, giving rise to an increase in the value of ZT [166–169]. Recent theoretical predictions have found noticeably low values for κ of around 1 W $(mK)^{-1}$ [170–172] compared to graphene, contrary to the experimental results, which show values around an order of magnitude higher [173–176].

7.7.3 Black phosphorous

Following the isolation of graphene using the Scotch tape method, which was successfully applied to the aforementioned TMDs, it was only a matter of time before the isolation of the most stable form of elemental phosphorous—the so-called black phosphorus (BP)—quickly followed. BP is a direct-bandgap semiconductor discovered and thoroughly studied more than a century ago in its bulk. Li *et al* were the first to synthesize [177] a few-nanometer-thick flake that was used to create a field-effect transistor with compelling characteristics such as an on/off ratio of 10^5 at room temperature and a field-effect mobility, μ, of about 1000 $cm^2\ V^{-1}\ s^{-1}$. Unlike graphene, but more similarly to some of the TMDs, black phosphorous presents noticeable in-plane anisotropy as a consequence of the armchair termination of the single layers, which is responsible for some of the most intriguing properties of this material in the context of its implementation in thermoelectric devices [178, 179]. In its bilayer form, the κ of BP differs by a factor two in the two privileged directions in its crystal structure, resulting in the highly anisotropic thermoelectric response of this material [180, 181]. The two privileged directions are armchair (AC)-like and zigzag (ZZ)-like [182, 183] and present κ values that have been found to range between κ_{AC} ~ 10 and κ_{ZZ} ~ 25 W $(mK)^{-1}$ [181] and between κ_{AC} ~ 10 and κ_{ZZ} ~ 50 W $(mK)^{-1}$, respectively [180]. Striking anisotropy has also been found in the S value of black phosphorous. While the bulk form of black phosphorous presents a robust Seebeck coefficient of S ~ +335 ± 10 μV K^{-1} measured at room temperature [184], there is a sizable discrepancy in the value of S in the two privileged directions when its thickness is reduced to few nanometers [181, 185], which is, nevertheless, enhanced to S values of around +510 μV K^{-1} [186] in the 2D limit. More interestingly, black phosphorous possesses an inverse anisotropy [187, 188] in its electrical conductivity compared to that present in κ. This double anisotropy and a moderate value of S are responsible for the low values of the figures of merit, ZT_{AC} ~ 0.0036 and ZT_{ZZ} ~ 0.0006, obtained for undoped samples. However, it has been theoretically predicted that the ZT values could be further enhanced by doping, intercalation, or even strain [182, 189–192], boosting them to ZT_{AC} ~ 2 and ZT_{ZZ} ~ 1.6, respectively, and putting this material in an advantageous position compared to other, more conventional 2D materials in the search for high electronic performance.

7.7.4 Layered binary IV–VI compounds

Chalcogenides from the IV–VI group can present both a layered structure in the growth direction and excellent thermoelectric characteristics. These compounds can consist of the following combinations of materials: Sn(S, Se, Te), Ge(S, Se, Te), Pb (S, Se, Te), $Sn(Se, S)_2$, and their alloys, some of which have remarkably high *ZT* values and hence great potential for their implementation in thermoelectric devices [193], as shown in figure 7.13. These high values of *ZT* [194–198] result from an inherent anisotropy in the effective mass tensor of the relevant charge carriers, leading to high Seebeck coefficients; this stems from a large DOS effective mass together with a high electrical conductivity caused by a small conductivity effective mass.

In the early years of their study, these monochalcogenide compounds exhibited a sizeable *ZT* of around 2.5 but with the major drawback of a remarkably high operational temperature, $T \sim 1000$ °C [198], making them unsuitable for use in standard thermoelectric devices. In the last few years, we have witnessed a reinvigoration in the field due to the use of their polycrystalline forms together

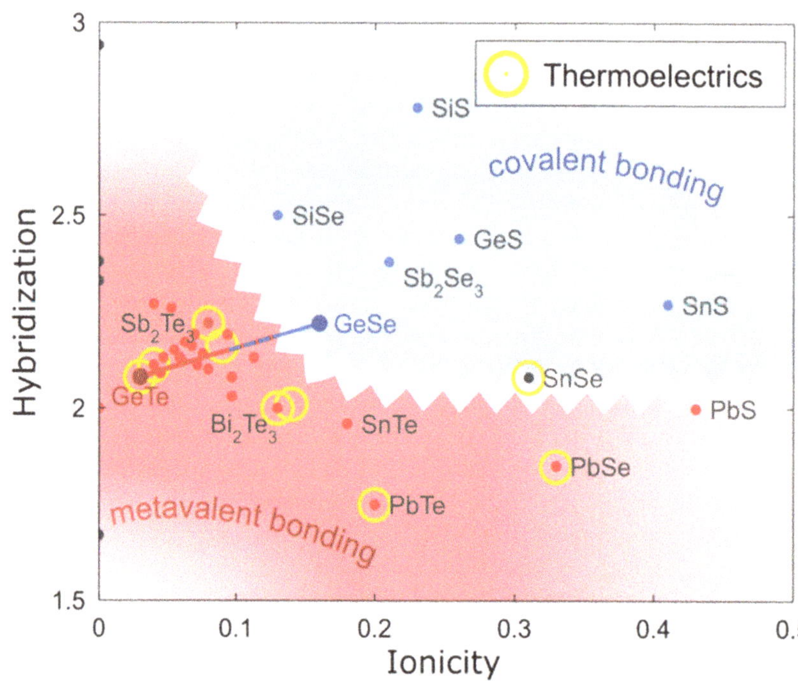

Figure 7.13. Map of various group V elements, IV–VI compounds, and V_2VI_3 compounds. The materials are sorted according to ionicity and s-p hybridization. Materials that exhibit metavalent bonding are shown in red, while materials that do not exhibit metavalent bonding are shown in blue. Yellow circles highlight single-phase materials with a recorded maximum ZT of more than 0.7. Interestingly, most thermoelectric materials are located in the red region of the map. This implies that these materials possess common intrinsic physical properties beneficial for thermoelectric performance [193] John Wiley & Sons. [© 2018 RWTH Aachen University. Published by WILEY-VCH Verlag GmbH & Co. KGaA, Weinheim].

with a substantial increment in the value of ZT up to 0.5–1.7 [199, 200]. Such forms operate at much lower temperatures than their crystalline counterparts but also possess much reduced electrical conductivity due to the presence of grain boundaries [201, 202]. Further experimental work intended to merge both the high electrical conductivity and the high ZT values of the crystalline and polycrystalline compounds, respectively, is being carried out.

7.7.5 MXenes

MXenes [203] are a newly discovered group of 2D materials synthesized for the first time in 2011 [204]. They are produced by rinsing the bulk MAX family in hydrofluoric acid followed by sonication of the compound. MAX phases [205] are a family of solids with layered hexagonal structures whose chemical compositions are traditionally known by the chemical formula $M_{n+1}AX_n$, where n is 1, 2, or 3, 'M' is an early transition metal (Sc, Ti, Zr, Hf, V, Nb, Ta, Cr, or Mo), 'A' is an element predominantly from group III (but it can also come from group VI), and 'X' is either nitrogen and/or carbon. In order to obtain layered MXenes, 'A' atoms are selectively etched by HF acid [204, 206, 207]. MXenes present excellent thermal stability in air with noticeable thermoelectric performances that outperform those of other 2D materials such as graphene. Kim *et al* [208] reported values for the Seebeck coefficient, S, of around $-27\ \mu V\ K^{-1}$ in $Mo_2Ti_2C_3T_x$ and $S = -30.5\ \mu V\ K^{-1}$ in Mo_2CT_x. Recently, Jing *et al* [209] showed a superior thermoelectric performance in ordered double transition-metal (TM)-based MXenes. They studied $Cr_2TiC_2T_2$ (where T is –OH or –F) and observed S values in the range of $>400\ \mu V\ K^{-1}$ in Cr_2TiC_2 and $S > 800\ \mu V\ K^{-1}$ in $Cr_2TiC_2F_2$. Moreover, it has been theoretically predicted that that functionalization of MXenes (by –F, –C, –O or –OH) will substantially modify the electron band structure of these materials and affect their thermoelectric response. As an example, ScC_2 functionalized with C presented an indirect gap of 1.84 eV, while its functionalization with –OH reduced the gap to 0.44 eV [210, 211]. The value for S has been reported to be as high as $S > 1000\ \mu V\ K^{-1}$ at room temperature for Sc_2CF_2 [210]. Hence, these layered compounds are expected to have the potential for great thermoelectric responses.

7.8 Applications and devices

Thermoelectric systems can work in three different modes of operation, i.e. power generation, Peltier refrigeration, and active cooling. The first mode, power generation, is purely based on the Seebeck effect; therefore heat is converted into electricity. This mode has direct application to solar cells and thermopower devices. The second mode, Peltier refrigeration, relies on the electrical pumping of heat from a cold source to a hot sink. Finally, in the latter mode, the active cooling mode, electrical current is basically reversed compared to that of Peltier refrigeration with the aim of pumping heat in its natural direction, namely, from a heat source to a cold drain. The first two modes of operation require materials and systems with high ZT values, whereas active cooling requires large thermoelectric PFs and therefore large thermal conductivity. Here, we briefly outline some of the devices and applications devoted mainly to the first and latter modes of operation. While 2D-like materials have typically been neglected for use in the power generation mode due to their inherently low *ZT* and enormous κ, great efforts have been made to integrate them further into these devices because they are naturally abundant, cheap, and typically biocompatible. In the case of active cooling, graphene and other 2D layered systems present two main advantages compared to standard thermoelectric materials such as topological insulators, as a result of their efficient passive (high κ) and active cooling (high PF) performances [212].

The typical structure of the nanodevices employed to measure the thermoelectric response of a low-dimensional material is presented in figure 7.14. While *S* can be obtained for samples directly deposited onto a substrate, the value of *ZT* must be determined using suspended devices in which the material to be studied is decoupled from the phononic contribution to κ. Figure 7.14 shows the minimal setup used to measure *S* and *ZT* in a NW InAs/InP quantum dot. A heater lies at one side of the device and electrical contacts are employed to evaluate *ZT* as follows. The heater induces a heat flow across the material and the electrical contacts serve as thermometers, since they are able to measure the temperature difference ΔT between

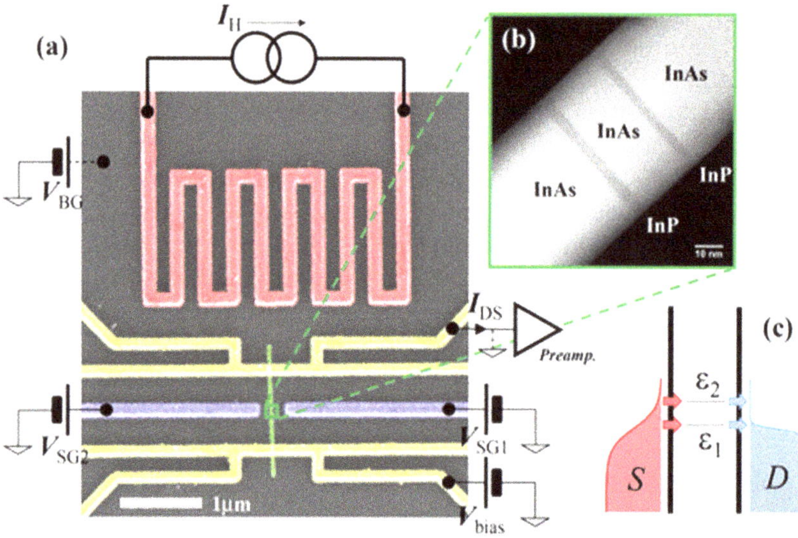

Figure 7.14. (a) A scanning electron micrograph of the typical setup used to measure the thermoelectric efficiency of a given device or material. A local heater (red) is used to establish a temperature difference ΔT between the two ends of a NW (green) embedded in an InAs/InP heterostructured quantum dot. The device is fabricated on top of a degenerately doped SiO_2/Si p++ substrate (gray), and a set of Ti/Au electrodes (yellow) can be used both as electrical contacts for the NW and as local resistive thermometers. The quantum dot's electronic configuration can be controlled using a pair of side gates (purple) or using the conductive substrate as a back-gate electrode. (b) A transmission electron micrograph of the heterostructured quantum dots. (c) Sketch of the energetics scheme: the quantum dot implements a multilevel system that can mediate heat and charge transport between the a source (S) electrode and a drain (D) electrode in the presence of thermal and electric biases. Two spin-degenerate levels ε_1 and ε_2 ($\Delta\varepsilon = \varepsilon_2 - \varepsilon_1$) play relevant roles in the regime studied in the experiment. Reprinted with permission from [80]. Copyright (2019) American Chemical Society.

both ends of the device. κ can then be inferred from the value obtained for ΔT and the heat delivered using a thermometric technique [213, 214].

This technique has been thoroughly implemented in order to measure a number of different low-dimensional materials in different configurations, for example: single NWs [213], quantum dots [215], superlattices [216], and Hall bars [217]. Structures based on graphene and graphene-like materials have been theoretically studied with the goal of increasing their ZT values [112, 115, 218]. Treating a graphene sheet in an oxygen plasma can enormously increase its Seebeck coefficient to values higher than 700 $\mu V\ K^{-1}$ [219, 220], and such sheets have direct applications in active cooling, as local hotspot cooling can be performed [212]. Different devices and applications involving on-chip graphene-based structures for cooling purposes use Peltier heating for carriers [221–223]; engineered graphene nanoribbons in which the thermal conductance can be reversed, which are therefore used as thermal conductance modulators [224]; and folded–unfolded monolayer graphene sheets in ordered 3D structures for thermal management at the nanoscale [225]. Sierra *et al* [129, 226] in their seminal works demonstrated that the spin-Seebeck effect [227–229] (spin current caused by a thermal gradient) is present in graphene structures near the charge neutrality point and that such a spin voltage might be increased by the thermal carriers present in the sample [230, 231]. These two works have triggered a novel research line of their own based on thermal spins in 2D materials; these envisage the production of, for example, spin-Seebeck diodes based on graphene [98, 232], BP [233], TMDs [234], or CrI_3 [235].

Thermoelectric materials have recently been used to develop efficient and wearable self-powered devices in photodetectors and for photovoltaic purposes. The Seebeck effect is the key ingredient, since it allows the device to generate a photocurrent from an induced temperature gradient created via absorbed light or an applied external voltage. Graphene and other 2D layered materials have been extensively implemented in photothermal devices [164, 236–238]; the seminal work from Gabor *et al* [239] showed that the heat that results from shining a laser on a single layer of graphene generates a thermoelectric voltage. High-sensitivity terahertz photodetectors based on the thermoelectric effect in graphene are already a reality [240, 241], showing a state-of-the-art value of 120 $pW\ Hz^{-1/2}$ for the tunable noise spectral density [241, 242] and sensitivities exceeding 10 $V\ W^{-1}$ with reference to the incident power. TMD-based devices operating through the photothermoelectric effect have arisen in the last few years, following their graphene- and BP-based counterparts. MoS_2 [164, 243] and WSe_2 [236] photoelectric and photothermoelectric devices have shown great potential when designed as double-gate p–n junctions, as shown in figure 7.15. In these devices, the photocurrent generated by the photothermoelectric effect (as a consequence of the presence of strong optical absorption regions in the sample) outperforms the pure photovoltaic effect. This response implies that a careful device design that produces an asymmetry in the thickness of WSe_2 will create a thermal gradient upon illumination that will be responsible for producing a sizable photocurrent. MXenes are starting to be employed for thermoelectric purposes. They can appear in combination with a standard topological insulator, thereby enhancing its thermoelectric power

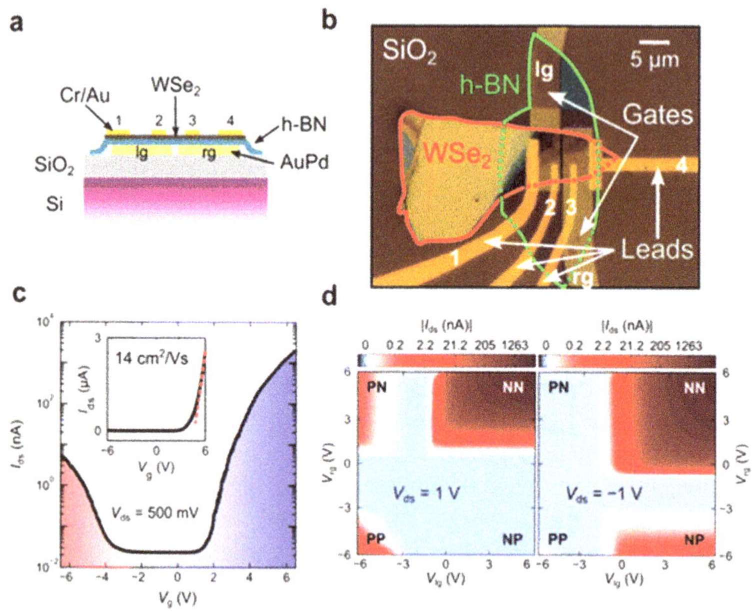

Figure 7.15. (a) A device schematic indicating the different components of a WSe_2 double-gate device. (b) Optical image of the WSe_2 flake (outlined in red) on top of the hBN (outlined in green) on the local gates. The leads, gates, and different flakes are labeled for clarity. (c) The measured current (I_{ds}) as the two local gates are linked and varied as a single back-gate V_g, shown on a logarithmic scale. The color gradient from red to blue highlights the transition from hole-doping to electron-doping as the gates are varied from negative to positive voltages. The inset shows the data on a linear scale. (d) Color maps of the magnitude of I_{ds} (on a logarithmic scale) as the voltages on the local gates are independently varied from −6 to 6 V with a fixed positive (left) and negative (right) V_{ds}. Reprinted with permission from [236]. Copyright (2014) American Chemical Society.

generation by the mutual interaction between both materials [244]. A noticeable increase in the power factor and a reduction in the thermal conductivity of $(Bi, Sb)_2Te_3$ have been shown when $Ti_3C_2T_x$ is incorporated into it, paving the way for a novel approach with which to increase the energy-conversion efficiency.

7.9 Summary

The growing energy consumption worldwide demands a paradigm shift in the way we exploit energy resources. We should be able to make fast progress toward environmentally friendly energy sources. But alongside the use of clean and renewable energies, we must seek ways to recover, at least partially, the energy that is wasted in the form of heat. This goal is deeply rooted in current thermoelectric research. The quest for new and innovative materials with enhanced heat-to-electricity conversion capabilities and large output powers is one of the best strategies for accomplishing this objective. Further, thermoelectric machines (heat engines and refrigerators) work with no moving parts and therefore show great reliability.

Bulk materials seem to be reaching their full potential for the design of more efficient thermoelectric devices based on conventional heat-to-electricity conversion processes. Therefore, a lot of effort should be devoted to proposing new physical mechanisms for energy harvesting in industrial and domestic facilities. In this regard, salient advances in nanotechnology have brought with them promising methods to exploit fundamental mechanisms and devise thermoelectric setups that have good performance. Several effects, such as quantum electron resonances and enhanced phonon scattering, can be combined together to significantly improve on the best figures of merit for macroscopic conductors. Moreover, thermoelectric quantities such as the Seebeck coefficient are sensitive probes for small changes in the local DOS, giving us information about the dominant carrier type (electron or hole) and its entropy and energy transfer during transport.

The present chapter comprises an overview of some of the fundamental concepts and breakthroughs in experimental and theoretical thermoelectric research at the nanoscale. Several physical effects arising in systems whose size is of the order of few nanometers might help to increase thermodynamic efficiency. On the one hand, quantum interference effects enable sharp resonances to appear that ultimately lead to high thermopower values in both the linear and nonlinear regimes of transport. On the other hand, reduced geometries greatly increase phonon scattering without affecting electrical conduction to a large extent. As a consequence, the contribution of the lattice to the heat current is lower than in bulk materials. The combination of both types of effect manifests itself as a marked increase in the thermoelectric figure of merit for nanodevices used in energy harvesting and cooling, especially for highly efficient microscale power generation and cryogenic refrigeration to very low temperatures.

Q & A

1. Bulk materials present better thermoelectric performance than 2D-based materials because of:
   (a) a much lower thermal transport;
   (b) their inherent asymmetry in the DOS because of quantum confinement;
   (c) none of the above.
2. Graphene presents sizable Nernst and Seebeck signals arising from:
   (a) its asymmetric DOS;
   (b) the presence of electron–hole puddles;
   (c) electron–electron interactions.
3. Graphene deposited onto alkyl-terminated hexaboron nitride would present an enhanced figure of merit due to:
   (a) an induced asymmetry in the DOS due to the poor lattice match;
   (b) a reduction of the phononic transmission from the substrate;
   (c) an increased carrier mobility within the graphene sheet.
4. Unlike graphene, BP has a negligible Seebeck coefficient as a consequence of:

(a) an extraordinary in-plane isotropy, which makes BP unsuitable for thermoelectric purposes;
(b) a high thermal conductance due to its dangling terminated bonds;
(c) none of the above—BP outperforms graphene in terms of Seebeck coefficient because of its marked in-plane anisotropy.

5. In order to unambiguously determine the figure of merit, the optimal approach would be:
   (a) to infer it from photothermoelectric experiments in which pure photovoltaic effects can be measured and neglected;
   (b) to suspend the device in order to decouple it from the phononic contribution;
   (c) to obtain low-noise electric measurements, keeping in mind that the local heater has to be in contact with the actual sample in order to reduce spurious phononic effects.
6. Calculate the figure of merit, ZT, for a device operating at 50% of the Carnot efficiency.
   (a) 0.5;
   (b) 1;
   (c) 8.
7. In an experiment, it is observed that a single-channel nanoconductor with Fermi energy E_F shows a linear conductance of $2e^2/h$ at zero temperature. Without performing any measurement, the thermopower is:
   (a) 0;
   (b) $\pi^2 k_B/3e$;
   (c) k_B/E_F.
8. When the energy level of a quantum dot aligns with the Fermi energy of the coupled terminals, the Seebeck coefficient is zero because:
   (a) the applied temperature difference is symmetric;
   (b) electron–electron interactions can be neglected;
   (c) electron–hole symmetry is not broken.
9. Fano resonances are beneficial for thermoelectrics since, in general, their transmission line shapes around the Fermi energy:
   (a) are symmetric;
   (b) are asymmetric;
   (c) attain zero values.
10. The high thermoelectric efficiency of silicon NWs compared with that of bulk silicon is attributed to:
   (a) enhanced phonon scattering;
   (b) enhanced electron scattering;
   (c) strong electron–phonon coupling.

Solutions:

1. (a), 2. (c), 3. (b), 4. (c), 5. (b), 6. (c), 7. (a), 8. (c), 9. (b), and 10. (a)

References

[1] 2019 *BP Statistical Review of World Energy*, 2019 https://www.bp.com/content/dam/bp/business-sites/en/global/corporate/pdfs/energy-economics/statistical-review/bp-stats-review-2019-full-report.pdf
[2] Ritchie H and Roser M 2014 Energy https://ourworldindata.org/energy
[3] He J and Tritt T M 2017 Advances in thermoelectric materials research: looking back and moving forward *Science* **357** 1369
[4] Wang X, Hong W and Liu B 2018 Carbon nanotube-based organic thermoelectric materials for energy harvesting *Polymers* **10** 1196
[5] Beretta D *et al* 2019 Thermoelectrics: from history, a window to the future *Mater. Sci. Eng.* R **138** 100501
[6] Goldsmid H J 2009 *Introduction to ThermoelectricitySpringer Series in Materials Science* (Berlin: Springer)
[7] Kim H S, Liu W, Chen G, Chu C-W and Ren Z 2015 Relationship between thermoelectric figure of merit and energy conversion efficiency *Proc. Natl Acad. Sci. USA* **112** 8205
[8] Hicks L D and Dresselhaus M S 1993 Effect of quantum-well structures on the thermoelectric figure of merit *Phys. Rev.* B **47** 12727
[9] Hicks L D and Dresselhaus M S 1993 Thermoelectric figure of merit of a one-dimensional conductor *Phys. Rev.* B **47** 16631
[10] Khitun A, Balandin A, Liu J L and Wang K L 2000 In-plane lattice thermal conductivity of a quantum-dot superlattice *J. Appl. Phys.* **88** 696
[11] Venkatasubramanian R, Siivola E, Colpitts T and O'Quinn B 2001 Thin-film thermoelectric devices with high room-temperature figures of merit *Nature* **413** 597
[12] Harman T C, Taylor P J, Walsh M P and LaForge B E 2002 Quantum dot superlattice thermoelectric materials and devices *Science* **297** 2229
[13] Balandin A A and Lazarenkova O L 2003 Mechanism for thermoelectric figure-of-merit enhancement in regimented quantum dot superlattices *Appl. Phys. Lett.* **82** 415
[14] Hochbaum A I, Chen R, Delgado R D, Liang W, Garnett E C, Najarian M, Majumdar A and Yang P 2008 Enhanced thermoelectric performance of rough silicon nanowires *Nature* **451** 163
[15] Boukai A I, Bunimovich Y, Tahir-Kheli J, Yu J-K, Goddard W A and Heath J R 2008 Silicon nanowires as efficient thermoelectric materials *Nature* **451** 168
[16] Dubi Y and Di Ventra M 2011 Colloquium: heat flow and thermoelectricity in atomic and molecular junctions *Rev. Mod. Phys.* **83** 131
[17] Benenti G, Casati G, Saito K and Whitney R S 2017 Fundamental aspects of steady-state conversion of heat to work at the nanoscale *Phys. Rep.* **694** 1
[18] Jonson M and Mahan G D 1980 Mott's formula for the thermopower and the Wiedemann–Franz law *Phys. Rev.* B **21** 4223
[19] Fano U 1961 Effects of configuration interaction on intensities and phase shifts *Phys. Rev.* **124** 1866
[20] Mahan G D and Sofo J O 1996 The best thermoelectric *Proc. Natl Acad. Sci. USA* **93** 7436
[21] Gómez-Silva G, Ávalos-Ovando O, Ladrón de Guevara M L and Orellana P A 2012 Enhancement of thermoelectric efficiency and violation of the Wiedemann–Franz law due to Fano effect *J. Appl. Phys.* **111** 053704
[22] Zheng J, Zhu M-J and Chi F 2012 Fano effect on the thermoelectric efficiency in parallel-coupled double quantum dots *J. Low Temp. Phys.* **166** 208

[23] Chen Z-G, Han G, Yang L, Cheng L and Zou J 2012 Nanostructured thermoelectric materials: current research and future challenge *Prog. Nat. Sci. Mater* **22** 535
[24] García-Suárez V M, Ferradás R and Ferrer J 2013 Impact of Fano and Breit–Wigner resonances in the thermoelectric properties of nanoscale junctions *Phys. Rev.* B **88** 235417
[25] Fu H-H, Gu L, Wu D-D and Zhang Z-Q 2015 Enhancement of the thermoelectric figure of merit in DNA-like systems induced by Fano and Dicke effects *Phys. Chem. Chem. Phys.* **17** 11077
[26] Sierra M A, Saiz-Bretín M, Domínguez-Adame F and Sánchez D 2016 Interactions and thermoelectric effects in a parallel-coupled double quantum dot *Phys. Rev.* B **93** 235452
[27] Wang R-N, Dong G-Y, Wang S-F, Fu G-S and Wang J-L 2016 Impact of contact couplings on thermoelectric properties of anti, Fano, and Breit–Wigner resonant junctions *J. Appl. Phys.* **120** 184303
[28] Mott N F and Davis E A 1979 *Electronic Processes in Non-Crystalline Materials* (Oxford: Clarendon)
[29] Landauer R 1957 Spatial variation of currents and fields due to localized scatterers in metallic conduction *IBM J. Res. Dev.* **1** 223
[30] Büttiker M 1986 Four-terminal phase-coherent conductance *Phys. Rev. Lett.* **57** 1761
[31] Ymry Y 1997 *Introduction to Mesoscopic Physics* (Oxford: Oxford University Press)
[32] Nazarov Y V and Blanter Y M 2009 *Quantum Transport: Introduction to Nanoscience* (Cambridge: Cambridge University Press)
[33] Butcher P N 1990 Thermal and electrical transport formalism for electronic microstructures with many terminals *J. Phys. Condens. Matter* **2** 4869
[34] Kubala B, König J and Pekola J 2008 Violation of the Wiedemann–Franz law in a single-electron transistor *Phys. Rev. Lett.* **100** 066801
[35] Lunde A M and Flensberg K 2005 On the Mott formula for the thermopower of non-interacting electrons in quantum point contacts *J. Phys. Condens. Matter* **17** 3879
[36] Van den Broeck C 2005 Thermodynamic efficiency at maximum power *Phys. Rev. Lett.* **95** 190602
[37] Curzon F L and Ahlborn B 1975 Efficiency of a carnot engine at maximum power output *Am. J. Phys.* **43** 22
[38] Matthews J, Sánchez D, Larsson M and Linke H 2012 Thermally driven ballistic rectifier *Phys. Rev.* B **85** 205309
[39] Matthews J, Battista F, Sánchez D, Samuelsson P and Linke H 2014 Experimental verification of reciprocity relations in quantum thermoelectric transport *Phys. Rev.* B **90** 165428
[40] Sánchez D and Serra L 2011 Thermoelectric transport of mesoscopic conductors coupled to voltage and thermal probes *Phys. Rev.* B **84** 201307
[41] Saito K, Benenti G, Casati G and Prosen T 2011 Thermopower with broken time-reversal symmetry *Phys. Rev.* B **84** 201306
[42] Brandner K, Saito K and Seifert U 2013 Strong bounds on onsager coefficients and efficiency for three-terminal thermoelectric transport in a magnetic field *Phys. Rev. Lett.* **110** 070603
[43] Sothmann B, Sánchez R and Jordan A N 2014 Thermoelectric energy harvesting with quantum dots *Nanotechnology* **26** 032001
[44] Sánchez R and Büttiker M 2011 Optimal energy quanta to current conversion *Phys. Rev.* B **83** 085428

[45] Entin-Wohlman O, Imry Y and Aharony A 2010 Three-terminal thermoelectric transport through a molecular junction *Phys. Rev.* B **82** 115314
[46] Thierschmann H, Sánchez R, Sothmann B, Arnold F, Heyn C, Hansen W, Buhmann H and Molenkamp L 2015 Three-terminal energy harvester with coupled quantum dots *Nat. Nanotechnol* **10** 854
[47] Büttiker M 1993 Capacitance, admittance, and rectification properties of small conductors *J. Phys. Condens. Matter* **5** 9361
[48] Sánchez D and López R 2013 Scattering theory of nonlinear thermoelectric transport *Phys. Rev. Lett.* **110** 026804
[49] Meair J and Jacquod P 2013 Scattering theory of nonlinear thermoelectricity in quantum coherent conductors *J. Phys. Condens. Matter* **25** 082201
[50] Hwang S-Y, Sánchez D, Lee M and López R 2013 Magnetic-field asymmetry of non-linear thermoelectric and heat transport *New J. Phys.* **15** 105012
[51] Erdman P A, Mazza F, Bosisio R, Benenti G, Fazio R and Taddei F 2017 Thermoelectric properties of an interacting quantum dot based heat engine *Phys. Rev.* B **95** 245432
[52] Whitney R S 2014 Most efficient quantum thermoelectric at finite power output *Phys. Rev. Lett.* **112** 130601
[53] López R, Lim J S and Sánchez D 2012 Fluctuation relations for spintronics *Phys. Rev. Lett.* **108** 246603
[54] Juergens S, Haupt F, Moskalets M and Splettstoesser J 2013 Thermoelectric performance of a driven double quantum dot *Phys. Rev.* B **87** 245423
[55] Zhou H, Thingna J, Hänggi P, Wang J-S and Li B 2005 Boosting thermoelectric efficiency using time-dependent control *Sci. Rep.* **5** 14870
[56] Ludovico M F, Battista F, von Oppen F and Arrachea L 2016 Adiabatic response and quantum thermoelectrics for ac-driven quantum systems *Phys. Rev.* B **93** 075136
[57] Li Z, Sun Q, Yao X D, Zhu Z H and Lu G Q 2012 Semiconductor nanowires for thermoelectrics *J. Mater. Chem.* **22** 22821
[58] Hung N, Nugraha A R T and Saito R 2019 Thermoelectric properties of carbon nanotubes *Energies* **12** 4561
[59] Domínguez-Adame F, Martín-González M, Sánchez D and Cantarero A 2019 Nanowires: a route to efficient thermoelectric devices *Physica* E **113** 213
[60] Hung N T, Hasdeo E H, Nugraha A R T, Dresselhaus M S and Saito R 2016 Quantum effects in the thermoelectric power factor of low-dimensional semiconductors *Phys. Rev. Lett.* **117** 036602
[61] Glassbrenner C J and Slack G A 1964 Thermal conductivity of silicon and germanium from 3 K to the melting point *Phys. Rev.* **134** A1058
[62] Thompson J C and Younglove B A 1961 Thermal conductivity of silicon at low temperatures *J. Phys. Chem. Solids* **20** 146
[63] Zhang T, Wu S, Xu J, Zheng R and Cheng G 2015 High thermoelectric figure-of-merits from large-area porous silicon nanowire arrays *Nano Energy* **13** 433
[64] Calaza C, Salleras M, Dávila D, Tarancón A, Morata A, Santos J D, Gadea G and Fonseca L 2015 Bottom-up silicon nanowire arrays for thermoelectric harvesting *Mater. Today Proc.* **2** 675
[65] Bennett N S, Wight N M, Popuri S R and Bos J-W G 2015 Efficient thermoelectric performance in silicon nano-films by vacancy-engineering *Nano Energy* **16** 350

[66] Xu Z, Zheng Q-R and Su G 2011 Thermoelectric properties of silicon carbide nanowires with nitride dopants and vacancies *Phys. Rev.* B **84** 245451
[67] Chen S *et al* 2020 Vacancy-based defect regulation for high thermoelectric performance in $Ge_9Sb_2Te_{12-x}$ compounds *ACS Appl. Mater. Interfaces* **12** 19664
[68] Xu S-Y *et al* 2012 Observation of a topological crystalline insulator phase and topological phase transition in $Pb_{1-x}Sn_xTe$ *Nat. Commun* **7** 12505
[69] Hsieh T H, Lin H, Li J, Duan W, Bansil A and Fu L 2012 Topological crystalline insulators in the SnTe material class *Nat. Commun.* **3** 982
[70] Ando Y and Fu L 2015 Topological crystalline insulators and topological superconductors: from concepts to materials *Annu. Rev. Condens. Matter Phys.* **6** 361
[71] Tchoumakov S, Jouffrey V, Inhofer A, Bocquillon E, Plaçais B, Carpentier D and Goerbig M O 2017 Volkov–Pankratov states in topological heterojunctions *Phys. Rev.* B **96** 201302
[72] Inhofer A *et al* 2017 Observation of Volkov–Pankratov states in topological HgTe heterojunctions using high-frequency compressibility *Phys. Rev.* B **96** 195104
[73] Xu N, Xu Y and Zhu J 2017 Topological insulators for thermoelectrics *npj Quantum Mater.* **2** 51
[74] Hinsche N F, Zastrow S, Gooth J, Pudewill L, Zierold R, Rittweger F, Rauch T, Henk J, Nielsch K and Mertig I 2015 Impact of the topological surface state on the thermoelectric transport in Sb_2Te_3 thin films *ACS Nano* **9** 4406
[75] Yang J 2017 Prospective thermoelectrics among topological insulators *Materials Aspect of Thermoelectricity* ed C Uher (Boca Raton, FL: CRC Press), 583–97 ch 21
[76] Huber T E, Adeyeye A, Nikolaeva A, Konopko L, Johnson R C and Graf M J 2011 Surface state band mobility and thermopower in semiconducting bismuth nanowires *Phys. Rev.* B **83** 235414
[77] Huber T E, Owusu K, Johnson S, Nikolaeva A, Konopko L, Johnson R C and Graf M J 2012 Thermoelectric prospects of nanomaterials with spin-orbit surface bands *J. Appl. Phys.* **111** 043709
[78] Humphrey T E and Linke H 2005 Reversible thermoelectric nanomaterials *Phys. Rev. Lett.* **94** 096601
[79] Josefsson M, Svilans A, Burke A M, Hoffmann E A, Fahlvik S, Thelander C, Leijnse M and Linke H 2018 A quantum-dot heat engine operating close to the thermodynamic efficiency limits *Nat. Nanotechnol* **13** 920
[80] Prete D, Erdman P A, Demontis V, Zannier V, Ercolani D, Sorba L, Beltram F, Rossella F, Taddei F and Roddaro S 2019 Thermoelectric conversion at 30 K in InAs/InP nanowire quantum dots *Nano Lett.* **19** 3033
[81] Nakpathomkun N, Xu H Q and Linke H 2010 Thermoelectric efficiency at maximum power in low-dimensional systems *Phys. Rev.* B **82** 235428
[82] Esposito M, Lindenberg K and Van den Broeck C 2009 Thermoelectric efficiency at maximum power in a quantum dot *Europhys. Lett.* **85** 60010
[83] Beenakker C W J and Staring A A M 1992 Theory of the thermopower of a quantum dot *Phys. Rev.* B **46** 9667
[84] Staring A A M, Molenkamp L W, Alphenaar B W, van Houten H, Buyk O J A, Mabesoone M A A, Beenakker C W J and Foxon C T 1993 Coulomb-blockade oscillations in the thermopower of a quantum dot *Europhys. Lett.* **22** 57

[85] Dzurak A S, Smith C G, Barnes C H W, Pepper M, Martín-Moreno L, Liang C T, Ritchie D A and Jones G A C 1997 Thermoelectric signature of the excitation spectrum of a quantum dot *Phys. Rev.* B **55** R10197

[86] Turek M and Matveev K A 2002 Cotunneling thermopower of single electron transistors *Phys. Rev.* B **65** 115332

[87] Boese D and Fazio R 2001 Thermoelectric effects in Kondo-correlated quantum dots *Europhys. Lett.* **56** 576

[88] Dong B and Lei X L 2002 Effect of the Kondo correlation on the thermopower in a quantum dot *J. Phys. Condens. Matter* **14** 11747

[89] Costi T A and Zlatic V 2010 Thermoelectric transport through strongly correlated quantum dots *Phys. Rev.* B **81** 235127

[90] Roura-Bas P, Tosi L, Aligia A A and Cornaglia P S 2012 Thermopower of an SU(4) Kondo resonance under an SU(2) symmetry-breaking field *Phys. Rev.* B **86** 165106

[91] Fahlvik Svensson S, Hoffmann E A, Nakpathomkun N, Wu P M, Xu H Q, Nilsson H A, Sánchez D, Kashcheyevs V and Linke H 2013 Nonlinear thermovoltage and thermocurrent in quantum dots *New J. Phys.* **15** 105011

[92] Sierra M A and Sánchez D 2014 Strongly nonlinear thermovoltage and heat dissipation in interacting quantum dots *Phys. Rev.* B **90** 115313

[93] Erlingsson S I, Manolescu A, Nemnes G A, Bardarson J H and Sánchez D 2017 Reversal of thermoelectric current in tubular nanowires *Phys. Rev. Lett.* **119** 036804

[94] Sierra M A, López R and Sánchez D 2017 Fate of the spin-½ Kondo effect in the presence of temperature gradients *Phys. Rev.* B **96** 085416

[95] Bergeret F S, Silaev M, Virtanen P and Heikkila T T 2018 Colloquium: nonequilibrium effects in superconductors with a spin-splitting field *Rev. Mod. Phys.* **90** 041001

[96] Machon P, Eschrig M and Belzig W 2013 Nonlocal thermoelectric effects and nonlocal Onsager relations in a three-terminal proximity-coupled superconductor–ferromagnet device *Phys. Rev. Lett.* **110** 047002

[97] Ozaeta A, Virtanen P, Bergeret F S and Heikkila T T 2014 Predicted very large thermoelectric effect in ferromagnet–superconductor junctions in the presence of a spin-splitting magnetic field *Phys. Rev. Lett.* **112** 057001

[98] Hwang S-Y, López R and Sánchez D 2016 Large thermoelectric power and figure of merit in a ferromagnetic–quantum dot–superconducting device *Phys. Rev.* B **94** 054506

[99] Hussein R, Governale M, Kohler S, Belzig W, Giazotto F and Braggio A 2019 Nonlocal thermoelectricity in a Cooper-pair splitter *Phys. Rev.* B **99** 075429

[100] Sánchez R, Burset P and Levy-Yeyati A 2018 Cooling by Cooper pair splitting *Phys. Rev.* B **98** 241414

[101] Nakanishi T and Kato T 2007 Thermopower of a quantum dot in a coherent regime *J. Phys. Soc. Jpn.* **76** 034715

[102] Finch C M, García-Suárez V M and Lambert C J 2009 Giant thermopower and figure of merit in single-molecule devices *Phys. Rev.* B **79** 033405

[103] Bergfield J P and Stafford C A 2009 Thermoelectric signatures of coherent transport in single-molecule heterojunctions *Nano Lett.* **9** 3072

[104] Wang X, Yang Y, Wu H-N and Gong W-J 2017 Thermoelectric and thermospin properties in a quantum ring with an embedded protein-like single-helical molecule *Appl. Phys. Lett.* **111** 063701

[105] Trocha P and Barnas J 2012 Large enhancement of thermoelectric effects in a double quantum dot system due to interference and Coulomb correlation phenomena *Phys. Rev.* B **85** 085408
[106] Vannucci L, Ronetti F, Dolcetto G, Carrega M and Sassetti M 2015 Interference-induced thermoelectric switching and heat rectification in quantum Hall junctions *Phys. Rev.* B **92** 075446
[107] Karlström O, Linke H, Karlström G and Wacker A 2011 Increasing thermoelectric performance using coherent transport *Phys. Rev.* B **84** 113415
[108] Munárriz J, Domínguez-Adame F and Malyshev A V 2011 Toward graphene-based quantum interference devices *Nanotechnology* **22** 365201
[109] Castro Neto A H, Guinea F, Peres N M R, Novoselov K S and Geim A K 2009 The electronic properties of graphene *Rev. Mod. Phys.* **81** 109
[110] Ouyang Y and Guo J 2009 A theoretical study on thermoelectric properties of graphene nanoribbons *Appl. Phys. Lett.* **94** 263107
[111] Rosales L, Núñez C D, Pacheco M, Latgé A and Orellana P A 2013 Graphene nanoribbon thermopower as a tool for molecular spectroscopy *J. Appl. Phys.* **114** 153711
[112] Sevinçli H and Cuniberti G 2010 Enhanced thermoelectric figure of merit in edge-disordered zigzag graphene nanoribbons *Phys. Rev.* B **81** 113401
[113] Mazzamuto F, Hung Nguyen V, Apertet Y, Caër C, Chassat C, Saint-Martin J and Dollfus P 2011 Enhanced thermoelectric properties in graphene nanoribbons by resonant tunneling of electrons *Phys. Rev.* B **83** 235426
[114] Abdullah N R 2018 Optical control of spin-dependent thermal transport in a quantum ring *Phys. Lett.* A **382** 1432
[115] Saiz-Bretín M, Malyshev A V, Orellana P A and Domínguez-Adame F 2015 Enhancing thermoelectric properties of graphene quantum rings *Phys. Rev.* B **91** 085431
[116] Munárriz Arrieta J 2014 *Modelling of Plasmonic and Graphene Nanodevices* (Berlin: Springer)
[117] Saiz-Bretín M, Malyshev A V, Domínguez-Adame F, Quigley D and Römer R A 2018 Lattice thermal conductivity of graphene nanostructures *Carbon* **127** 64
[118] http://lammps.sandia.gov
[119] Plimpton S 1995 Fast parallel algorithms for short-range molecular dynamics *J. Comput. Phys.* **117** 1
[120] Stuart S J, Tutein A B and Harrison J A 2000 A reactive potential for hydrocarbons with intermolecular interactions *J. Chem. Phys.* **112** 6472
[121] Pop E, Varshney V and Roy A K 2012 Thermal properties of graphene: fundamentals and applications *MRS Bull.* **37** 1273
[122] Yang P, Tang Y, Yang H, Gong J, Liu Y, Zhao Y and Yu X 2013 Thermal management performance of bent graphene nanoribbons *RSC Adv.* **3** 17349
[123] Li K-M, Xie Z-X, Su K-L, Luo W-H and Zhang Y 2014 Ballistic thermoelectric properties in double-bend graphene nanoribbons *Phys. Lett.* A **378** 1383
[124] Aradi B, Hourahine B and Frauenheim T 2007 DFTB+, a sparse matrix-based implementation of the DFTB method *J. Phys. Chem.* A **111** 5678
[125] Saiz-Bretín M, Medrano Sandonas L, Gutierrez R, Cuniberti G and Domínguez-Adame F 2019 Impact of device geometry on electron and phonon transport in graphene nanorings *Phys. Rev.* B **99** 165428
[126] Milošević I, Kepčija N, Dobardžić E, Damnjanović M, Mohr M, Maultzsch J and Thomsen C 2010 Electron–phonon coupling in graphene *Int. J. Mod. Phys.* B **24** 655

[127] Novoselov K S, Geim A K, Morozov S V, Jiang D, Zhang Y, Dubonos S V, Grigorieva I V and Firsov A A 2004 Electric field effect in atomically thin carbon films *Science* **306** 666
[128] Zhang G and Zhang Y-W 2017 Thermoelectric properties of two-dimensional transition metal dichalcogenides *J. Mater. Chem.* C **5** 7684
[129] Sierra J F, Neumann I, Cuppens J, Raes B, Costache M V and Valenzuela S O 2018 Thermoelectric spin voltage in graphene *Nat. Nanotechnol* **13** 107
[130] Markov M and Zebarjadi M 2019 Thermoelectric transport in graphene and 2D layered materials *Nanoscale Microscale Thermophys. Eng.* **23** 117
[131] Li D, Gong W, Chen Y, Lin J, Khan Q, Zhang Y, Li Y, Zhang H and Xie H 2020 Recent progress of two-dimensional thermoelectric materials *Nano-Micro Lett.* **12** 36
[132] Cao Y, Fatemi V, Fang S, Watanabe K, Taniguchi T, Kaxiras E and Jarillo-Herrero P 2018 Unconventional superconductivity in magic-angle graphene superlattices *Nature* **556** 43
[133] Nie X, Zhao L, Deng S, Zhang Y and Du Z 2019 How interlayer twist angles affect in-plane and cross-plane thermal conduction of multilayer graphene: a non-equilibrium molecular dynamics study *Int. J. Heat Mass Transf* **137** 161
[134] Ouyang W, Qin H, Urbakh M and Hod O 2020 Controllable thermal conductivity in twisted homogeneous interfaces of graphene and hexagonal boron nitride *Nano Lett.* **20** 7513
[135] Farhangfar S 2011 Size-dependent thermoelectricity in nanowires *J. Phys. D: Appl. Phys.* **44** 125403
[136] Balandin A A 2011 Thermal properties of graphene and nanostructured carbon materials *Nat. Mater* **10** 569
[137] Nika D L and Balandin A A 2012 Thermal properties of graphene: fundamentals and applications *J. Phys. Condens. Matter* **24** 233203
[138] Xu Y, Li Z and Duan W 2014 Thermal and thermoelectric properties of graphene *Small* **10** 2182
[139] Morozov S V, Novoselov K S, Katsnelson M I, Schedin F, Elias D C, Jaszczak J C and Geim A K 2008 Giant intrinsic carrier mobilities in graphene and its bilayer *Phys. Rev. Lett.* **100** 016602
[140] Balandin A A, Ghosh S, Bao W, Calizo I, Teweldebrhan D, Miao F and Lau C-N 2008 Superior thermal conductivity of single-layer graphene *Nano Lett.* **8** 902
[141] Faugeras C, Faugeras B, Orlita M, Potemski M, Nair R R and Geim A K 2010 Thermal conductivity of graphene in Corbino membrane geometry *ACS Nano* **4** 1889
[142] Wei P, Bao W, Lau C-N and Shi J 2009 Anomalous thermoelectric transport of Dirac particles in graphene *Phys. Rev. Lett.* **102** 166808
[143] Zuev Y M, Chang W and Kim P 2009 Thermoelectric and magnetothermoelectric transport measurements of graphene *Phys. Rev. Lett.* **102** 096807
[144] Hwang E H, Rossi E and Das Sarma S 2009 Theory of thermopower in two-dimensional graphene *Phys. Rev.* B **80** 235415
[145] Ghahari F, Xie H-Y, Taniguchi T, Watanabe K, Foster M S and Kim P 2016 Enhanced thermoelectric power in graphene: violation of the Mott relation by inelastic scattering *Phys. Rev. Lett.* **116** 136802
[146] Duan J, Wang X, Lai X, Li G, Watanabe K, Taniguchi T, Zebarjadi M and Andrei E Y 2016 High thermoelectric power factor in graphene/hBN devices *Proc. Natl Acad. Sci. US* A **113** 14272

[147] Crossno J *et al* 2016 Observation of the Dirac fluid and the breakdown of the Wiedemann–Franz law in graphene *Science* **351** 1058
[148] Checkelsky J G and Ong N P 2010 Thermopower and Nernst effect in graphene in a magnetic field *Phys. Rev.* B **80** 081413
[149] Ghosh S, Bao W, Nika D L, Subrina S, Pokatilov E P, Lau C-N and Balandin A A 2010 Dimensional crossover of thermal transport in few-layer graphene *Nat. Mater.* **9** 555
[150] Nam S-G, Ki D-K and Lee H-J 2010 Thermoelectric transport of massive Dirac fermions in bilayer graphene *Phys. Rev.* B **82** 245416
[151] Wang Q H *et al* 2012 Understanding and controlling the substrate effect on graphene electron-transfer chemistry via reactivity imprint lithography *Nat. Chem.* **4** 724
[152] Haskins J, Kınacı A, Sevik C, Sevinçli H, Cuniberti G and Çağın D 2011 Control of thermal and electronic transport in defect-engineered graphene nanoribbons *ACS Nano* **5** 3779
[153] Dollfus P, Nguyen V H and Saint-Martin J 2015 Thermoelectric effects in graphene nanostructures *J. Phys. Condens. Matter* **27** 133204
[154] Seol J H, Jo I, Moore A L, Lindsay L, Aitken Z H, Pettes M T, Li C X, Yao Z, Huang R and Broido D 2010 Two-dimensional phonon transport in supported graphene *Science* **328** 213
[155] Chen C-C, Li Z, Shi L and Cronin S B 2016 Thermoelectric transport across graphene/hexagonal boron nitride/graphene heterostructures *Nano Res.* **8** 666
[156] Casolo S, Martinazzo R and Tantardini G F 2011 Band engineering in graphene with superlattices of substitutional deffects *J. Phys. Chem.* C **115** 3250
[157] Mu X, Wu X, Zhang T, Go D B and Luo T 2011 Thermal transport in graphene-oxide from ballistic extreme to amorphous limit *Sci. Rep.* **4** 3909
[158] Anno Y, Imakita Y, Takei K, Akita S and Arie T 2017 Enhancement of graphene thermoelectric performance through defect engineering *2D Mater.* **4** 025019
[159] Alomar M I and Sánchez D 2014 Thermoelectric effects in graphene with local spin–orbit interaction *Phys. Rev.* B **89** 115422
[160] Inglot M, Dugaev V K and Barnás J 2015 Thermoelectric and thermospin transport in a ballistic junction of graphene *Phys. Rev.* B **92** 085418
[161] Mak K F, Lee C, Hone J, Shan J and Heinz T F 2010 Atomically thin MoS_2: a new direct-gap semiconductor *Phys. Rev. Lett.* **105** 136805
[162] Tongay S, Zhou J, Ataca C, Lo K, Matthews T S, Li J, Grossman J C and Wu J 2012 Thermally driven crossover from indirect toward direct bandgap in 2D semiconductors: $MoSe_2$ versus Mo2 *Nano Lett.* **12** 5576
[163] Braga D, Gutiérrez Lezama I, Berger H and Morpurgo A F 2012 Quantitative determination of the band gap of WS_2 with ambipolar ionic liquid-gated transistors *Nano Lett.* **12** 5218
[164] Buscema M, Barkelid M, Zwiller V, van der Zant H S J, Steele G A and Castellanos-Gomez A 2013 Large and tunable photothermoelectric effect in single-layer MoS_2 *Nano Lett.* **13** 358
[165] Kumar S and Schwingenschlögl U 2015 Thermoelectric response of bulk and monolayer $MoSe_2$ and WSe_2 *Chem. Mater.* **27** 1278
[166] Hong J, Lee C, Park J-S and Shim J H 2016 Control of valley degeneracy in MoS_2 by layer thickness and electric field and its effect on thermoelectric properties *Phys. Rev.* B **93** 035445

[167] Pu J, Kanahashi K, Cuong N T, Chen C-H, Li L-J, Okada S, Ohta H and Takenobu T 2016 Enhanced thermoelectric power in two-dimensional transition metal dichalcogenide monolayers *Phys. Rev.* B **94** 014312

[168] Wang R-N, Dong G-Y, Wang S-F, Fu G-S and Wang J-L 2017 Variations of thermoelectric performance by electric fields in bilayer MX_2 (M = W, Mo; X = S, Se) *Phys. Chem. Chem. Phys.* **19** 5797

[169] Özbal G, Senger R T, Sevik C and Sevinçli H 2019 Ballistic thermoelectric properties of monolayer semiconducting transition metal dichalcogenides and oxides *Phys. Rev.* B **100** 085415

[170] Liu X, Zhang G, Pei Q-X and Zhang Y-W 2013 Phonon thermal conductivity of monolayer MoS_2 sheet and nanoribbons *Appl. Phys. Lett.* **103** 133113

[171] Wickramaratne D, Zahid F and Lake R K 2014 Electronic and thermoelectric properties of few-layer transition metal dichalcogenides *J. Chem. Phys.* **140** 124710

[172] Ding Z, Jiang J-W, Pei Q-P and Zhang Y-W 2015 In-plane and cross-plane thermal conductivities of molybdenum disulfide *Nanotechnology* **26** 065703

[173] Sahoo S, Gaur A P S, Ahmadi M, Guinel M J-F and Katiyar R S 2013 Temperature-dependent Raman studies and thermal conductivity of few-layer MoS_2 *J. Phys. Chem.* C **117** 9042

[174] Yan R, Simpson J R, Bertolazzi S, Brivio J, Watson M, Wu X, Kis A, Luo T, Hight Walker A R and Xing H G 2014 Thermal conductivity of monolayer molybdenum disulfide obtained from temperature-dependent Raman spectroscopy *ACS Nano* **8** 986

[175] Massicotte M, Schmidt P, Vialla F, Watanabe K, Taniguchi T, Tielrooij K J and Koppens F H L 2016 Photo-thermionic effect in vertical graphene heterostructures *Nat. Commun.* **7** 12174

[176] Yuan P, Li C, Xu S, Liu J and Wang X 2017 Interfacial thermal conductance between few to tens of layered-MoS_2 and c-Si: effect of MoS_2 thickness *Acta Mater.* **122** 152

[177] Li L, Yu Y, Ye G J, Ge Q, Ou X, Wu H, Feng D, Chen X H and Zhang Y 2014 Black phosphorus field-effect transistors *Nat. Nanotechnol* **9** 372

[178] Churchill H O H and Jarillo-Herrero P 2015 Phosphorus joins the family *Nat. Nanotechnol* **9** 330

[179] Carvalho A, Wang M, Zhu X, Rodin A S, Su H and Castro Neto A H 2016 Phosphorene: from theory to applications *Nat. Rev. Mater* **1** 16061

[180] Luo Z, Maasen J, Deng Y, Du Y, Garrelts R P, Lundstrom M S, Ye P D and Xu X 2015 Anisotropic in-plane thermal conductivity observed in few-layer black phosphorus *Nat. Commun.* **6** 8572

[181] Lee S *et al* 2015 Anisotropic in-plane thermal conductivity of black phosphorus nanoribbons at temperatures higher than 100 K *Nat. Commun.* **6** 8573

[182] Fei R, Faghaninia A, Soklaski R, Yan J-A, Lo C and Yang L 2014 Enhanced thermoelectric efficiency via orthogonal electrical and thermal conductances in phosphorene *Nano Lett.* **14** 6393

[183] Medrano Sandonas L, Teich D, Gutierrez R, Lorenz T, Pecchia A, Seifert G and Cuniberti G Anisotropic thermoelectric response in two-dimensional puckered structures *J. Phys. Chem.* C **120** 18841

[184] Flores E, Ares J R, Castellanos-Gomez A, Barawi M, Ferrer I J and Sánchez C 2015 Thermoelectric power of bulk black-phosphorus *Appl. Phys. Lett.* **106** 022102

[185] Liu H, Choe H S, Chen Y, Suh J, Ko C, Tongay S and Wu J 2017 Variable range hopping electric and thermoelectric transport in anisotropic black phosphorus *Appl. Phys. Lett.* **111** 102101

[186] Saito Y, Iizuka T, Koretsune T, Arita R, Shimizu S and Iwasa Y 2016 Gate-tuned thermoelectric power in black phosphorus *Nano Lett.* **16** 4819

[187] Xia F, Wang H and Jia Y 2014 Rediscovering black phosphorus as an anisotropic layered material for optoelectronics and electronics *Nat. Commun.* **5** 4458

[188] Qiao J, Kong X, Yang F, Hu Z-X and Ji W 2014 High-mobility transport anisotropy and linear dichroism in few-layer black phosphorus *Nat. Commun.* **5** 4475

[189] Zhang J, Liu H J, Cheng L, Wei J, Liang J H, Fan D D, Shi J, Tang X F and Zhang Q J 2014 Phosphorene nanoribbon as a promising candidate for thermoelectric applications *Sci. Rep.* **4** 6452

[190] Zhang J, Liu H J, Cheng L, Wei J, Liang J H, Fan D D, Jiang P H, Suna L and Shi J 2016 High thermoelectric performance can be achieved in black phosphorus *J. Mater. Chem.* C **4** 991

[191] Qin G, Yan Q-B, Qin Z, Yue S-Y, Cui H-J, Zheng Q-R and Su G 2014 Hinge-like structure induced unusual properties of black phosphorus and new strategies to improve the thermoelectric performance *Sci. Rep.* **4** 6946

[192] Lv H Y, Lu W J, Shao D F and Sun Y P 2014 Enhanced thermoelectric performance of phosphorene by strain-induced band convergence *Phys. Rev.* B **90** 085433

[193] Cagnoni M, Führen D and Wuttig M 2018 Thermoelectric performance of IV–VI compounds with octahedral-like coordination: achemical-bonding perspective *Adv. Mater.* **30** 1801787

[194] Dresselhaus M S, Chen G, Tang M Y, Yang R, Lee H, Wang D Z, Ren Z F, Fleurial J-P and Gogna P 2007 New directions for low-dimensional thermoelectric materials *Adv. Mater.* **19** 1043

[195] Snyder G J and Toberer E R 2008 Complex thermoelectric materials *Nat. Mater.* **7** 105

[196] Wang H, Pei Y, LaLonde A D and Snyder G J 2011 New directions for low-dimensional thermoelectric materials *Adv. Mater.* **23** 1366

[197] Szczech J R, Higgins J M and Jin S 2011 Enhancement of the thermoelectric properties in nanoscale and nanostructured materials *J. Mater. Chem.* **21** 4037

[198] Zhao L-D, Lo S-H, Zhang Y, Sun H, Tan G, Uher C, Wolverton C, Dravid V P and Kanatzidis M G 2014 Ultralow thermal conductivity and high thermoelectric figure of merit in SnSe crystals *Nature* **508** 373

[199] Zhao L-D, Chang C, Tan G and Kanatzidis M G 2016 SnSe: a remarkable new thermoelectric material *Energy Environ. Sci.* **9** 3044

[200] Chen Z G, Shi X L, Zhao L-D and Zhou J 2018 High performance SnSe thermoelectric materials: progress and future challenge *Prog. Mater. Sci.* **97** 293

[201] He J, Xu J, Liu G-Q, Shao H, Tan X, Liu Z, Xu J, Jiang H and Jiang J 2016 Enhanced thermopower in rock-salt SnTe–CdTe from band convergence *RSC Adv.* **6** 32189

[202] Ma Z, Wang C, Lei J, Zhang D, Chen Y, Wang J, Cheng Z and Wang Y 2019 High thermoelectric performance of SnTe by the synergistic effect of alloy nanoparticles with elemental elements *ACS Appl. Energy Mater* **2** 7354

[203] Gogotsi Y and Anasori B 2019 The rise of MXenes *ACS Nano* **13** 8491

[204] Naguib M, Kurtoglu M, Presser V, Lu J, Niu J, Heon M, Hultman L, Gogotsi Y and Barsoum M W 2011 Two-dimensional nanocrystals produced by exfoliation of Ti_3AlC_2 *Adv. Mater.* **11** 4248
[205] Barsoum M W 2000 The $M_{N+1}AX_N$ phases: a new class of solids: thermodynamically stable nanolaminates *Prog. Solid State Chem.* **28** 201
[206] Khazaei M, Arai M, Sasaki T, Estili M and Sakka Y 2014 Two-dimensional molybdenum carbides: potential thermoelectric materials of the MXene family *Phys. Chem. Chem. Phys.* **16** 7841
[207] Lukatskaya M R, Mashtalir O, Ren C E, Dall'Agnese Y, Rozier P, Taberna P L, Nagui M, Simon P, Barsoum M W and Gogotsi Y 2013 Cation intercalation and high volumetric capacitance of two-dimensional titanium carbide *Science* **341** 1502
[208] Kim H, Anasori B, Gogotsi Y and Alshareef H N 2017 Thermoelectric properties of two-dimensional molybdenum-based MXenes *Chem. Mater.* **29** 6472
[209] Jing Z, Wang H, Feng W, Xiao B, Ding Y, Wu K and Chen Y 2019 Superior thermoelectric performance of ordered double transition metal MXenes: $Cr_2TiC_2T_2$ (T = –OH or –F) *J. Phys. Chem. Lett.* **10** 5721
[210] Kumar S and Schwingenschlögl U 2016 Thermoelectric performance of functionalized Sc_2C MXenes *Phys. Rev.* B **94** 035405
[211] Sarikurt S, Çakır D, Keçeli M and Sevik C 2018 The influence of surface functionalization on thermal transport and thermoelectric properties of MXene monolayers *Nanoscale* **10** 8859
[212] Zebarjadi M 2015 Electronic cooling using thermoelectric devices *Appl. Phys. Lett.* **106** 203506
[213] Small J P, Perez K M and Kim P 2003 Modulation of thermoelectric power of individual carbon nanotubes *Phys. Rev. Lett.* **91** 256801
[214] Hippalgaonkar K, Huang B, Chen R, Sawyer K, Ercius P and Majumdar A 2010 Fabrication of microdevices with integrated nanowires for investigating low-dimensional phonon transport *Nano Lett.* **10** 4341
[215] Yu J-K, Mitrovic S, Tham D, Varghese J and Heath J R 2010 Reduction of thermal conductivity in phononic nanomesh structures *Nat. Nanotechnol* **5** 718
[216] Zeng G, Zide J M O, Kim W, Bowers J E and Gossard A C 2007 Cross-plane Seebeck coefficient of ErAs:InGaAs/InGaAlAs superlattices *J. Appl. Phys.* **101** 035502
[217] Jia Z, Li C, Li X, Shi J, Liao Z, Yu D and Wu W 2016 Thermoelectric signature of the chiral anomaly in Cd_3As_2 *Nat. Commun* **7** 13013
[218] Chen Y, Jayasekera T, Calzolari A, Kim K W and Nardelli M B 2010 Thermoelectric properties of graphene nanoribbons, junctions and superlattices *J. Phys. Condens. Mat* **22** 372202
[219] Xiao N *et al* 2011 Enhanced thermopower of graphene films with oxygen plasma treatment *ACS Nano* **5** 2749
[220] Ryu G H, Lee J, Kang D, Jo H J, Shin H S and Lee Z 2017 Effects of dry oxidation treatments on monolayer graphene *2D Mater.* **4** 024011
[221] Grosse K L, Bae M-H, Lian F, Pop E and King W P 2011 Nanoscale Joule heating, Peltier cooling and current crowding at graphene–metal contacts *Nat. Nanotechnol.* **6** 287
[222] Vera-Marun I J, van den Berg J J, Dejene F K and van Wees B J 2016 Direct electronic measurement of Peltier cooling and heating in graphene *Nat. Commun.* **7** 11525

[223] Harzheim A, Evangeli C, Kolosov O V and Gehring P 2020 Direct mapping of local Seebeck coefficient in 2D material nanostructures via scanning thermal gate microscopy *2D Mater.* **7** 041004
[224] Ouyang T, Chen Y, Xie Y, Stocks G M and Zhong J 2011 Thermal conductance modulator based on folded graphene nanoribbons *Appl. Phys. Lett.* **99** 233101
[225] Xu W, Qin Z, Chen C-T, Kwag H R, Ma Q, Sarkar A, Buehler M J and Gracias D H 2017 Ultrathin thermoresponsive self-folding 3D graphene *Sci. Adv.* **3** e1701084
[226] Sierra J F, Neumann I, Costache M V and Valenzuela S O 2015 Hot-carrier Seebeck effect: diffusion and remote detection of hot carriers in graphene *Nano Lett.* **15** 4000
[227] Uchida K, Takahashi S, Harii K, Koshibae W, Ando K, Maekawa S and Saitoh E 2008 Observation of the spin Seebeck effect *Nature* **455** 778
[228] Bauer G E W, Saitoh E and van Wees B J 2012 Spin caloritronics *Nat. Mater.* **11** 391
[229] Adachi H *et al* 2012 Spin Seebeck insulator *Nat. Mater.* **9** 894–7
[230] Berciaud S, Han M Y, Mak K F, Brus L E, Kim P and Heinz T F 2010 Electron and optical phonon temperatures in electrically biased graphene *Phys. Rev. Lett.* **104** 227401
[231] Betz A C, Jhang S H, Pallecchi E, Ferreira R, Fe've G, Berroir J and Plaçais B 2012 Supercollision cooling in undoped graphene *Nat. Phys.* **9** 109
[232] Fu H-H, Wu D-D, Gu L, Wu M and Wu R 2015 Design for a spin-Seebeck diode based on two-dimensional materials *Phys. Rev.* B **92** 045418
[233] Farghadan R and Ildarabadi F 2020 Gate-voltage induced giant spin Seebeck effect in phosphorene nanoribbons *Phys. Rev.* B **102** 035430
[234] Yang X, Tan F, Dong Y, Yua H and Liu Y 2019 Transition metal-containing molecular devices: controllable single-spin negative differential thermoelectric resistance effects under gate voltages *Phys. Chem. Chem. Phys.* **21** 5243
[235] Marfoua B, Khan I and Hong J 2020 Giant spin seebeck effect in two-dimensional ferromagnetic CrI_3 monolayer *Nanotechnology* **31** 455404
[236] Groenendijk D J, Buscema M, Steele G A, de Vasconcellos S M, Bratschitsch R, van der Zant H S J and Castellanos-Gomez A 2014 Photovoltaic and photothermoelectric effect in a double-gated WSe_2 device *Nano Lett.* **14** 5846
[237] Xu X, Gabor N M, Alden J S, van der Zande A M and McEuen P L 2010 Photothermoelectric effect at a graphene interface junction *Nano Lett.* **10** 562
[238] Hong T, Chamlagain B, Lin W, Chuang H-J, Pan M, Zhou Z and Xu Y-Q 2014 Polarized photocurrent response in black phosphorus field-effect transistors *Nanoscale* **6** 8978
[239] Gabor N M, Song J C W, Ma Q, Nair N L, Taychatanapat T, Watanabe K, Taniguchi T, Levitov L S and Jarillo-Herrero P 2011 Hot carrier–assisted intrinsic photoresponse in graphene *Science* **334** 648
[240] Cai X *et al* 2014 Sensitive room-temperature terahertz detection via the photothermoelectric effect in graphene *Nat. Nanotechnol* **9** 814
[241] Viti L, Cadore A R, Yang X, Muench J E, Watanabe K, Taniguchi T, Satej J, Ferrari A C and Vitiello M S 2020 Thermoelectric graphene photodetectors with sub-nanosecond response times at terahertz frequencies *Nanophotonics* **10** 89
[242] Viti L, Purdie D G, Lombardo A, Ferrari A C and Vitiello M S 2020 HBN-encapsulated, graphene-based, room-temperature terahertz receivers, with high speed and low noise *Nano Lett.* **20** 3169–77

[243] Wu J, Schmidt H, Amara K K, Xu X, Eda G and Özyilmaz B 2014 Large thermoelectricity via variable range hopping in chemical vapor deposition grown single-layer MoS_2 *Nano Lett.* **14** 2730
[244] Lu X *et al* 2020 High-efficiency thermoelectric power generation enabled by homogeneous incorporation of MXene in $(Bi,Sb)_2Te_3$ matrix *Adv. Eng. Mater* **10** 1902986

Chapter 8

Electro- and photochromic oxide-related materials and applications

José Montero Amenedo

Chromogenic materials are materials that can change their optical properties in a reversible and persistent way in response to an external stimulus [1]. The external stimulus can be, for example, an electrical current, light of an adequate wavelength and intensity, a change in temperature, or the presence of certain gases. Accordingly, chromogenic materials can be classified as electrochromic, photochromic, thermochromic, or gasochromic, respectively.

Chromogenic materials can have many technological uses [2], including their uses in ophthalmic lenses [3], rear-view mirrors in cars [4], displays [5], etc. However, in the context of energy applications—the topic of this book—the implementation of chromogenic devices in smart (dynamic) windows stands out [1, 2].

A smart window is a particular type of window able to modify its optical properties according to the changing weather conditions. For this reason, a smart window is able to provide superior thermal and visual comfort and larger energy savings than any standard static window. Smart windows are, therefore, expected to become instrumental in the decarbonization of the building sector [6].

8.1 Introduction

Our modern lifestyles keep us longer and longer inside buildings. Indeed, according to recent studies, we spend over 90% of our lives indoors [7]. Thus, globally, up to 40% of primary energy is consumed in buildings, mostly for heating and cooling [8]. During 2008, for example, this accounted for 39% of global greenhouse gas emissions related to energy processes [9]. Increasing the energy efficiency of buildings is, therefore, a crucial carbon-abatement strategy. Fortunately, the building sector has a large potential for significantly reducing greenhouse gas emissions compared to other major emitting sectors [10].

doi:10.1088/978-0-7503-3531-7ch8

© IOP Publishing Ltd 2023

Figure 8.1. (a) A palloza located in Piornedo, Lugo, Spain. Photo courtesy of Alberto Ramón González Acedo. (b) House 10, Ångström Laboratory, Uppsala, Sweden.

The energy performance of a building is highly conditioned by its envelope, i.e. its external walls, roofs, ceilings, windows, and doors. In terms of energy efficiency, the fenestration is the weakest point, being the source of undesirable energy flows. In other words, windows often allow heat to leak out on cold days and allow too much heat from the Sun to enter on warm days. A solution used to overcome this problem is to reduce the size of the windows to the minimum: figure 8.1(a) shows a *palloza*, an ancient traditional house found in mountain areas in north-western Spain. These traditional houses, which have a possible Celtic origin, were intended to protect their occupants—people and cattle—from harsh weather conditions. The windows of a palloza, if any, were tiny openings in the walls [11].

Obviously, reducing the areas of windows, or not including windows at all, is not a very attractive solution to the energy-efficiency problem. Natural light and visual contact with the surroundings have a direct correlation with our wellbeing and are very much appreciated by building residents. Studies show that employees working in offices with big windows are happier and more productive [12] and that hospitalized patients recover faster from surgery if they enjoy natural light and visual contact with the environment [13]. Indeed, modern architectural trends tend to incorporate larger and larger windows [14], as shown by the example in figure 8.1(b). For all these reasons, it is not possible to simply neglect the energy challenge that large-area fenestration poses. Instead, we need to find strategies that can harness the energy flow that passes through it [14, 15].

8.1.1 Three parameters

We need to quantify the energy performance of a window in order to be able to improve it. In this context, windows are generally characterized by three parameters [16, 17]:

i. **The visible transmittance (VT)**, that is, the amount of light in the visible portion of the spectrum that passes through a glazing material.
ii. **The total heat transfer coefficient (*U*-value)**, which represents the heat flow per unit of time, window area, and difference in temperature between the

indoor and outdoor air. It includes conductive, convective, and radiative heat transfer.

iii. **The solar transmittance coefficient (*g*-value)**, i.e. the fraction of the solar energy directly transmitted through the glazing[1].

At this stage, it is possible to discuss what the different parameters (VT, *U*-value, and *g*-value) look like for an ideal energy-efficient window [16, 17]:

- **The VT**: since the objective of any window is to provide visual contact with the environment and natural light, the VT must be as high as possible, except if the possibility of glare and visual discomfort exists. In this case, lower values of VT may be required.
- **The *U*-value**: a low *U*-value is always desirable, because a window, as a fundamental part of the building's envelope, has to provide good insulation.
- **The *g*-value**: choosing an adequate *g*-value is not straightforward: in warm regions, a low *g*-value is desirable; in this way the solar energy that arrives indoors is limited, which results in a lower cooling load. In cold regions, on the other hand, a high *g*-value can be beneficial; in this way the extra solar energy admitted reduces the heating load. In mid latitudes, characterized by cold winters and hot summers, it is problematic to define an adequate *g*-value. In this case, the energy savings provided by a high *g*-value during the winter will be compensated by the larger cooling load during the summer. The reverse situation is true for a low *g*-value, which is suitable during the summer but results in a higher heating load during the winter [16]. The solution to this problem is solved by a smart window that can provide different *g*-values (and VTs) depending on the meteorological conditions. Note that the use of static windows that have constant *g*-values leaves a large energy savings potential untapped, especially in the mid latitudes. It is in the development of *smart* windows that chromogenic materials come into play [14].

8.1.2 Spectrally selective coatings

In a window, heat transfer can be minimized by the use of double and triple glazing combined with vacuum, Ar filling, etc [15]. These strategies, which aim to minimize the heat transferred by conduction, are not related to chromogenic materials and hence are not discussed here. Nor do we discuss the energy performance associated with the window frame, airtightness, etc. We instead focus on the approaches for controlling heat transfer by radiation in the window glazing, which is where electrochromic and photochromic technologies can play an important role.

The strategy for dealing with radiative exchange in a window relies on spectrally selective coatings [15]. Such coatings block some regions of the electromagnetic spectrum while remaining transparent to others. In figure 8.2, the spectral

[1] The *g*-value is commonly used in Europe [17]. In the USA, the solar heat gain coefficient (SHGC) is considered instead [16]. The SHGC takes into account the energy directly transmitted through the glazing (as does the *g*-value), plus the energy absorbed by the window and released to the building interior in the form of heat.

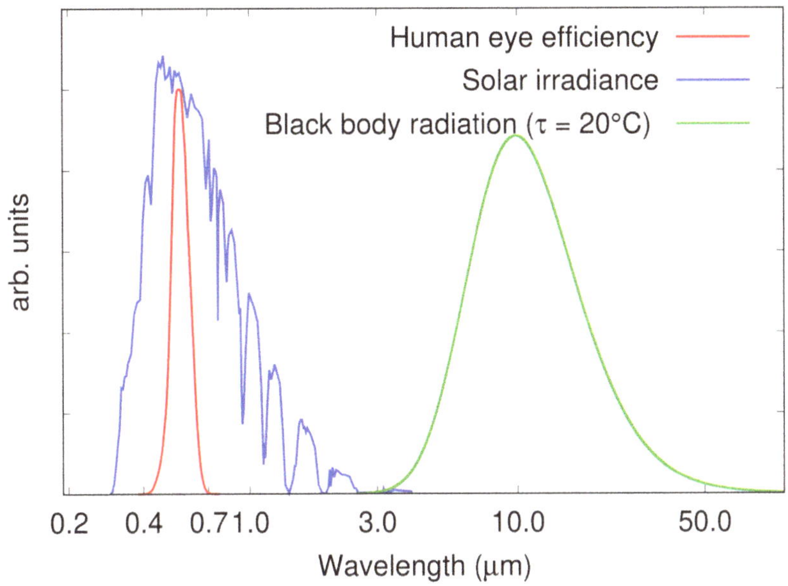

Figure 8.2. Solar irradiance at sea level [18], the relative luminous efficiency of the human eye [19] (assuming photopic vision, i.e. good illumination conditions), and the blackbody radiation of a body at 20° C, all presented as a function of wavelength λ. For simplicity, the y-axis has been set to arbitrary units.

distribution of the solar irradiance (after traveling through the atmosphere) [18], the radiation of a blackbody at a temperature τ equal to 20° C, and the luminous efficiency of the human eye (photopic vision) [19] are presented as a function of wavelength λ. Only the spectral distributions of those curves are needed for our discussion, therefore figure 8.2 has been simplified by setting the y-axis to arbitrary units.

From figure 8.2 we can infer that the visible portion of the solar radiation lies in the wavelength range $0.4\mu\text{m} \leqslant \lambda \leqslant 0.7\mu\text{m}$. The solar radiation outside this range does not contribute to better illumination nor to any improvement in the visual contact with the environment. In addition to the visible radiation, the solar radiation includes ultraviolet (UV) radiation, $\lambda < 0.4$ μm, and infrared (IR) radiation, $0.7\mu\text{m} \leqslant \lambda \leqslant 3.0\mu\text{m}$. A good window must block all the UV radiation, not only because UV degrades and fades the colors of the furniture and objects that we keep indoors, but also because it is harmful for our eyes and skin.

The thermal radiation emitted by a body at room temperature is represented in figure 8.2 by the blackbody radiation at temperature $\tau = 20$ °C. In order to calculate the actual spectral distribution of the thermal radiation emitted by a particular object, including the window glazing, the equivalent blackbody radiation has to be multiplied by the emittance E corresponding to the object. According to Kirchhoff's Law, [15]:

$$A(\lambda) = E(\lambda), \tag{8.1}$$

where $A(\lambda)$ is the absorptance of the object. In our case, equation (8.1) is only relevant in the spectral range between 3.0 μm and 50.0 μ m, due to the range of temperatures expected, see figure 8.2. Since the overlap between the thermal radiation and the solar radiation is negligible[2], it is possible to manipulate the ambient thermal radiation without affecting the solar radiation.

Taking into account the spectral regions presented in figure 8.2, we are ready to define what the spectral selectivity of a window glazing material must look like in order to achieve adequate VT, U-, and g-values:

- Achieving high VT implies that the transmittance $T(\lambda) = 1$ when $0.4\ \mu\text{m} \leqslant \lambda \leqslant 0.7\ \mu\text{m}$.
- In order to reduce the radiative heat transfer and consequently the U-value, a window is expected to present $E(\lambda) = 0$ in the spectral range of $3.0\ \mu\text{m} \leqslant \lambda \leqslant 50.0\ \mu\text{m}$. By applying energy conservation, we get:

$$T(\lambda) + R(\lambda) + A(\lambda) = 1, \tag{8.2}$$

  where $R(\lambda)$ is the reflectance. By taking account of equation (8.1), we get:

$$T(\lambda) + R(\lambda) = 1. \tag{8.3}$$

  We need $T(\lambda) = 0$, because otherwise the window will still allow heat transfer by radiation, thus:

$$R(\lambda) = 1. \tag{8.4}$$

- In cold regions, a high g-value is required, and hence a window has to be transparent, $T(\lambda) = 1$ to the infrared solar radiation $0.7\ \mu\text{m} \leqslant \lambda \leqslant 3.0\ \mu\text{m}$. On the other hand, in warm regions, a low g-value is desirable, and therefore the $0.7\ \mu\text{m} \leqslant \lambda \leqslant 3.0\ \mu\text{m}$ range has to be reflected, $R(\lambda) = 1$.

In summary, two types of coatings have to be defined, one for cold weather (low emittance or low-E) and another for warm weather (solar control). An ideal low-E coating presents the following optical characteristics [14, 15]:

$$\begin{aligned} T(\lambda) &= 1 \text{ for } 0.4 < \lambda < 3.0\ \mu\text{m} \\ R(\lambda) &= 1 \text{ for } 3.0 < \lambda < 50.0\ \mu\text{m}, \end{aligned} \tag{8.5}$$

whereas the optical characteristics of an ideal solar control coating are [14, 15]:

$$\begin{aligned} T(\lambda) &= 1 \text{ for } 0.4 < \lambda < 0.7\ \mu\text{m} \\ R(\lambda) &= 1 \text{ for } 0.7 < \lambda < 50.0\ \mu\text{m}. \end{aligned} \tag{8.6}$$

[2] In fact, the Sun radiates like a blackbody at $\tau = 5505\ ^\circ\text{C}$. Indents observed in the solar irradiance curve at sea level, as shown in figure 8.2, are caused by atmospheric absorption processes.

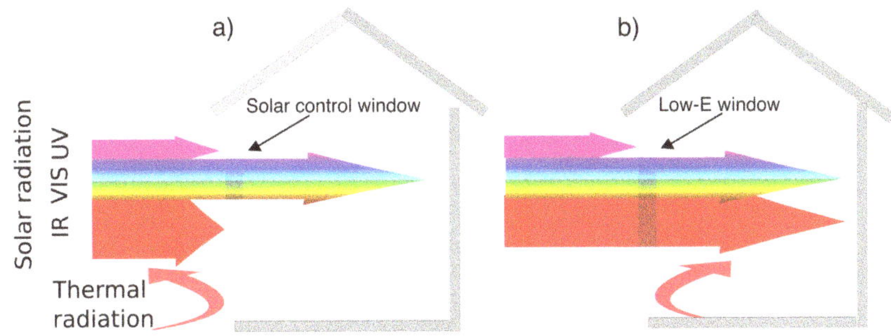

Figure 8.3. (a) The optical selectivity of a solar control window, (b) that of a low-E window. A solar control window is desirable during warm weather, whereas a solar control coating is adequate for cold weather.

Note that if the solar radiation is too strong, to avoid glare and visual discomfort, it may be convenient to use a solar control coating in which $T(\lambda) < 1$ in the visible region. The information provided by equations (8.5) and (8.6) is represented graphically in figure 8.3, where the spectral selectivity for a solar control window (panel (a)) and a low-E window (panel (b)) are depicted.

In mid latitudes, it is desirable to have a smart window that can behave like a low-E window in the winter and like a solar control window in the summer, thus providing large energy savings. Today's commercial smart windows are generally based on electrochromic devices and are described in section 8.2. In addition, recent innovations in materials science have the potential to transform photochromic windows into a reality; see section 8.3.

8.2 Electrochromism

Figure 8.4 shows the archetypical arrangement of an electrochromic device as described by Granqvist [20]. This device consists of a total of five layers, all squeezed between two glass panes which act as a scaffold. These five layers are: two transparent electrodes, an ion storage coating, an electrolyte containing small ions —e.g. H^+ or Li^+—and the electrochromic coating itself.

In short, an electrochromic device such as the one depicted in figure 8.4 works as follows: a voltage of suitable polarity applied to the electrodes drives the ions into the electrochromic layer. As the positively charged ions are intercalated, charge-compensating electrons fill the lowest unoccupied electronic states of the electrochromic material, thus ensuring charge neutrality. Note that the structure of the device presented in figure 8.4 resembles that of a battery. Both ions and electrons can move in the ion storage and electrochromic layers, but only ions are transported in the electrolyte. As a consequence, an electrochromic device exhibits open-circuit memory, that is, a voltage is only needed to induce an optical change.

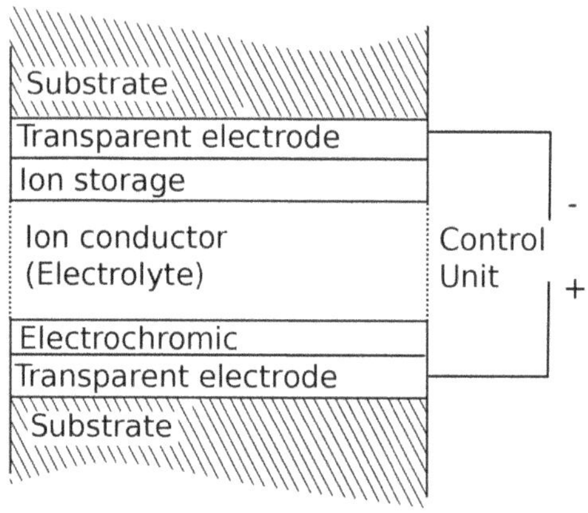

Figure 8.4. The classic arrangement of an electrochromic device, after Granqvist [20]. The thicknesses of the different layers are not to scale.

As charge-compensating electrons are inserted as a response to the intercalation of ions, the electrochromic layer gradually darkens, depending on the number of intercalated electrons. Due to the open-circuit memory of the device, the application of the voltage can be interrupted at any stage of intercalation, leading to different colouration degrees.

When the polarity of the voltage is inverted, the intercalated ions (and hence the charge-compensating electrons) abandon the electrochromic layer, which bleaches back to its original transparent state.

In addition to the five layers described in figure 8.4, a low-E coating—a must in any modern window technology—needs to be conveniently placed in relation to the electrochromic device [21]. Otherwise, after absorbing sunlight and getting hot, the electrochromic device radiates heat indoors. In addition, an electrochromic device needs a control unit, i.e. the electronics that enable the interaction of the user with the device.

8.2.1 The electrochromic device

These are the usual materials that compose an electrochromic device such as the one depicted in figure 8.4:

- Transparent electrodes: these are commonly based on transparent conducting oxides [20], which, as their name indicates, are oxide materials exhibiting transparency in the visible region as well as electrical conductivity. Transparent conducting oxides are materials that combine a wide optical bandgap of more than 3.0 eV—which enables transparency in the visible region—with shallow donor levels—which provide electrical conductivity [22]. There is a limit to the conductivity of these materials. This is because an *overlarge* charge carrier concentration N leads to low transmittance in the visible region. Achieving both transparency and electrical conductivity requires a compromise in which N is high enough for the material to exhibit electrical conductivity, but low enough that the plasma frequency is located far from the visible region. Normally, ITO ($Sn:In_2O_3$) is chosen as the transparent electrode material due to its easy fabrication by magnetron sputtering and excellent electrical and optical properties. Alternatives to the use of indium—which is expensive and considered a critical material—include FTO ($F:SnO_2$) and others [22].

- The electrochromic layer: although several transition-metal oxides (Ti, Nb, Ta, Mo, and V oxides), as well as some organic compounds, present *cathodic* electrochromic properties, i.e. they darken upon ion intercalation, the usual choice for smart window applications is tungsten oxide, WO_3 [20]. Amorphous WO_3 films provide a network of nanometer-sized tunnels that are favorable for the easy intercalation of ions [20]. WO_3 films can easily be prepared by sputtering [23].
- The ion storage layer: ideally, the ion storage layer has to be based on a *anodic* electrochromic material, i.e. a material that darkens upon ion de-intercalation. In this way, the properties of the ion storage layer and the electrochromic layer complement each other: as the ions are de-intercalated from the ion storage layer, move across the electrolyte, and are intercalated into the electrochromic layer, both the ion storage layer and the electrochromic layer darken. The reverse situation is true: when a reverse voltage is applied and the ions are de-intercalated from the electrochromic layer and intercalated in the ion storage layer, both layers bleach. The archetypical anodic electrochromic material is NiO [20].
- Electrolyte: the electrolyte has to be an ion conductor, but it also has to prevent the conduction of electrons [20]. Electrolytes containing small anions such as Li^+ and H^+ are advantageous for intercalation in the ion storage and electrochromic layers. The typical electrolyte is based on a solution of lithium perchlorate in propylene carbonate [24], although there are other possible alternatives including solid-state electrolytes [25].
- Substrate: glass is the usual substrate. However, the use of PET and/or other flexible substrates is possible [20]. Deposition onto flexible substrates enables roll-to-roll fabrication [26], which reduces production and transport costs. In addition, electrochromic devices produced on plastic substrates can be used to retrofit existing windows. Electrochromic foil can be laminated in glass for incorporation into a conventional window.

8.2.2 Coloring mechanism

From now on, we assume that Li^+ is the ion to be intercalated into the WO_3 electrochromic film. As mentioned above, when a number x of Li^+ ions is intercalated, x electrons (e^-) are inserted into the conduction band of the WO_3 coating, ensuring charge neutrality. Thus, the coloring reaction can be schematically represented as follows [14]:

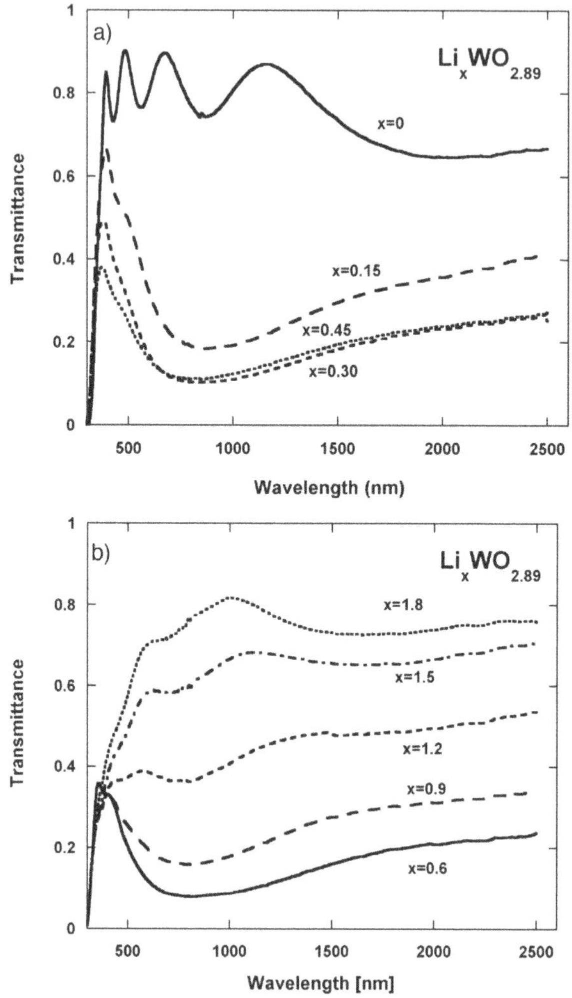

Figure 8.5. Transmittance curves as a function of wavelength for an amorphous $Li_xWO_{2.89}$ thin film deposited onto an ITO-coated glass substrate. The transmittances corresponding to x values ranging from 0 to 0.46 and from 0.6 to 1.8 are presented in panels (a) and (b), respectively. Panels (a) and (b) are reprinted from reference [24], with the permission of AIP Publishing.

$$WO_3 + xLi^{+} + xe^{-} \rightarrow Li_xWO_3. \qquad (8.7)$$

Therefore, x is the Li-to-W atomic ratio, i.e. $x = Li/W$; x is also equal to the fraction of inserted electrons per W atom, as is evident from equation (8.7).

Figure 8.5 shows $T(\lambda)$ curves for different values of x in (slightly substoichiometric) tungsten oxide as reported by Berggren *et al* [24]. Lower intercalation levels $0 \leqslant x \leqslant 0.45$ are presented in panel (a), while higher values $0.6 \leqslant x \leqslant 1.8$ are shown in panel (b). Note that the transmittance decreases drastically from the case corresponding to the non-intercalated film $x = 0.0$ up to roughly $x = 0.5$. For

$x > 0.5$, the situation is reversed and the transmittance increases again as x increases.

In the site saturation (SS) theory developed by Denesuk and Uhlman [27], as a consequence of Li^+ intercalation, charge-compensating electrons are accommodated in tungsten W^{6+} sites, which are consequently reduced to W^{5+}. Assuming a W^{5+} site in position i and a W^{6+} site in position j, the inserted electron in site i can jump to site j by absorbing a photon of adequate energy [28]:

$$W_i^{5+} + W_j^{6+} \xrightarrow{\text{photon}} W_i^{6+} + W_j^{5+}. \tag{8.8}$$

In the SS model, the process presented in equation (8.8) is responsible for the electrochromic coloring of WO_3. Thus, the increase in optical density (OD) caused by this process is proportional to P, where P is the probability associated with the process presented in equation (8.8) [28]:

$$\mathrm{OD}(\lambda, x) = \xi(\lambda)P(x). \tag{8.9}$$

$\xi(\lambda)$ is the optical absorption strength at wavelength λ. An increase in the OD of the material means an increase in its optical absorption and hence a decrease in $T(\lambda)$. Note that P depends on the intercalation level x as follows [28]:

$$P(x) = x(1 - x) = x - x^2. \tag{8.10}$$

Equations (8.10) and (8.9) qualitatively explain the evolution of $T(\lambda)$ of a Li_xWO_3 film at different intercalation levels up to $x = 1.0$, as seen in figure 8.5: when x is small, there are plenty of W^{6+} sites, but barely any W^{5+} sites, so the probability of absorbing a photon is very low and thus $T(\lambda)$ remains high (the OD remains low). As Li^+ and electrons enter the film, more and more W^{5+} sites are created, while at the same time there are still plenty of W^{6+} sites, so $T(\lambda)$ decreases rapidly (the OD increases). This situation culminates at $x = 0.5$, at which there are equal numbers of W^{5+} and W^{6+} sites, and hence the probability of absorbing a photon reaches a maximum, which means that $T(\lambda)$ and the OD reach a minimum and a maximum, respectively. As x increases above 0.5, the probability of absorbing a photon is reduced because, although there are plenty of W^{5+} sites, there are fewer and fewer W^{6+} sites. As a consequence, $x > 0.5$ means an increase in $T(\lambda)$ as the OD decreases. This situation is sustained until $x = 1$, where $P(x) = 0$.

In the SS theory, x cannot be larger than 1.0, i.e. a maximum of one inserted electron is allowed per W atom. This is a problem because, as we know experimentally, values of x lager than 1.0 can be achieved; see figure 8.5(b). On this basis, Berggren *et al* [24] introduced the extended site saturation (ES) theory by allowing values of x up to 2.0. In other words, the ES theory allows up to two electrons to be inserted per W atom. Therefore, according to the ES model W^{4+}, W^{5+}, and W^{6+} sites are possible, enabling transitions from W^{4+} sites to W^{5+} and W^{6+} sites. Nevertheless, Rojas and Niklasson [28] demonstrated that only the transitions involving W^{5+} and W^{6+} sites are relevant in explaining the optical properties, at least in the range of wavelengths studied. However, it is important to remark that the presence of W^{4+} sites affects the probability associated with the W^{5+} and W^{6+} transitions [28]:

$$P(x) = x(1 - x/2)^3, \tag{8.11}$$

which is different from the probability obtained for the SS case in equation (8.10). Thus, the presence of W^{4+} sites does affect the optical properties, albeit indirectly.

The results predicted by the SS and ES models need to be compared to experimental data in order to assess their suitability. For this purpose, we need to calculate a parameter that can easily be obtained experimentally, for example, the differential coloration efficiency K [28]:

$$K(x, \lambda) = \frac{\mathrm{d}\,\mathrm{OD}(\lambda, x)}{\mathrm{d}q}, \tag{8.12}$$

where q is the inserted change per unit area. K can be experimentally obtained from the OD through equation (8.12), using spectrophotometric measurements performed at different intercalation levels [28]. On the other hand, by combining equations (8.9) and (8.12) we get:

$$K(x, \lambda) = \xi(\lambda)\frac{\mathrm{d}P(x)}{\mathrm{d}q}. \tag{8.13}$$

Taking into account that $q = Gx$, where G is a constant, we can write:

$$K(x, \lambda) = C(\lambda)\frac{\mathrm{d}P(x)}{\mathrm{d}x}, \tag{8.14}$$

where $C(\lambda) = G\xi(\lambda)$. By combining equations (8.14) and (8.10), we get the differential coloration efficiency for the SS theory K^{SS}:

$$K^{SS}(x, \lambda) = C(\lambda)[1 - 2x]. \tag{8.15}$$

Equation (8.15) is defined in the range of $0.0 \leqslant x \leqslant 1.0$.

Analogously, the differential coloration efficiency for the ES theory K^{ES} can be obtained by combining equations (8.14) and (8.11):

$$K^{ES}(x, \lambda) = C(\lambda)[(1/2)(x - 2)^2(1/2 - x)]. \tag{8.16}$$

Equation (8.16) is defined in the range of $0.0 \leqslant x \leqslant 2.0$. Note that although the terms corresponding to transitions from W^{4+} have been neglected[3] in equation (8.16), their effects are still present. This is the case because contemplating the possibility of W^{4+} sites affects the probability of transition from W^{5+} to W^{6+}.

Rojas and Niklasson [28] compared the differential coloration efficiency predicted by the SS and ES models with the experimental coloration efficiency obtained experimentally by chronopotentiometry, see figure 8.6. As can be observed, the ES

[3] The extended version of equation (8.16), which also includes terms corresponding to the transitions from W^{4+} sites to W^{5+} and W^{6+} sites can be found in reference [28].

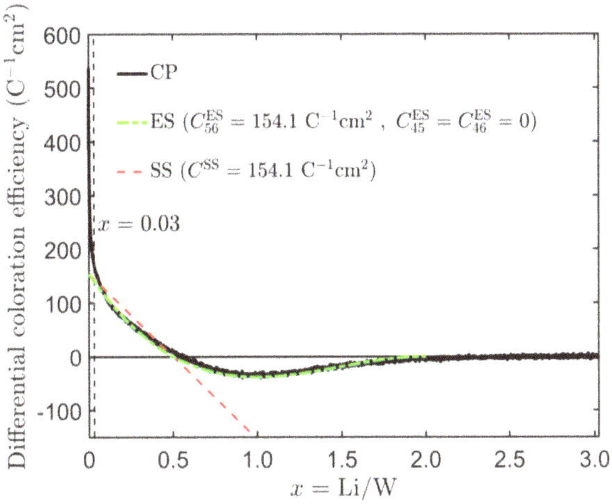

Figure 8.6. Differential coloration efficiency at $\lambda = 810$ nm as a function of the intercalation level, x. The experimental curve is compared with curves calculated according to the SS and ES models. Reprinted from [28], with the permission of AIP Publishing.

model reproduces the experimental data very well and is better able to explain the processes involved in the electrochromic colouration of WO_3-based films than the SS model.

8.2.3 Energy savings

The advantages of using a smart electrochromic window rather than a conventional static window can easily be understood qualitatively: as discussed above, during the winter, when solar gains are desirable, an electrochromic window can be set to its bleached state, providing a high *g* value [29]. During the summer, when solar gains represent a burden, an electrochromic window can be set to a colored state that exhibits a low *g* value [29]. In addition, the VT can be regulated, both increasing the visual comfort of the users and providing energy savings [25].

However, when it comes to *quantifying* the energy-saving potential of an electrochromic window, things can get a little bit more complicated. This is because the energy savings that can be achieved not only depend on the characteristics of the electrochromic window itself but also on the geographic location of the building, its orientation, use (offices, residential, ···), etc. In addition, the correct use of the device, i.e. selecting a suitable *g*-value/VT at the right time, plays an important role [25]. The window may be smart, but its use has to be intelligent. Furthermore, the energy savings obtained depend on which type of static window is used as a baseline. For example, the estimation of the energy-saving potential provided by an electrochromic device is lower than that of a solar control window and larger than that of a simple single-pane window fabricated from clear float glass.

Table 8.1, adapted from Cannavale *et al* [25], summarizes some approaches to the quantification of the energy savings provided by electrochromic fenestration as

Table 8.1. Energy savings provided by different electrochromic smart window technologies as found in the literature. Adapted from [25]. CC BY 4.0.

Reference	Technology	Energy savings*
Azens and Granqvist [30].	Solid electrolyte, cathodic (WO_3) and anodic (Ni-based oxide) EC materials for laminated glazing	170 kWh (m^2 year)$^{-1}$.
Piccolo [31].	Fully solid-state EC device based on cathodic (WO_3) and anodic (Ni-based oxide) materials	Load reductions of 50% compared to the performance of a clear float glass.
Jonsson and Roos [32].	N/A	EC windows always result in a lower cooling requirement than static alternatives. Between 100 kWh (m^2 year)$^{-1}$ and 200 kWh (m^2 year)$^{-1}$, depending on operational mode.
Aste *et al* [33].	EC glazing properties from the International Glazing Database (IGDB)	Virtual test cell compared to automated external venetian blinds properly used (−7.3% for yearly cooling loads) and base case clear glass (−40% for yearly cooling loads).
Sbar *et al* [34].	EC device manufactured by SAGE Electrochromics, Inc.	Global energy saving of 57% compared to static single-pane glazing and 37% compared to American Society of Heating, Refrigerating and Air-Conditioning Engineers (ASHRAE)-compliant glazing in the USA climate (Minneapolis).
Tavares *et al* [35].	EC device manufactured by SAGE Electrochromics, Inc.	From 20 kWh (m^2 year)$^{-1}$ to 37 kWh (m^2 year)$^{-1}$ (per square meter of glazed area).
Piccolo *et al* [36].	Solid-state device. Cathodic WO_3, anodic Ni-based oxide, and PEO-PEGMA:Li polymer electrolyte.	Location: Messina (Italy). −20% to −40% compared to a double-glazed unit.
Cannavale *et al* [37].	Single-substrate, solid-state device with substrate/ITO/WO_3/Nafion/ITO architecture.	Overall yearly energy savings as high as 40 kWh (m^2 year)$^{-1}$.
Tallberg *et al* [38]	EC device manufactured by SAGE Electrochromics, Inc.	Lowest energy demand in Madrid (Spain) with total delivered energy of 69% compared to the reference case.

* Energy savings are often given in units KWh(m^2year)$^{-1}$, i.e., annual energy savings per glazing area.

found in the literature. As shown in table 8.1, very diverse results can be obtained, depending on the conditions of the study/technology considered. Nevertheless, all the studies agree that the energy savings provided by an electrochromic window are by no means negligible.

8.3 Photochromism

The history and evolution of photochromic materials is deeply intertwined with their most successful commercial product. We are talking, of course, of the photochromic ophthalmic lens, which has been used for decades in the fabrication of photochromic spectacles [39].

During the 1960s, when photochromic silver halide glasses were introduced by Armistead and Stookey [40], ophthalmic lenses were made of glass and hence inorganic photochromic materials, silver halides, rode a wave of popularity [39]. During the 1990s, as glass lenses were gradually replaced by lighter and less brittle polymeric organic lenses, silver halides—which have to be prepared in bulk glass—were replaced by organic photochromic compounds [39, 41].

Photochromic spectacles are based on a fantastic idea, i.e. eyeglasses that turn into sunglasses by the action of strong sunlight and that have the ability to bleach back to the original transparent state when the illumination ceases, e.g. when the user walks indoors. Why not apply the same idea to windows? A photochromic window would help to regulate the ingress of solar radiation, increasing the visual and thermal comfort of the users as well as providing large energy savings. Unfortunately, the photochromic materials available today are unsuitable for such an application, either because they cannot be implemented in large-area float glass (silver halides) or because they are not durable enough (organic compounds) [41]. We will discuss the lack of adequate photochromic materials for smart windows further in section 8.3.2.

8.3.1 $\mathcal{T}$-type and positive photochromism

A photochromic material changes reversibly from a state $\mathcal{A}$ to a state $\mathcal{B}$ when illuminated by light of a suitable wavelength and intensity. The optical properties of states $\mathcal{A}$ and $\mathcal{B}$ are different in the visible region, thus the photochromic transition from $\mathcal{A}$ to $\mathcal{B}$ is perceived by the human eye as a change in color or visual appearance. For a material to be considered photochromic, the transformation from $\mathcal{A}$ to $\mathcal{B}$ has to be fully reversible, as schematically represented in the equation below [42]:

$$\mathcal{A} \underset{\text{Recovery}}{\overset{h\nu}{\rightleftharpoons}} \mathcal{B}, \tag{8.17}$$

where the term $h\nu$, i.e. Planck's constant times the frequency of the impinging light, represents electromagnetic energy. Following the definitions given by Towns [42], a photochromic material is considered to be $\mathcal{T}$-type if the recovery from $\mathcal{B}$ to $\mathcal{A}$ occurs by thermal relaxation. Not all photochromic materials recover thermally; some of them require illumination at longer wavelengths to recover the bleached state, which

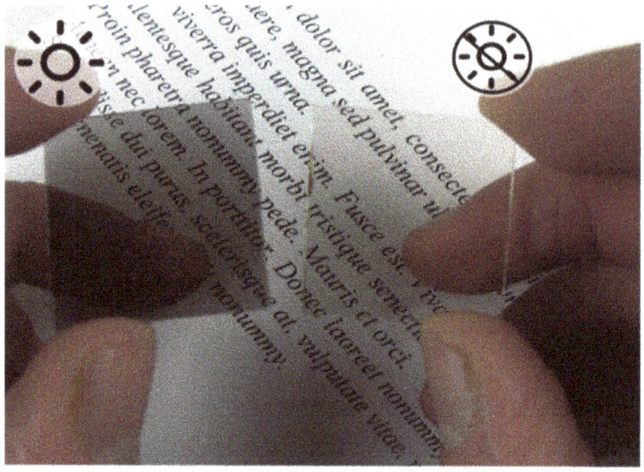

Figure 8.7. Yttrium oxyhydride in the photodarkened state, i.e. after illumination (left-hand side), and in the bleached state (right-hand side).

is known as $\mathcal{P}$-type photochromism [42], or exhibit a mix of $\mathcal{P}$- and $\mathcal{T}$-type photochromisms. In addition, $\mathcal{T}$-type photochromic materials are said to exhibit *positive* photochromism if state $\mathcal{B}$ presents a darker visual appearance than state $\mathcal{A}$ [42]. It goes without saying that most of the photochromic materials used industrially nowadays are $\mathcal{T}$-type and positive, the materials used to fabricate photochromic spectacles being the most obvious case [42].

The photochromic effect can be characterized by different parameters, including the *photochromic contrast* ΔVT defined as:

$$\Delta \mathrm{VT} = \mathrm{VT}_{\mathrm{clear}} - \mathrm{VT}_{\mathrm{dark}}, \tag{8.18}$$

where $\mathrm{VT}_{\mathrm{clear}}$ and $\mathrm{VT}_{\mathrm{dark}}$ denote the VT values in the clear (bleached) state and photodarkened (after illumination) states, respectively. Note that ΔVT typically depends on the illumination intensity and illumination time. Other parameters and factors to take into account when considering the possible practical application of a photochromic material are the darkening/bleaching times as well as durability features, including resistance to fatigue (durability when subjected to bleaching/darkening cycles) and the UV dosage that the material can take without losing its photochromic properties. We note that the darkening/bleaching times depend on the ambient temperature; this is the reason that photochromic spectacles require longer to darken/bleach in cold weather [43].

As an example of a $\mathcal{T}$-type positive photochromic material, in figure 8.7 the author shows two pieces of yttrium oxyhydride, a material that belongs to an emerging class of inorganic photochromic materials, rare-earth (RE) oxyhydrides, which can be deposited without intentional substrate heating by magnetron sputtering onto a wide range of substrates including plastic or, as shown in figure 8.7, float glass. The sample on the left side is in the photodarkened state (after illumination), whereas the sample on the right side remains in the clear state.

8.3.2 Photochromic windows

A photochromic window seems like a good idea for improving energy efficiency and comfort; yet, photochromic windows cannot be found on the market today. The main reason for this is the lack of adequate photochromic materials, as we discuss next.

As mentioned above, the popularity of inorganic silver halides decreased in favor of organic photochromic compounds as polymeric lenses became more and more popular: the fabrication of photochromic glass based on silver halides entails the dispersion of halide nanoparticles in a glass matrix, a method that is incompatible with the use of polymeric lenses [39].

Unfortunately, the substitutes for silver halides adopted by the ophthalmic industry—organic photochromic dyes—do not exhibit the excellent durability and fatigue resistance of their inorganic counterparts [44]. Organic compounds compensate for their much shorter life spans by offering an easier fabrication method, a reduction in production costs and a lighter and less brittle final product. According to the leading manufacturers of photochromic lenses, organic-based photochromic products have a life span of two years of continuous use [43]. This life span clearly hinders the application of such photochromic technology in windows and façades. In glaring contrast, silver halides never fatigue and can exhibit the same photochromic contrast and bleaching speed even after half a million darkening/bleaching cycles [44].

So, why not use silver halides for architectural applications? Despite more or less successful attempts to address the problem [45], photochromic glass based on silver halides cannot be produced by the usual float glass fabrication method, the common approach taken by the glazing industry for the fabrication of large-area glass panes.

With silver halides and organic compounds off the table, today's alternatives are photochromic transition-metal oxides and, notably, RE photochromic oxyhydrides.

8.3.3 Photochromism in inorganic photochromic materials

Figure 8.8 shows three examples of $T(\lambda)$ curves before illumination (the clear state) and after illumination (the dark state), corresponding to different inorganic photochromic materials: silver chloride glass, according to reference [46] in panel (a); a thin film composed of 5 nm-diameter tungsten oxide nanoparticles reported in reference [47] in panel (b); and an yttrium oxyhydride film [48] in panel (c). Both the tungsten oxide and the oxyhydride thin films were deposited onto glass. The relative luminous efficiency of the human eye has been added for reference in figure 8.8 [19].

In the particular case of the materials represented in figure 8.8, VT_{clear} = 85.1%, 50.2%, and 78.1% for silver chloride, tungsten oxide, and yttrium oxyhydride, respectively. Analogously, the photochromic contrast ΔVT = 53.7%, 3.7%, and 51.7% for silver chloride, tungsten oxide, and yttrium oxyhydride. Given the different experimental conditions (illumination time, illumination wavelength, etc.) and the intrinsic characteristics of the different samples (thickness, composition, whether it consists of nanostructures, etc.), it is not possible to quantitatively compare ΔVT nor VT for the different materials shown in figure 8.8. Nevertheless, it

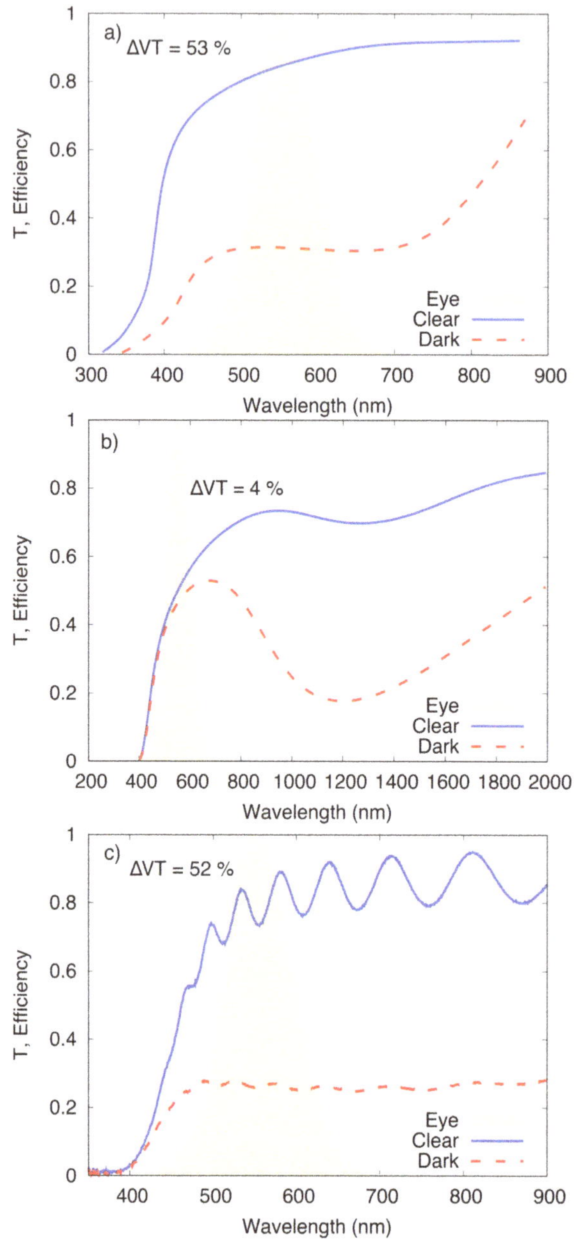

Figure 8.8. The transmittances of different inorganic photochromic materials before (clear) and after (dark) illumination. The relative luminous efficiency of the human eye has been added to the figures as a reference. (a) A glass based on silver halides. Reprinted from [46], Copyright (2016), with permission from Elsevier. (b) A tungsten oxide film composed of WO_{3-y} nanoparticles 5 nm in diameter. Reproduced from [47] with permission of The Royal Society of Chemistry. (c) An yttrium oxyhydride thin film. Adapted from [48]. CC BY 4.0.

is possible to make some qualitative observations: we note that the largest photochromic contrast in silver chloride takes place in the visible region, while the contrast becomes smaller at longer wavelengths (in the near-infrared). On the other hand, for the tungsten oxide nanostructured film, the larger contrast takes place in the near-infrared at $\lambda = 1200$ nm, though it is poor in the visible region [47]. Observe also how the decrease in $T(\lambda)$ in the visible part of the spectrum takes place evenly in yttrium oxyhydride and silver chloride, allowing the film to keep a neutral color appearance. Both the silver halide and the oxyhydride show excellent VT and ΔVT.

What is the reason for the darkening of the different inorganic photochromic compounds depicted in figure 8.8? We next analyze the photochromic mechanism of the most representative inorganic photochromic materials: silver halides, transition-metal oxides, and RE oxyhydrides.

8.3.3.1 Silver halides

The model proposed to explain the photochromic effect in silver halides is well established and can be found in, e.g. Araujo [44] and elsewhere [49, 50].

Silver-halide-based photochromic glass consists of a dispersion of silver halide (e.g. silver chloride or bromide) crystals containing Cu^+ impurities within a glass matrix (borosilicate) [40]. When this glass is illuminated with light of an adequate energy (blue or UV light), electrons are transferred from the chloride to the silver ions [44]:

$$Ag^+ + Cl^- \xrightarrow{h\nu} Ag^0 + Cl^0. \tag{8.19}$$

Since the reverse of the process presented in equation (8.19) is possible and indeed very quick, in principle, no optical change would be expected (because the glass would immediately return to its original state). However, freshly formed Cl^0 can take away one electron from a neighboring Cu^+ impurity [44]:

$$Cl^0 + Cu^+ \rightarrow Cl^- + Cu^{2+}. \tag{8.20}$$

In other words, each copper impurity can trap a hole. This event extends the lifetime of the Ag^0 species, which were created according to equation (8.19), ultimately leading to the formation of metallic silver domains within the glass matrix. These metallic domains are responsible for the change in the optical properties of the material after illumination; see figure 8.8(a).

However, the holes are not trapped forever in the copper impurities. Thus, the bleaching process takes place through the reaction [44]:

$$Cu^{2+} + Ag^0 \rightarrow Cu^+ + Ag^+. \tag{8.21}$$

For the process presented in equation (8.21) to take place, Cu^{2+} and Ag^0 species must be close to each other in the lattice [44]. As a consequence, the bleaching process is slower than the darkening process. Since bleaching and darkening processes compete, this limitation is the one which ultimately enables the darkening of the glass under illumination. When the illumination is interrupted, no more Ag^0

species are formed and the glass recovers its initial transparent state through the process presented in equation (8.21) [44].

8.3.3.2 Transition-metal oxides

Photochromic transition-metal oxides include tungsten oxide, molybdenum oxide, and others (niobium oxide, titanium dioxide, and zinc oxide) [51, 52].

In transition-metal oxides, photochromic behavior is commonly explained through the double-charge injection (DCI) model [53]. In the DCI model, holes, h^+, generated in the oxide upon illumination with sufficiently energetic light react with adsorbed water, producing protons, H^+:

$$H_2O + 2h^+ \rightarrow 2H^+ + O\uparrow . \tag{8.22}$$

These protons are inserted into the film, where they react with the oxide in question; e.g. in the case of tungsten oxide, the reaction results in the formation of a hydrogen tungsten bronze [54]:

$$WO_3 + xH^+ + xe^- \rightarrow H_xWO_3 \tag{8.23}$$

Note that the process described by equation (8.23) is identical to that described in equation (8.7), except that in this case, the inserted ion is H^+ instead of Li^+, and hence the DCI model explains the photodarkening of the film by the intervalence charge transfer between W^{5+} and W^{6+} species as presented in equation (8.8).

The reverse of the process described in equation (8.22) is very fast, therefore for the darkening of WO_3 to take place, a quick separation of O and H^+ species must occur. This can happen in two ways: by the release of oxygen gas to the atmosphere or by the trapping of oxygen atoms at oxygen vacancy sites [55]. Indeed, perfectly stoichiometric WO_3 films present almost no measurable photochromic effect. Thus, it is more preferable to refer to photochromic tungsten oxide as WO_{3-y}, in this way highlighting the deviation from the WO_3 stoichiometry.

Concerns about the DCI model were first raised by Bechinger, Bange, and others [55–58]. The criticism was based, among other things, on the fact that no measurable changes in hydrogen content were detected in the photodarkened films compared to films in the transparent or bleached states. In other words, the hydrogen was already present in the films in the clear state before illumination.

Bechinger *et al* [58] proposed an alternative model to the DCI model. In this case, the starting point is an oxygen-deficient tungsten oxide film in which *n* molecules of water are absorbed before illumination [56]. Very much as in the DCI model, some of these adsorbed water molecules are broken into protons and oxygen by the action of photogenerated holes in the oxide:

$$WO_{3-y} \cdot nH_2O \xrightarrow{h\nu} WO_{3-y} \cdot (n - x/2)H_2O + xH^+ + xe^- + \frac{x}{2}O\uparrow . \tag{8.24}$$

Up until equation (8.24), the model proposed by Bechinger *et al* is very similar to the DCI model. However, in this case, the proton is assumed to capture oxygen from the lattice, replenishing the initial water content:

$$WO_{3-y} \cdot (n - x/2)H_2O + xH^+ + xe^- \rightarrow WO_{3-y-x/2} \cdot nH_2O. \tag{8.25}$$

By combining equations (8.24) and (8.25), we get:

$$WO_{3-y} \cdot nH_2O \xrightarrow{h\nu} WO_{3-y-x/2} \cdot nH_2O + \frac{x}{2}O \uparrow . \tag{8.26}$$

According to Bechinger's model, light-induced oxygen release is the cause of the photochromic behavior seen in tungsten oxide. Note that (i) $WO_{3-y-x/2}$ includes W atoms in different oxidation states: $W^{6+}_{1-y-x}W^{5+}_{x}W^{4+}_{y}O_{3-y-x/2}$ (very much as in the ES model discussed in section 8.2.2), and (ii), 'oxygen release' does not necessarily mean that oxygen effectively leaves the film. As in the DCI model, the released oxygen is able to remain in interstitial positions [51].

As proven experimentally, and in agreement with equation (8.26), the photochromic effect in tungsten oxide is strongly affected by the surrounding atmosphere: when a photodarkened tungsten film is kept in vacuum or in an inert atmosphere, it recovers only partially and very slowly [57, 59]. Accordingly, it is possible to identify two bleaching channels, a quick one associated with atmospheric oxygen and a slower one associated with interstitial oxygen [57]. A similar dependence on the availability of oxygen for adequate bleaching is found in other photochromic compounds involving tungsten, for example, in phosphotungstic acid–nanocellulose composites [60].

8.3.3.3 Rare-earth oxyhydrides

In recent years, a new family of materials has irrupted into the inorganic photochromic scene. This family is based on RE oxyhydrides [61], where RE = Y, La, Gd, Dy, Er, etc. Oxyhydrides are extraordinary materials that present fascinating possibilities. We may think of an oxyhydride as an oxide in which many of the oxide ligands have been replaced by hydride anions [62]. However, the presence of two anion species radically transforms the chemistry of the oxide parent compound [62]. First of all, we find an extra flexibility in the building blocks that form the material (i.e. coordination geometries, including the positions of anions and the ratios of different species), resulting in different crystal field splitting, tunability of optical properties, and H^- anion diffusion (hopping), among other features [62].

The photochromic mechanism of RE oxyhydrides is still open to debate. Using variable-angle spectroscopic ellipsometry, we have demonstrated that the photochromic behavior in oxyhydrides is caused by the light-induced formation of metallic domains in the oxyhydride semiconducting matrix [63]. At the same time, light causes the contraction of the lattice [48]. Once the illumination ceases, samples bleach back to the original transparent state only if they are kept in an oxygen-containing atmosphere (air, dry air, or oxygen), but remain dark if kept in an inert atmosphere (N_2) [48], pointing to sample–ambient exchange much like that of tungsten oxide.

At this point, it is important to remark on the similarities between photochromism in transition-metal oxides and RE oxyhydrides: (i) in both cases, in addition to oxygen and hydrogen atoms, the material contains metallic atoms that can easily change oxidation state, e.g. W^{4+}, W^{5+}, and W^{6+} (transition-metal oxides) or Y^{2+} and Y^{3+} (RE oxyhydrides); (ii) both groups of materials, after being photodarkened, require of the presence of oxygen in the environment to recover. In this sense, we can assume a photochromic mechanism involving oxygen release, similar to that described by equation (8.26). Optical modeling predicts that a very small number of oxygen atoms need to be released [63]. Furthermore, released atoms do not need to abandon the film and can fill interstitial positions in the REHO structure. To give some numbers, a 0.6% volume fraction of an oxygen-deficient (not oxygen-free!) $REHO_{1-x}$ phase is predicted to cause a substantial darkening of the film [63]. These numbers are in agreement with those observed in photodarkened WO_3 [56].

Nevertheless, consensus has not been reached regarding the photochromic mechanism of RE oxyhydrides; some authors suggest that structural rearrangements enabled by H^- anion diffusion must be at the heart of the photochromic behavior of REHO [64]. However, H^- diffusion alone cannot explain the differences in bleaching speed as a function of oxygen content in the surrounding atmosphere.

8.4 Summary

Windows are the Achilles' heels of buildings in terms of energy efficiency. Quite often, windows allow the ingress of excessive solar energy during the summer and *leak heat* during the winter. Conventional solutions to this problem are based on spectrally selective static windows, which either can provide good performance during the winter (low-E) or during the summer (solar control). However, by using static windows, we are missing a huge potential for achieving lager energy savings. Thus, dynamic smart windows are a game-changing technology, as they can admit solar radiation when there is a heating demand and reject solar radiation when there is a cooling demand, providing energy savings during both the summer and the winter. Smart windows can be based on chromogenic technologies, such as electrochromism and photochromism.

The usual electrochromic device is composed of five layers: two transparent electrodes, an ion storage film, an electrolyte, and an electrochromic film. The darkening of such a device is achieved by applying a voltage that can displace ions (typically Li^+) from the ion storage film and into the electrochromic film (typically WO_3). The insertion of Li^+ ions is accompanied by the intercalation of charge-compensating electrons that can reduce some W^{6+} species to W^{5+}. According to the SS theory, the darkening of the electrochromic film is, therefore, caused by charge transfer between the W^{5+} and W^{6+}oxidation states. However, the SS model cannot explain the behavior observed when the Li/W ratio is larger than 1.0. In contrast, the ES theory—which considers not only the W^{6+} and W^{5+} cases but also the oxidation state W^{4+}—successfully explains the behavior of electrochromic films in a wider intercalation range of up to Li/W = 2.0. In both models, the application of a reverse voltage is able to de-intercalate the Li^+ ions from the electrochromic film, causing the bleaching of the device.

In addition to electrochromism, photochromism is becoming a promising technology for the development of smart windows. Photochromic materials have been used for decades in the fabrication of eyeglasses. Unfortunately, the traditional photochromic materials used by the ophthalmic industry are either not durable enough (in the case of organic dyes) or are not easily fabricated in large-area glass panes (in the case of silver halides). Alternatives for the fabrication of photochromic windows are transition-metal oxides (such as WO_3) and, most importantly, RE oxyhydrides. Both transition-metal oxides and rare-earth oxyhydrides can be fabricated on an industrial scale by sputtering, for example, and are inorganic—hence, in principle, they are more durable than their organic counterparts.

A typical silver-halide-based photochromic glass consists of a dispersion of silver halide (e.g. silver chloride) nano-crystals. Under illumination, some of the Ag^+ species present are reduced to Ag^0, creating clusters of metallic silver domains responsible for the darkening of the glass under illumination. Once illumination ceases, the bleaching process involves the return of Ag^0 to Ag^+.

Photochromism in transition-metal oxides, e.g. WO_3, is commonly explained by the DCI model, in which hydrogen, a product of the reaction of adsorbed water with photogenerated holes, is inserted into the oxide, forming a hydrogen tungsten bronze. As a consequence, W^{6+} to W^{5+} coexist, causing a darkening of the film in a similar way to that of the SS model. Some authors have proposed a different model, in which hydrogen, instead of forming a tungsten bronze, combines with oxygen from the lattice, replenishing the reserves of adsorbed water. As a consequence, under illumination, tungsten appears in the oxidation states W^{6+}, W^{5+}, and W^{4+}, a behavior which is similar to that predicted by the ES theory. The latter model explains why photodarkened tungsten oxide does not bleach without an external supply of oxygen.

Although their photochromic mechanism is not yet well established, RE oxyhydrides share several features with transition-metal oxides, including the prominent role of hydrogen and the need for oxygen in the surrounding atmosphere for the adequate bleaching of photodarkened films.

8.5 Questions

1. The façade of the new House 10 of the Ångström Laboratory (Uppsala, Sweden) is almost fully covered by static glazing, figure 8.1(b). Most of the glazing installed in House 10 has the following characteristics:

- *U*-value: 0.5 W $(m^2\ K)^{-1}$
- *g*-value: 26.7%
- VT: 52.7%

Write a short paragraph commenting on these values. Use the following questions as a guide: would you say this is low-E or solar control glazing? How good are the

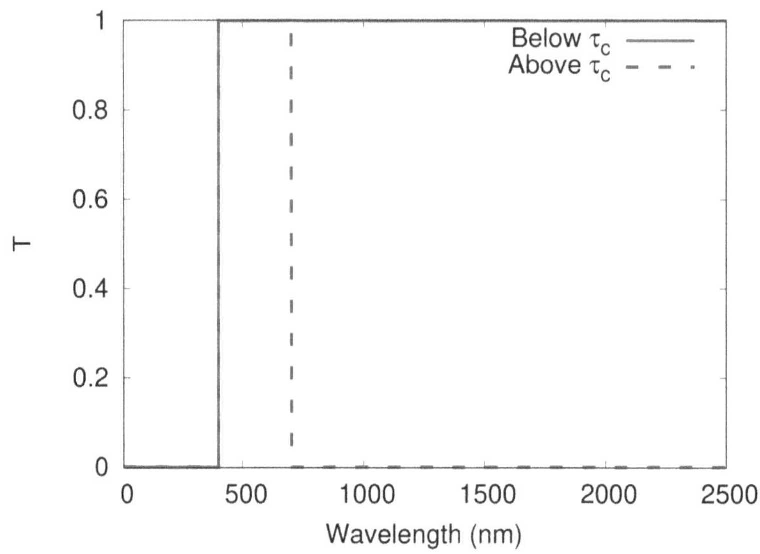

Figure 8.9. Transmittance above and below the critical temperature τ_c of an imaginary thermochromic material.

U-value and *g*-value, given the location of the building? Regarding the VT, what does it look like inside House 10 during the winter? And during the summer? Why do you think the architects have chosen this type of glazing? Note: before answering these questions, you may need to gather some information regarding the climate in Uppsala.

2. Imagine you can fabricate a smart glazing technology exhibiting any parameter values that you want. What set of parameter values (U, g, and VT) would your imaginary glass exhibit during the winter? And during the summer? Justify your answer.

3. An imaginary thermochromic coating is transparent in the near-infrared spectral region below a critical temperature τ_c, but blocks infrared radiation when heated to temperatures higher than τ_c. Irrespective of the temperature, this coating is transparent in the visible region. The $T(\lambda)$ of this material is presented in figure 8.9.

Describe the application of such a coating for the development of a smart (switchable) window able to control the ingress of solar energy, depending on the temperature. Assume that during the winter, the temperature remains below τ_c and that it remains above τ_c during the summer. Hint: compare figure 8.9 with figure 8.2. Describe the behavior of the coating during summer and winter in the different spectral regions: ultraviolet, visible, and near-infrared. Explain the benefit of using this coating in a window.

4. Vanadium dioxide VO_2 is a thermochromic material, i.e. its optical properties change when heated to temperatures higher than a critical temperature τ_c. $T(\lambda)$

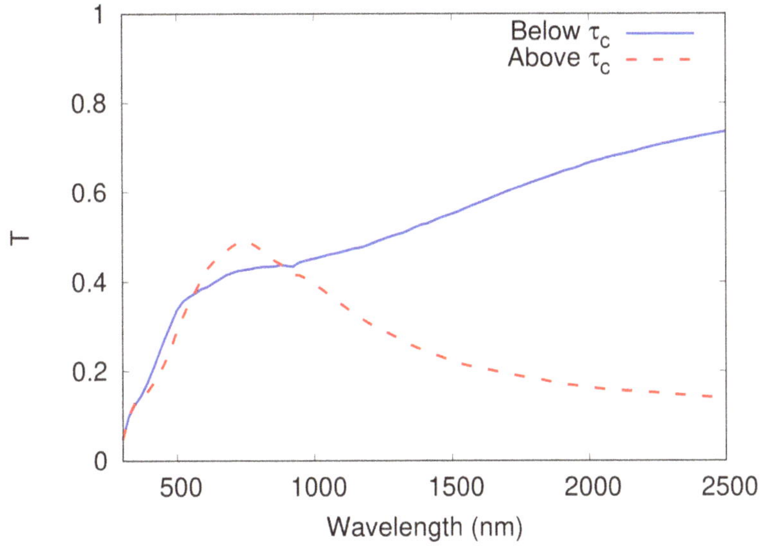

Figure 8.10. Transmittance above and below the critical temperature τ_c of a VO_2 thin film (50 nm thick) deposited onto glass by reactive magnetron sputtering. The film was obtained by following the fabrication procedure described by Montero *et al* [65].

values below and above the critical temperature τ_c of a 50 nm-thick VO_2 film deposited onto a glass substrate are shown in figure 8.10.

(a) Is the optical transmittance of this VO_2 film in the visible region adequate for a window? Justify your answer.

(b) As can be observed in figure 8.10, the maximum transmittance modulation provided by the VO_2 coating takes place at around 2500 nm and longer wavelengths. What is the nature of the solar irradiance in this spectral region? Is this beneficial in terms of energy efficiency?

5. Now, instead of a thin film, let us consider a dispersion of VO_2 nanoparticles in a dielectric (transparent) matrix, the refractive index of which is ~1.52, i.e. the same as that of glass. In this case, the $T(\lambda)$ values below and above τ_c are as presented in figure 8.11.

The idea of using a dispersion of VO_2 nanoparticles instead of a VO_2 thin film was first introduced by Li *et al* [66]. This approach is known as *nanothermochromism*. You can find more about the challenges posed by the implementation of VO_2 coatings in smart windows in reference [67].

(a) Considering $T(\lambda)$ in the visible region and with a window application focus, why would it be advantageous to use a dispersion of VO_2 nanoparticles rather than a VO_2 thin film?

(b) Is the nanothermochromic approach also advantageous in the near-infrared region?

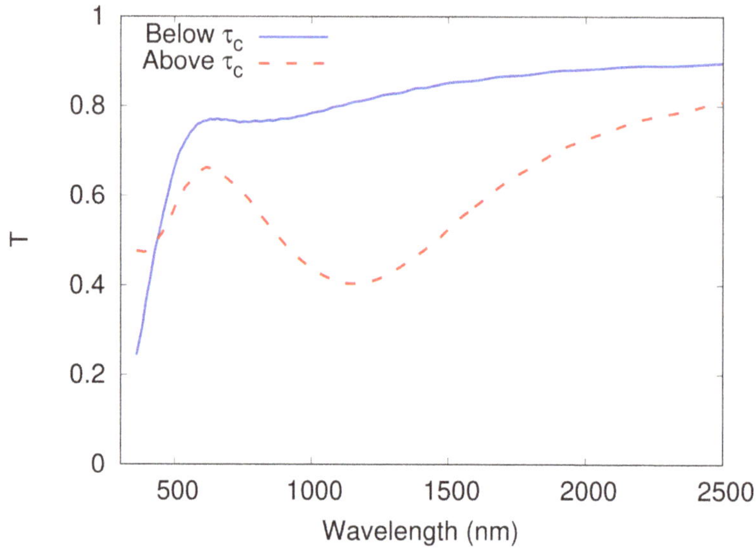

Figure 8.11. Transmittance above and below the critical temperature τ_c of a dispersion of VO_2 nanoparticles in a glass matrix. Calculated using the effective-medium approximation of Maxwell Garnett following the work by Li *et al* [66].

References

[1] Lampert C M and Granqvist C G Sr. (ed) 1990 *Large-Area Chromogenics: Materials and Devices for Transmittance Control Proceedings* 10304 vol IS4 (Bellingham, WA: The International Society for Optical Engineering) https://www.spiedigitallibrary.org/conference-proceedings-of-spie/10304.toc

[2] Lampert C M 2004 Chromogenic smart materials *Mater. Today* **7** 28–35

[3] Corns S N, Partington S M and Towns A D 2009 Industrial organic photochromic dyes *Color. Technol.* **125** 249–61

[4] Baucke F G K 1991 Electrochromic applications *Mater Sci. Eng.* B **10** 285–92

[5] Gugole M, Olsson O, Xiong K, Blake J C, Montero J, Pehlivan I B, Niklasson G A and Dahlin A 2020 High-contrast switching of plasmonic structural colors: inorganic versus organic electrochromism *ACS Photonics* **7** 1762–72

[6] Granqvist C G, Arvizu M A, Pehlivan I B, Qu H-Y, Wen R-T and Niklasson G A 2018 Electrochromic materials and devices for energy efficiency and human comfort in buildings: a critical review *Electrochim. Acta* **259** 1170–82

[7] Klepeis N E, Nelson W C, Ott W R, Robinson J P, Tsang A M, Switzer P, Behar J V, Hern S C and Engelmann W H 2001 The national human activity pattern survey (NHAPS): a resource for assessing exposure to environmental pollutants *J. Expo. Anal. Environ. Epidemiol* **11** 231–52

[8] Energy Technology Perspectives 2017 *Technical Report* (Paris: International Energy Agency) https://www.iea.org/reports/energy-technology-perspectives-2017

[9] Global Alliance for Buildings and Construction, International Energy Agency and the United Nations Environment Programme 2019. 2019 Global status report for buildings and construction: towards a zero-emission, efficient and resilient buildings and construction sector

[10] Richter B *et al* 2008 How America can look within to achieve energy security and reduce global warming *Rev. Mod. Phys.* **80** S1–109
[11] Palanco P and Gómez Olazábal L 1976 *Arquitetura gallega en León: Las pallozas* (Madrid: Facultad de Filosofía y Letras, Universidad Autónoma de Madrid) http://hdl.handle.net/10486/7899
[12] Lottrup L, Stigsdotter U K, Meilby H and Claudi A G 2015 The workplace window view: a determinant of office workers' work ability and job satisfaction *Landsc. Res.* **40** 57–75
[13] Ulrich R S 1984 View through a window may influence recovery from surgery *Science* **224** 420–1
[14] Smith G B and Granqvist C G 2010 *Green Nanotechnology: Solutions for Sustainability and Energy in the Built Environment* (Boca Raton, FL: CRC Press)
[15] Granqvist Claes G 2007 Transparent conductors as solar energy materials: a panoramic review *Sol. Energy Mater. Sol. Cells* **91** 1529–98
[16] U.S. Environmental Protection Agency and U.S. Department of Energy 2021. ENERGY STAR®, www.energystar.gov (Accessed: Jan 2021)
[17] Glass for Europe 2021. Building codes: minimum performance requirements for windows, www.glassforeurope.com (Accessed: Jan 2021)
[18] Annual Book of ASTM Standards 2008 *Standard tables of Reference, Solar Spectral Irradiances: Direct Normal and Hemispherical on 37° Tilted Surface* (Philadelphia, PA: American Society for Testing and Materials) Chapter ASTM G173-03
[19] Wyszecki G and Stiles W 2000 *Color Science: Concepts and Methods, Quantitative Data and Formulae* 2nd edn (New York: Wiley) https://www.wiley.com/en-us/Color+Science%3A+Concepts+and+Methods%2C+Quantitative+Data+and+Formulae%2C+2nd+Edition-p-9780471399186
[20] Granqvist C G 1995 *Handbook of Inorganic Electrochromic Materials* (Amsterdam: Elsevier Science)
[21] Piccolo A and Simone F 2015 Performance requirements for electrochromic smart window *J. Build. Eng.* **3** 94–103
[22] Ginley D S, Hosono H and Paine D C 2010 *Handbook of Transparent Conductors* (New York: Springer)
[23] Atak G, Pehlivan I B, Montero J, Granqvist C G and Niklasson G A 2021 Electrochromic tungsten oxide films prepared by sputtering: optimizing cycling durability by judicious choice of deposition parameters *Electrochim. Acta* **367** 137233
[24] Berggren L, Jonsson J C and Niklasson G A 2007 Optical absorption in lithiated tungsten oxide thin films: experiment and theory *J. Appl. Phys.* **102** 083538
[25] Cannavale A, Ayr U, Fiorito F and Martellotta F 2020 Smart electrochromic windows to enhance building energy efficiency and visual comfort *Energies* **13** 1449
[26] Granqvist C G, Pehlivan I B and Niklasson G A 2018 Electrochromics on a roll: web-coating and lamination for smart windows *Surf. Coat. Technol.* **336** 133–8
[27] Denesuk M and Uhlmann D R 1996 Site-saturation model for the optical efficiency of tungsten oxide-based devices *J. Electrochem. Soc.* **143** L186–8
[28] Rojas-González E A and Niklasson G A 2020 Differential coloration efficiency of electrochromic amorphous tungsten oxide as a function of intercalation level: comparison between theory and experiment *J. Appl. Phys.* **127** 205101
[29] Casini M 2018 Active dynamic windows for buildings: a review *Renew. Energy* **119** 923–34

[30] Azens A and Granqvist C G 2003 Electrochromic smart windows: energy efficiency and device aspects *J. Solid State Electrochem.* **7** 64–8

[31] Piccolo A 2010 Thermal performance of an electrochromic smart window tested in an environmental test cell *Energy Build.* **42** 1409–17

[32] Jonsson A and Roos A 2010 Evaluation of control strategies for different smart window combinations using computer simulations *Sol. Energy* **84** 1–9

[33] Aste N, Compostella J and Mazzon M 2012 Comparative energy and economic performance analysis of an electrochromic window and automated external venetian blind *Energy Procedia* **30** 404–13

[34] Sbar N L, Podbelski L, Yang H M and Pease B 2012 Electrochromic dynamic windows for office buildings *Int. J. Sustain. Built Environ* **1** 125–39

[35] Tavares P F, Gaspar A R, Martins A G and Frontini F 2014 Evaluation of electrochromic windows impact in the energy performance of buildings in Mediterranean climates *Energy Policy* **67** 68–81

[36] Piccolo A, Marino C, Nucara A and Pietrafesa M 2018 Energy performance of an electrochromic switchable glazing: experimental and computational assessments *Energy Build.* **165** 390–8

[37] Cannavale A, Martellotta F, Cossari P, Gigli G and Ayr U 2018 Energy savings due to building integration of innovative solid-state electrochromic devices *Appl. Energy* **225** 975–85

[38] Tallberg R, Jelle B P, Loonen R, Gao T and Hamdy M 2019 Comparison of the energy saving potential of adaptive and controllable smart windows: a state-of-the-art review and simulation studies of thermochromic, photochromic and electrochromic technologies *Sol. Energy Mater. Sol. Cells* **200** 109828

[39] Van Gemert B 2000 The commercialization of plastic photochromic lenses: a tribute to John Crano *Mol. Cryst. Liq. Cryst. Sci. Technol.* A **344** 57–62

[40] Armistead W H and Stookey S D 1964 Photochromic silicate glasses sensitized by silver halides: their characteristic of changing color reversibly, in combination with other properties, suggests many uses *Science* **144** 150–4

[41] Ferrara M and Bengisu M 2014 *Materials that Change Colour* (Cham: Springer)

[42] Towns A D 2016 Industrial photochromism *Applied PhotochemistryLecture Notes in Chemistry* 92 (Cham: Springer), 227–79 ch 5

[43] Essilor 2022 *Technical Report* https://www.essilorusa.com/products/transitions/faq (Accessed: Jan 2021)

[44] Araujo R J 2003 Photochromic glasses *Encyclopedia of Physical Science and Technology* ed R A Meyers 3rd edn (New York: Academic), 49–56

[45] Landa K A, Landa L M, Longobardo A V and Thomsen S 2004 Photochromic float glasses and methods of making the same *United States Patent* https://image-ppubs.uspto.gov/dirsearch-public/print/downloadPdf/6711917 No. US 6711917B2

[46] Karmakar B 2016 Functionality of reversible glass nanocomposites and their applications *Glass Nanocomposites* ed B Karmakar, K Rademann and A L Stepanov (Norwich, NY: William Andrew Publishing), 11 265–78

[47] Bourdin M, Salek G, Fargues A, Messaddeq S, Messaddeq Y, Cardinal T and Gaudon M 2020 Investigation on the coloring and bleaching processes of WO_{3-x} photochromic thin films *J. Mater. Chem.* C **8** 9410–21

[48] Baba E M, Montero J, Strugovshchikov E, Zayim E and Karazhanov S 2020 Light-induced breathing in photochromic yttrium oxyhydrides *Phys. Rev. Mater* **4** 025201

[49] Brinkmann M 2012 Optical Materials and Their Properties *Springer Handbook of Lasers and Optics* (Berlin: Springer), 253–399

[50] Hoffmann H J 1990 Photochromic glass *Large-Area Chromogenics: Materials and Devices for Transmittance Control*10304 ed C M Lampert and C-G Granqvist (Bellingham, WA: International Society for Optics and Photonics, SPIE), 92–107

[51] Vollath D 2008 *Nanomaterials: An Introduction to Synthesis, Properties and Applications* (New York: Wiley), 195–200 https://www.wiley.com/en-gb/Nanomaterials%3A+An+Introduction+to+Synthesis%2C+Properties+and+Applications%2C+2nd+Edition-p-9783527671861 pp, ch 6

[52] He T and Yao J-N 2004 Photochromism in transition-metal oxides *Res. Chem. Intermed* **30** 459–88

[53] Faughnan B W and Crandall R S 1980 Electrochromic displays based on WO3 *Display Devices* (Berlin: Springer), 5 181–211

[54] Gavrilyuk A I 1998 Photoinjection of hydrogen in solids *Ionics* **4** 372–82

[55] Bechinger C, Burdis M S and Zhang J-G 1997 Comparison between electrochromic and photochromic coloration efficiency of tungsten oxide thin films *Solid State Commun.* **10** 753–6

[56] Bange K 1999 Colouration of tungsten oxide films: a model for optically active coatings *Sol. Energy Mater. Sol. Cells* **58** 1–131

[57] Bechinger C, Oefinger G, Herminghaus S and Leiderer P 1993 On the fundamental role of oxygen for the photochromic effect of WO_3 *J. Appl. Phys.* **7** 4527–33

[58] Zhang J G, Benson D K, Tracy C E, Deb S K, Czanderna A W and Bechinger C 1997 Chromic mechanism in amorphous WO_3 films *J. Electrochem. Soc.* **144** 2022–6

[59] Miyazaki H, Ishigaki T and Suzuki T 2016 Effect of film thickness and air atmosphere on photochromic properties of wo_3-based composite films *Bull. Chem. Soc. Jpn.* **89** 20–3

[60] Santos M V *et al* 2021 Self-supported smart bacterial nanocellulose–phosphotungstic acid nanocomposites for photochromic applications *Front. Mater.* **8** 668835

[61] Cornelius S, Colombi G, Nafezarefi F, Schreuders H, Heller R, Munnik F and Dam B 2019 Oxyhydride nature of rare-earth-based photochromic thin films *J. Phys. Chem. Lett.* **10** 1342–8

[62] Kageyama H, Hayashi K, Maeda K, Attfield J P, Hiroi Z, Rondinelli J M and Poeppelmeier K R 2018 Expanding frontiers in materials chemistry and physics with multiple anions *Nat. Commun.* **9** 772

[63] Montero J, Martinsen F A, García-Tecedor M, Karazhanov S Z, Maestre D, Hauback B and Marstein E S 2017 Photochromic mechanism in oxygen-containing yttrium hydride thin films: an optical perspective *Phys. Rev.* B **95** 201301

[64] Colombi G, Stigter R, Chaykina D, Banerjee S, Kentgens A P M, Eijt S W H, Dam B and de Wijs G A 2022 Energy, metastability, and optical properties of anion-disordered ro_xh_{3-2x} $(r = \mathrm{Y, La})$ oxyhydrides: a computational study *Phys. Rev.* B **105** 054208

[65] Montero J, Ji Y-X, Li S-Y, Niklasson G A and Granqvist C G 2015 Sputter deposition of thermochromic VO_2 films on In_2O_3:Sn, SnO_2, and glass: structure and composition versus oxygen partial pressure *J. Vac Sci. Technol.* B **33** 031805

[66] Li S-Y, Niklasson G A and Granqvist C G 2010 Nanothermochromics: calculations for VO_2 nanoparticles in dielectric hosts show much improved luminous transmittance and solar energy transmittance modulation *J. Appl. Phys.* **108** 063525

[67] Li S Y, Niklasson G A and Granqvist C G 2012 Thermochromic fenestration with VO_2-based materials: three challenges and how they can be met *Thin Solid Films* **520** 3823–8

IOP Publishing

Sustainable Nanomaterials for Energy Applications

Ana Cremades and David Maestre

Chapter 9

Nanosized phosphors for white light-emitting diodes

Sara Ferrara and Rubén D Costa

This chapter reviews the state of the art of the materials design and device applications of nanosized oxides as color down-converting filters, i.e. phosphors, highlighting the challenges and progresses in white light-emitting diodes (WLEDs) and organic–inorganic white hybrid LEDs (WHLEDs). Although metal-based oxide phosphors have been widely used in different lighting applications, such as three-band fluorescent lamps and plasma display panels, this chapter focuses on the most relevant achievements related to single-point phosphor-converted WLEDs and WHLEDs. It is divided in four sections: first, a general introduction to the working mechanism of light-emitting diode (LED) and hybrid LED (HLED) technologies and the main figures of merit related to the production of white light. Second, a critical review of the state of the art and the need for future developments is provided. This leads to the third section, which reviews the foremost examples of metal oxide nanophosphors based on YAG:Ce nanoparticles and core@shell nanoparticles bearing metal coordination complexes, metal–halide perovskites, and polymers as luminescent cores and TiO_2, ZrO_2, SiO_2, etc. as protective shells. Finally, a conclusion and outlook are discussed in depth. Overall, this chapter aims to provide both a didactic instrument for those who are approaching the field of phosphor converters for LEDs and an overview for readers interested in the design and limitations of nanosized phosphors.

9.1 Introduction

Nowadays, there are two major players in white artificial illumination, namely white organic light-emitting diodes (WOLEDs) and white light-emitting diodes (WLEDs). The former have spread in both industrial and scientific communities [1]. However, the multilayered structure based on several organic semiconductors, which is needed for charge injection, charge transport/blocking, and light emission

doi:10.1088/978-0-7503-3531-7ch9

© IOP Publishing Ltd 2023

still represents a significant limitation on its wide commercialization [2, 3]. Although there are large-area devices featuring luminous efficiencies of 50–60 lm W^{-1} and stabilities of thousands of hours [3], it is, indeed, highly desirable to develop (i) simple architectures to enhance charge transport and recombination as well as light out-coupling, (ii) stable blue emitters, and (iii) easy and low-cost fabrication processes to ensure their large-scale application [2, 3]. The leading alternative is all-inorganic white light-emitting diodes (IWLEDs) that are dominating the lighting market. IWLEDs are highly efficient UV- or blue-emitting inorganic chips covered by color down-converting coatings that partially convert the high-energy chip emission into visible light [1, 3]. The most widely used color down-converters are based on yellow-emitting rare-earth-doped crystals, such as cerium-doped yttrium aluminum garnet ($Y_3Al_5O_{12}:Ce^{3+}$; YAG:Ce). They feature excellent photo- and thermal stabilities, high photon flux saturations, and photoluminescent quantum yields (ϕ) > 70%.

After blue-emitting gallium nitride (GaN) chip technology was developed, research into emitters that could be used with color down-converting filters acquired the utmost relevance [4]. Overall, luminescent nanocrystals (NCs) have attracted great interest due to (i) their easy synthesis that can be used to tune color emission, (ii) their excellent photoluminescent features (ϕ > 80%), and (iii) easy manipulation and compatibility in polymer matrices [5]. The first investigations of luminescent NCs were carried out using semiconductors such as InP [6] or Si [7], transition-metal-doped chalcogenides [8], and N/V-doped diamond [9] and were then extended to rare-earth-doped oxides [10–14] and halides [15, 16]. In addition, optimized systems have been further studied for a wide and various range of possible applications, such as labels for the dynamic tracking of biological species [17] and transparent or nanostructured light-emitting devices [14, 18].

With regard to luminescent nanoparticles, the optimization of their emission features has been focused on understanding the influence of size and surface effects on their radiative and non-radiative processes. For instance, a decrease in the phosphor size to a few tenths of a nanometer reduces light-scattering effects, while surface defects and surface reactivity both control emission deactivation via trapping effects. In addition, this family of emitters does not exhibit other specific size effects, since the excited states are mainly localized on the sensitizers and/or activator centers, which usually exhibit emission properties similar to those of the bulk materials [19, 20]. Thus, it is all-important to determine the minimum nanoparticle size that retains bulk-like photoluminescent properties. Here, the main limitation is synthesis procedures that require high-temperature conditions, leading to aggregated powders. This can be controlled by lowering the calcination temperature or using porogen agents such as surfactants or polymers to limit the sintering of nanoparticles and/or crystal growth. In addition, many groups have focused on the synthesis of well-dispersed crystalline nanoparticles through colloid chemistry routes (YVO_4:Eu, Sm,Dy [21, 22], $LaPO_4$:Eu,Ce,Tb [23], and $CePO_4$:Tb [23]). This has made possible the precipitation of the desired compound in its crystalline phase through reactions between precursors in solution at moderate temperature (<300 °C). Many rare-earth-doped phosphors have been successfully synthetized in nanocrystalline form (<50 nm) and different shapes, exhibiting good values of ϕ [24]. In general, oxide phosphors feature high values of ϕ and thermal stability as well as simple preparation methods, but reliance on rare-earth materials implies high production costs as well as uncertain sustainability. Finally, the availability of chemical routes that can tune the color emission toward the low-energy part of the visible spectrum still remains the main roadblock. This strongly limits the LED color quality available using these phosphors [25–28]. This scenario implies high luminous efficiencies and stabilities at the cost of acceptable color quality in terms of, for example, low color rendering index (CRI) values (ca. 70).

In view of all of the aforementioned, further efforts have been made to produce hybrid inorganic/organic WLED (HWLED) architectures for use in high-quality lighting systems [4, 29, 30]. State-of-the-art HWLEDs achieve luminous efficiencies of up to 120 lm W^{-1} and high color quality, with x/y Commission Internationale de l'Eclairage (CIE) color coordinates of 0.30–3/0.30–3, CRI values above 80–90, and correlated color temperatures (CCTs) ranging from 2700 K to 6500 K. Their major drawback is the limited device stability, which reaches a maximum of a few hundred hours [30–33]. Here, the current leading examples of nanosized hybrid phosphors are: (i) metal complexes embedded in silica nanoparticles, (ii) metal–halide perovskites stabilized by metal oxide shells (MHP@MOx), and (iii) metal oxide–polymer nanoparticles.

The aim of this chapter is to present the most representative advances in the above nanosized phosphors, focusing on their design, main photophysical features, and their use as color conversion layers for WLEDs and WHLEDs. This is complemented by a previous section introducing the main concepts of the fabrication, mechanisms, and characterization of white-emitting devices. This is followed by a short description of the state of the art of WLEDs and WHLEDs, highlighting their most critical aspects for their future development. Finally, a conclusion and outlook section provides the most relevant aspects to be tackled for the future development of nanosized phosphors. This provides the reader with an overview of how these phosphors have contributed to the development of artificial white lighting, establishing themselves as leading examples of color down-converting filters.

9.2 Basic concepts

9.2.1 The architectures and electroluminescent mechanisms of WLEDs

In 1993, Nakamura, Senoh, and Mukai demonstrated a double-heterostructure (DH) blue-emitting diode (p-GaN/n-InGaN/n-GaN) [34]. At that time, research was quite active in the field of II–VI semiconductors. Blue–green and blue injection laser diodes (LDs) had already been developed [35–37]. At the same time, gallium nitride (GaN) was attracting much attention as a semiconductor and important progress was rapidly being made in terms of crystal quality, p-type control, and the formation of thin films [38]. In this context, Nakamura and coworkers selected InGaN material as the active layer for blue emission. The background story of the invention of this efficient InGaN technology is well described in the Nobel Lecture

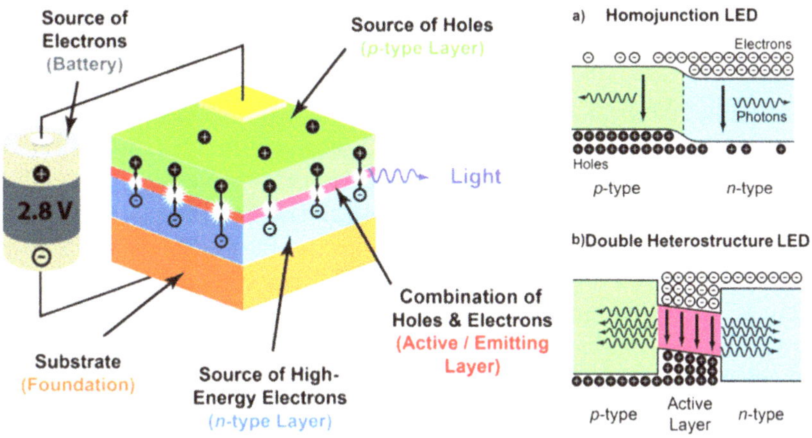

Figure 9.1. Left: schematic representation of a DH thin-film LED. Right: schematics of the energy band structures of (a) a homojunction LED and (b) a DH LED [39] John Wiley & Sons. [© 2015 WILEY-VCH Verlag GmbH & Co. KGaA, Weinheim].

given by Shuji Nakamura, in which he also clarifies the working mechanism of the optoelectronic device—figure 9.1 [39]. This consists of semiconductor materials sandwiching an active emitting film. High-energy electrons are abundant in the n-type semiconductor layer, while the p-type semiconductor has an abundance of available holes, in which electrons may reside at a lower energy level. The energy difference between the high- and low-energy electron states is referred to as the bandgap of the material. Such a system is called a DH because every junction between two materials that have different bandgaps is called a heterostructure; see figure 9.1. For DH LEDs, the bandgap of the active layer is smaller than that of the n- and p-type layers. When a forward bias is applied to a DH LED using a direct current source, electrons and holes are injected into the active layer from the n- and p-type layers, respectively. They undergo radiative recombination in the aforementioned active layer, with the consequent emission of photons that have an energy approximately equal to the bandgap of the active-layer material. As a result, it is sufficient to modify the bandgap of the active layer to produce photons at different energies [39].

Before the development of the DH LEDs, p–n homojunction LEDs without an active layer were first demonstrated by Amano *et al* in 1989 [38]. The comparison between this technology and DH LEDs is made easier if we consider the internal quantum efficiency (IQE). The IQE represents the efficiency expressed as the number of injected carriers that is converted into generated photons. An exciton can experience a radiative process or two non-radiative processes, namely, the Shockley–Read–Hall (SRH) and Auger processes, which depend on the carrier concentration. For instance, p–n homojunction LEDs operating at low carrier densities exhibit a prominent SRH process, leading to poor IQE. As a result of the p–n structure, both carriers are confined within the active region, which is typically around 3–200 nm thick. This confinement results in an increased concentration at the same current density, enhancing the probability of radiative recombination.

Indium gallium nitride (InGaN) was identified as the ideal candidate for the active layer, since the bandgap of this material shrinks through addition of indium to the GaN, providing both the confinement of carriers in the active layers and the ease of tuning the color by varying the amount of In. The first blue DH LED (440 nm) with a p-GaN/n-InGaN/n-GaN structure featured a Si-doped InGaN active layer (20 nm) [34]. A year later (1994), the first blue DH LED (p-GaN/p-AlGaN/Zn-doped InGaN/n-GaN) was commercialized. Here, an electron-blocking layer (p-AlGaN) was introduced [40]. Nakamura further demonstrated high-brightness blue, green, and yellow LEDs with InGaN quantum-well (QW) structures in 1995 [41], considering QWs to be DH systems with a very thin active layer, usually on the order of a few nanometers. This means that if the middle layer is thin enough, it acts as a potential well with discrete energy values. Ultraviolet (UV) InGaN/AlGaN DH LEDs were developed in 1998 using a single quantum well (SQW) InGaN active layer [42]. Finally, multi-QW structures (consisting of at least two QWs) of different bandgaps were developed for low-energy LEDs. High-quality InGaN LEDs based on conventional planar InGaN/GaN multi-QW structures are currently grown along the polar *c*-axes of wurtzite crystal structures. The growth of InGaN/GaN QWs with a high indium content usually results in high internal electrostatic (piezoelectric) fields due to a high density of defects [45, 46]. The internal fields in the InGaN wells spatially separate the electron and hole wave functions (the quantum confined Stark effect, QCSE), making it hard to obtain highly efficient longer-wavelength LEDs. For blue InGaN LEDs with low indium contents, the carrier localization phenomenon and the use of ultrathin QW structures (about 2–4 nm in commercial LEDs) can alleviate the effects of high defect density and the QCSE. Unfortunately, these solutions are not applicable in the case of high-In-content InGaN quantum wells. It is clear that the great success in the field of III nitride optoelectronics is mainly limited to InGaN-based blue emitters on polar GaN grown on *c*-plane sapphire. Even though there have been some impressive results in the development of green emitters, green (LEDs) on conventional *c*-plane GaN generally show inferior properties compared to their blue and violet counterparts. For longer wavelengths, such as yellow and red, progress is even slower. It is also extremely difficult to employ other III–V semiconductors such as (Al,Ga,In)P to fabricate efficient emitters in the green and yellow spectral region, thus confirming the so-called 'green/yellow gap' [47].

Following the success of the high-efficiency and high-power blue LED, Nichia Chemical Corporation commercialized the first white LEDs by combining a blue InGaN QW DH LED [41, 43] with a yellow yttrium aluminum garnet (YAG, $Y_3Al_5O_{12}$)-based phosphor. To date, the most common LED chips available on the market are still based on those materials. A great advantage of $In_xGa_{1-x}N$ ($0 \leqslant x \leqslant 1$) semiconductors is the possibility of tuning the bandgap from the near-infrared (0.6 eV, InN) to the near-UV (3.4 eV, GaN) region, including the entire visible spectrum. However, this potential is limited by the dramatic drop in InGaN emission efficiency at longer wavelengths. Therefore, efficient InGaN LEDs are only available in the high-frequency regions of the spectrum. As a result, white LEDs are typically realized using color down-converting filters or phosphor [44].

In the above, we already referred to the use of blue LED technology coupled with a down-converting phosphor as a successful strategy for the production of white light. In general, there are three ways to generate white light using LED technology —figure 9.2 [48]:

- mixing red-, green-, and blue-emitting (RGB) LEDs
- using a UV LED chip to excite red-, green-, and blue-emitting phosphors
- using a blue LED chip to excite either a yellowish-orange-emitting phosphor or a mixture of green- and red-emitting phosphors

The RGB LED approach is not the most effective, as RGB LEDs require different driving currents for each emitting chip; they also have low color quality, as they are prone to show variations of the CCT caused by the quick degradation of red LEDs compared to that of blue LEDs [49]. In addition, we already discussed the issues related to the production of commercial green and red LEDs using GaN technology, leading the scientific and industrial communities to prefer a

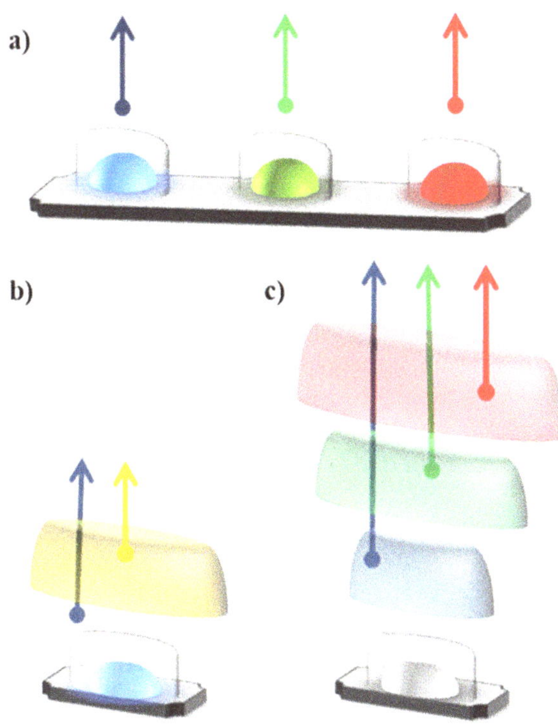

Figure 9.2. A schematic representation of WLED architectures: (a) the RGB approach, which combines different blue, green, and red LEDs, (b) WLEDs in which a blue chip excites a yellowish-orange-emitting phosphor, (c) WLEDs with a UV LED chip that excites RGB phosphors.

phosphor-converted LED (pc-LED) approach. The white light produced by following the UV LED + RGB phosphor approach has usually resulted in the highest efficiency, CRI, and color stability at different driving currents. In addition, the CCT can be controlled by changing the amounts of the phosphors and/or the thicknesses of the color filters. Nevertheless, blending different phosphors results in expensive devices, while there is still only a handful of stable and efficient red phosphors [50]. Thus, devices based on the architecture that consists of a blue LED and a yellowish-orange phosphor benefit from low costs, ease of fabrication, and good efficiency, but compromise the color quality—see figure 9.2 [48].

In addition to the photoluminescent properties of a phosphor (emission band shape, ϕ, τ, etc.), their performance in terms of thermal- and photostability under continuous irradiation at ambient conditions is of the utmost relevance. The photostability of the emitters has been investigated by exposing the coatings to continuous irradiation using high-power, UV- or blue-emitting LED chips. The typical designs include remote and on-chip approaches, depending on the distance between the color filter and the emitting chip. The overall emission spectra (electro- and photoluminescent spectra) and the temperature of the coating are monitored over time until the lifetime of the device is reached. This could be either the time taken to reach the half of the initial emission intensity of the color filter or a percentage loss in device efficiency (L_{90} or L_{50}) along with a variation in the color quality. Finally, thermal stability involves the self-stability of the coating at high temperatures under conditions of both self-heating upon continuous excitation and without irradiation stress.

9.2.2 Photoluminescence in inorganic/organic phosphors

Luminescence is defined as the emission of electromagnetic radiation in an excess of thermal radiation. It may help the reader if we point out that luminescence and incandescence are two distinct processes. Incandescence is related to the emission of light by solids due to their high temperature. Luminescence arises when the atoms of a solid are energized without appreciable heating of the bulk material and then release the excess of energy in the form of ultraviolet, visible, or infrared radiation. In LEDs, the phosphors are excited by the radiative emission of LED chips; therefore, their luminescence is referred to as photoluminescence.

9.2.2.1 Inorganic phosphors

Inorganic phosphors normally consist of a crystalline host lattice doped with a relatively small amount of rare-earth or transition-metal ions, called activators [51]. The activator ions introduce energy levels that can be populated either by direct excitation or indirectly by energy transfer and are responsible for the luminescence [52]. In general, the luminescent process is divided into absorption, transfer, and radiative relaxation. Energy absorption does not necessarily take place at the activator ion itself, but can occur at a random place in the lattice. This implies an efficient energy transfer from the lattice to the luminescent center via exciton transfer, a resonance process, and the reabsorption of photons emitted by another activator ion or sensitizer. There are two types of activator ion, which exhibit weak (e.g. lanthanide ions, Ln^{3+}) and strong (Mn^{2+}, Ce^{3+}, and complex anions, such as MoO_4^{2-} or NbO_4^{3-}) interactions with the host lattice. The occurrence of energy transfer loss via crystal surface defects and lattice defects is one of the major limitations in terms of ϕ. Luminescent materials be highly crystalline to achieve high light output. In addition, the surface area of the crystals has to be minimized, which means that their surface has to be neat.

9.2.2.2 Organic phosphors

In 1997, Schneider and coworkers reported the first WHLEDs to use luminescent polymers as down-converters. They consisted of either organic emitters (conjugated polymers, small molecules, proteins, etc.) or inorganic or metallo-organic emitters (perovskites, quantum dots (QDs), metal complexes, etc.) embedded in an organic matrix (polymers, MOFs, biogenic matrices). An organic color filter coupled with an inorganic LED chip is the basis of the so-called hybrid LED. If a component of biological origin is present, the device is referred to as a Bio-HLED.

The down-conversion process in an organic phosphor is basically related to the photoluminescent features of the emitter. Therefore, it is necessary to have a good overlap between the emission spectrum of the LED chip and the excitation spectrum of the emitter embedded in the coating matrix (the latter should be inert with respect to spectroscopic and optical features). In addition, the solid matrix plays an important role in preventing the aggregation of the emitters — vide infra.

9.2.3 *x*/*y* CIE color coordinates and diagram

The International Commission on Illumination or CIE is an international body that defines universal standards in the field of light, color, vision, and lighting. The Commission has proposed three models, namely CIE 1931, CIE 1960, and CIE 1976, even though the latter two are just modifications of the former. The chromaticity coordinates x, y, z can be calculated from the tristimulus values X, Y, and Z as follows:

$$x = X/(X + Y + Z)$$
$$y = Y/(X + Y + Z)$$
$$z = Z/(X + Y + Z).$$

Given that $x + y + z = 1$, it is sufficient to indicate two of the three coordinates. Usually, x and y CIE are used to provide the x/y CIE color diagram—figure 9.3. A dominant wavelength (λ_D) can be pinpointed on the horseshoe curve on the crossing point of a line passing through the color point (C) and the color of the illuminant (I). A complementary wavelength can be determined on the opposite side of the horseshoe curve, prolonging the line over the illuminant point. The purity of the excitation is the ratio between the segment connecting the illuminant point and the color point and the total segment connecting the illuminant to the dominant wavelength. It can be reported as a purity percentage by multiplying the result by 100.

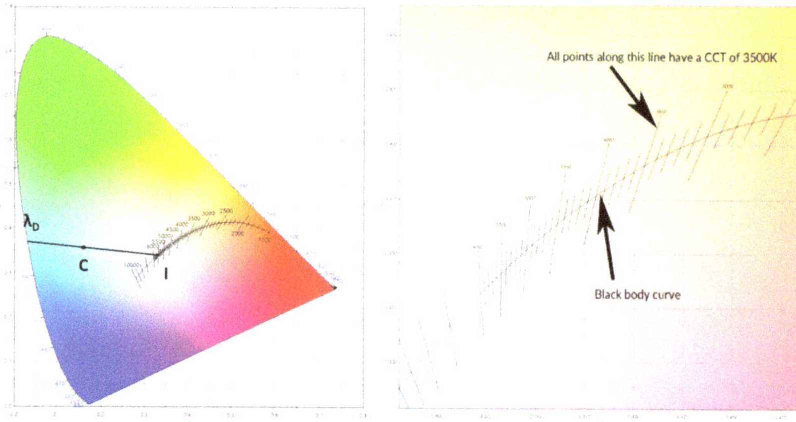

Figure 9.3. CIE 1931 color diagram (left) and a magnification of the black-body curve (right). The right-hand panel is reproduced from and © 2023 Waveform Lighting, LLC. All rights reserved. https://www.waveformlighting.com/tech/calculate-color-temperature-cct-from-cie-1931-xy-coordinates.

9.2.4 Correlated color temperature

A white light source is usually defined using a correlated color temperature, which basically describes the appearance of a light source by correlating it with the temperature of an ideal black-body radiator — see Figure 9.3. Most light sources do not exhibit a color that exactly lies on the black-body curve; therefore, certain amounts of deviation are accepted by the different standards — i.e. the vertical lines crossing the black-body curve in figure 9.3 [53]. Nevertheless, if the deviation is so large that the light source is not considered white, the obtained number should not be considered meaningful [53].

Depending on the CCT values, the white light source can be defined as warm or cold light. Warm LEDs produce white light with a significant low-energy component, while cold LEDs feature a strong bluish component. Thus, we can consider CCT values between 2500 K and 3500 K for warm white and over 5000 K for cold white.

The formula used to approximate the CCT starting from CIE 1931 x, y values is shown below:

$$n = (x - 0.3320)/(0.1858 - y);$$
$$\text{CCT} = 437 \times n^3 + 3601 \times n^2 + 6861 \times n + 5517.$$

The above equation clarifies the dependence of the CCT on the CIE coordinates.

9.2.5 Color rendering index

The CRI represents the ability of a light source to give a true reflection of the colors of an object illuminated by the light source when compared to a reference light such as the Sun [54]. The CRI is a single, quantified number that can assume a maximum value of 100. This means that the highest value of 100 represents the maximum

Figure 9.4. An intuitive representation of the perception of colors when an apple is irradiated with a light source featuring different CRI values. Reproduced from Benèl BV. https://www.benel.eu/en/advice/studio-photography/cri-value.

ability to reproduce the colors of illuminated objects, which is attributed to daylight — see Figure 9.4. This is a useful way to compare the color quality of artificial lamps to that of sunlight. The higher the CRI exhibited by a light source, the closer the light emitted by the source is to daylight. Usually, CRI values over 90 are considered excellent, while values of less than 80 are generally considered poor.

It must be mentioned that changes in the reference source cause variation in the calculated CRI. Usually, a source that has the chromatic coordinates (x, y) localized at (0.33, 0.33) is considered to be pure white [55]. This is located at the *equal-energy point*, which means the point where $x=y=z=0.33$, and it is named the standard illuminant, E. A CCT of 5455 K is attributed to this point, even though this is actually the CCT of the so-called standard illuminant D_{55}, due to the fact that the aforementioned point does not fall in the Planckian locus because it is not a black-body radiator and it is therefore not possible to evaluate its real CCT [55].

At this point, it should be of interest to give a short overview of the standard illuminants—figure 9.5. The first three illuminants, namely A, B, and C, were introduced in 1931, with the aim of representing average incandescent light (tungsten-filament lamp), direct sunlight, and average daylight, respectively. The illuminants belonging to the D series represent natural daylight. D_{50}, D_{55}, and D_{65} are the most commonly cited. Illuminant E and those of the F series are an equal-energy radiator that acts as a sort of theoretical reference and illuminants referring to several types of fluorescent lightning, respectively. The reader is referred to [55] for more information.

To summarize, the figures of merit of illuminants A–E are indicated below—figure 9.5:

Illuminant A: $(x,y) = (0.4476, 0.4074)$ – 2865K
Illuminant B: $(x,y) = (0.3484, 0.3516)$ – 4870 K
Illuminant C: $(x,y) = (0.3101, 0.3162)$ – 6770 K
Illuminant D_{65}: $(x,y) = (0.3128, 0.3292)$ – 6500 K
Illuminant E: $(x,y) = (0.3333, 0.3333)$

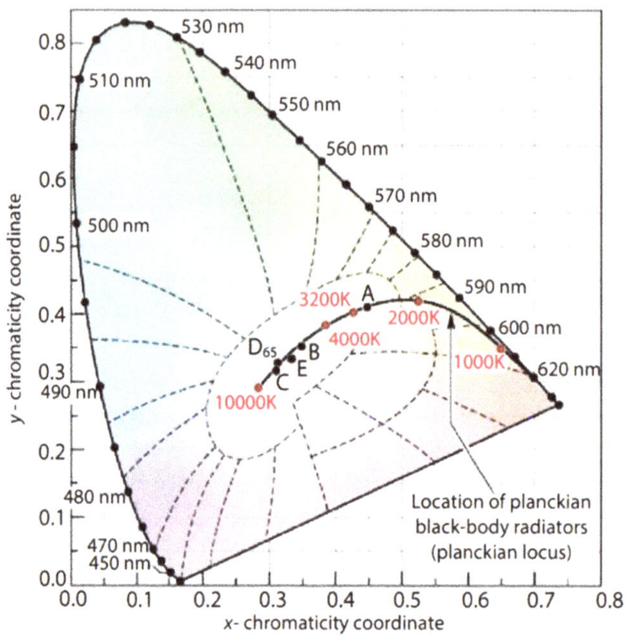

Figure 9.5. Illuminants in the CIE chromaticity diagram [55] John Wiley & Sons. [© 2020 Scrivener Publishing LLC].

9.2.6 Luminous efficiency

The luminous efficiency of a light source is defined as the luminous flux divided by the electrical input power (lm W^{-1}) [56]. Therefore, in the following, we refer to this figure of merit using both terms, i.e. luminous efficiency and power efficiency.

$$\text{Power efficiency} = \text{Luminous flux (lm)}/(\text{Current (A)}\times\text{Voltage (V)})$$

The lumen is defined as 1/683 W of monochromatic green light at a frequency of 540×10^{12} Hz, which means at a wavelength of 555 nm, where the human eye is most sensitive. The maximum value in the presence of the complete conversion of energy at 555 nm would be 683 lm W^{-1}. The luminous efficiency usually remains far below this ideal value, given that the maximum values reported are around 300 lm W^{-1} [48]. Please note that a high CRI value requires proper spectral dispersion in the visible range, with a consequent decrease in the luminous efficiency to far less than the ideal value. Therefore, the performance of a light source always has to address a trade-off between the CRI and the luminous efficiency [48].

9.3 The state of the art and critical aspects of WLEDs and WHLEDs

While we always refer to pc-LEDs—figure 9.2, WLEDs and WHLEDs are characterized as devices that contain all-inorganic and organic phosphors, respectively, in addition to the inorganic emitting chip—see section 9.2.

Among all-inorganic phosphors, oxide-based materials doped with rare-earth ions have shown good luminescent properties (high ϕ as well as absorption extinction coefficients in the blue and UV regions). The most famous is YAG:Ce ($Y_3Al_5O_{12}$ doped with Ce (III)). However, the use of this type of phosphor involves three major shortcomings. On one hand, the phosphor represents up to 20%–30% of the total cost of a WLED, but this share is continuously increasing due to the limited resources of rare-earth ion elements controlled by China [58]. On the other hand, there are important concerns related to the visual and nonvisual effects of blue light on human health [59]. In 1966, Noell *et al* first reported that blue light can cause damage to rod cells [60], while in 2001 Dawson *et al* confirmed that blue light damages the retinas of primates [61], and Algvere *et al* (2006) reported on age-related maculopathy and the hazardous impact of blue light [62]. Moreover, Brainard *et al* measured the nonvisual bio-spectral response curve, i.e. the circadian rhythm function, to demonstrate the influence of light at different wavelengths on the circadian rhythm of the human body [63]. Finally, rare-earth-doped phosphors are prone to emission-temperature quenching. In fact, when the power consumption increases, the p–n junction temperature can reach 400 K, at which, for example, thermal-emission quenching of Ce (III) luminescent centers occurs. The reader can refer to the indicated contribution for an overview of YAG:Ce phosphors [64]. Other nitride and oxynitride phosphors have also attracted attention. Unfortunately, they require critical synthesis conditions, namely high temperature and high N_2 pressure. In addition, there are no clear guidelines for controlling their crystal structures. Finally, red-emitting commercial phosphors for white UV-LEDs are also scarce, due to their poor self-stability under ambient conditions. For a comprehensive overview, we recommend the following reviews [48, 65].

Despite the wide use of inorganic phosphors, they raise significant concerns related to high costs, sustainability, the lack of efficient recycling methods, color-tuning difficulties in the low-energy region, thermal quenching, and non-uniform dispersion into commercial packaging matrices. Thus, organic phosphors are becoming more and more attractive in terms of low costs, sustainability, and easy tuning [66, 67]. They usually possess higher absorption extinction coefficients in the near-UV and UV regions that can also be tuned through specific molecular design [68]. Furthermore, the choice of an organic material avoids an industry dependency on costly rare-earth elements as the demand for solid-state lighting grows [69]. Finally, the short excited-state lifetimes compared to those of inorganic phosphors allows the use of LEDs in, for example, visible light communication devices [70]. Encouraged by these advantages, research has been focusing on HLEDs as an eco-friendly lighting solution [71].

Although there are several ways to prepare organic phosphors, the most used consists of UV-curable or thermally curable sealing reagents employed with the emitter to produce thin films on glass or on top of the LED packaging. The emitter is mechanically dispersed in the polymer or resin matrix [72]. Originally, conjugated polymers and small molecules were the most studied families of emitter for HLEDs [73, 74]. However, the degradation of the emitter upon phosphor fabrication, phase separation over time from the polymer matrices, and strong quenching due to the absorption of water and oxygen over time have been identified as the most limiting factors [75]. Other issues are related to the degradation due to thermal stress, as high-power HWLEDs reach operating temperatures of ca. 70 °C at the phosphor (photoinduced heat generation) [55, 76–79]. This heat needs to be dissipated to the environment, but polymer matrices usually show low thermal conductivities (0.1/0.2 W m^{-1} K^{-1}), leading to quenching of the phosphor emission [79, 80]. Separation (glass or air) between the LED chip and the phosphor (i.e. a remote configuration) is typically used to minimize heat generation and enhance heat dissipation. Finally, an increase in either the driving current or the phosphor thickness results in higher luminous output, but there is a maximum layer thickness for a given concentration of each emitter, beyond which the luminous flux decreases due to optical losses [81]. Finally the stability under irradiation stress implies the formation of radicals due to polymer degradation and/or the presence of absorbed water and oxygen, resulting in photoinduced degradation of the emitter. In addition, aging processes can occur when treating, processing, storing, and using polymer matrices: (i) thermal aging, (ii) chemical aging, (iii) oxidative aging, and (iv) light-induced degradation [82]. Photooxidative degradation is typically more critical with UV light and at high temperatures. In the HLED field, the best strategy for dealing with this is via the choice of pumping source; blue LEDs are preferred [83, 84]. Moreover, the removal of oxygen atoms from the host matrix results in a notable retardation of photooxidative degradation [55, 85–87]. For example, the addition to the matrix of hydroxyl group species has been quite successful [87].

Thus, the development of organic emitters has recently focused on: (i) stable coordination complexes [88], inorganic and graphitic QDs [84, 89, 90] and perovskites [91], and (ii) a further protection of the organic emitter using metal–organic frameworks (MOFs) [32, 92, 93], metal oxide shells (@MOx) [94], and proteins [95]. Among them, this chapter focuses on inorganic/organic nanoparticle-based phosphors for WLEDs. The synthesis of micro- and nanophosphors as well as their characterization and functionalization form one of the key research fields for highly efficient devices. This is related to the challenge of the easy synthesis of phosphors featuring high ϕ and a nanosize that reduces scattering events. A decrease in the size of phosphor particles to several nanometers should eliminate light losses related to scattering events, as these particle sizes are much shorter than the excitation light wavelength. However, most rare-earth-doped nanophosphors show poor luminescent properties because of their more prominent surface defects and aggregation issues. Therefore, we first focus on the design and use of YAG:Ce nanoparticles as the most representative inorganic phosphors in section 9.4 and review the advances in nanosized organic phosphors based on (i) metal complexes embedded in silica nanoparticles, (ii) metal–halide perovskites stabilized by metal oxide shells (MHP@MOx), and (iii) metal oxide–polymer nanoparticles in section 9.5.

9.4 Inorganic phosphors based on YAG:Ce nanoparticles

YAG:Ce is a useful yellow-emitting phosphor with a short excited-state lifetime (60 ns), high ϕ (>70%), and significant chemical stability and reliability [96, 97]. The LED field would certainly profit from the large-scale production of nanosized YAG:Ce. For instance, it would give a great boost to microLED fabrication, which promises to revolutionize flat-panel displays by directly incorporating red, green, and blue emitters into a full-color pixel, thereby avoiding the losses associated with liquid-crystal displays and achieving the brightness and color quality typical of OLEDs [98]. Finally, nanosized YAG:Ce is expected to be an interesting material for tagging [99] and imaging [100, 101] as well as for advanced ceramics designs for better transparency, absorption, and light extraction [102].

9.4.1 The synthesis of YAG:Ce nanoparticles

The synthesis of nanocrystalline YAG:Ce follows standard wet chemistry methods, such as precipitation reactions, sol–gel methods, and solvothermal reactions—see figure 9.6.

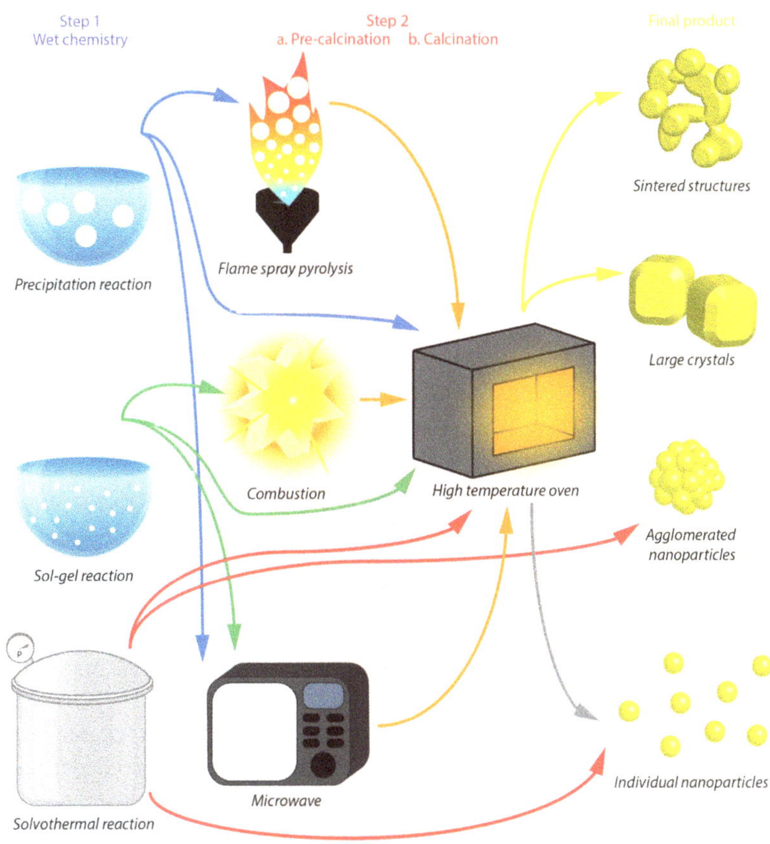

Figure 9.6. Schematic representation of the synthesis routes used to obtain YAG:Ce nanoparticles. Reprinted with permission from [98]. Copyright (2020) American Chemical Society.

Usually, a flame spray pyrolysis step, a combustion, or a microwave treatment is applied before a calcination step in a high-temperature oven [98]. When precipitation synthesis is used, either flame spray pyrolysis [103–107] or pre-combustion in a microwave oven [108, 109] and direct combustion at high temperature [110–115] is used. The final product can be either a sintered structure [112, 114, 116] or a large crystal [103–105]. When starting from sol–gel reactions, the YAG precursor can either undergo a preventive combustion treatment [117–120] (eventually in a microwave oven [121–123]), or it can be directly subjected to high-temperature calcination [124]. Sintered structures [117, 119, 120, 125–127] and large crystals [121, 122] are obtained. The solvothermal reaction is the only one-step synthesis method, and it leads to either agglomerated nanoparticles [128–131] or individual nanoparticles [132, 133].

As described in the literature, all the aforementioned techniques have been used with great success. In this sense, there is not a most desirable synthesis method. The solvothermal method has a major selling point, as it comprises a one-step reaction. However, there are several difficulties related to undesired agglomeration and aggregation, which hamper the production of individual nanoparticles.

9.4.2 Devices based on nano YAG:Ce: issues and strengths

The main characteristics to take into account are (i) the absorption extinction coefficient, (ii) the particle size and morphology, and (iii) the photo- and thermal stabilities. Here, the spectroscopic features are influenced by (i) the high crystallinity, (ii) a homogenous Ce-dopant distribution, and (iii) the presence of Ce in the desired oxidation state. In particular, the Ce concentration must be controlled, since high doping concentrations can lead to concentration quenching and self-absorption. For microsized YAG crystals, the optimum concentration is between 0.1% and 3%, while for nanoparticles, the optimal concentration is between 0.4% and 4%. Higher concentrations lead to an eventual distortion of the crystal structure.

In addition, the ϕ of YAG:Ce nanoparticles is around 50%. Although this is common in luminescent nanoparticles due to presence of surface defects, the formation of radiative excitons in YAG:Ce nanoparticles is localized on the dopant ions. Thus, the emission features are more tolerant of the nanoparticle surface. Moreover, in active ion-doped materials, changing the size of the host nanoparticle should not change the optical properties. The origin of the reduced ϕ is, indeed, still under debate, since it has no clear relationship with the synthetic method. The most plausible answers are the formation of Ce (IV) traces during the synthesis and the inhomogeneous distribution of Ce atoms, leading to (i) concentration quenching and (ii) lower crystallinity due to the low calcination temperatures required to achieve the desired small size of the particles—see figure 9.7.

Finally, the photostability of YAG:Ce nanoparticles under device operational conditions is also an issue. While the photostability of microsized YAG:Ce crystals is excellent in devices ($>10^5$ h), devices that use YAG:Ce nanoparticles feature average lifetimes of hundreds of hours. For instance, Kamiyama *et al* observed an increase in Ce (IV) concentration when highly doped YAG:Ce nanoparticles were exposed to blue light under ambient conditions due to the photooxidation of Ce (III)

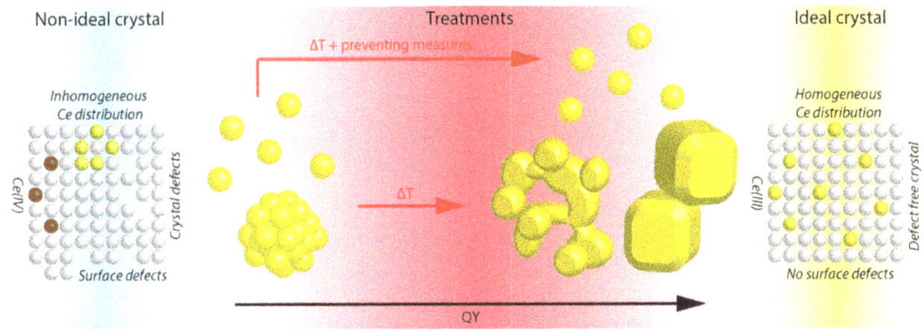

Figure 9.7. Schematic representation of ϕ-limiting factors in YAG:Ce nano particles. Reprinted with permission from [98]. Copyright (2020) American Chemical Society..

to Ce (IV). In particular, this could be related to the presence of Ce (III) near the surface. Thus, a synthesis leading either to a lower concentration of Ce (III) next to the surface and less contact with oxygen or to a higher crystallinity of the material could alleviate the photo-bleaching issue [134]. In fact, some encouraging reports can be found in the literature in which poly(vinyl alcohol) (PVA) was used as a matrix to suppress photooxidation, leading to only reversible initial drop in the photoluminescent intensity upon 450 nm light irradiation. Other polymers, such as polyvinylpyrrolidone and polyvinylbutyral, have shown less ideal behaviors [135]. The aspects presented here and deeper considerations are extensively described in a recent comprehensive review of nanosized YAG:Ce published in 2020 [98].

9.5 Other metal-oxide-based nanophosphors

9.5.1 Organometallic@MOx emitters in HLEDs

The embedding of emitters such as phosphorescent iridium (III) complexes in silica nanoparticles was proposed in 2010 by Lee and collaborators [136]. This work was the birth of phosphorescent silica-based phosphors. Interestingly, iridium (III) complexes have some drawbacks related to their poor solubility in environmentally friendly solvents and strong sensitivity to oxygen in air due to their long-lived triplet excited state [137]. Thus, a silica protective matrix was of high interest [138–140]. Moreover, the silica nanoparticles form stable suspensions in water and show excellent chemical and mechanical stabilities as well as easy functionalization via modification of the silica surface [141].

The typical strategy used to incorporate organic emitters into a silica matrix is the chemical coupling of emitters functionalized with reactive organosilicon reagents to form, for example, thiourea linkage [142, 143]. However, the external decoration of silica nanoparticles leads to unstable phosphors, as the protection thus provided from the surrounding environment is ineffective. Thus, it is more convenient to use the functionalized emitters as a seed for the growth of surrounding silica nanoparticles via allylation and hydrosilylation to form a trialkoxysilane derivative of the Ir(III) complex, which, after controlled hydrolysis by the Stöber method, forms the doped silica nanoparticle with a nonporous amorphous form—see figure 9.8 [144]. The authors also showed the high versatility of this approach by preparing

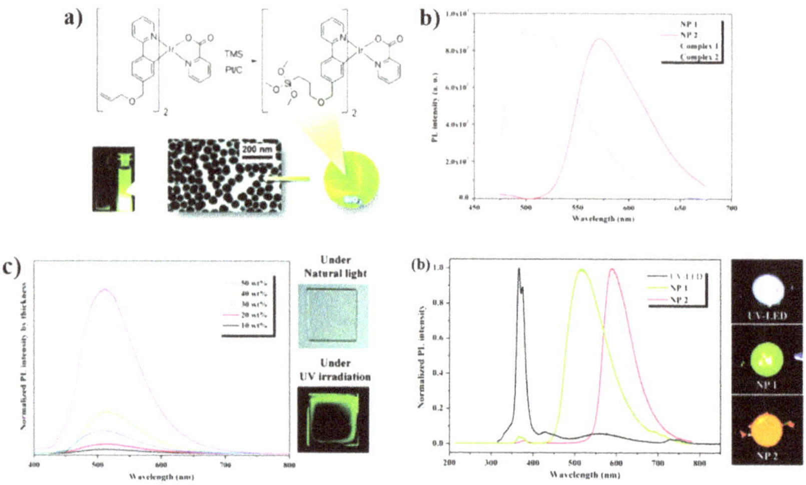

Figure 9.8. (a) A schematic representation of the work on silica nanoparticles by Lee *et al.* (b) Emission spectra of complexes 1 and 2 in $CHCl_3$ and NP1 and NP2 in ethanol at room temperature (RT). (c) Dependence of the emission spectra of NP1–polymer nanocomposite thin films on the NP1 content. Inset: photos of NP1–polymer nanocomposite films on quartz under natural and UV light irradiation. (d) Emission spectra of NP1–polymer and NP2–polymer nanocomposites on UV LED devices. Inset: photos of the nanocomposites in the devices. Reprinted with permission from [136]. Copyright (2010) American Chemical Society.

green- and red-emitting Ir(III)-complex@SiO_2 with similar success. The hybrid organometallic–silica nanoparticles showed slightly different photophysical features to those of iridium (III) complexes in solution (wavelength maxima at 501 and 571 nm, ϕ of 40% and 75%, and excited-state lifetimes of 10 ns for both green- and red-emitting nanoparticles, respectively). With respect to green emitters, the ϕ was three times higher than that of the complex in solution, indicating minor contact between the complex and the oxygen quencher when encapsulated and a decrease in vibration relaxation and non-radiative decay. Indeed, the emission band shape is strongly blue-shifted in the nanoparticles compared to that of the complex in solution—see figure 9.8(b). In the case of the red emitters, no differences were noticed between the emission features of the complex in solution and in the silica nanoparticle—figure 9.8(b). In addition, both Ir(III)-complex@SiO_2 nanoparticles featured more stable phosphorescence at higher temperatures. For example, the green nanoparticles were treated at 120 °C for 18 h and then further heated to 200 °C for 1 h without significant changes in their phosphorescent intensity. Thus, they prepared a 50% mass ratio of green and red Ir(III)-complex@SiO_2 composite films using a PVA polymer matrix. This was prepared by dissolving PVA in the nanoparticle water suspension and then sonicating it for 1h. To prepare the films (80 nm), the aforementioned solutions were spin cast at 2000 rpm for 30 s on quartz substrates. In order to fabricate the device, the polymer thin film was cast on the lens of a UV LED (no information about the wavelength). Although the authors did not provide an in-depth analysis of the hybrid LED (they did not disclose the device stability and/or color quality parameters), this study opened up a new way to use organometallic@MOx nanocomposites as phosphors for hybrid LEDs. Indeed, this has constituted a significant source of inspiration for HLEDs using perovskite nanoparticles, as discussed in the next paragraph.

In 2019, Costa and Berenguer [94] proposed a radically new way to prepare white-emitting organometallic@SiO_2 based on the formation of a previous organometallic dot formed via the kinetically controlled condensation of three emitting iridium (III) complexes decorated with alkoxysiloxane groups at the periphery. This white-emitting dot was used as a seed to subsequently form mesoporous silica nanoparticles. The synthetic route is displayed in figure 9.9. The first sol–gel step involved the pre-hydrolysis of the three alkoxysiloxane derivatives B, G, and R (at a mass ratio of 62.5 B : 32 G : 5.5 R) in an ethanol–water mixture under kinetic control, in the presence of hexadecyltrimethylammoniumbromide (CTAB) to control the growth of the organometallic dots formed, each of which was around 5 nm in diameter (denoted by OD_W). The OD_Ws were then subjected to a second 24 h sol–gel condensation step with tetraethyl orthosilicate (TEOS), leading to white-emitting organometallo-silica nanoparticles (denoted by NP_W).

The approach employed consists of the so-called sol–gel coordination chemistry based on the co-condensation of an emissive organometallic complex bearing alkoxysiloxane terminal groups with a silica source, such as TEOS. This technique can be employed to produce luminescent organometallo-silica materials, even if they are well diffused to enable the incorporation of single organometallic complexes [145, 146].

The organometallo-silica nanoparticles NP_B,G,R and their corresponding dots OD_B,G,R display similar emission properties to those observed for the corresponding B, G, and R complexes in the solid state—see figure 9.10. This indicates that the ODs are generated without any significant structural change to their

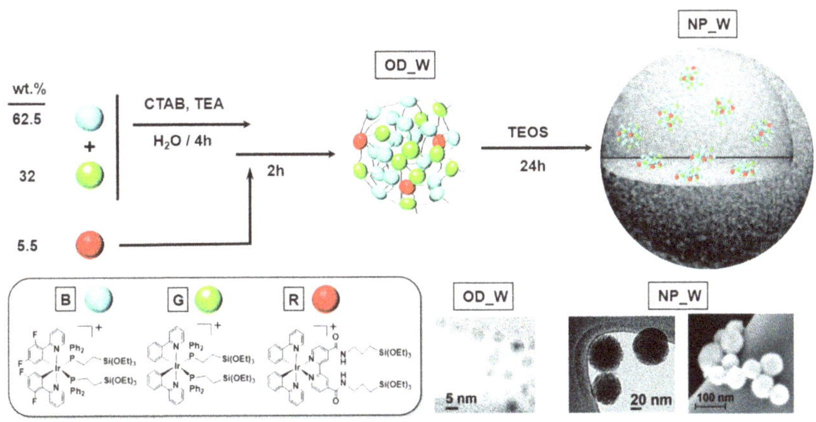

Figure 9.9. A schematic representation of the synthesis of white-emitting organometallic dots (OD_W) and nanoparticles (NP_W) using blue- (B), green- (G), and red-emitting (R) alkoxysilane-based Ir (III) complexes. The pictures show transmission electron microscopy (TEM) photographs of both types of nanomaterial and field-emission scanning electron microscopy of NP_W (FESEM). Reproduced from [94] with permission of The Royal Society of Chemistry.

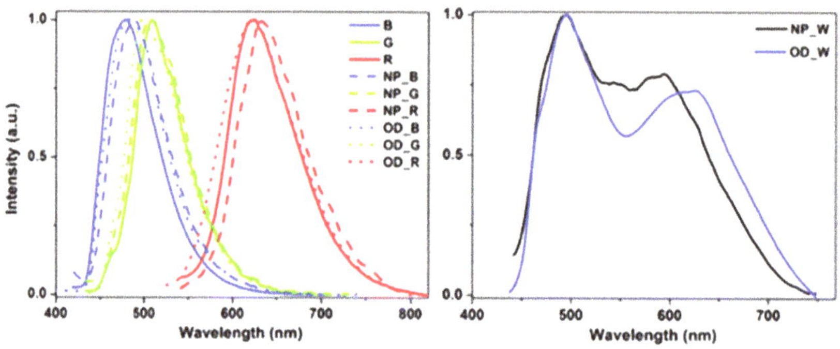

Figure 9.10. Left: emission spectra (λ_{exc} = 365 nm for B and G; λ_{exc} = 470 nm for R) at RT of the B, G, and R complexes in the solid state (solid line), their respective monochromatic emitter ODs in suspension (dotted line), and NPs in the solid state (dashed line). Right: emission spectra (λ_{exc} = 390 nm) at RT of the white-emitting OD_W and NP_W in suspension and the solid state, respectively. Reproduced from [94] with permission of The Royal Society of Chemistry.

constituent complexes. Notably, although their ϕ values are smaller than those found for the pure complexes, the NP_B,G,R nanoparticles show ϕ values of up to 20%.

In contrast, NP_W nanoparticles, which are formed from a mixture of B, G, and R complexes with the appropriate mass contents exhibit white emission associated with a broad band that has maxima at 490 and 595 nm along with a tail extending up to 750 nm—see figure 9.10.

The authors reported the occurrence of an energy transfer process from the high-energy-emitting complexes to the red-emitting one inside the NP_W. Further confirmation was obtained upon comparing the τ values of NP_W monitored at 480 nm (0.57 μs), 510 nm (0.64 μs), and 595 nm (0.63 μs) with those of NP_B (0.99 μs), NP_G (0.89 μs), and NP_R (0.12 μs) at their respective emission maxima. The energy transfer process is confirmed by the slight τ decrease of the blue and green emitters and a clear τ increase of the red one. This energy transfer is only feasible in a short-range process, such as the 5 nm size of the OD_W formed with the three complexes in the NP_W sample, and it cannot happen in the homogeneous mixture of NP_B,G,R, in which each single emitter is localized inside the silica nanoparticles, which are 50–70 nm in diameter.

The authors fabricated a HWLED by dispersing the NP_W in a rubber-like polymer matrix composed of linear and branched polyethylene oxide derivatives. The coating was prepared with an optimized thickness of around 1–1.5 mm and put on a UV LED chip emitting 400 nm light. Remarkably, these devices showed a broad emission spectrum with a maximum at 590 nm, perfectly matching that of the natural sunlight measured in Madrid in December 2017—figure 9.11. This device was fully characterized, demonstrating *x*/*y* CIE color coordinates of 0.34/0.33, a CRI of 85 and a CCT of 5143 K, in spite of its poor luminous efficiency of 2.5 lm W^{-1} (<5% decrease). Moreover, the white-emitting nanoparticles featured excellent color quality, demonstrating a stability of around thousands of hours, specifically

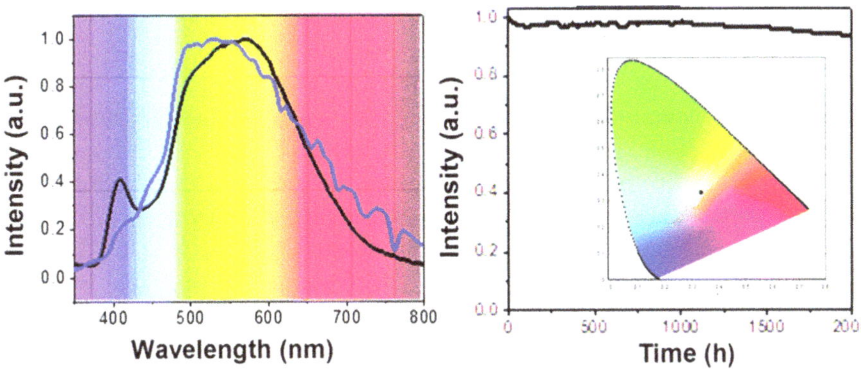

Figure 9.11. Left: electroluminescence spectrum of single-component WHLEDs compared with the visible part of the sunlight spectrum (measured in Madrid, 12/12/2017 at 13:00). Right: emission intensity of the WHLEDs and the *x*/*y* CIE color coordinates (inset) over time at 15 mA of applied current. Reproduced from [94] with permission of The Royal Society of Chemistry.

more than 2000 h with an extrapolated lifetime of 10 000 h (50% of the initial brightness). Once again, the use of white NPs instead of monochromatic NPs is fundamental in order to obtain white light, since the lack of an efficient energy transfer between monochromatic NPs does not permit the production of white light.

These studies represent two milestones, since Lee's work opened up a new scenario for the employment of organometallic complexes embedded in nanoparticles, and Costa's work refined this approach, leading to white-emitting nanoparticles with high ϕ. However, the reader should take into account that, in spite of the excellent photoluminescent performances of its complexes, iridium scarcity in the Earth's crust still represents an issue. Therefore, other sustainable emitters must be investigated.

9.5.2 Metal–halide perovskites@MOx in HLEDs

Metal–halide perovskites (MHP) have the general formula ABX_3, where A is a monovalent and either organic (methylammonium or MA^+) or inorganic (Cs^+) cation, B is a divalent cation (Pb^{+2}) and X is the halide ion. When A is an organic cation, the perovskite is referred to as a hybrid.

The first MHP QDs with the formula $APbX_3$ (A = Cs^+, MA^+ or $CH(NH_2)^{2+}$ (formamidinium or FA^+); X = Cl^-, Br^-, I^-) were prepared by Møller *et al* [147]. They had a halide composition and a size-dependent luminescent behavior that covered the whole visible spectrum range (400–700 nm), relatively low fluorescence blinking, narrow emission, and a facile synthesis method [148–150]. These advantages make them the ideal candidates as color down-converting materials for LEDs [151–153]. However, they were unstable, exhibiting the prominent formation of lead halide species upon exposition to moisture, oxygen, or polar solvents. To overcome this issue, many efforts have been devoted to stabilizing MHPs via modifying their external ligands, core–shell designs, encapsulation in mesoporous silica or polymers, etc [154–156]. Since the best-performing materials are $CsPbX_3$ NCs further discussions will be focused on them.

Ding and Song prepared highly stable and water-soluble monodisperse $CsPbX_3@SiO_2$ nanocomposites [157]. They first prepared all-inorganic $CsPbX_3$ QDs ($CsPbCl_3$, $CsPbCl_{3-x}Br_x$, $CsPbBr_3$, $CsPbBr_{3-x}I_x$, and $CsPbI_3$) through a modified hot-injection method. Monodisperse $CsPbCl_3@SiO_2$, $CsPbBr_3@SiO_2$, and $CsPbI_3@SiO_2$ nanocomposites with an average diameter of 120 nm were prepared via the modified Stöber method. In detail, $CsPbX_3$ (X = Cl, Br, I) QDs were encapsulated into monodisperse SiO_2 nanoplates to form highly luminescent $CsPbX_3@SiO_2$ nanocomposites inlaid with several types of $CsPbX_3$—see figure 9.12. The most remarkable photoluminescent features were ϕ values of 84% for $CsPbI_3@SiO_2$ and 45% for $CsPbBr_3@SiO_2$. After coating, their stability exposed in the air was largely improved for all the $CsPbX_3$ (X = Cl, Br, I) composites, which

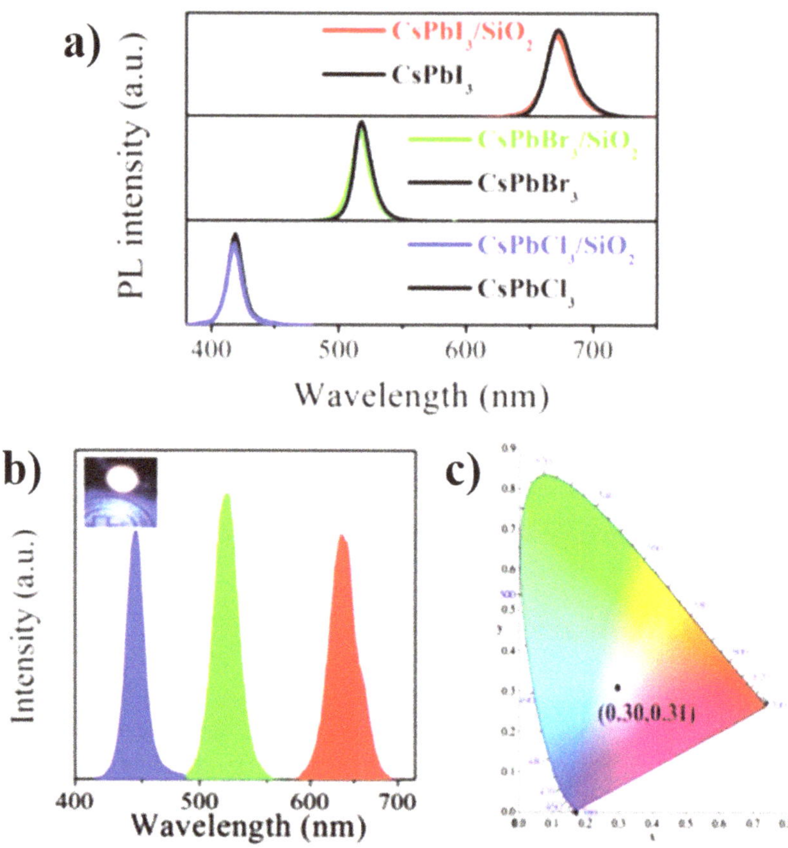

Figure 9.12. (a) Emission spectra of the $CsPbX_3$ perovskite quantum dots PQDs and $CsPbX_3$ $QDs@SiO_2$ (X = Cl, Br, I). (b) Emission spectra of the white LED based on a blue-emitting InGaN LED chip with (inset) a photograph of the device; (c) *x*/*y* CIE color coordinates of the white LED device. Reproduced from [157] © IOP Publishing. Reproduced with permission. All rights reserved.

did not change their emission bands over a period of days due to the suppression of the anion-exchange decomposition processes. Moreover, the authors demonstrated that the $CsPbX_3@SiO_2$ nanocomposites were more stable in water than naked $CsPbX_3$ QDs (>4 h vs 1 h, respectively). In order to fabricate the WHLEDs, $CsPbBr_3@SiO_2$ (λ_{em} = 523 nm and ϕ = 82%) and $CsPb(Br_{0.3}/I_{0.7})_3@SiO_2$ (λ_{em} = 634 nm and ϕ = 73%) powders were dispersed in a solution of poly(methyl methacrylate) (PMMA) in toluene. This mixture was stirred for 2 h in a vacuum glove box. The WHLEDs were then prepared by coating the prepared viscous mixture onto a commercially available blue-emitting chip (441 nm). After the color down-conversion layer was dry (the thickness was not reported), on-chip WHLEDs were obtained. The *x*/*y* CIE color coordinates were 0.30/0.31, which nicely corresponded to the emission bands of the emitting chip and both $CsPbBr_3@SiO_2$ and $CsPb(Br_{0.3}/I_{0.7})_3@SiO_2$ emitters—see figure 9.12. The luminous efficiency was 35.32 lm W^{-1} and the device stability was >40 h. This is remarkable, if we take into account that WHLEDs prepared with similar naked $CsPbX_3$ emitters featured average stabilities of less than 1 h [158].

In 2017, Li and collaborators successfully incorporated $CsPbBr_3$ QDs into a double-shell@SiO_2/Al_2O_3 monolith through a simple sol–gel process in toluene. These were referred to as QD-SAMs [159]. They incorporated ca. 18 wt.% of $CsPbBr_3$ and showed a surprisingly high water stability, a ϕ value of ca. 90%, and a narrow emission width of 25 nm. In contrast to previous works, the synthesis used di-sec-butoxyaluminoxytriethoxysilane (DBATES, $(sec\text{-}BuO)_2$-Al–O–$Si(OEt)_3$) as the single molecular precursor for the SiO_2/Al_2O_3 matrix, which contains two alkoxide functionalities and a typical Al–O–Si linkage — see Figure 9.13 [160]. This

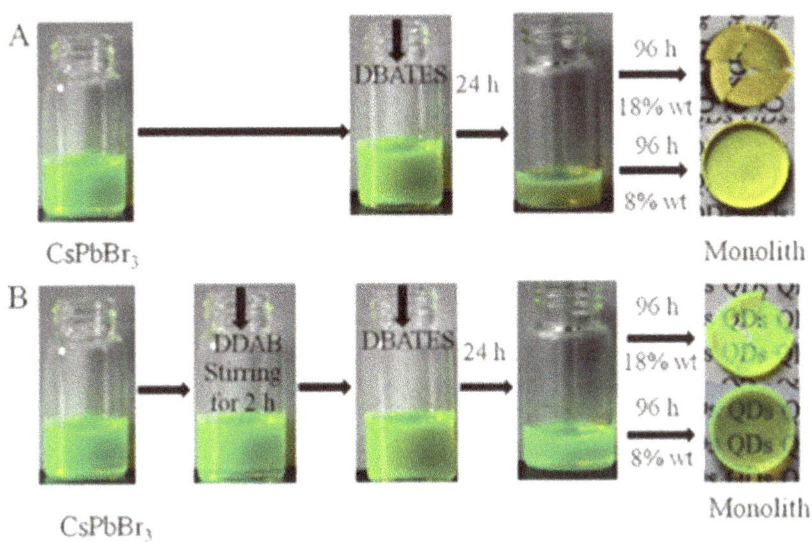

Figure 9.13. The synthesis steps used to obtain $CsPbBr_3$ QDs with (A) and without (B) DDAB treatment and the corresponding monoliths at different concentrations of QDs [159] John Wiley & Sons. [© 2017 Wiley-VCH Verlag GmbH & Co. KGaA, Weinheim].

is supports preparation of the SiO_2/Al_2O_3 monolith without the need to add a catalyst, even if a ligand exchange step with didodecyl dimethyl ammonium bromide (DDAB) was necessary to stabilize the $CsPbBr_3$ QDs and retain the photoluminescence features of the QDs once encapsulated. The major limitation of this approach is the lack of control over the morphology of both coatings, resulting in a composite-like material.

The $CsPbBr_3$ QD-SAM was annealed at different temperatures under vacuum conditions. It was observed that the emission intensity of $CsPbBr_3$ QD-SAM was unchanged after 1 h of treatment at 50 °C. If the annealing temperature was increased to 100 °C, the emission intensity of $CsPbBr_3$ QDs-SAM decreased to 70% of its initial intensity—see figure 9.14.

Since one of the aims is not to modify the properties of the QDs after they are encapsulated in metal oxide shells, some features of the $CsPbBr_3$ QDs were monitored for reference. In order to maintain the ϕ of the material and strengthen the cross-linking degree of the material, the QD-SAM was annealed at 50 °C for 1 h and then tested using 470 nm irradiation. The emission of the $CsPbBr_3$ QD-SAM powder was maintained at approximately 90% of its initial intensity after 300 h of operation, while the emission of the $CsPbBr_3$ QDs dropped sharply at the start of illumination and decreased to approximately 10% of its initial intensity after 72 h, demonstrating the effectiveness of the encapsulation.

Li and collaborators tested the operational stabilities of the aforementioned samples blended in polydimethylsiloxane (PDMS) on the blue LED chip (455 nm), which was operated at 5 mA and 2.7 V at the ambient temperature. In detail, to prepare the HLED, the desired amount of QD-SAM in toluene solution was added to PDMS. The resulting mixture was then heated to 40 °C for 1 h to remove the toluene and any bubbles. The mixture was then dropped onto a blue chip and thermally cured for 3 h at 80 °C in an oven. An analogous procedure was followed for the reference $CsPbBr_3$ QDs. The luminous efficacies of the resulting devices were

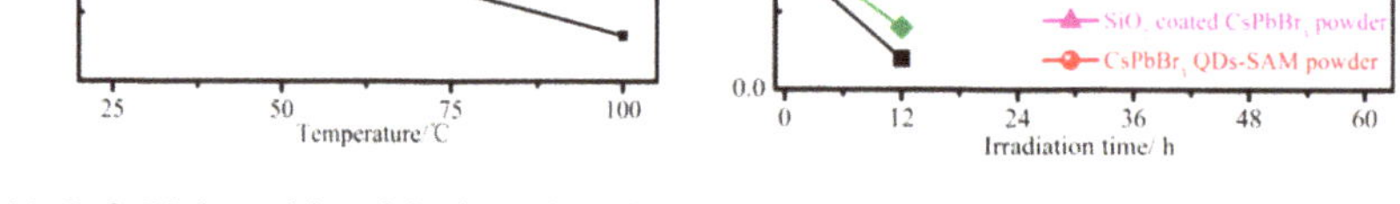

Figure 9.14. Left: PL intensities of $CsPbBr_3$ QDs-SAM and $CsPbBr_3$ QDs after thermal annealing at different temperatures for 1 h under vacuum conditions. Right: the photostability of $CsPbBr_3$ QDs-SAM powder, SiO_2 coated $CsPbBr_3$ QD powder, DDAB-treated $CsPbBr_3$ QD, and bare QDs sealed with polydimethylsiloxane (PDMS) on an LED chip (20 mA, 2.7 V) [159] John Wiley & Sons. [© 2017 Wiley-VCH Verlag GmbH & Co. KGaA, Weinheim].

80.77 lm W^{-1} for $CsPbBr_3$ QD-SAM powder and 36.15 lm W^{-1} for $CsPbBr_3$ QDs, which further demonstrated the superiority of QDs-SAM over pristine QDs. Compared to the bare $CsPbBr_3$ QDs, the $CsPbBr_3$ QDs-SAM exhibited superior photostability against blue-light radiation, because the compact SiO_2 and Al_2O_3 acted as a robust protective layer to protect the QDs from attack by oxygen and moisture.

Yin *et al* demonstrated an effective sol–gel process for modifying the surface of perovskite QDs at the interface between water and a nonpolar solvent, realizing highly stable monodisperse $CsPbX_3$@SiO_2 and $CsPbBr_3$@Ta_2O_5—figure 9.15 [161]. In short, the synthesis consisted of adding tetramethoxysilane (TMOS) or tantalum (V) ethoxide (TTEO) and acetylacetone, respectively, to a Cs_4PbBr_6 nanocrystal/hexane solution. Water was then added under vigorous oscillation in a vortex, and the solution was then kept for 12 h under ambient conditions. The following centrifugation yielded the product by discarding the precipitates.

Due to the surface passivation derived from the water treatment and silica modification, the $CsPbX_3$@SiO_2 emitters exhibited excellent photoluminescent features (ϕ of around 70% and τ of 19.8 ns) and significant photoluminescent stability during underwater (>7 days), humid air (>4 days), and constant irradiation (>10 h under a 375 nm LED at a constant current of 200 mA, 117 mW cm^{-2}) stress scenarios. As a reference, the ϕ of $CsPbX_3$ NCs obtained via water-triggered and hot-injection methods was strongly reduced under the same

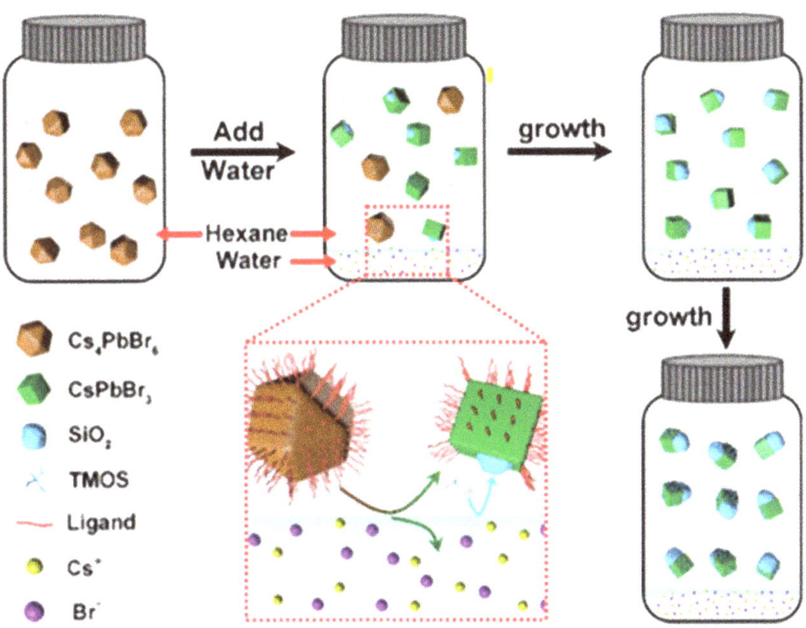

Figure 9.15. Illustration of the formation pathway of $CsPbBr_3/SiO_2$ proposed by Yin *et al.* Reprinted with permission from [161]. Copyright (2017) American Chemical Society.

stress conditions. Importantly, the discrete morphology is ideal for the formation of high-quality thin films due to its excellent dispersibility and small size. By combining green $CsPbBr_3@SiO_2$ emitters with red-emitting CdSe nanocrystal PMMA films on blue-emissive GaN LED chips, they further demonstrated the potential applications of this approach in WHLEDs. In detail, they separately dissolved $CsPbBr_3@SiO_2$ and CdSe nanocrystals (NCs) in a CH_3Cl solution of PMMA. $CsPbBr_3@SiO_2$–PMMA films and CdSe–PMMA solutions were then cast on a blue GaN chip via a multilayered approach. At an applied current of 5 mA, the WHLED showed *x*/*y* CIE color coordinates of 0.3/0.32 that were associated with CRI values of 63.

Recently, Duan and coworkers prepared green-emitting $CsPbBr_3$ NCs by the hot injection (140 °C) of a preformed cesium oleate solution into a $PbBr_2$ solution, both in octadecene (ODE) [91]. The $PbBr_2$ solution contained oleic acid (OA) and (3-aminopropyl)triethoxysilane (APTES) as capping agents in order to control perovskite NCs. After fast cooling at RT, the reaction medium was opened to the air and stirred at RT for 3 h. Air moisture induced the formation of an initial silica layer around the MHPs by the slow hydrolysis of the terminal alkoxysilane groups of the APTES. Following this step, zirconium (IV) tert-butoxide (ZTB) was added as a zirconia source. The mixture was further stirred for 3 h under ambient conditions, achieving the formation of the final SiO_2/ZrO_2 coating protecting the MHPs. Finally, the reaction mixture was centrifuged and the obtained solids were washed out with hexane and air dried — see Figure 9.16.

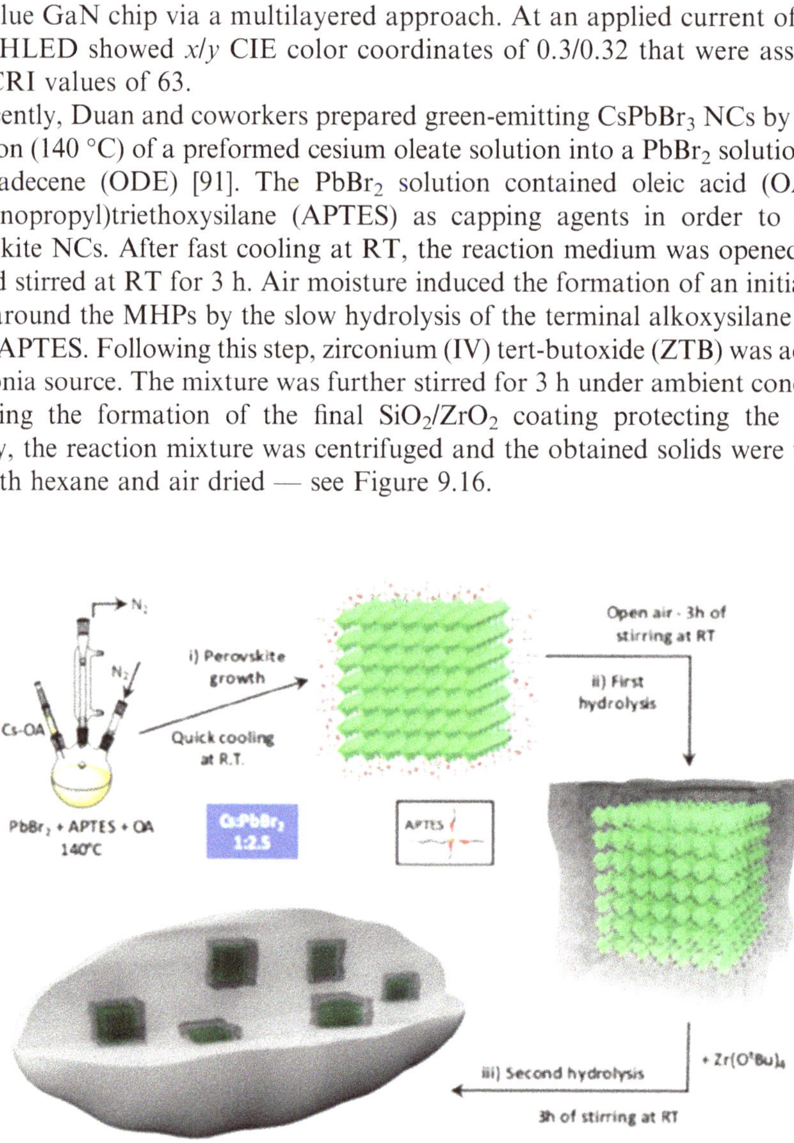

Figure 9.16. Schematic representation of the sequential one-pot synthesis of the hybrid $CsPbBr_3@SiO_2/ZrO_2$ composite (Cs-OA refers to the solution of cesium oleate in oleic acid) [91] John Wiley & Sons. [© 2020 The Authors. Published by Wiley-VCH GmbH].

The emission band centered at 524 nm was associated with a ϕ of ≈55% and an average excited-state lifetime of 30 ns. Starting from 1 mg of powder, $CsPbBr_3@SiO_2/ZrO_2$-based color filters for phosphor-converted HLEDs (pc-HLEDs) were prepared by embedding the nanoparticles into a PMMA matrix. These filters showed a highly intense green emission band centered at 524 nm associated with a ϕ of 65% and an average excited-state lifetime of 31 ns. They featured a stability of over 30 days under ambient storage conditions. The thermal stability was also investigated by monitoring eventual changes in the emission spectrum upon increasing the temperature from RT to 120 °C and then cooling the filters to RT. The filters exhibited exponential temperature-induced emission quenching, reaching a ≈60% intensity loss at 70 °C and an almost substitute quantitative with total quenching at 120 °C.

HLEDs were prepared by covering the blue-emitting chip of a commercial LED (3 W, 440 nm) using films of PMMA with $CsPbBr_3@SiO_2/ZrO_2$ in a remote configuration—figure 9.17. Devices with $CsPbBr_3@SiO_2/ZrO_2$/PMMA filters showed a quantitative blue-to-green conversion, regardless of the applied current range (5–200 mA). In particular, the pc-HLED showed a narrow green emission centered at 524 nm, corresponding to that of the hybrid $CsPbBr_3@SiO_2/ZrO_2$ nanoparticles. The luminous efficiency reached maximal values of 70–75 lm W^{-1} at currents falling in the range of 10–100 mA. An extreme applied current of 200 mA is unsuitable for this device, since it leads to a decrease in the power efficiency to 50 lm W^{-1}.

This work claims that the main benefits of ZrO_2 shells are: (i) a highly transparent coating in the whole visible range, (ii) easy synthesis in a one-pot method, achieved using a sequential sol–gel process and gentle drying under ambient conditions, and (iii) a highly compact and chemically inert layer that effectively protects the MHPs under ambient storage as well as temperature and water stress scenarios. These benefits result in one of the most stable and efficient pc-HLEDs reported to date, which achieved ≈200 h and 75 lm W^{-1} as well as ≈700 h and 70 lm W^{-1} while operating under ambient conditions at 100 and 10 mA, respectively.

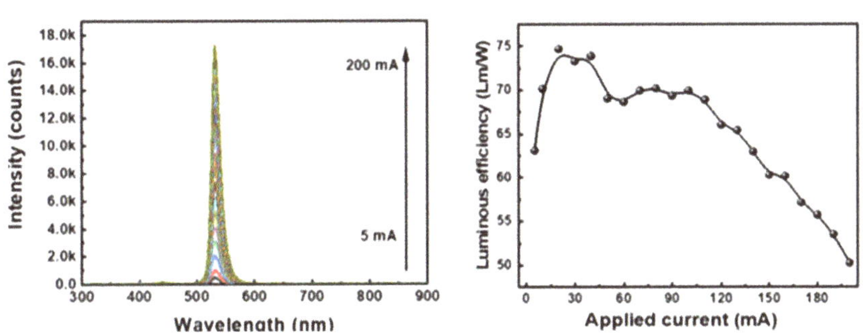

Figure 9.17. Changes in the emission spectra (left) and luminous efficiency (right) of devices with $CsPbBr_3@SiO_2/ZrO_2$/PMMA filters placed on a 440 nm LED chip operating at different applied currents [91] John Wiley & Sons. [© 2020 The Authors. Published by Wiley-VCH GmbH].

9.5.3 Polymer@MOx phosphors for WHLEDs

Hui and coworkers prepared four white LEDs using fluorescent nanosphere opal coatings as color down-converters [162]. They employed several nanosphere sizes with diameters ranging from 190 nm to 2 μm incorporating fluorescent dyes with the aim of applying prepared coatings containing mixtures of green and red-orange nanospheres onto GaN 470 nm LED chips—see figure 9.18. They chose green nanospheres internally dyed with green fluorescent protein (GFP) and orange–red nanospheres containing propidium iodide (PI). The former can produce broadband green emission with two spectral peaks a 525 and 560 nm when excited at 468 nm. The absorption of the orange–red nanospheres reaches a maximum at 558 nm, and they have fluorescent peaks at 570 and 610 nm.

Spherical particles have higher packing densities and can self-assemble into an ordered hexagonal array structure, forming regular and uniform coating layers for efficient spectral conversion because the ordered nanospheres provide large surface areas for absorption and fluorescence. The authors already provided the details of the array preparation in a previous work [163]. In detail, the microsphere mixture, suspended in deionized water, was added to a second solution containing the

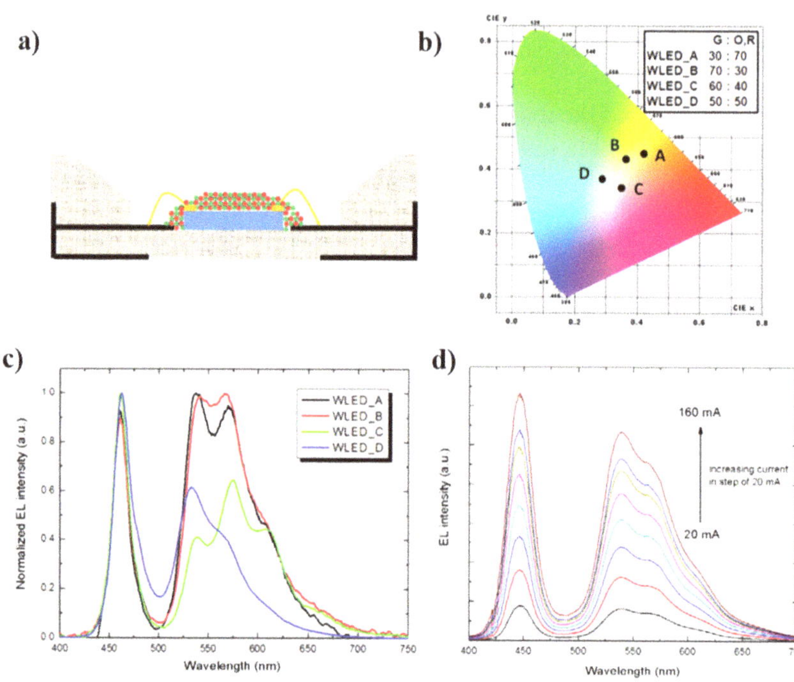

Figure 9.18. (a) A schematic diagram of the fluorescent nanosphere-coated device. (b) A CIE (1931) diagram showing the coordinates of fluorescent nanosphere-coated LEDs A–D and the nanosphere compositions. (c) The emission EL spectra of fluorescent nanosphere-coated devices A–D. (d) The electroluminescence spectra of device D at injection currents from 20 to 160 mA. Reproduced from [162]. © IOP Publishing. Reproduced with permission. All rights reserved.

n-dodecyl-sodium-sulfate surfactant, prior to drop coating. The solution was deposited onto the surface of a GaN substrate and then placed in an oven to remove excess water. The evaporation of water was reported to be preliminary to the instauration of capillary forces that drew the microspheres together, with consequent crystallization into an ordered hexagonal close-packed pattern on the substrate—figure 9.18. In this case, the green and orange–red nanospheres with diameters of 390 nm were mixed in various proportions and applied to the surfaces of LED chips as color conversion coatings consisting of around 10–14 layers of nanospheres closely packed into a hexagonal lattice.

The authors prepared four WHLEDs with *x*/*y* CIE color coordinates of 0.42/0.45, 0.37/0.43, 0.34/0.34, and 0.28/0.37 using different nanosphere compositions—see figure 9.18. These devices featured CCT values of 3587 K, 4778 K, 5271 K, and 13 000 K and luminous efficiencies of 40.5 lm W^{-1}, 57.7 lm W^{-1}, 63.1 lm W^{-1}, and 67.2 lm W^{-1}, respectively. The devices were tested at different applied currents—see figure 9.18, demonstrating their versatility by virtue of the absence of a significant shift in the EL spectra when the driving conditions were increased to 160 mA.

Unfortunately, the field of hybrid polymeric nanoparticles has not yet spread to WHLEDs. However, interesting studies are paving the way. Very recently, Daniele and Mesenelli's groups presented a study of hybrid inorganic/organic ZnO/poly-acrilic acid (PAA) nanocomposites as rare-earth-free phosphors for WHLEDs, showing great versatility in terms of chromaticity [164]. They did not provide significant data about operative WHLEDs, but they proposed the introduction of dopants as radiative centers in the lattice ZnO nanoparticles and the use of thin polymer coatings to shield the nanoparticles. In short, they studied the effect of dopants of different valence states and concentrations to tune the emission color without affecting the ϕ of hybrid polymer/ZnO nanocomposites. This was achieved for Ce^{4+}, Ag^{+}, Bi^{3+}, Ni^{2+} (which yielded almost identical ϕ values with redshifted emission) and Cr^{3+}, Cu^{2+}, Al^{3+} and Fe^{3+} (which yielded an improvement in the ϕ values with redshifted emission). As shown in figure 9.19, a small ionic radius and a large valence of the precursor are necessary to improve ϕ. However, these are not sufficient, since some dopants with a 2+ valence and a relatively small ionic radius (Mn and Co) did not improve the ϕ of the ZnO nanohybrids. Finally, the nature of the dopant plays a key role during the synthesis, which was modified for each material to provide comparable nanoparticles (polymer mesospheres containing similar metal-doped ZnO NCs). This includes not only the nanoparticle growth process with regards to size and morphology (the smaller the NCs the higher the value of ϕ) [165] but also the complexation of PAA in solution [166, 167]. This modifies the pH of the solution and the conformation of PAA chains, leading to a change in the interaction between PAA and Zn ions in the solution.

9.6 Conclusions and perspectives

In this chapter, our objectives were to (i) make the reader appreciate the extensive variety of strategies proposed in the field of inorganic and organic phosphors for white-emitting LEDs, and (ii) to present a selection of recent innovative materials in

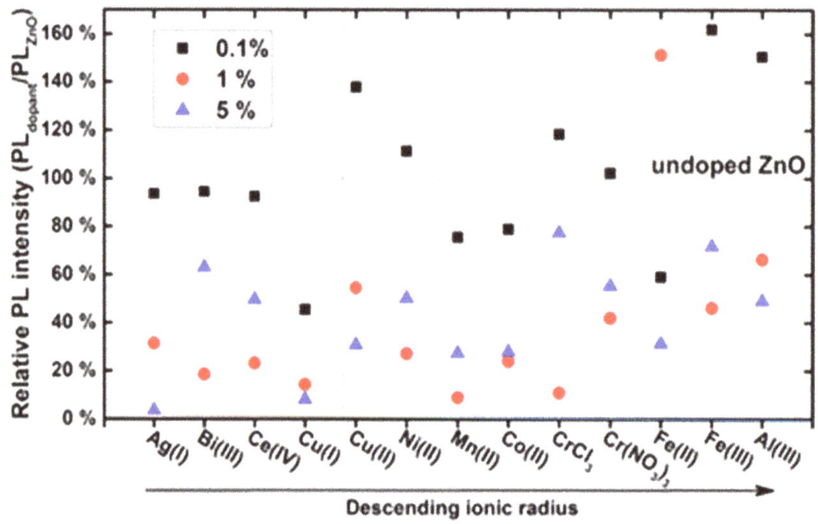

Figure 9.19. ϕ values of doped ZnO nanohybrids normalized to the ϕ value of an undoped ZnO nanohybrid. Reprinted from [164]. CC BY 4.0.

the context of nanosized phosphors. The efforts of the scientific community are moving toward the implementation of nanostructured emitters for illumination devices in an attempt to overcome some of the traditional issues associated with high-cost synthesis, low self-stability under ambient and device operational conditions, low stabilities with regards to temperature and irradiation stresses, etc. Therefore, a major commitment is emerging within the field of nanoemitters for LEDs, and this chapter may be a source of inspiration for those who are entering this field.

In this context, we have shown how the synthesis of YAG:Ce nanoparticles has already been extensively investigated without achieving low- cost, efficient, and easy synthesis. Even if greener microwave-assisted techniques seem to be a plausible solution, preliminary tough steps will be required. If single-step reactions are proposed, difficulties in controlling morphology and doping are always highlighted. This is critical for device performance, as Ce-doping is not homogenous, while Ce^{4+} formation could be the reason for the degradation of these nanoparticles. These still constitute the main limitations in this field.

The field of luminescent hybrid nanoparticles is much younger. The implementation of complexes in nanoparticles does not suffer from the issue of coating stability, as the coatings can remain stable even for thousands of hours. However, all the examples have focused on iridium (III) complexes, which are scarce materials. In addition, the ϕ values are still moderate, at around 20%. Thus, WHLEDs feature high stabilities (>1000 h) and low efficiencies (<5 lm W^{-1}). In contrast, devices based on perovskite@MOx exhibit very good luminous efficiencies (75 lm W^{-1}), but their device lifetimes and self-stabilities hardly reach 1000 h. In the field of polymer nanoparticles, in-depth scientific investigation is still lacking. It is necessary to move from the preliminary studies described in this chapter to more concrete techniques for developing luminescent polymer nanoparticles. Thus, further efforts must be made to use sustainable and efficient luminescent cores based on d^{10}-copper complexes, lead-free perovskites, and conjugated polymers. If this can be realized, it is expected to make a great impact in the mid-term.

Overall, the previous considerations prove that, almost thirty years after the demonstration of the first DH LEDs by Nakamura *et al*, this field is still able to take advantage of new, fruitful ideas, reinventing itself in the new clothes of nanoparticle technology and thus confirming that it has not yet reached maturity.

Acknowledgments

S. F. and R. D. C. acknowledge the European Union's Horizon 2020 research and innovation FET-OPEN program under grant agreement ARTIBLED No. 863170, the ERC-Co InOutBioLight No. 816856.

Summary

This chapter has reviewed the state of the art in material designs and device applications of nanosized oxides as color down-converting filters, i.e. phosphors, highlighting the challenges and progresses in WLEDs and organic–inorganic WHLEDs. Although metal-based oxide phosphors have been widely used in different lighting applications, such as three-band fluorescent lamps and plasma display panels, this chapter has focused on the most relevant achievements related to single-point phosphor-converted WLEDs and WHLEDs. It was presented in four sections. The first and the second provided a general introduction to the working mechanism of LED and HLED technologies and the main figures of merit related to the production of white light. In the next sections, a critical revision of the state of the art and the perspectives for future development was provided.

Overall, WLEDs can be produced via three main approaches, namely: (a) by combining different blue, green, and red LEDs; (b) by coupling a blue chip with a yellowish-orange emitting phosphor; or (c) by combining a UV LED chip with RGB phosphors. Phosphor-converted LEDs are currently the commonest way to produce high-quality white light with the required values of CCT, CRI, and luminous efficiency. Research is strongly active in the field of organic phosphors, in which the development of organic emitters has recently focused on: (i) stable coordination complexes, inorganic and graphitic QDs, and perovskites; and (ii) the further protection of organic emitters using MOFs, metal oxide shells (@MOx), and proteins. The synthesis of micro- and nanophosphors as well as their characterization and functionalization are some of the key points to be investigated in the development of highly efficient devices. The state of the art of metal oxide nanophosphors is characterized by YAG:Ce nanoparticles, core@shell nanoparticles bearing coordination metal complexes, metal–halide perovskites, and polymers as the luminescent core protected by TiO_2, ZrO_2, or SiO_2 shells.

Questions

1. Describe the electroluminescent mechanism of WHLEDs.
2. Discuss the phosphor-converted approach for the fabrication of LEDs as well the related photoluminescent processes.
3. Contextualize the use of nanosized phosphors in the current WHLED scenario.
4. Which are the main figures of merit related to WHLEDs?
5. What requirements should a white light satisfy?
6. Define the stability issues related to LED devices.
7. Which are the main challenges in the implementation of nano YAG:Ce in LEDs?
8. Provide a chronological overview of the use of metal complexes in phosphors.
9. Which strategies have been used to stabilize metal–halide perovskites in HLEDs?
10. Summarize the current status of polymer@MOx phosphors for WHLEDs.

References and further reading

[1] Kido J, Hongawa K, Okuyama K and Nagai K 1994 White light-emitting organic electroluminescent devices using the poly(*N*-vinylcarbazole) emitter layer doped with three fluorescent dyes *Appl. Phys. Lett.* **64** 815–7

[2] Chen J, Zhao F and Ma D 2014 Hybrid white OLEDs with fluorophors and phosphors *Mater. Today* **17** 175–83

[3] Zhao D, Qin Z, Huang J and Yu J 2017 Progress on material, structure and function for tandem organic light-emitting diodes *Org. Electron.* **51** 220–42

[4] Fresta E, Fernández-Luna V, Coto P B and Costa R D 2018 Merging biology and solid-state lighting: recent advances in light-emitting diodes based on biological materials *Adv. Funct. Mater.* **28** 1707011

[5] Holder E, Tessler N and Rogach A L 2008 Hybrid nanocomposite materials with organic and inorganic components for opto-electronic devices *J. Mater. Chem.* **18** 1064–78

[6] Xu S, Kumar S and Nann T 2006 Rapid synthesis of high-quality InP nanocrystals *J. Am. Chem. Soc.* **128** 1054–5

[7] Wilson W L, Szajowski P F and Brus L E 1993 Quantum confinement in size-selected, surface-oxidized silicon nanocrystals *Science* **262** 1242–4

[8] Counio G, Esnouf S, Gacoin T and Boilot J P 1996 CdS:Mn nanocrystals in transparent xerogel matrices: synthesis and luminescence properties *J. Phys. Chem.* **100** 20021–6

[9] Treussart F *et al* 2006 Photoluminescence of single colour defects in 50 nm diamond nanocrystals *Physica* B **376–7** 926–9

[10] Huignard A, Gacoin T and Boilot J P 2000 Synthesis and luminescence properties of colloidal YVO_4:Eu phosphors *Chem. Mater.* **12** 1090–4

[11] Haase M, Riwotzki K, Meyssamy H and Kornowski A 2000 Synthesis and properties of colloidal lanthanide-doped nanocrystals *J. Alloys Compd.* **303–4** 191–7

[12] Buissette V, Giaume D, Gacoin T and Boilot J P 2006 Aqueous routes to lanthanide-doped oxide nanophosphors *J. Mater. Chem.* **16** 529–39

[13] Louis C *et al* 2005 Nanosized hybrid particles with double luminescence for biological labeling *Chem. Mater.* **17** 1673–82
[14] Kasuya R, Kawano A, Isobe T, Kuma H and Katano J 2007 Characteristic optical properties of transparent color conversion film prepared from YAG:Ce^{3+} nanoparticles *Appl. Phys. Lett.* **91** 3–6
[15] Heer S, Kömpe K, Güdel H U and Haase M 2004 Highly efficient multicolour upconversion emission in transparent colloids of lanthanide-doped $NaYF_4$ nanocrystals *Adv. Mater.* **16** 2102–5
[16] Stouwdam J W, Hebbink G A, Huskens J and Van Veggel F C J M 2003 Lanthanide-doped nanoparticles with excellent luminescent properties in organic media *Chem. Mater.* **15** 4604–16
[17] Michalet X *et al* 2005 Quantum Dots for Live Cells, in Vivo Imaging, and Diagnostics *Science* **307** 538–44
[18] Pénard A L, Gacoin T and Boilot J P 2007 Functionalized sol–gel coatings for optical applications *Acc. Chem. Res.* **40** 895–902
[19] Stouwdam J W and Van Veggel F C J M 2002 Near-infrared emission of redispersible Er^{3+}, Nd^{3+}, and Ho^{3+} doped LaF_3 nanoparticles *Nano Lett.* **2** 733–7
[20] Sudarsan V, Van Veggel F C J M, Herring R A and Raudsepp M 2005 Surface Eu^{3+} ions are different than 'bulk' Eu^{3+} ions in crystalline doped LaF_3 nanoparticles *J. Mater. Chem.* **15** 1332–42
[21] Huignard A, Buissette V, Laurent G, Gacoin T and Boilot J P 2002 Synthesis and characterizations of YVO_4:Eu colloids *Chem. Mater.* **14** 2264–9
[22] Riwotzki K and Haase M 1998 Wet-chemical synthesis of doped colloidal nanoparticles: YVO_4:Ln (Ln = Eu, Sm, Dy) *J. Phys. Chem.* B **102** 10129–35
[23] Riwotzki K, Meyssamy H, Kornowski A and Haase M 2000 Liquid-phase synthesis of doped nanoparticles: colloids of luminescing $LaPO_4$:Eu and $CePO_4$:Tb particles with a narrow particle size distribution *J. Phys. Chem.* B **104** 2824–8
[24] Gai S, Li C, Yang P and Lin J 2014 Recent progress in rare earth micro/nanocrystals: soft chemical synthesis, luminescent properties, and biomedical applications *Chem. Rev.* **114** 2343–89
[25] Cho J, Park J H, Kim J K and Schubert E F 2017 White light-emitting diodes: history, progress, and future *Laser Photonics Rev.* **11** 1600147
[26] Tolhurst T M *et al* 2016 Electronic structure, bandgap, and thermal quenching of $Sr[Mg_3SiN_4]:Eu^{2+}$ in comparison to $Sr[LiAl_3N_4]:Eu^{2+}$ *Adv. Opt. Mater.* **4** 584–91
[27] Dai P P *et al* 2016 A single Eu^{2+}-activated high-color-rendering oxychloride white-light phosphor for white-light-emitting diodes *Light: Sci. Appl.* **5** e16024
[28] Cheng W, Rechberger F and Niederberger M 2016 Three-dimensional assembly of yttrium oxide nanosheets into luminescent aerogel monoliths with outstanding adsorption properties *ACS Nano* **10** 2467–75
[29] Schlotter P, Schmidt R and Schneider J 1997 Luminescence conversion of blue light emitting diodes *Appl. Phys.* A **64** 417–18
[30] Fernández-Luna V, Coto P B and Costa R D 2018 When fluorescent proteins meet white light-emitting diodes *Angew. Chemie Int. Ed* **57** 8826–36
[31] Findlay N J *et al* 2014 An organic down-converting material for white-light emission from hybrid LEDs *Adv. Mater.* **26** 7290–4

[32] Sun C Y *et al* 2013 Efficient and tunable white-light emission of metal–organic frameworks by iridium-complex encapsulation *Nat. Commun.* **4** 1–8
[33] Cui Y *et al* 2015 Dye encapsulated metal–organic framework for warm-white LED with high color-rendering index *Adv. Funct. Mater.* **25** 4796–802
[34] Nakamura S, Senoh M and Mukai T 1993 p-GaN/n-InGaN/n-GaN double-heterostructure blue-light-emitting diodes *Jpn. J. Appl. Phys.* **32** L8–11
[35] Haase M A, Qiu J, DePuydt J M and Cheng H 1991 Blue-green laser diodes *Appl. Phys. Lett.* **59** 1272–4
[36] Jeon H *et al* 1992 Blue and green diode lasers in ZnSe-based quantum wells *Appl. Phys. Lett.* **60** 2045–7
[37] Xie W *et al* 1992 Room temperature blue light emitting p–n diodes from Zn(S,Se)-based multiple quantum well structures *Appl. Phys. Lett.* **60** 1999–2001
[38] Amano H, Kito M, Hiramatsu K and Akasaki I 1989 p-Type conduction in Mg-doped GaN treated with low-energy electron beam irradiation (LEEBI) *Jpn. J. Appl. Phys.* **28** L2112–4
[39] Nakamura S 2015 Background story of the invention of efficient InGaN blue-light-emitting diodes (nobel lecture) *Angew. Chemie Int. Ed* **54** 7770–88
[40] Nakamura S, Mukai T and Senoh M 1994 Candela-class high-brightness InGaN/AlGaN double-heterostructure blue-light-emitting diodes *Appl. Phys. Lett.* **64** 1687–9
[41] Nakamura S, Senoh M, Iwasa N and Nagahama S I 1995 High-brightness InGaN blue, green and yellow light-emitting diodes with quantum well structures *Jpn. J. Appl. Phys.* **34** L797–9
[42] Mukai T, Morita D and Nakamura S 1998 High-power UV InGaN/AlGaN double-heterostructure LEDs *J. Cryst. Growth* **189** 778–81
[43] Nakamura S *et al* 1995 Superbright green InGaN single-quantum-well-structure light-emitting diodes *Jpn. J. Appl. Phys.* **34** L1332–5
[44] Lin H W, Lu Y J, Chen H Y, Lee H M and Gwo S 2010 InGaN/GaN nanorod array white light-emitting diode *Appl. Phys. Lett.* **97** 073101
[45] Fiorentini V, Bernardini F, Della S F, Di Carlo A and Lugli P 1999 Effects of macroscopic polarization in III–V nitride multiple quantum wells *Phys. Rev.* B **60** 8849–58
[46] Waltereit P *et al* 2000 Nirtride semiconductors free of electrostatic fields for efficient white light-emitting diodes *Nature* **406** 865–8
[47] Wang T 2016 Topical review: development of overgrown semi-polar GaN for high efficiency green/yellow emission *Semicond. Sci. Technol.* **31** 093003
[48] Ye S, Xiao F, Pan Y X, Ma Y Y and Zhang Q Y 2010 Phosphors in phosphor-converted white light-emitting diodes: recent advances in materials, techniques and properties *Mater. Sci. Eng.* R **71** 1–34
[49] Shur M S and Žukauskas A 2005 Solid-state lighting: toward superior illumination *Proc. IEEE.* **93** 1691–703
[50] Osborne R A *et al* 2020 New red phosphor ceramic K_2SiF_6:Mn^{4+} *Opt. Mater.* **107** 110140
[51] Lin Y C, Karlsson M and Bettinelli M 2016 Inorganic phosphor materials for lighting *Top. Curr. Chem.* **374** 374–421
[52] Jüstel T, Nikol H and Ronda C 1998 New developments in the field of luminescent materials for lighting and displays *Angew. Chem. Int. Ed.* **37** 3084–103

[53] Calculate color temperature (CCT) from CIE 1931 *xy* coordinates, Waveform Lighting https://waveformlighting.com/tech/calculate-color-temperature-cct-from-cie-1931-xy-coordinates
[54] What is CRI? The ultimate guide to the Color Rendering Index, Waveform Lighting https://waveformlighting.com/tech/what-is-cri-color-rendering-index
[55] Mosca M, Macaluso R and Crupi I 2020 Hybrid inorganic–organic white light emitting diodes *Polymers for Light-Emitting Devices and Displays* (New York: Wiley) 197–262
[56] Mehr M Y, Van Driel W D and Zhang G Q 2014 Colour shift in remote phosphor based LED products *2014 IEEE 64th Electronic Components and Technology Conference (ECTC)* (Piscataway, NJ: Institute of Electrical and Electronics Engineers Inc.) 1477–81
[57] Ronda C R, Jüstel T and Nikol H 1998 Rare earth phosphors: fundamentals and applications *J. Alloys Compd.* **275–7** 669–76
[58] Pattison M *et al* 2020 *DOE BTO Lighting RandD Program* (Washington, DC: U.S. Department of Energy Office of Energy efficiency and Renewable Energy) https://www.energy.gov/sites/prod/files/2020/01/f70/ssl-rd-opportunities2-jan2020.pdf
[59] Chaopu Y *et al* 2018 Change of blue light hazard and circadian effect of LED backlight displayer with color temperature and age *Opt. Express* **26** 27021–32
[60] Noell W K, Walker V S, Kang B S and Berman S 1966 Retinal damage by light in rats *Invest. Ophthalmol. Vis. Sci.* **5** 450–73 https://iovs.arvojournals.org/article.aspx?articleid=2128202
[61] Ham W T, Mueller H A and Sliney D H 1976 Retinal sensitivity to damage from short wavelength light *Nature* **260** 153–5
[62] Algvere P V, Marshall J and Seregard S 2006 Age-related maculopathy and the impact of blue light hazard *Acta Ophthalmol. Scand.* **84** 4–15
[63] Brainard G C *et al* 2001 Action spectrum for melatonin regulation in humans: evidence for a novel circadian photoreceptor *J. Neurosci.* **21** 6405–12
[64] Tucureanu V, Matei A and Avram A M 2015 Synthesis and characterization of YAG:Ce phosphors for white LEDs *Opto-electron. Rev.* **23** 239–51
[65] Xia Z and Liu Q 2016 Progress in discovery and structural design of color conversion phosphors for LEDs *Prog. Mater Sci.* **84** 59–117
[66] Kim H J, Jin J Y, Lee Y S, Lee S H and Hong C H 2006 An efficient luminescence conversion LED for white light emission, fabricated using a commercial InGaN LED and a 1,8-naphthalimide derivative *Chem. Phys. Lett.* **431** 341–5
[67] Kwak S-Y, Yang S, Kim N R, Kim J H and Bae B-S 2011 Thermally stable, dye-bridged nanohybrid-based white light-emitting diodes *Adv. Mater.* **23** 5767–72
[68] Zhang L, Li B, Lei B, Hong Z and Li W 2008 A triphenylamine derivative as an efficient organic light color-conversion material for white LEDs *J. Lumin.* **128** 67–73
[69] Yue Z *et al* 2013 Hybrid GaN/organic white light emitters with aggregation induced emission organic molecule *Opt. Mater. Express* **3** 1906
[70] Laurand N *et al* 2012 Colloidal quantum dot nanocomposites for visible wavelength conversion of modulated optical signals *Opt. Mater. Express* **2** 250
[71] Jin J Y, Kim H G, Hong C H, Suh E K and Lee Y S 2007 White light emission from a blue LED, combined with a sodium salt of fluorescein dye *Synth. Met.* **157** 138–41
[72] Di Martino D *et al* 2014 Straightforward fabrication of stable white LEDs by embedding of inorganic UV-LEDs into bulk polymerized polymethyl-methacrylate doped with organic dyes *Sci. Rep.* **4** 1–5

[73] Lai C-F, Chang C-J, Hsieh C-L, Chen Y-L and Tuan C-S 2013 Highly bright broadband red light produced by fluorescence polymer/InGaN hybrid light-emitting diodes *Opt. Lett.* **38** 4082
[74] Huyal I O, Koldemir U, Ozel T, Demir H V and Tuncel D 2008 On the origin of high quality white light emission from a hybrid organic/inorganic light emitting diode using azide functionalized polyfluorene *J. Mater. Chem.* **18** 3568–74
[75] Yazdan Mehr M *et al* 2020 Degradation of optical materials in solid-state lighting systems *Int. Mater. Rev.* **65** 102–28
[76] Luo X, Hu R, Liu S and Wang K 2016 Heat and fluid flow in high-power LED packaging and applications *Prog. Energy Combust. Sci.* **56** 1–32
[77] Arik M, Becker C A, Weaver S E and Petroski J 2004 Thermal management of LEDs: package to system *3rd Int. Conf. on Solid State Lighting* ed I T Ferguson, N Narendran, S P DenBaars and J C Carrano vol 5187 (Bellingham, WA: SPIE), 64 pp
[78] Petroski J 2002 Thermal challenges facing new-generation light-emitting diodes (LEDs) for lighting applications *Solid State Lighting II* ed I T Ferguson, N Narendran, S P DenBaars and Y-S Park vol 4776 (Bellingham, WA: SPIE), 215 pp
[79] Espasa A *et al* 2020 Long-living and highly efficient bio-hybrid light-emitting diodes with zero-thermal-quenching biophosphors *Nat. Commun.* **11** 1–10
[80] Kim Y H *et al* 2017 A zero-thermal-quenching phosphor *Nat. Mater.* **16** 543–50
[81] Tan C M, Singh P, Zhao W and Kuo H-C 2018 Physical limitations of phosphor layer thickness and concentration for white LEDs *Sci. Rep.* **8** 2452
[82] Emanuel N M *et al* 1998 Definition of terms and classification of processes relating to aging polymers *Int. J. Polym. Mater. Polym. Biomater* **41** 7–22
[83] Xiang H F, Yu S C, Che C M and Lai P T 2003 Efficient white and red light emission from GaN/tris-(8-hydroxyquinolato) aluminum/platinum(II) meso-tetrakis(pentafluorophenyl) porphyrin hybrid light-emitting diodes *Appl. Phys. Lett.* **83** 1518–20
[84] Yousif E and Haddad R 2013 Photodegradation and photostabilization of polymers, especially polystyrene: review *SpringerPlus* **2**
[85] Sokołowska J, Czajkowski W and Podsiadły R 2001 The photostability of some fluorescent disperse dyes derivatives of coumarin *Dyes Pigm.* **49** 187–91
[86] Enko B *et al* 2013 *J. Phys. Chem.* A **117** 8873–82 Singlet oxygen-induced photodegradation of the polymers and dyes in optical sensing materials and the effect of stabilizers on these processes
[87] Nowakowska M, Smoluch M and Sendor D 2001 The effect of cyclodextrins on the photochemical stability of 7-amino-4-methylcoumarin in aqueous solution *J. Inclusion Phenom. Macrocyclic Chem.* **40**
[88] Ye Y *et al* 2018 Application of an orange–yellow emitting cationic iridium(III) complex in GaN-based warm white light-emitting diodes *J. Mater. Sci. Mater. Electron.* **29** 1554–61
[89] Tetsuka H, Nagoya A and Asahi R 2015 Highly luminescent flexible amino-functionalized graphene quantum dots@cellulose nanofiber-clay hybrids for white-light emitting diodes *J. Mater. Chem.* C **3** 3536–41
[90] Zhou D *et al* 2015 Surface ligand dynamics-guided preparation of quantum dots-cellulose composites for light-emitting diodes *ACS Appl. Mater. Interfaces* **7** 15830–9
[91] Duan Y *et al* 2020 Meeting high stability and efficiency in hybrid light-emitting diodes based on SiO_2/ZrO_2 coated $CsPbBr_3$ perovskite nanocrystals *Adv. Funct. Mater.* **30** 2005401

[92] Gong Q *et al* 2014 Solution processable MOF yellow phosphor with exceptionally high quantum efficiency *J. Am. Chem. Soc.* **136** 16724–7
[93] Lu Y and Yan B 2014 Lanthanide organic–inorganic hybrids based on functionalized metal–organic frameworks (MOFs) for a near-UV white LED *Chem. Commun.* **50** 15443–6
[94] Ezquerro C *et al* 2019 White-emitting organometallo-silica nanoparticles for Sun-like light-emitting diodes *Mater. Horiz.* **6** 130
[95] Fernández-Luna V *et al* 2020 Biogenic fluorescent protein-silk fibroin phosphors for high performing light-emitting diodes *Mater. Horizons* **7** 1790–800
[96] Xia Z and Meijerink A 2017 Ce^{3+}-doped garnet phosphors: composition modification, luminescence properties and applications *Chem. Soc. Rev.* **46** 275–99
[97] Bachmann V, Ronda C and Meijerink A 2009 Temperature quenching of yellow Ce^{3+} luminescence in YAG:Ce *Chem. Mater.* **21** 2077–84
[98] Berends A C, van de Haar M A and Krames M R 2020 YAG:Ce^{3+} phosphor: from micron-sized workhorse for general lighting to a bright future on the nanoscale *Chem. Rev.* **1209** 13461–79
[99] Asakura R, Isobe T, Kurokawa K, Aizawa H and Ohkubo M 2006 Tagging of avidin immobilized beads with biotinylated YAG:Ce3+ nanocrystal phosphor *Anal. Bioanal. Chem.* **386** 1641–7
[100] Dong B *et al* 2012 Non-photobleaching YAG:Ce nanoparticles for optical imaging with blue excitation *RSC Adv.* **2** 3897
[101] Wang Q *et al* 2015 Optical properties of Ce^{3+}–Nd^{3+} co-doped YAG nanoparticles for visual and near-infrared biological imaging *Spectrochim. Acta* A **149** 898–903
[102] Bechtel H, Schmidt P, Busselt W and Schreinemacher B S 2008 Lumiramic: a new phosphor technology for high performance solid state light sources *8th Int. Conf. on Solid State Lighting*7058 ed I T Ferguson, T Taguchi, I E Ashdown and S-J Park (Bellingham, WA: SPIE), 70580E
[103] Zhou Y H *et al* 2003 Morphology control and luminescence properties of YAG:Eu phosphors prepared by spray pyrolysis *Mater. Res. Bull.* **38** 1289–99
[104] Purwanto A *et al* 2008 High luminance YAG:Ce nanoparticles fabricated from urea added aqueous precursor by flame process *J. Alloys Compd.* **463** 350–7
[105] Baranwal R, Villar M P, Garcia R and Laine R M 2001 Flame spray pyrolysis of precursors as a route to nano-mullite powder: powder characterization and sintering behavior *J. Am. Ceram. Soc.* **84** 951–61
[106] You Y, Qi L, Li X and Pan W 2013 Preparation of YAG nano-powders via an ultrasonic spray co-precipitation method *Ceram. Int.* **39** 3987–92
[107] Marchal J, John T, Baranwal R, Hinklin T and Laine R M 2004 Yttrium aluminum garnet nanopowders produced by liquid-feed flame spray pyrolysis (LF-FSP) of metalloorganic precursors *Chem. Mater.* **16** 822–31
[108] Wang J, Xu H, Wang Y and Yue Y 2006 Effect of sulfate ions on YAG powders synthesized by microwave homogeneous precipitation *J. Rare Earths* **24** 284–7
[109] Yong W, Peng Y, Hongyan X and Jieqiang W 2006 Synthesis of Ce:YAG phosphor via homogeneous precipitation under microwave irradiation *J. Rare Earths* **24** 183–6
[110] Xu X *et al* 2012 Synthesis of monodispersed spherical yttrium aluminum garnet (YAG) powders by a homogeneous precipitation method *J. Am. Ceram. Soc.* **95** 3821–6
[111] Pickup H and Burke H 1992 Synthesis of yttrium aluminum garnet by reverse strike precipitation *J. Mater. Res.* **7** 706–11

[112] Matsushita N, Tsuchiya N, Nakatsuka K and Yanagitani T 1999 Precipitation and calcination processes for yttrium aluminum garnet precursors synthesized by the urea method *J. Am. Ceram. Soc.* **82** 1977–84

[113] Vrolijk J W G A, Willems J W M M and Metselaar R 1990 Coprecipitation of yttrium and aluminium hydroxide for preparation of yttrium aluminium garnet *J. Eur. Ceram. Soc.* **6** 47–51

[114] Li X, Liu H, Wang J, Zhang X and Cui H 2004 Preparation and properties of YAG nano-sized powder from different precipitating agent *Opt. Mater.* **25** 407–12

[115] Xu G, Zhang X, He W, Liu H and Li H 2006 The study of surfactant application on synthesis of YAG nano-sized powders *Powder Technol.* **163** 202–5

[116] Lv Y *et al* 2012 Synthesis of nano-sized and highly sinterable Nd:YAG powders by the urea homogeneous precipitation method *Powder Technol.* **217** 140–7

[117] Guo K *et al* 2010 Morphology investigation of yttrium aluminum garnet nano-powders prepared by a sol–gel combustion method *J. Alloys Compd.* **500** 34–8

[118] Veith M *et al* 1999 Low temperature synthesis of nanocrystalline $Y_3Al_5O_{12}$ (YAG) and Ce-doped $Y_3Al_5O_{12}$ via different sol–gel methods *J. Mater. Chem.* **9** 3069–79

[119] Guo K *et al* 2010 Influence of fuels on the morphology of undoped $Y_3Al_5O_{12}$ and photoluminescence of $Y_3Al_5O_{12}$:Eu^{3+} prepared by a combustion method *Mater. Res. Bull.* **45** 1157–61

[120] Zhang L *et al* 2012 Citrate sol–gel combustion preparation and photoluminescence properties of YAG:Ce phosphors *J. Rare Earths* **30** 289–96

[121] Abd H R *et al* 2018 Effect of annealing time of YAG:Ce^{3+} phosphor on white light chromaticity values *J. Electron. Mater.* **47** 1638–46

[122] Hassan Z, Abd H R, Alsultany F H, Omar A F and Ahmed N M 2019 Investigation of sintering temperature and Ce^{3+} concentration in YAG:Ce phosphor powder prepared by microwave combustion for white-light-emitting diode luminance applications *Mater. Chem. Phys.* **229** 22–31

[123] Vaidhyanathan B and Binner J G P 2006 Microwave assisted synthesis of nanocrystalline YAG *J. Mater. Sci.* **41** 5954–7

[124] Aboulaich A *et al* 2015 In situ synthesis of a highly crystalline Tb-doped YAG nano-phosphor using the mesopores of silica monoliths as a template *J. Mater. Chem.* C **3** 5041–9

[125] Lu C H, Hong H C and Jagannathan R 2002 Sol–gel synthesis and photoluminescent properties of cerium-ion doped yttrium aluminium garnet powders *J. Mater. Chem.* **12** 2525–30

[126] Sun Z *et al* 2004 Preparation and characterization of Co^{2+}-doped $Y_3Al_5O_{12}$ nano-crystal powders by sol–gel technique *J. Cryst. Growth* **260** 171–5

[127] Zhang J J, Ning J W, Liu X J, Pan Y B and Huang L P 2003 Synthesis of ultrafine YAG:Tb phosphor by nitrate–citrate sol–gel combustion process *Mater. Res. Bull.* **38** 1249–56

[128] Dantelle G *et al* 2018 A new solvothermal method for the synthesis of size-controlled YAG: Ce single-nanocrystals *RSC Adv.* **8** 26857–70

[129] Asakura R *et al* 2007 Effects of citric acid additive on photoluminescence properties of YAG:Ce^{3+} nanoparticles synthesized by glycothermal reaction *J. Lumin.* **127** 416–22

[130] Nyman M, Shea-Rohwer L E, Martin J E and Provencio P Nano-YAG:Ce mechanisms of growth and epoxy-encapsulation

[131] Odziomek M, Chaput F, Lerouge F, Sitarz M and Parola S 2017 Highly luminescent YAG: Ce ultra-small nanocrystals, from stable dispersions to thin films *J. Mater. Chem.* C **5** 12561–70

[132] Aboulaich A *et al* 2012 Rapid synthesis of Ce^{3+}-doped YAG nanoparticles by a solvothermal method using metal carbonates as precursors *New J. Chem.* **36** 2493–500

[133] Wiatrowska A, Keur W and Ronda C 2017 Single and multicomponent garnets nanoparticles—synthesis and luminescence *J. Lumin.* **189** 9–18

[134] Kamiyama Y *et al* 2010 Photostability of YAG : Ce3 + Nanophosphors Synthesized by Glycothermal Method *J. Electrochem. Soc.* **157** J149

[135] Sasaki Y 2016 Preparation, Photoluminescence, and Photostability of Transparent Composite Films of Glycothermally Synthesized YAG:Ce3+ Nanoparticles for White LED *ECS J. Solid State Sci. Technol.* **5** R3049

[136] Kim O H, Ha S W, Kim J I and Lee J K 2010 Excellent photostability of phosphorescent nanoparticles and their application as a color converter in light emitting diodes *ACS Nano* **4** 3397–405

[137] Lo K K W, Ng D C M and Chung C K 2001 First examples of luminescent cyclometalated iridium(III) complexes as labeling reagents for biological substrates *Organometallics* **20** 4999–5001

[138] Burns A, Ow H and Wiesner U 2006 Fluorescent core–shell silica nanoparticles: towards 'lab on a particle' architectures for nanobiotechnology *Chem. Soc. Rev.* **35** 1028–42

[139] Wang F, Tan W B, Zhang Y, Fan X and Wang M 2006 Luminescent nanomaterials for biological labelling *Nanotechnology* **17** R1

[140] Rampazzo E, Bonacchi S, Montalti M, Prodi L and Zaccheroni N 2007 Self-organizing core–shell nanostructures: spontaneous accumulation of dye in the core of doped silica nanoparticles *J. Am. Chem. Soc.* **129** 14251–6

[141] Ow H *et al* 2005 Bright and stable core–shell fluorescent silica nanoparticles *Nano Lett.* **5** 113–7

[142] van Blaaderen A and Vrij A 1993 Synthesis and characterization of monodisperse colloidal organo-silica spheres *J. Colloid Interface Sci.* **156** 1–18

[143] Rossi L M, Shi L, Quina F H and Rosenzweig Z 2005 Stöber synthesis of monodispersed luminescent silica nanoparticles for bioanalytical assays *Langmuir* **21** 4277–80

[144] Ha S-W, Camalier C E, Beck G R Jr and Lee J-K 2009 New method to prepare very stable and biocompatible fluorescent silica nanoparticles *Chem. Commun.* 2881–3

[145] Rico M *et al* 2012 A stable luminescent hybrid mesoporous copper complex–silica *Chem. Commun.* **48** 8883–5

[146] Valenti G *et al* 2016 Variable doping induces mechanism swapping in electrogenerated chemiluminescence of $Ru(bpy)_3^{2+}$ core–shell silica nanoparticles *J. Am. Chem. Soc.* **138** 15935–42

[147] MØLLER C H R K N 1958 Crystal structure and photoconductivity of caesium plumbohalides *Nature* **182** 1436

[148] Zhang F *et al* 2015 Brightly luminescent and color-tunable colloidal $CH_3NH_3PbX_3$ (X = Br, I, Cl) quantum dots: potential alternatives for display technology *ACS Nano* **9** 4533–42

[149] Mondal N, De A and Samanta A 2018 All-inorganic perovskite nanocrystal assisted extraction of hot electrons and biexcitons from photoexcited CdTe quantum dots *Nanoscale* **10** 639–45

[150] Yettapu G R *et al* 2016 Terahertz conductivity within colloidal $CsPbBr_3$ perovskite nanocrystals: remarkably high carrier mobilities and large diffusion lengths *Nano Lett.* **16** 4838–48
[151] Lin C C *et al* 2018 Water-resistant efficient stretchable perovskite-embedded fiber membranes for light-emitting diodes *ACS Appl. Mater. Interfaces* **10** 2210–5
[152] Cha W, Kim H J, Lee S and Kim J 2017 Size-controllable and stable organometallic halide perovskite quantum dots/polymer films *J. Mater. Chem.* C **5** 6667–71
[153] Li G *et al* 2016 Solvent-polarity-engineered controllable synthesis of highly fluorescent cesium lead halide perovskite quantum dots and their use in white light-emitting diodes *Adv. Funct. Mater.* **26** 8478–86
[154] Huang S *et al* 2016 Enhancing the stability of $CH_3NH_3PbBr_3$ quantum dots by embedding in silica spheres derived from tetramethyl orthosilicate in 'waterless' toluene *J. Am. Chem. Soc.* **138** 5749–52
[155] Sun C *et al* 2016 Efficient and Stable White LEDs with Silica-Coated Inorganic Perovskite Quantum Dots *Adv. Mater.* **28** 10088–94
[156] Xuan T *et al* 2017 Highly stable $CsPbBr_3$ quantum dots coated with alkyl phosphate for white light-emitting diodes *Nanoscale* **9** 15286–90
[157] Ding N *et al* 2018 Highly stable and water-soluble monodisperse $CsPbX_3/SiO_2$ nanocomposites for white-LED and cells imaging *Nanotechnology* **29** 345703
[158] Bidikoudi M, Fresta E and Costa R D 2018 White perovskite based lighting devices *Chem. Commun.* **54** 8150–69
[159] Li Z, Kong L, Huang S and Li L 2017 Highly luminescent and ultrastable $CsPbBr_3$ perovskite quantum dots incorporated into a silica/alumina monolith *Angew. Chem. Int. Ed.* **129** 8246–50
[160] Lemaire A, Rooke J C, Chen L H and Su B L 2011 Direct observation of macrostructure formation of hierarchically structured meso-macroporous aluminosilicates with 3D interconnectivity by optical microscope *Langmuir* **27** 3030–43
[161] Hu H *et al* 2018 Interfacial synthesis of highly stable $CsPbX_3$/oxide janus nanoparticles *J. Am. Chem. Soc.* **140** 406–12
[162] Xia G, Qiao-Li L and Chong L 2008 Polychromatic light-emitting diodes with a fluorescent nanosphere opal coating *Nanotechnology* **19** 355203
[163] Hui K N, Lai P T and Choi H W 2008 Spectral conversion with fluorescent microspheres for light emitting diodes *Opt. Express* **16** 13
[164] Zhang Y *et al* 2019 Doping of ZnO inorganic–organic nanohybrids with metal elements *Sci. Rep.* **9** 1–10
[165] Felbier P *et al* 2014 Highly luminescent ZnO quantum dots made in a nonthermal plasma *Adv. Funct. Mater.* **24** 1988–93
[166] Prochowicz D, Sokołowski K and Lewiński J 2014 Zinc hydroxides and oxides supported by organic ligands: synthesis and structural diversity *Coord. Chem. Rev.* **270–1** 112–26
[167] Nishikiori H, Natori D, Ebara H, Teshima K and Fujii T 2016 Zinc complex formation of organic ligands on zinc oxide and titanium dioxide *J. Photochem. Photobiol.* A **327** 51–7

Chapter 10

Oxide electronic devices

David Rogers, Ferechteh Teherani, Eric Sandana and Philippe Bove

The topic of oxide-based electronic devices encompasses an enormous range of materials and applications. A number of these have already been addressed in chapters 1–9, so this chapter focuses on two specific topics that have made significant progress over the past few years: amorphous oxide semiconductors (AOSs) and their use in flexible electronics (section 10.1) along with the use of gallium oxide in power electronic devices (section 10.2). In both cases, we introduce the underlying materials development and provide some examples of applications and state-of-the-art device realizations. These include various device architectures along with a discussion of the need, status, and potential for p–n based devices.

10.1 Oxide-based flexible electronics

Oxide-based flexible electronics is a vast and dynamic field with multiple technologies/applications and thousands of innovative device demonstrations to date. In this section we give an overview of the current status of this domain for the general reader. For fuller details and explanations of specific device functioning/performance, we refer the reader to a number of comprehensive recent review papers [1–8] and the bibliography below.

10.1.1 Amorphous oxide semiconductors for flexible electronics

Although the definition of flexibility varies from application to application (e.g. from rollability in displays/solar cells to bendability in electronic paper (e-paper), foldability in cell phone screens, and conformability/stretchability/twisting in electronic skin), the flexibility of a material is generally inversely proportional to the material's thickness. Hence, thin films and thin film transistors (TFTs) have become quintessential components for use in flexible electronics.

Many different kinds of material have been investigated for use as channels in flexible TFTs. Hydrogenated amorphous silicon (a-Si:H) thin films have been widely used as driving elements in displays [9, 10]. Compared to crystalline silicon

doi:10.1088/978-0-7503-3531-7ch10 © IOP Publishing Ltd 2023 (c-Si), a-Si:H offers better flexibility and better uniformity of its properties (linked to the lack of grain boundaries and related issues). a-Si:H also has a lower fabrication temperature (<300 °C), which makes it compatible with inexpensive heat-sensitive flexible substrates such as plastic/polymers. a-Si:H is opaque, however, offers poor stretchability, exhibits instability under high gate bias (due to the creation of dangling bonds and charge trapping) and its carrier mobility (and thus its operational speed in TFTs) is rather limited. It also generally requires a vacuum-based, hydrogen plasma exposure in order to show acceptable mobilities and on–off ratios.

There have also been many studies of organic TFTs on account of their inherent combination of flexibility and stretchability [11–14], their low weight, and their large-scale/low-temperature fabrication potential (<120 °C) along with the large range of potential molecules/properties to choose from (including the availability of both n- and p-type forms). Under certain circumstances, organic semiconductors have also been shown to be healable [15] and biodegradable [16–18]. However, like a-Si:H, organic semiconductors also exhibit relatively low mobilities and often have stability challenges [19, 20].

Two-dimensional materials form another very promising flexible electronics materials platform [21, 22]. They offer inherent flexibility/transparency along with excellent materials quality and can tolerate extreme deformation. They also comprise a large portfolio of potential materials with highly tunable properties. The most explored 2D flexible devices are based on graphene, transition-metal dichalcogenides (e.g. MoS_2, WS_2, $MoSe_2$ and WSe_2), hexagonal BN (h-BN) and black phosphorous. Various high-performance devices including fast field-effect transistors (FETs), sensitive sensors and supercapacitors have been reported. Moreover, fabrication by exfoliation, low-temperature, large-area chemical vapor deposition (CVD) growth on heat-sensitive polymer substrates and roll-to-roll production of large-area, flexible graphene electrodes have already been demonstrated. Durability and the reproducible production of continuous/homogeneous layers with stringent thickness variation tolerances remain challenges, however.

Nanostructures are also a candidate basis for flexible electronics [23, 24] (e.g. silicon nanosprings [25], ZnO nanowires [26] and single-wall carbon nanotubes [27]). These offer intrinsic advantages such as excellent flexibility and transparency as well as the potential for low-temperature fabrication, high carrier mobilities and widely tuneable property sets. Their manufacture is complex/costly, however, and there are ongoing issues with uniformity, scalability, and reproducibility [23].

Recently, a new kind of flexible oxide semiconductor has emerged: AOSs [28]. Like a-Si:H, AOSs offer superior flexibility, better uniformity of their properties and the possibility of much lower-temperature fabrication compared to crystalline oxides or crystalline silicon (c-Si) [29]. AOSs also avoid the need for the plasma-assisted deposition that is required for a-Si:H and thus offer the potential for simplified production using only lower-cost non-vacuum-based approaches (e.g. solution processing). Moreover, AOSs show excellent stability/reproducibility of properties during and after repetitive bending, so they are compatible with flexible substrates/electronics. Compared with a-Si:H, AOSs can also exhibit enhanced carrier mobility (probably due to reduced carrier trapping [30]), superior processability in air and improved thermodynamic stability [31].

These AOSs are generally ternary or quaternary alloys. This is because most binary oxides (e.g. ZnO and In_2O_3) tend to form crystalline films, even when deposited at room temperature, whereas the mixing of two or more cations with different ionic charges and sizes promotes the formation of an amorphous phase and suppresses crystallization. Hence, certain multielement alloys have been found to yield AOSs with crystallization temperatures of >500 °C [29].

Amorphous InZnO (a-IZO) is currently considered to be the ternary AOS flexible transparent conductor system of reference [31, 32]. The resistivity of a-IZO can be varied very widely (from 10^{-4} to 10^{8} Ω cm) by controlling the cation composition and cation/oxygen stoichiometry [33]. a-IZO can also exhibit a mobility of more than 50 $cm^2\ V^{-1}\ s^{-1}$, and Sn doping can produce even higher mobilities (>100 $cm^2\ V^{-1}\ s^{-1}$ [34]) along with transparencies that are comparable with state-of-the-art crystalline transparent conductors such as indium-doped tin oxide (ITO) [35].

10.1.2 AOS-based field-effect transistors on flexible substrates

A new generation of FETs employing AOS-based channels has recently emerged. The AOSs used as channels for FETs need as low a carrier concentration as possible in order to achieve stable performance [36]—ideally <~10^{14} cm^{-3} [37]. This can be achieved by introducing an extra component into a-IZO in order to suppress carriers. Gallium (Ga) is the most common carrier suppressor adopted for the a-IZO system [37, 38], hence amorphous InGaZnO (a-IGZO) has become the reference in AOS TFT channel materials. However, increased gallium doping suppresses the relatively high carrier mobility (>20 $cm^2\ V^{-1}\ s^{-1}$ [39]) of In-rich a-IGZO [40]; thus, a compositional trade-off is required.

In 2004, Nomura *et al* demonstrated a flexible a-IGZO-based Metal–Insulator–Semiconductor FET (MISFET) on a polymer substrate for the first time [31]. Figure 10.1 shows a schematic of the device. Figure 10.1 shows a schematic of the device. The transparent channel layer, gate dielectric, and electrodes (gate, drain, and source) were all deposited at room temperature using pulsed laser deposition (PLD). Figure 10.2 shows the transfer and output characteristics of the front-gated AOS MISFET shown in figure 10.1 before and after bending. The transfer characteristics are rectifying, confirming transistor behavior in the device. The turn-on voltage (V_{ON}) is about 0 V, which suggests that neither the electron nor the trap densities in the channel are too high. The device shows a relatively large drain current (I_d) with good pinch-off and a subsequent plateau, which indicates that the channel is almost depleted of carriers. The saturation mobility is ~6–9 $cm^2\ V^{-1}\ s^{-1}$, the leakage current is about 10^{-10} A and the on/off ratio is about 10^3. The characteristics were similar during and after bending down to a radius of curvature of 30 mm.

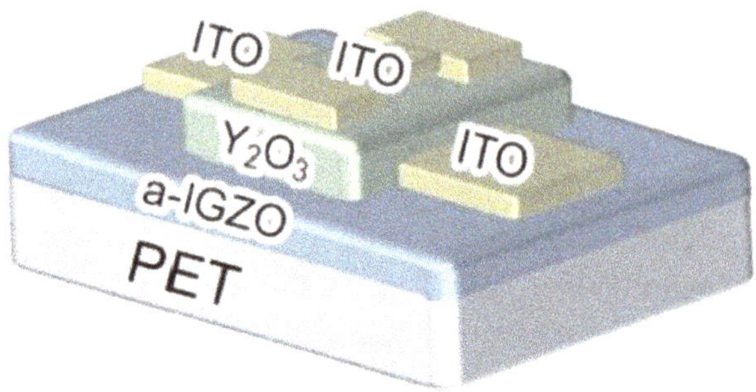

Figure 10.1. A front-gated MISFET with a polyethylene terephthalate (PET) substrate, an a-IGZO channel (30 nm thick), a Y_2O_3 gate (140 nm thick), and an ITO electrode (40 nm thick) [31] (2004) (Copyright © 2004, Macmillan Magazines Ltd.). With permission of Springer.

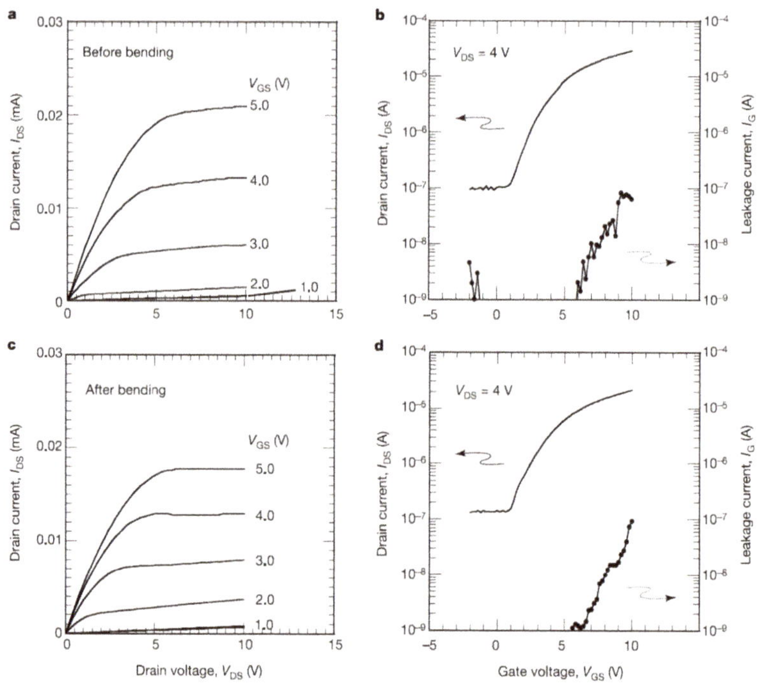

Figure 10.2. Output (left) and transfer (right) characteristics (both before (top) and after bending (bottom)) for the back-gated oxide MISFET of figure 10.1 [31] (2004) (Copyright © 2004, Macmillan Magazines Ltd). With permission of Springer.

A key advantage of such transistors is that they are (low leakage) enhancement mode (E-mode) devices (they consume no power in the off state, i.e. when the gate–source voltage (V_{gs}) = 0 V). This is preferable to the depletion mode (D-mode) alternative, since circuit design is simpler and power dissipation is lower (because the off state is the power-down state).

Following the work of Nomura *et al*, flexible a-IGZO-based TFT devices have been demonstrated on a number of other polymer substrates including polyimide

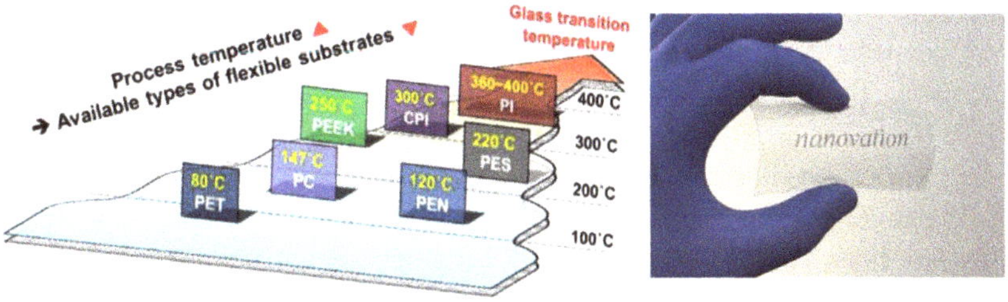

Figure 10.3. (left) Maximum process temperatures for common flexible polymer substrates: PET, PI, polycarbonate (PC), polyether ether ketone (PEEK), PES, colorless PI (CPI), and PEN [6] John Wiley & Sons. [© 2019 WILEY-VCH Verlag GmbH & Co. KGaA, Weinheim]. (right) An image of a flexible/transparent PET substrate (www.nanovation.com).

(PI) [41, 42], polyethylene naphthalate (PEN) [43] and polyestersulfone (PES) [44]. These devices have achieved much higher on–off ratios (up to $>10^8$), higher I_d (into the mA range) and channel mobilities of up to about 16 $cm^2\ V^{-1}\ s^{-1}$ [45] (compared to $\mu < 1\ cm^2\ V^{-1}\ s^{-1}$ for comparable devices using a-Si:H channels). Of particular note in studies of TFTs on ultrathin parylene substrates is that a-IGZO-based TFTs can function reproducibly after repeated bending down to radii of curvature as low as 50 μm [46].

In addition to polymers and plastics, paper has attracted increasing attention for use as a compliant substrate for flexible TFTs. Paper offers the advantages of being robust, ubiquitous, lightweight, low-cost, renewable and recyclable/biodegradable [47–49]. Paper has a relatively rough surface, however, and it is hygroscopic. Therefore, a coating (e.g. cyclotene [50], SiO_2 [51, 52] or epoxy acrylate copolymer [53]) is usually applied in order to both prevent water/solvent absorption during TFT fabrication and reduce the surface roughness.

Although polymer/plastic/paper substrates are cheap, have large areas and are compatible with low-temperature wet chemical TFT fabrication, they all impose significant processing temperature limitations (see figure 10.3) which have turned out to be a problem if the maximal electrical performance is required from AOS channels. To obtain maximal performance, a thermal annealing step (ideally at a temperature >300 °C) is usually required [54–58].

An alternative low-temperature approach used to realize the necessary property optimization for a-IGZO TFTs is laser annealing [59]. In laser annealing, a short burst of laser energy (an excimer laser pulse typically tens of ns in duration) is locally absorbed by the top few hundred nanometers of the layers and thus avoids detrimental substrate heating. This method has shown promising results and is also much faster than thermal annealing.

Another approach that has been reported to avoid thermal damage to the substrate during annealing is microwave annealing (MWA) [58]. In MWA, the AOS volumetrically absorbs electromagnetic energy and converts it into heat. Because the sample is directly heated by energy transfer rather than heat transfer, MWA is a cost-effective heat treatment method with no energy wastage. In addition, the heat ramp-up and ramp-down times of MWA are relatively rapid, which significantly reduces thermal budgets. Min *et al* [58] reported that after a 2 min 1800 W MWA treatment, layer properties were significantly improved and PI substrates did not show the discoloration, cracking, or change in optical transmittance commonly observed as a result of thermal annealing.

Another approach that has been demonstrated to avoid substrate thermal damage is annealing under high (2 MPa) O_2 gas pressure at reduced temperature (100 °C) [60]. This has been found to give a-IGZO property improvements equivalent to a those of conventional 300 °C thermal annealing in air.

Alternatively, the annealing step can be directly incorporated through the adoption of different kinds of flexible substrate which are inherently more temperature resistant than polymers. For instance, flexible foils of metals such as aluminum, copper, gold and silver exhibit both excellent mechanical robustness and the potential to support higher process temperatures [61]. Moreover, the ductility of such metal foils accommodates an extremely small radius of curvature and thus opens up applications with higher strain, such as thermal and pressure sensors integrated onto artificial skin [62].

Metals are not suitable for applications requiring transparency, however, and flexible thin glass (a few tens of microns thick) is often adopted for this. High-performance transparent/flexible TFTs based on annealed a-IGZO have been demonstrated on thin glass [63].

10.1.3 Current and emerging flexible electronics applications

Figure 10.4 shows estimates of the evolution of the flexible electronics market (broken down by component) from 2019 to 2027 (adapted from Allied Market Research (www.alliedmarketresearch.com)).

The total market was estimated to be \$23.6 billion in 2019 and projected to reach \$42.5 billion by 2027, based on a compound annual growth rate (CAGR) of 7.4% from 2020 to 2027. This growth estimate was justified by a number of factors, including a general rise in demand for portable and smart electronics products, growth in government funding for flexible electronics and sales of printed electronics plus Internet of Things devices.

With over 90% of the market share, the current leading application is flexible displays (comprising active-matrix organic light-emitting diodes (AM-OLEDs), active-matrix liquid crystal displays (AM-LCDs), and e-paper) (figure 10.5).

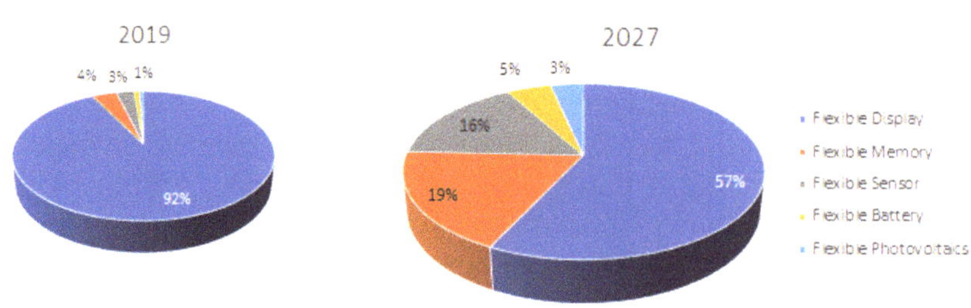

Figure 10.4. An estimate of the flexible electronics market broken down by component (adapted from Allied Market Research (www.alliedmarketresearch.com)).

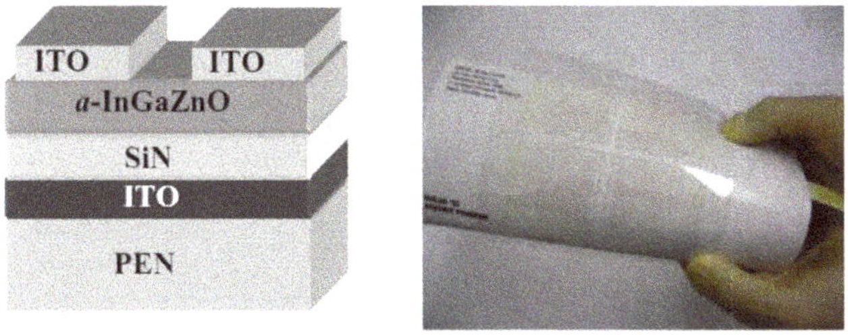

Figure 10.5. A schematic and an image of a transparent/flexible a-IGZO TFT active-matrix display backplane fabricated on a PEN film [64] John Wiley & Sons.

a-IGZO-based FETs are already well established for use as backplane transistors in rigid AM-OLED and AM-LCD screens. These 'select' transistors toggle between transmitting or blocking the backlight for each pixel. The higher mobility of AOS-based FETs offers increased refresh rates compared those of the competing a-Si:H-based devices, and their wide bandgap means that AOS-based FETs: (a) have reduced light sensitivity (lower noise), (b) have less degradation, (c) can support higher device currents (and thus higher driving powers), and (d) show improved transparencies, compared with their opaque a-Si:H counterparts, thus allowing more backlight to be transmitted, yielding a brighter and more efficient (lower power consumption) display. This also means that the aperture ratio can be reduced, so as to yield smaller pixels and higher resolution. The transparency also makes AOS channels compatible with use as front-driven 'select' FETs for e-paper [64].

According to Applied Markets Research, flexible memories will be the next biggest flexible electronics application. Indeed, many flexible electronic systems will require flexible memories with low power consumption and robust bending performance. A number of such systems have been proposed, including a conductive-bridging random access memory based on a Cu/TiW/a-Ga_2O_3/Pt layer stack on PI [56], a flexible charge-trap-type a-IGZO memory TFT on a PEN substrate [65], and a selective floating-gate non-volatile memory transistor employing cellulose paper as both the gate dielectric and the substrate [66, 67] (see figure 10.6).

The next application, in terms of market share, is flexible sensors, which are either based on separate sensing elements which activate AOS TFT switches, or sensing performed directly by an AOS-based sensor (i.e. without an associated AOS TFT). These low-cost, wide-area flexible sensors have many application areas, such as pH sensing [68–71] (figure 10.7), x-ray detection [72–74], physiological/biosensing [75, 76], temperature/pressure sensing [77, 78], photosensing [79], UV sensing [80, 81] (figure 10.8), and chemical and ionic sensing [82–84]. Wearable/implantable sensors for health monitoring/care [88, 89] are of particular interest, as are electronic skin sensors for medical robotics [90].

The final two current market applications identified by Allied Market Research (flexible photovoltaics and flexible batteries) do not currently involve AOS TFTs.

Beyond these current commercial applications, there have also been promising demonstrations of unipolar flexible electronic circuits integrating a-IGZO TFTs

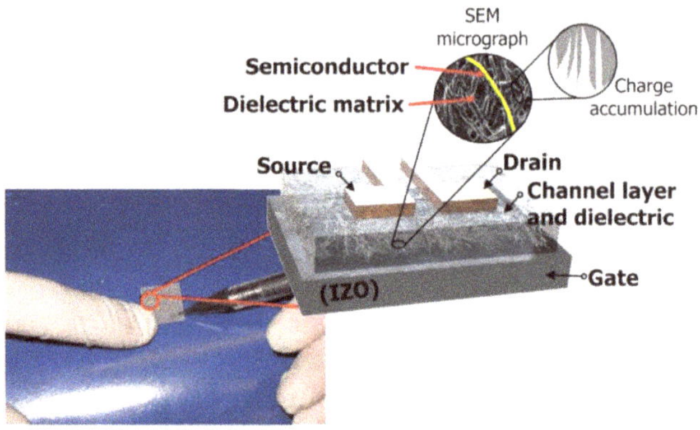

Figure 10.6. An image and a schematic of a selective floating-gate non-volatile memory a-IGZO TFT employing cellulose paper as both the gate dielectric and the substrate. © [2008] IEEE. Reprinted, with permission, from [67].

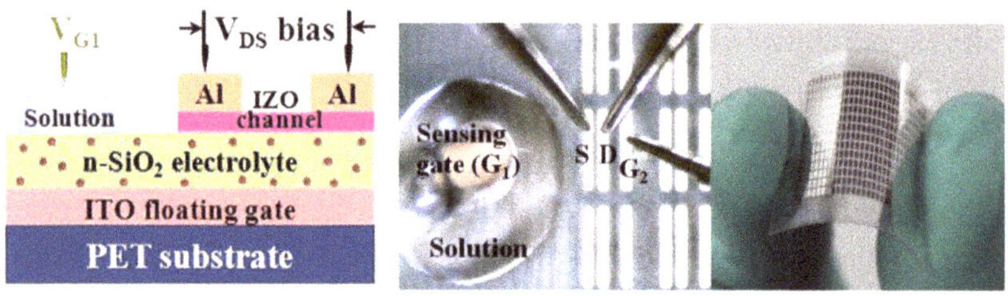

Figure 10.7. Schematic and images of a flexible AOS-channel-TFT-enabled pH sensor. Reproduced from [70].

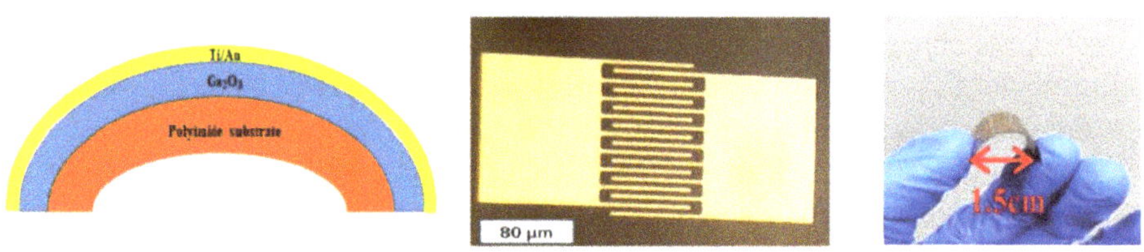

Figure 10.8. Schematic and images of a flexible solar blind metal–semiconductor–metal photodetector based on a-Ga_2O_3 (one of the few binary oxides to show good stability of the amorphous phase up to several hundred °C). © [2019] IEEE. Reprinted, with permission, from [80]. The performance of such a-Ga_2O_3 based UV photodetectors is reported to be comparable to, or even better than, those of their crystalline counterparts [85–88].

[91], including operational amplifiers and logic circuits [92, 93], neuromorphic chips [70], RFID transponder chips [43, 94], clock generators [95] and analog/RF circuits [96] (figure 10.9).

These have extended to flexible complementary circuits [97], including recyclable CMOS inverter architectures on paper substrates [98] and combinations of p-organic

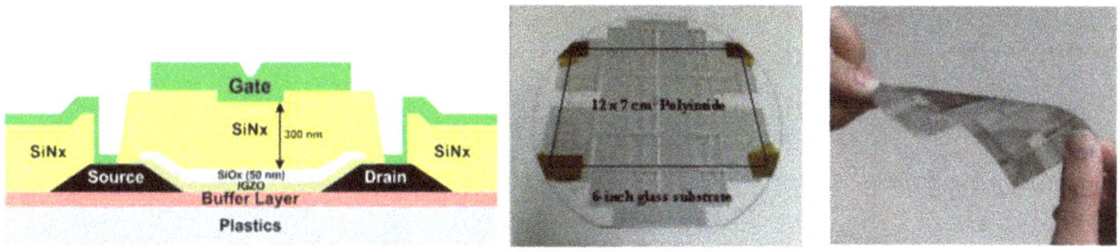

Figure 10.9. Schematic and images of flexible a-IGZO based pseudo CMOS circuitry on a PI substrate (images before and after lift-off from a rigid glass support) © [2016] IEEE. Reprinted, with permission, from [95].

and n-AOS TFTs [99–105]; examples include the use of p-type self-assembled-monolayer organic TFTs in combination with n-type a-IGZO-based TFTs [10] or hybrid complementary inverters composed of p-type pentacene and n-type a-IGZO TFTs [97].

10.1.4 Future prospects for AOS-based flexible electronics

With regards to the future, several significant tendencies can be identified.

First of all, there is a trend toward the use of non-toxic, cheaper, and non-resource-critical (sustainable) materials in the AOS channel. Hence, In- and Ga-free alternatives are being explored. a-ZnSnO and a-MgZnO, in particular, are showing promising results [106].

Next, there are significant innovations with regards to the dielectric. In metal–insulator–semiconductor (MIS) transistors, the device resistance is controlled by the electric field, which is generated by the gate voltage in the capacitor formed by the MIS structure. Therefore, the quality/uniformity of the insulating dielectric and the smoothness/abruptness of the semiconductor/dielectric interface are critical. A variety of gate insulators (including SiO_2, SiN_x, $SiON_x$,Y_2O_3, HfO_2, Al_2O_3 [107], (Ba,Sr)TiO_3 [108], and organics) have been examined for use in AOS TFTs, and all gave reasonable performances [29]. This tolerance toward various dielectrics was attributed partly to the unipolar nature of these devices (making the valence band offset between the gate insulator and the AOS less critical) and partly to the highly ionic nature of the AOS chemical bonds (which leads to fewer defects at the gate insulator/AOS interfaces) [29].

As well as such conventional inorganic and organic dielectric gates, electrolyte gates have also been investigated [109–111]. In these electrolytes, the electric field across the gate displaces anions and cations towards the oppositely charged electrodes, thus forming an electric double layer (EDL) which resembles the charged plates of a parallel-plate capacitor. This short-range charge separation gives a boost in capacitance, which is favorable for lower-power TFT operation [112]. Various EDL-AOS TFTs have been developed of late using ionic liquids, polymer electrolytes, ionic gels and inorganics as EDL dielectrics [113–115]. It has generally been concluded that they hold particular promise for biological sensing, low-cost flexible displays and logic circuitry applications.

Electrodes are another area of innovation. In addition to the requirement for flexibility, the gate, source, and drain electrodes in flexible TFTs should have low specific electrical resistance and good mechanical stability. Metals are the usual choice for this. If the application requires transparency, however, metals are unsuitable, and the most widely adopted alternative is crystalline transparent conductive oxides (TCOs). However, the mechanical stability of crystalline oxides under repeated bending remains an issue, so there is a trend toward the adoption of AOS-based electrodes [116–119].

The physical and chemical vapor deposition techniques that are currently the standard for the commercial fabrication of high-performance flexible AOS TFTs require vacuum equipment which is relatively expensive and complex. Therefore, the potential for solution-based fabrication of flexible AOS TFTs at atmospheric pressure has generated huge interest. Indeed, solution-based fabrication offers many advantages, including high throughput, wide-area synthesis and improved control of compositional homogeneity. It has been estimated that fabrication costs could be lowered by >50% using printing approaches because such processes do not require vacuum, gases, or lithography [120]. Moreover, it has been demonstrated that a laser annealing step can be compatible with high-throughput roll-to-roll printing [121] via wide-area annealing in which laser beams are shaped into a line extending across the width of the roll.

Another active development front is that of novel device architectures. For instance, oxide-based metal–semiconductor FETs (MESFETs) [122] have been demonstrated to operate in a lower voltage regime (± 1 V) than that of MISFETs. Thus, they should be better suited to driving display pixels. A method for obtaining fully transparent MESFET devices (including all contacts) has also been developed.

Bipolar devices could offer a number of advantages over FETs for flexible electronics. These include lower reverse current, lower voltage/power operation, faster response, smaller footprint, less current crowding, higher noise margin, higher logic swing output, higher circuit integration density, simpler architecture, and greater ease of fabrication. There is a problem in this regard, however, in that the majority of oxides and, in particular, those that currently form the backbone of AOS-based flexible electronics (ZnO, Ga_2O_3, In_2O_3, and SnO_2) only exist as unipolar materials. Indeed, the vast majority of oxide semiconductors (in either crystalline or amorphous form) are only available in n-type form. This is because the creation of holes is often limited by: (i) a lack of shallow acceptors; (ii) the high formation energy of native acceptors that produce holes, such as cation vacancies; and (iii) the relatively low formation energy of background donors that compensate acceptors, e.g. native defects (oxygen vacancies and/or cation interstitials) and common impurities.

There are, nevertheless, some oxide semiconductors that are natively p-type (e.g. NiO, SnO, delafossites, Cu_2O). Unfortunately, these also lack their n-type counterparts, so AOS homojunction p–n devices have not emerged.

In the absence of p–n AOS homojunctions, the community has turned its efforts to the creation of heterojunction bipolar devices. In order for these to function well, the n- and p-type layers need to have appropriate bandgaps, band alignments, carrier concentrations, and mobilities. There is a problem in achieving p-type oxide semiconductors with similar mobilities to those of n-type ones, however. This arises because of the intrinsic features of the electronic configuration. Even when a shallow acceptor is available (e.g. the Ni vacancy in the case of cubic NiO) and a relatively high concentration of holes is possible, the valence band maximum (VBM), is usually comprised mainly of anisotropic and localized oxygen 2*p* orbitals, which result in a large effective mass, hopping hole transport, and relatively low mobility.

There have been some demonstrations of heterojunction bipolar AOS devices, however. These include fully amorphous bipolar oxide diodes based on n-zinc tin oxide/p-zinc $cobalt_2$ $oxide_4$ (ZTO/ZCO) junctions with a rectification $>10^6$ and an ideality close to one [123].

The low-cost fabrication and control of stable, reproducible, and homogeneous wide-area p–n AOS heterojunctions with sharp interfaces remains a challenge, however.

10.2 Gallium-oxide-based power electronics

Around half of the energy used by humankind is in the form of electricity and this figure is increasing rapidly, due, in no small part, to the accelerating adoption of renewable energy technologies and the trend toward the electrification of a huge panoply of everyday objects [124]. It is a little-known fact that more than 30% of this electricity passes through at least one power electronic device during its generation, transmission, and final use [125]. Moreover, it is forecast that this could grow to 80% over the next decade due to the increasing use of power devices in applications such as power management, electric vehicles, photovoltaic inverters, RF switching, and power supplies (e.g. servers and IT equipment).

According to Allied Market Research (www.alliedmarketresearch.com) the global power electronics market size was estimated to be about $23.3 billion in 2019 and to grow to $36.6 billion by 2027 (i.e. a CAGR of 5.7% from 2020 to 2027).

Today, silicon-based devices form the basis of the vast majority of these power electronics components (>95% in 2019, according to Informatech). Although silicon-based devices work well and are relatively low cost, they are comparatively inefficient/lossy and are plagued by intrinsic running temperature and power-handling limitations when compared to devices based on wide-bandgap (WBG) semiconductors (i.e. $E_g > 3$ eV).

Hence, a lot of effort has been made over the past few decades to replace silicon-based ($E_g = 1.12$ eV) power electronics devices with WBG silicon carbide (SiC) and gallium nitride (GaN) based devices. WBG devices have developed enormously over this period and there are now commercial components offering much higher energy efficiencies, power-handling capacities, and maximum running temperatures (i.e. with fewer cooling requirements) than those of the incumbent silicon-based components [126]. WBG semiconductors still suffer from some important limitations, however, such as wafer scale-up issues, limited breakdown voltages (V_{br}), and relatively high costs.

Recently, the ultimate performance frontier in the field of power electronics has been challenged by a new generation of power devices based on ultra-wide-bandgap

WBG	Bandgap E_g [eV]	Permitt. ε_r [ε_0]	Mobility μ_n [cm²/Vs]	Crit. Field E_c [MV/cm]	BFOM x10⁶ [V²/Wcm²]	Ther. Cond. k [W/mK]
Si	1.12	11.9	1240	0.3	8.8	145
4H-SiC	3.20	9.7	980	3.1	6270	350
GaN	3.40	10.4	1000	4.9	27900	140
β-Ga_2O_3	4.90	10.0	150	10.3	36300	27
Diamond	5.50	5.7	2000	13.0	554000	3450
AlN	6.00	9.8	426	15.4	336000	319
c-BN	6.40	7.1	825	17.5	695000	2145

Figure 10.10. A comparison of the critical characteristics of WBG and UWBG power electronics materials. Reproduced with permission from [127]. © 2018 Society of Photo-Optical Instrumentation Engineers (SPIE).

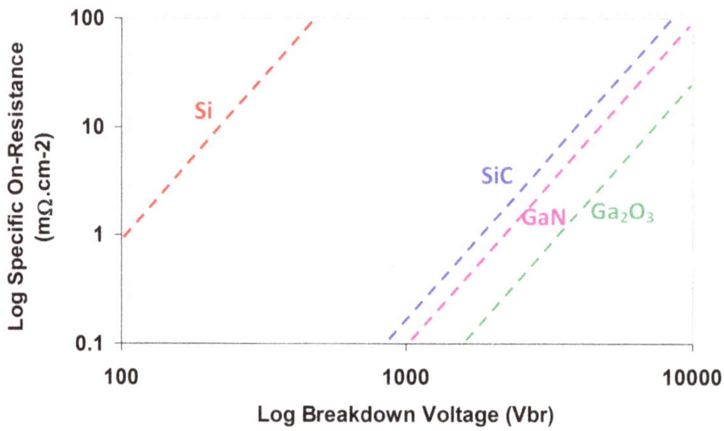

Figure 10.11. Specific on-resistance (R_{on}) vs breakdown voltage (V_{br}) for common (ultra) wide-bandgap semiconductors [128]. It should be noted that recent theoretical work [129] suggests that these R_{on} vs V_{br} curves should be revised to account for the effects of incomplete dopant ionization, impact ionization coefficients, and dielectric constants. This approach suggests that V_{br} should, in fact, be significantly higher for Ga_2O_3, due largely to a particularly low impact ionization coefficient. Moreover, it was also recently suggested that the critical electric field of Ga_2O_3 could actually be as large as 13.2 MV cm^{-1} if residual donors are efficiently removed. Reproduced with permission from [128]. © 2021 Society of Photo-Optical Instrumentation Engineers (SPIE).

(UWBG) semiconductors (E_g > 4.5 eV), which have the theoretical potential to further reduce power losses and upshift both the power-handling capacity and the operating temperature (figures 10.10 and 10.11).

Although diamond has very promising characteristics on paper, it has been investigated for use in power electronics for over 40 years, and there has been relatively limited progress in fabricating devices, scaling up native substrates, and obtaining high-quality wide-area epitaxy. Recently, other UWBG materials, such as gallium oxide (Ga_2O_3), aluminum nitride (AlN), and boron nitride (BN), have started to yield compelling device demonstrations. Unfortunately, AlN has proven difficult to dope, while BN epitaxy remains relatively poor and native substrates are not available. Hence, much attention for next-generation power devices has turned to Ga_2O_3 [130, 131].

10.2.1 An overview of β-Ga_2O_3

There has recently been a sharp rise in the number of publications focused on Ga_2O_3 (figure 10.12). This exponential increase in research interest is due to its distinctive and engineerable property set, combined with a number of recent breakthroughs, including the development of relatively large single-crystal substrates (up to 6 inches in diameter) [132–135], the demonstration of bandgap engineering (from 4.9 up to 6.2 eV) [136, 137] (figure 10.13) and the demonstration of tunable n-type doping with reasonably high mobilities [138, 139] (figure 10.14).

Indeed, it was demonstrated over the past decade that β-Ga_2O_3 could be n-doped from low intrinsic background levels (approaching 10^{13} carriers per cm^3) [141] up to degenerate transparent conductor levels (over 10^{20} carriers per cm^3) [140]. Known shallow donors include Si [142], Ge and Sn [143] (substituted on the Ga site) and F or Cl (substituted on the O site) [144].

These properties/potentialities have generated considerable interest in the adoption of β-Ga_2O_3 in a number of important technological applications ranging from solar transparent electrodes [145] and gas sensors [146] to UVC photodetectors [147, 148], LEDs [149] and power/switching electronics.

Many techniques have been adopted for the epitaxial growth of Ga_2O_3 (including molecular-beam epitaxy (MBE), atomic layer deposition (ALD), halide vapor phase epitaxy (HVPE), metal–organic chemical vapor deposition (MOCVD) and PLD) [140, 150–156], and excellent epitaxial quality has been obtained.

Heteroepitaxial β-Ga_2O_3 thin films have also been grown using non-native substrates such as a-, r-, m-, and c-sapphire, silicon, MgO, and yttrium-stabilized

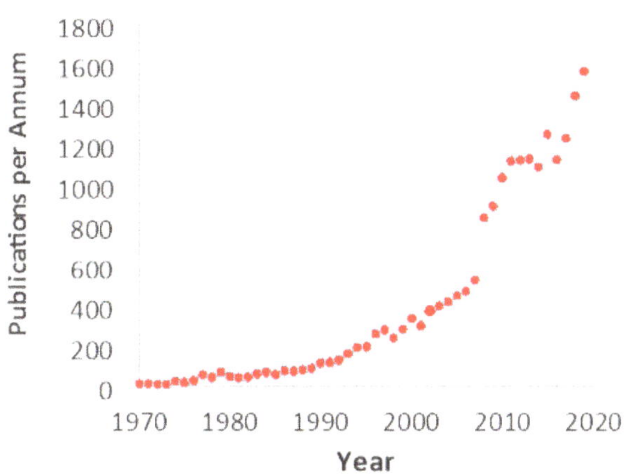

Figure 10.12. The number of publications per annum as a function of year (based on a search in the Scopus database for 'Ga_2O_3'). The majority of the application studies of Ga_2O_3 have focused on monoclinic β-Ga_2O_3, which is the most thermodynamically stable of the five common UWBG polymorphs (α-, β-, γ-, δ-, κ- and ε).

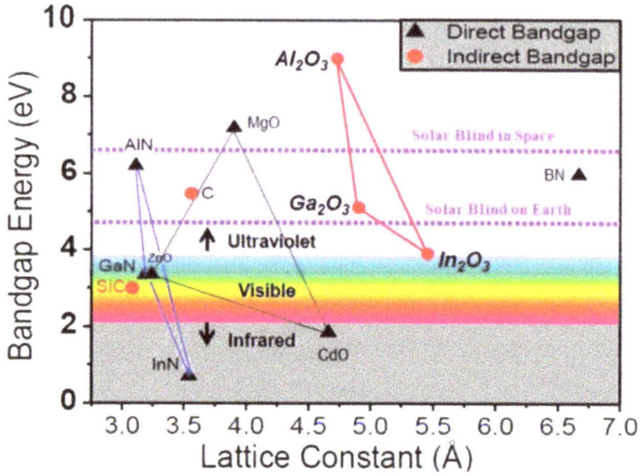

Figure 10.13. An illustration of the bandgap engineering range possible with Ga_2O_3 (through the addition of In or Al) in contrast to the possibilities offered by alternative WBG and UWBG semiconductors. Reproduced with permission from [137]. © 2021 Society of Photo-Optical Instrumentation Engineers (SPIE).

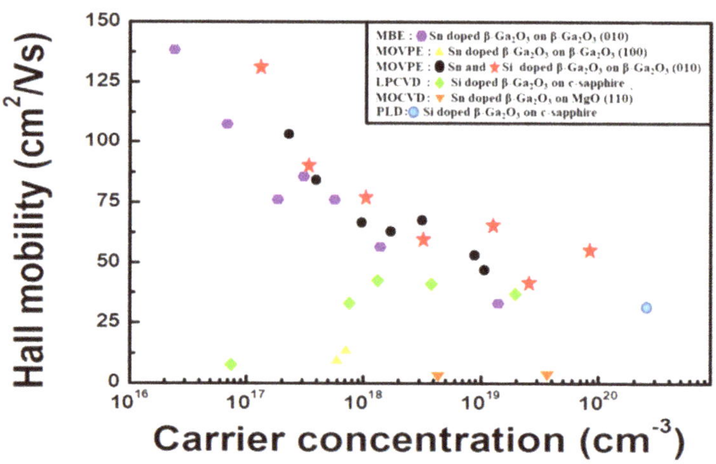

Figure 10.14. Illustration of n-type doping levels and carrier mobility ranges possible with epitaxial β-Ga_2O_3 layers. Adapted with permission from [138, 140]. © 2018 Society of Photo-Optical Instrumentation Engineers (SPIE).

zirconia (YSZ) [156–159]. However, when β-Ga_2O_3 thin films are grown on non-native substrates, it has proven more challenging to control the n-type doping concentration due to the occurrence of a higher density of defects, such as stacking faults and threading dislocations caused by lattice and thermal expansion mismatches between the non-native substrate and the β-Ga_2O_3 thin layers. Hence, the vast majority of power electronics studies have been based on homoepitaxial isotype devices grown on semi-insulating β-Ga_2O_3 substrates (generally obtained by Fe or Mg doping).

Although the growth has advanced significantly, it is still challenging to fabricate high-quality epilayers that are thick enough to resist high voltages and keep the on-resistance (R_{on}) low while maintaining high enough growth rates to remain cost-effective. Moreover, high crystal quality is essential for device integrity under high-voltage operation and a smooth surface morphology is critical because sharp, particle-free interfaces are necessary in order to obtain trap-free Ga_2O_3/dielectric interfaces and thin (but continuous) dielectric layers which can support very high electric fields.

10.2.2 An overview of Ga_2O_3-based power devices

Two overarching families of β-Ga_2O_3 power devices have been developed: FET-based switches and Schottky barrier diode (SBD)-based rectifiers.

Figure 10.15 shows the first reported FET-based power device produced using β-Ga_2O_3 [160].

There have subsequently been considerable ongoing advances in the field and a whole range of progressively more sophisticated device architectures and epitaxial realizations have been combined in a drive to increase V_{br} and reduce R_{on}. These include Si-doped channels, ohmic contacts formed by implantation [161, 162], and field-plated metal–oxide–semiconductor FET (MOSFET) architectures, which allowed increased V_{br} [163].

Further significant developments which advanced the twin targets of low R_{on} and high V_{br} in FETs were the development of high-performance fin-array field-effect transistors (FinFETs) (in which fins are etched into the epilayer in order to confine charge to the vertical channel fin and achieve high current density) [164–167] and current-aperture vertical transistors (CAVETs) (in which charge is confined to a vertical channel by a p-type current-blocking layer (CBL)) [168, 169] (see figures 10.16 and 10.17).

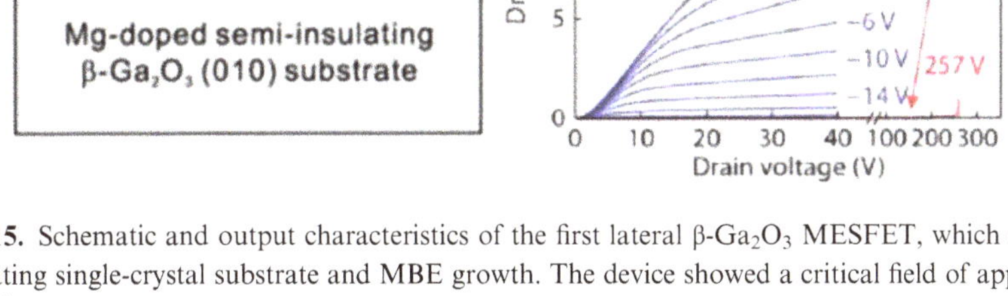

Figure 10.15. Schematic and output characteristics of the first lateral β-Ga_2O_3 MESFET, which employed a semi-insulating single-crystal substrate and MBE growth. The device showed a critical field of approximately 8 MV cm^{-1} and a V_{br} of 257 V. Reprinted with permission from [160], Copyright (2012) by the American Physical Society.

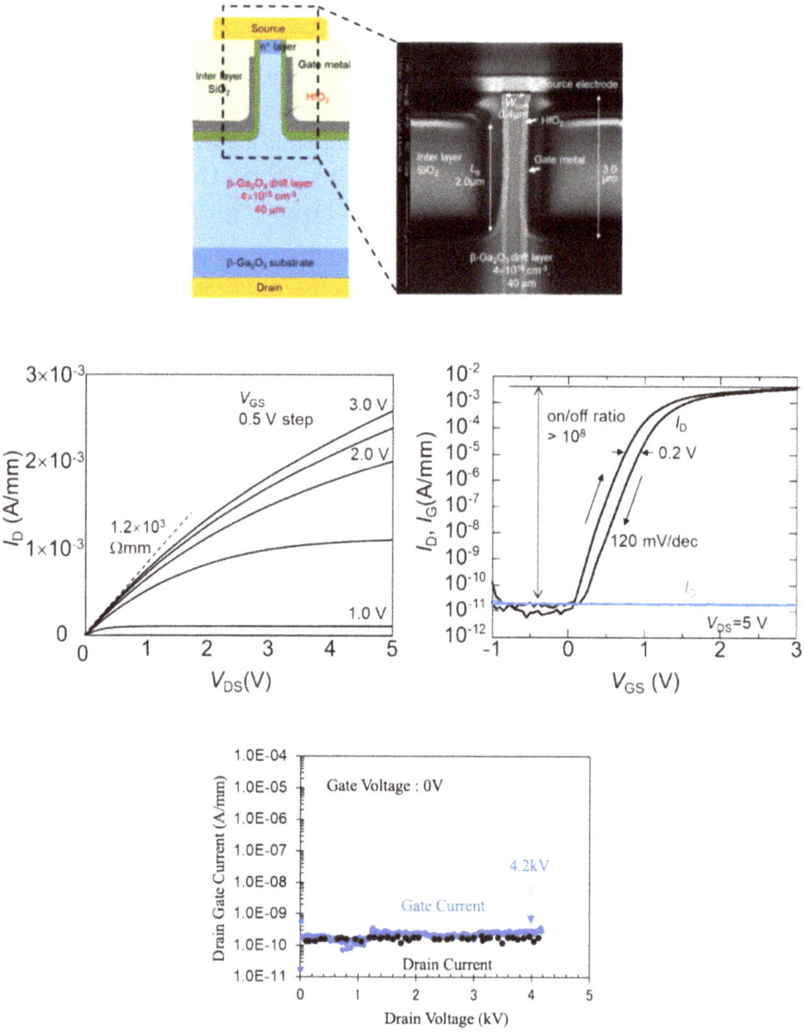

Figure 10.16. A device schematic, a scanning electron microscopy (SEM) image, and performance characteristics of a state-of-the-art Ga_2O_3 power FET. The vertical FinFET architecture enabled normally off operation (without using a p-type conductive layer) by narrowing the mesa width to submicron dimensions. The device employed an innovative 'high k' HfO_2 dielectric [170] (so as to reduce the electric field in the gate insulator). Its characteristics included a V_{br} of 4.6 kV and a gate leakage current that was below the detection limit. Reproduced from [167]. Copyright © Novel Crystal Technology, Inc.

These vertical finFET and CAVET architectures also offer the advantage of operating in the E-mode. The E-mode is 'normally off' and thus offers better energy efficiency and safer high-voltage operation than those of FETs which operate in the 'normally on' D-mode.

Figure 10.18 shows the first reported SBD power device [171] to be made using β-Ga_2O_3. Since then, ever more impressive specs have been achieved for β-Ga_2O_3 power SBDs. Such rectifying devices perform better when thinner epitaxial layers

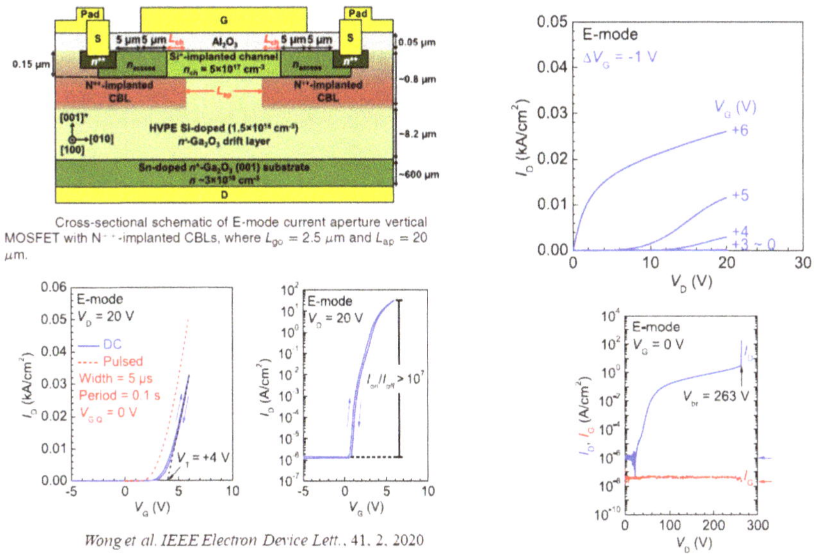

Figure 10.17. Schematic and performance characteristics for a state-of-the-art Ga_2O_3 CAVET. © [2020] IEEE. Reprinted, with permission, from [169].

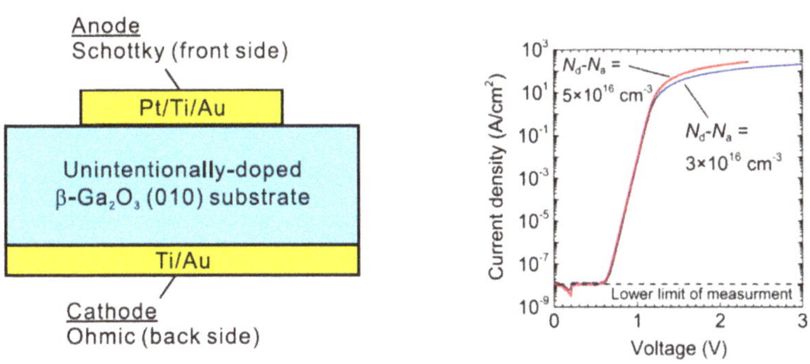

Figure 10.18. A schematic and the *J–V* characteristics of the first reported SBD-based Ga_2O_3 device based on an (010) oriented β-Ga_2O_3 single-crystal substrate grown using the floating-zone method. © [2013] IEEE. Reprinted, with permission, from [171].

are employed so as to sustain a given blocking voltage. Therefore, device terminations are very important and various device innovations such as field plates (FPs) [172], implanted edge terminations (ETs) [173], implanted guard-ring FPs [174], thermally oxidized terminations, beveled mesas [175] and trench structures [176] have been incorporated in order to reduce electric field crowding at the diode edges. Figure 10.19 shows the schematic of a state-of-the-art vertical-trench SBD [176].

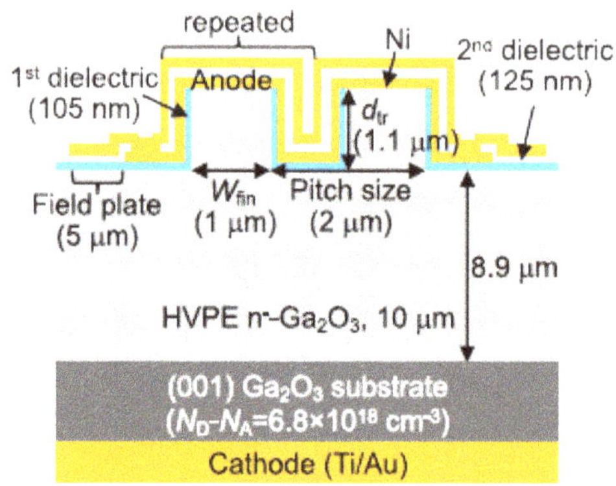

Figure 10.19. A schematic of a vertical-trench SBD device with a V_{br} of 2.9 kV and a 0.95 GW cm^{-2} Baliga figure of merit (BFOM).© [2020] IEEE. Reprinted, with permission, from [176].

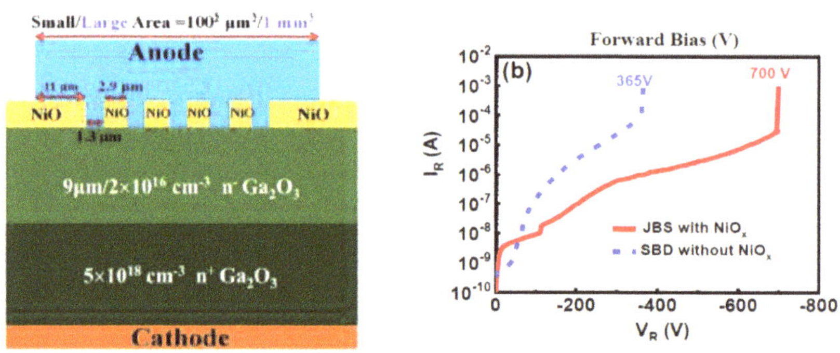

Figure 10.20. A schematic and performance *I–V* characteristics of a state-of-the-art Ga_2O_3 JBS diode formed using NiO as the p-type grid. © [2021] IEEE. Reprinted, with permission, from [178].

Another key SBD innovation is the development of β-Ga_2O_3-based junction barrier Schottky (JBS) diodes [177]. A JBS diode can combine the high V_{br} and low reverse leakage current of a p–n diode with the low forward voltage drop of an SBD. Figure 10.20 shows a schematic and *I–V* characteristics of a state-of-the-art Ga_2O_3-based JBS diode [178]. The *I–V* characteristics with and without NiO show the importance of the NiO grid for the device performance.

10.2.3 The prospects of Ga_2O_3-based power electronics

Figure 10.21 shows a mapping of the current perspective on the key comparative advantages and disadvantages of the main semiconductors used as the basis of power electronics. From this graphic, it can clearly be seen that while Ga_2O_3 dominates in terms of bandgap energy and breakdown field, it has very significant disadvantages in terms of electron mobility and thermal conductivity.

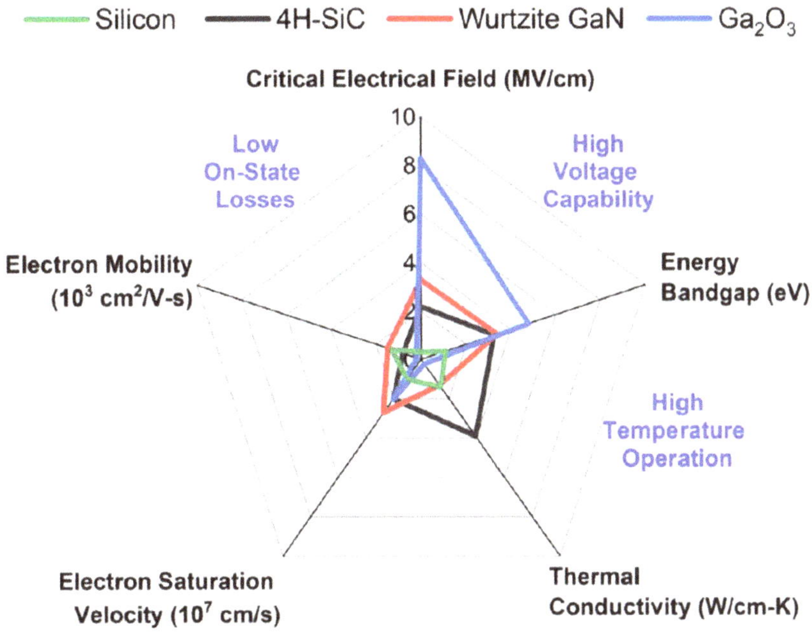

Figure 10.21. A graphic contrasting key β-Ga_2O_3 materials properties with those of incumbent power electronics materials. Reproduced from [179]. AIP Publishing.

With regards to improving electron mobility (which decreases the 'on' resistance), recent papers suggest that it may be possible to produce 2D electron gases (2DEGs) with enhanced carrier mobility at (Al)Ga_2O_3/Ga_2O_3 interfaces, and that high electron mobility transistors may thus be a possibility going forward [180] (figure 10.22).

The second issue, thermal management, is a big outstanding challenge for Ga_2O_3 because the bulk substrate's thermal conductivity is smaller than even that of sapphire. Hence, thermal resistance becomes a challenge during device operation in that heat accumulates at high voltages, leading to performance drift and accelerated aging. Several potential solutions (employing either top, bottom, or both types of cooling) have been proposed, such as direct epitaxial growth on a higher-thermal-conductivity substrate (e.g. SiC or diamond) or integrating Ga_2O_3 with a heat sink either by a flip chip approach (figure 10.23), a top bonding approach, and/or by thinning the Ga_2O_3 substrate away and bonding the wafer to a heat sink [181–187] (figure 10.23).

Assuming that these mobility and thermal management issues can be resolved, Mastro *et al* [135] predict that β-Ga_2O_3-based power devices could find competitive advantage in the higher-power, lower-frequency space of the overall market (figure 10.24). This will nevertheless depend on whether β-Ga_2O_3 can compete on cost with SiC. Recent market studies have suggested that β-Ga_2O_3 may well achieve a significant fabrication cost advantage over SiC going forward [188] (SiC power devices are currently cheaper to manufacture than GaN-based power devices)

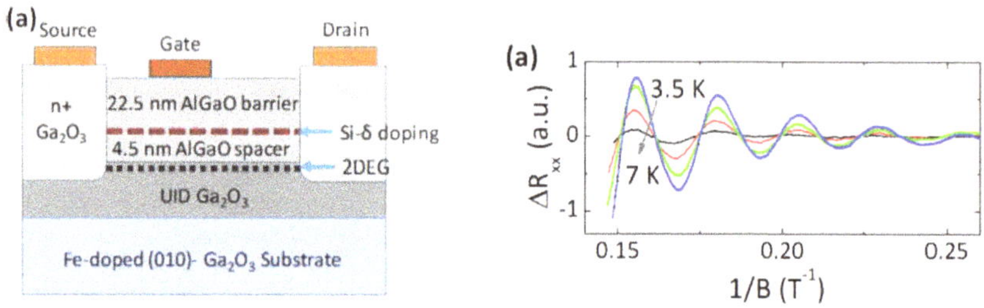

Figure 10.22. Device schematic and the corresponding temperature-dependent magnetoresistive measurements confirming the formation of a 2DEG at the $AlGa_2O_3/Ga_2O_3$ interface. Reprinted with permission from [180], Copyright (2018) by the American Physical Society.

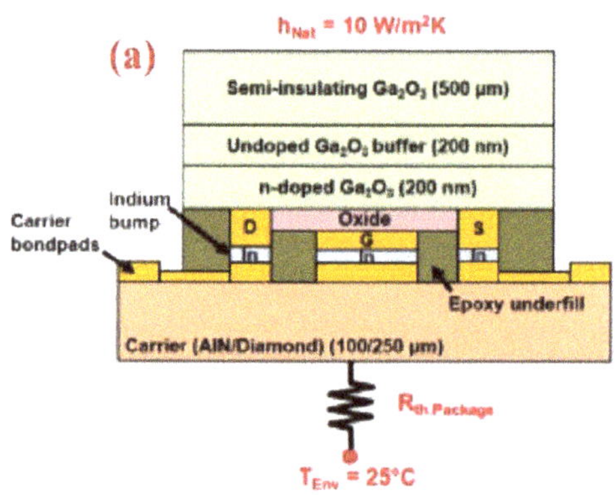

Figure 10.23. Illustration of a flip chip configuration used to bond a Ga_2O_3 power device to a heat sink. © [2019] IEEE. Reprinted, with permission, from [187].

(figure 10.25). This is largely based on the fact that single-crystal β-Ga_2O_3 substrates can be manufactured in relatively large formats by inexpensive melt growth processes [132–134]. For the moment, however, β-Ga_2O_3 substrates cost almost two orders of magnitude more than the equivalent sapphire substrates and they are not yet available in production volumes (the current manufacturing base is very small since it relies exclusively on a limited R&D market).

Hence, there is significant interest and work dedicated to the epitaxy of β-Ga_2O_3 on sapphire substrates and, as was the case in the development of GaN LEDs, the improvement in β-Ga_2O_3 epitaxial quality on sapphire has exceeded expectations. Moreover, it has been understood that sapphire brings several advantages (other than cost and industrial availability in large formats/volumes) that may allow it to find a place in future β-Ga_2O_3 power electronics devices. First of all, sapphire substrates have higher thermal conductivities than those of β-Ga_2O_3 substrates, which allows for better intrinsic heat management. Next, the epitaxial mismatch with sapphire allows strain engineering, which can improve the doping potential, stabilize different/non-equilibrium phases/polytypes/orientations, and extend the range of bandgap engineering. Furthermore, β-Ga_2O_3 can readily be transferred

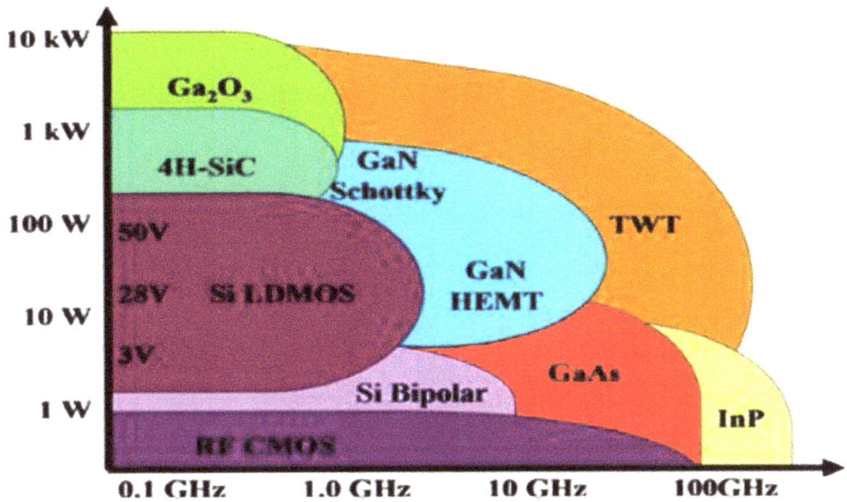

Figure 10.24. A map of the technologies that are considered to be ultimately superior in the power/frequency space. Reproduced with permission from [135]. © IOP Publishing 2017. Reproduced with permission. All rights reserved.

off sapphire onto another substrate by means of laser lift-off (LLO). This approach employs a laser beam fired through the sapphire substrate so as to ablate the Ga_2O_3/sapphire interface and thus release the epilayer from the substrate (not possible with Ga_2O_3 substrates)—see figure 10.26. Such an approach is widely used for high-brightness GaN LED fabrication and introduces several advantages, including improved heat evacuation (the new host substrate can be a heat sink) and the potential for back-contacting the β-Ga_2O_3 (which could allow vertical current flow in the device and thus facilitate higher currents due to less current crowding and less local heating compared with lateral current flow devices), simpler device architectures and smaller device footprints (i.e. more devices per wafer and thus lower costs).

In spite of the materials potential of Ga_2O_3 and these impressive advances in device architecture, realization and performance, Ga_2O_3-based power devices still lag significantly behind state-of-the-art GaN-based devices in their performance [189, 190].

Indeed, p–n junction architectures, such as those employed in GaN-based power devices, are a key factor in minimizing Ga_2O_3 device R_{on} while keeping V_{br} high. Thus, the p-type doping capability is of the utmost importance in the quest for ultimate performance. It is generally accepted, however, that useful levels of majority shallow acceptor (p-type) doping in β-Ga_2O_3 are unlikely to be achieved because of a combination of factors, including the relatively low absolute energy level of the valence band, the lack of an identified shallow acceptor, the relatively high effective masses of holes at the top of the valence band, the propensity for the self-trapping of holes and the relatively low formation energy of the oxygen vacancy background donor (that compensates acceptors) [135, 144, 191]. There have,

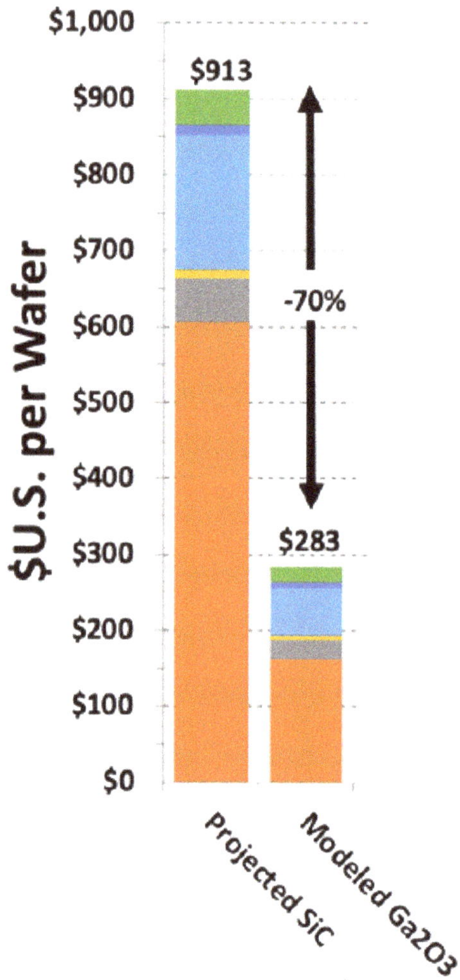

Figure 10.25. A projected cost comparison of SiC- and β-Ga_2O_3-based power device fabrication. Reproduced with permission from [188]. © 2020 Society of Photo-Optical Instrumentation Engineers (SPIE).

nevertheless, been several reports describing relatively resistive layers with predominant p-type doping in nominally undoped [192, 193], H-doped [194] and N-doped [195] β-Ga_2O_3. In the latter case, p-type β-Ga_2O_3 lateral MOSFETs were reported.

The lack of a low-resistance p-type layer for the anode has prevented the demonstration of a homojunction p–n β-Ga_2O_3 diode thus far. p–n heterojunction diodes have, nevertheless, been realized by integrating n-type Ga_2O_3 with p-type semiconductors such as Cu_2O [196] or NiO [197–200] (figure 10.27). Moreover, groundbreaking reports suggest that exploration of the metastable Ga_2O_3 phases and other Ga_2O_3-related compounds offers intriguing possibilities for p–n-junction-based power electronics going forward. For instance, the Japanese company Flosfia has chosen the corundum α-Ga_2O_3 phase as a basis for the development of power

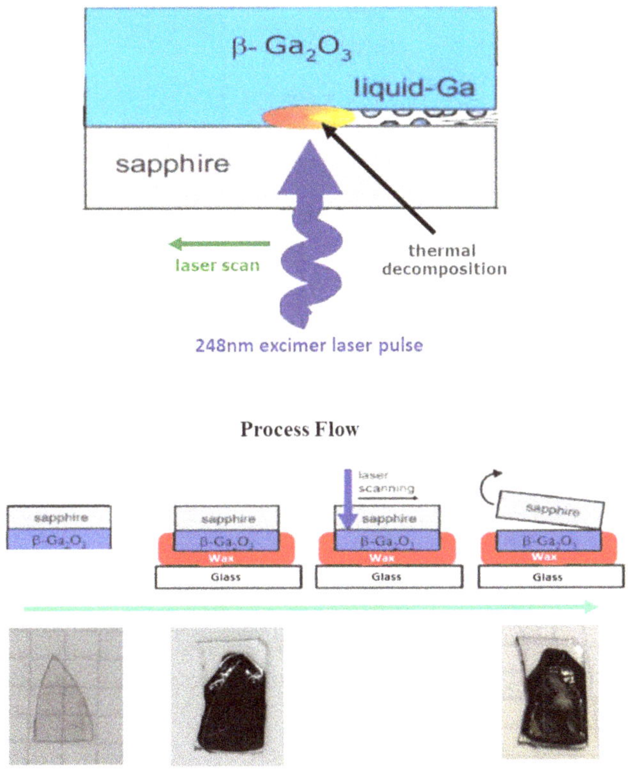

Figure 10.26. An illustration (above) and feasibility study (below) of the LLO process, which involves firing a laser beam through the sapphire substrate so as to ablate the Ga_2O_3/sapphire interface and release the Ga_2O_3 epilayer from the thermally insulating substrate using a 248 nm excimer laser (www.nanovation.com).

electronics. In 2015, they achieved a world-leading specific R_{on} for an SBD (of 0.1 mΩ cm^2). More recently, they have promoted p–n heterojunction power electronics that uses iridium oxide (Ir_2O_3) as a p-type counterpart to n-type α-Ga_2O_3 (www.flosfia.com).

In another metastable phase approach, Razeghi *et al* [201] recently reported a demonstration of both p- and n-type orthorhombic κ-Ga_2O_3. High levels of both n- and p-type shallow doping were obtained using amphoteric Si doping, as evidenced by both Hall measurements (a p-type resistivity of ~0.007 Ω.cm, a carrier concentration of ~1.7×10^{20} cm^{-3}, and a mobility of ~5.7 cm^2 V^{-1} s^{-1}) and the realization of both p-type and n-type ring FETs (figure 10.28).

Finally, a Ga_2O_3-related oxide, zinc gallate ($ZnGa_2O_4$), has been put forward as another candidate UWBG (~5 eV) oxide semiconductor for use in future power electronics [202, 203]. $ZnGa_2O_4$ has been reported to have a higher breakdown field than that of Ga_2O_3, and the spinel nature of $ZnGa_2O_4$ is expected to offer the advantage of superior dopability (based on the inherent diversity in cation coordination possibilities [204]). Moreover, some recent work has indicated that $ZnGa_2O_4$ may have the potential for both n- and p-type doping [205]. Hence,

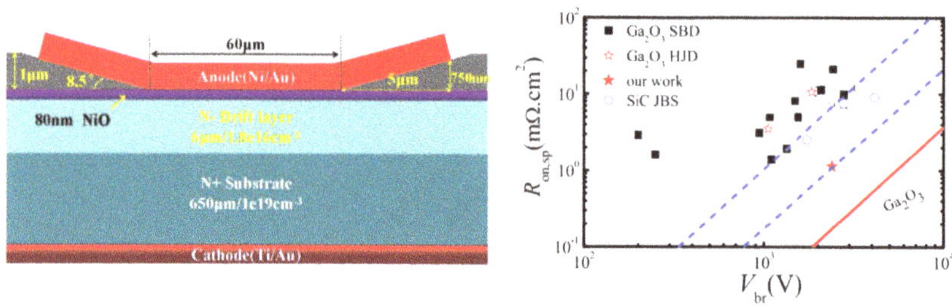

Figure 10.27. A schematic and a state-of-the-art performance comparison for a vertical p-NiO/n-Ga_2O_3 heterojunction diode with a V_{br} of 2.4 kV and a specific R_{on} of 1.1 mΩ cm^2. © [2022] IEEE. Reprinted, with permission, from [199].

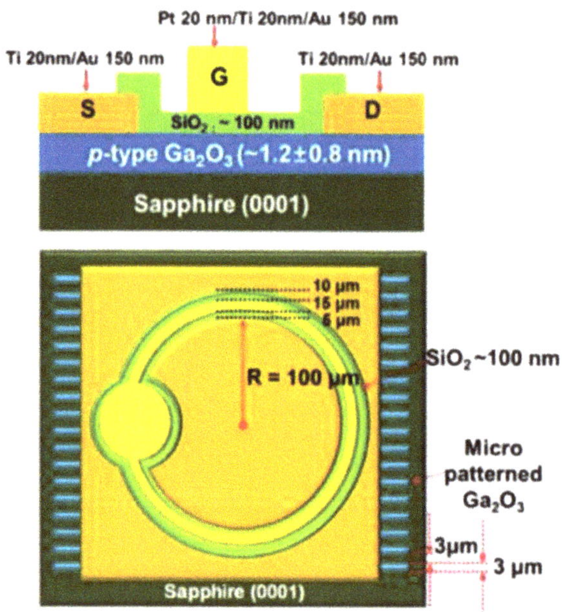

Figure 10.28. A schematic and I–V curves (at various V_{gs}) of a κ-Ga_2O_3-based 2D ring FET confirming p-type behavior. The inset shows an optical microscope image of the fabricated device. Reproduced from [201].

$ZnGa_2O_4$ may hold the potential to enable homoepitaxial bipolar power devices to be produced.

Acknowledgments

The authors would like to thank Dr Ekaterina Chikoidze (of the CNRS-GEMAC, Versailles, France), Dr Amador Pérez-Tomás (of ICN2, Barcelona, Spain), and Dr Marko Tadjer (of US Naval Research Laboratories, Washington DC) for useful discussions.

References and further reading

[1] Corzo D, Tostado-Blázquez G and Baran D 2020 Flexible electronics: status, challenges and opportunities *Front. Electron.* **1** 594003

[2] He F, Qin Y, Wan L, Su J, Lina Z, Zhang J, Chang J, Wu J and Hao Y 2020 Metal oxide heterojunctions for high performance solution grown oxide thin film transistors *Appl. Surf. Sci.* **527** 146774

[3] He Y, Wang X, Gao Y, Hou Y and Wan Q 2018 Oxide-based thin film transistors for flexible electronics *J. Semicond* **39** 011005

[4] Hosseini M and Nawrocki R 2021 A review of the progress of thin-film transistors and their technologies for flexible electronics *Micromachines* **12** 655

[5] Lorenz M *et al* 2016 The 2016 oxide electronic materials and oxide interfaces roadmap *J. Phys. D: Appl. Phys.* **49** 433001

[6] Park J, Kang B and Kim H 2020 A review of low-temperature solution-processed metal oxide thin-film transistors for flexible electronics *Adv. Funct. Mater* **30** 1904632

[7] Rogers D J, Teherani F H, Sandana V E and Bove P 2021 Zinc-oxide-based electronics and photonics *Encyclopedia of Materials: Technical Ceramics and Glasses* 3 (Oxford: Elsevier), 241–53

[8] Zhang Y, Mei Z, Liang H and Du X 2017 Review of flexible and transparent thin-film transistors based on zinc oxide and related materials *Chin. Phys.* B **26** 047307

[9] Lee J K, Lim Y S and Park C *et al* 2010 a-Si:H thin-film transistor driven flexible color E-paper display on flexible substrates *IEEE Electron Device Lett.* **31** 833

[10] Huang J J, Chen Y P and Huang Y S *et al* 2009 A 4.1-inch flexible QVGA AMOLED using a microcrystalline-Si:H TFT on a polyimide substrate *SID Symp. Dig. Tech. Pap* **40** 866

[11] Liu K, Ouyang B, Guo X, Guo Y and Liu Y 2022 Advances in flexible organic field-effect transistors and their applications for flexible electronics *npj Flex. Electron.* **6** 1

[12] Fukuda K, Takeda Y and Yoshimura Y *et al* 2014 Fully-printed high performance organic thin-film transistors and circuitry on one-micron-thick polymer films *Nat. Commun* **5** 4147

[13] Yi H T, Payne M M and Anthony J E *et al* 2012 Ultra-flexible solution-processed organic field-effect transistors *Nat. Commun* **3** 1259

[14] Guo X, Xu Y and Ogier S *et al* 2017 Current status and opportunities of organic thin-film transistor technologies *IEEE Trans. Electron Devices* **64** 1906

[15] Oh J Y, Rondeau-Gagné S, Chiu Y-C, Chortos A, Lissel F and Wang G-J N *et al* 2016 Intrinsically stretchable and healable semiconducting polymer for organic transistors *Nature* **539** 411–5

[16] Kim S J, Jeon D B and Park J H *et al* 2015 Nonvolatile memory thin film transistors using biodegradable chicken albumen gate insulator and oxide semiconductor channel on eco-friendly paper substrate *ACS Appl. Mater. Interfaces* **7** 4869

[17] Bettinger C J and Bao Z 2010 Biomaterials-based organic electronic devices *Polym. Int* **59** 563–7

[18] Li K, Liu X, Chn S, Pan W and Zhang J 2019 A flexible solid-state supercapacitor based on graphene/polyaniline paper electrodes *J. Energy Chem.* **32** 166–73

[19] Dimitrakopoulos C D and Mascaro D J 2001 Organic thin-film transistors: a review of recent advances *IBM J. Res. Dev* **45** 11

[20] Lin P and Yan F 2012 Organic thin-film transistors for chemical and biological sensing *Adv. Mater* **24** 34

[21] Gao L 2017 Flexible device applications of 2D semiconductors *Small* **13** 1603994

[22] Jiang D, Liu Z, Xiao Z, Qian Z, Sun Y, Zeng Z and Wang R 2022 Flexible electronics based on 2D transition metal dichalcogenides *J. Mater. Chem.* A **10** 89
[23] Peng B, Zhao F, Ping J and Ying Y 2020 Recent advances in nanomaterial-enabled wearable sensors: material synthesis, sensor design, and personal health monitoring *Small* **16** 2002681
[24] Shrivas K, Ghosale A, Bajpai P K, Kant T, Dewangan K and Shankar R 2020 Advances in flexible electronics and electrochemical sensors using conducting nanomaterials: a review *Microchem.* J **156** 104944
[25] Xue Z *et al* 2017 Deterministic line-shape programming of silicon nanowires for extremely stretchable springs and electronics *Nano Lett.* **17** 7638–46
[26] Rong P, Ren S and Yu Q 2019 Fabrications and applications of ZnO nanomaterials in flexible functional devices—a review *Crit. Rev. Anal. Chem.* **49** 336–49
[27] Che Y, Chen H, Gui H, Liu J, Liu B and Zhou C 2014 Review of carbon nanotube nanoelectronics and macroelectronics *Semicond. Sci. Technol* **29** 073001
[28] Hosono H 1996 Working hypothesis to explore novel wide band gap electrically conducting amorphous oxides and examples *J. Non-Cryst. Solids* **198–200** 165–9
[29] Kamiya T and Hosono H 2010 Material characteristics and applications of transparent amorphous oxide semiconductors *NPG Asia Mater* **2** 15
[30] Stewart K A, Yeh B-S and Wager J F 2016 Amorphous semiconductor mobility limits *J. Non-Cryst. Solids* **432** 169–99
[31] Nomura K, Ohta H, Takagi A, Kamiya T, Hirano M and Hosono H 2004 Room-temperature fabrication of transparent flexible thin-film transistors using amorphous oxide semiconductors *Nature* **432** 488–92
[32] Kim M G, Kanatzidis M G, Facchetti A and Marks T J 2011 Low-temperature fabrication of high-performance metal oxide thin-film electronics via combustion processing *Nat. Mater* **10** 382–8
[33] Kwon J Y and Jeong J K 2015 Recent progress in high performance and reliable n-type transition metal oxide-based thin film transistors *Semicond. Sci. Technol* **30** 024002
[34] Kim M G, Kim S K, Ha Y G, Kanatzidis M G, Facchetti A and Marks T J 2010 High-performance solution-processed amorphous zinc–indium–tin oxide thin-film transistors *J. Am. Chem. Soc* **132** 10352–64
[35] Sashabayashi T, Ito N, Nishimura E, Kon M, Song P K, Utsumi K, Kaijo A and Shigesato Y 2003 Comparative study on structure and internal stress in tin-doped indium oxide and indium–zinc oxide films deposited by r.f. magnetron sputtering *Thin Solid Films* **445** 219–23
[36] Kwon Y, Li Y and Heo Y *et al* 2004 Enhancement-mode thin-film field-effect transistor using phosphorus-doped (Zn,Mg)O channel *Appl. Phys. Lett.* **84** 2685–7
[37] Nomura K, Takagi A, Kamiya T, Ohta H, Hirano M and Hosono H 2006 Amorphous oxide semiconductors for high-performance flexible thin-film transistors *Jpn. J. Appl. Phys.* **45** 4303–8
[38] Barquinha P, Pereira L, Goncalves G, Martins R and Fortunato E 2008 The effect of deposition conditions and annealing on the performance of high-mobility GIZO TFTs *Electrochem. Solid-State Lett.* **11** 248–52
[39] Hosono H 2006 Ionic amorphous oxide semiconductors: material design, carrier transport, and device application *J. Non-Cryst. Solids* **352** 851–8
[40] Orita M, Ohta H, Hirano M, Narushima S and Hosono H 2001 Amorphous transparent conductive oxide $InGaO_3(ZnO)_m$ ($m < 4$): a Zn 4s conductor *Phil. Mag.* B **81** 501–15

[41] Mativenga M, Choi M H and Choi J W *et al* 2011 Transparent flexible circuits based on amorphous-indium–gallium–zinc–oxide thin film transistors *IEEE Electron Device* **32** 170
[42] Li H U and Jackson T N 2015 Oxide semiconductor thin film transistors on thin solution-cast flexible substrates *IEEE Electron Device Lett.* **36** 35
[43] Tripathi A K, Myny K and Hou B *et al* 2015 Electrical characterization of flexible InGaZnO transistors and 8-b transponder chip down to a bending radius of 2 mm *IEEE Trans. Electron Devices* **62** 4063
[44] Oh H, Cho K and Park S *et al* 2016 Electrical characteristics of bendable a-IGZO thin-film transistors with split channels and topgate structure *Microelectron. Eng* **159** 179
[45] Park M J, Yun D J and Ryu M K 2015 Improvements in the bending performance and bias stability of flexible InGaZnO thin film transistors and optimum barrier structures for plastic polyethylene naphthalate substrates *J. Mater. Chem.* C **3** 4779
[46] Salvatore G A, Münzenrieder N and Kinkeldei T *et al* 2014 Waferscale design of lightweight and transparent electronics that wraps around hairs *Nat. Commun* **5** 2982
[47] Martins R F, Ahnood A and Correia N *et al* 2013 Recyclable, flexible, low-power oxide electronics *Adv. Funct. Mater* **23** 2153
[48] Martins R, Ferreira I and Fortunato E 2011 Electronics with and on paper *Phys. Status Solidi* **5** 332
[49] Gaspar D, Fernandes S and De Oliveira A *et al* 2014 Nanocrystalline cellulose applied simultaneously as the gate dielectric and the substrate in flexible field effect transistors *Nanotechnology* **25** 094008
[50] Lim W, Douglas E and Norton D *et al* 2010 Low-voltage indium gallium zinc oxide thin film transistors on paper substrates *Appl. Phys. Lett.* **96** 053510
[51] Wu G D, Zhang J and Wan X 2014 Junctionless coplanar-gate oxidebased thin-film transistors gated by Al_2O_3 proton conducting films on paper substrates *Chin. Phys. Lett.* **31** 108505
[52] Sun J, Jiang J and Lu A *et al* 2010 One-volt oxide thin-film transistors on paper substrates gated by SiO_2-based solid electrolyte with controllable operation modes *IEEE Trans. Electron Devices* **57** 2258
[53] Choi N, Khan S A and Ma X *et al* 2011 Amorphous oxide thin film transistors with methyl siloxane based gate dielectric on paper substrate *Electrochem. Solid-State Lett.* **14** H247
[54] Jeong J K 2011 The status and perspectives of metal oxide thin film transistors for active matrix flexible displays *Semicond. Sci. Technol* **26** 034008
[55] Xu H, Luo D and Li M *et al* 2014 A flexible AMOLED display on the PEN substrate driven by oxide thin-film transistors using anodized aluminium oxide as dielectric *J. Mater. Chem.* C **2** 1255
[56] Gan K-J, Liu P-T, Chien T-C, Ruan D-B and Sze S M 2019 Highly durable and flexible gallium oxide based conductive-bridging random access memory *Sci. Rep.* **9** 14141
[57] Gelinck G H, Huitema H E A and Van Veenendaal E *et al* 2004 Flexible active-matrix displays and shift registers based on solution-processed organic transistors *Nat. Mater* **3** 106
[58] Min J-G and Cho W-J 2021 Sol–gel composites-based flexible and transparent amorphous indium gallium zinc oxide thin-film synaptic transistors for wearable intelligent electronics *Molecules* **26** 7233
[59] Nathan A and Chalamala B R 2005 Special issue on flexible electronics technology, part 1: systems and applications *Proc. IEEE* **93** 1235

[60] Kim W-G, Tak Y J, Ahn B D, Jung T S, Chung K-B and Kim High-pressure Gas H J 2016 Activation for amorphous indium–gallium–zinc–oxide thin-film transistors at 100 °C *Sci. Rep.* **6** 23039

[61] Mahmoudabadi F, Ma X and Hatalis M K *et al* 2014 Amorphous IGZO TFTs and circuits on conformable aluminum substrates *Solid-State Electron* **101** 57

[62] Park I J, Jeong C Y and Cho I T *et al* 2012 Fabrication of amorphous InGaZnO thin-film transistor-driven flexible thermal and pressure sensors *Semicond. Sci. Technol* **27** 105019

[63] Lee G J, Kim J and Kim J H *et al* 2014 High performance, transparent a-IGZO TFTs on a flexible thin glass substrate *Semicond. Sci. Technol* **29** 035003

[64] Ito M, Kon M and Ishizaki M *et al* 2005 A flexible active-matrix TFT array with amorphous oxide semiconductors for electronic paper *IDW/AD '05 Proceedings* 5 (Tokyo, Japan, San Jose CA: Institute of Image Information and Television Engineers ; Society for Information Display), 845–6

[65] Kim S J, Park M J and Yun D J *et al* 2016 High performance and stable flexible memory thin-film transistors using In–Ga–Zn–O channel and ZnO charge-trap layers on poly (ethylene naphthalate) substrate *IEEE Trans. Electron Devices* **63** 1557

[66] Martins R, Fortunato E, Barquinha P, Pereira L, Correia N, Gonçalves G, Ferreira I and Fortunato E 2008 Write-erase and read paper memory transistor *Appl. Phys. Lett.* **93** 2035011–3

[67] Fortunato E, Correia N, Barquinha P, Pereira L, Gonçalves G and Martins M 2008 High-performance flexible hybrid field-effect transistors based on cellulose fiber paper *IEEE Elec. Dev. Lett.* **29** 988–90

[68] Qin M, Guo H, Dai Z, Yan X and Ning X 1607 Advances in flexible and wearable pH sensors for wound healing monitoring *J. Semicond* **40** 11

[69] Smith J T, Shah S S and Goryll M *et al* 2014 Flexible ISFET biosensor using IGZO metal oxide TFTs and an ITO sensing layer *IEEE Sens.* J **14** 937

[70] Liu N, Zhu L Q and Feng P *et al* 2015 Flexible sensory platform based on oxide-based neuromorphic transistors *Sci. Rep.* **5** 18082

[71] Shah S, Smith J, Stowell J and Blain Christen J 2015 Biosensing platform on a flexible substrate *Sensors Actuators* **B 210** 197

[72] Liang H, Cui S, Su R, Guan P, He Y, Yang L, Chen L, Zhang Y, Mei Z and Du X 2019 Flexible X-ray detectors based on amorphous Ga_2O_3 thin films *ACS Photonics* **6** 351

[73] Lujan R and Street R 2012 Flexible X-ray detector array fabricated with oxide thin-film transistors *IEEE Electron Device Lett.* **33** 688

[74] Smith J T, Couture A J and Stowell J R *et al* 2014 Optically seamless flexible electronic tiles for ultra large-area digital X-ray imaging *IEEE Trans Compon. Packag. Manuf. Technol.* **4** 1109

[75] He R, Liu H, Niu Y, Zhang H, Genin M and Xu F 2022 Flexible miniaturized sensor technologies for long-term physiological monitoring *npj Flex. Electron.* **6** 20

[76] Li Y, Liu Y and Yun F *et al* 2014 Self-assembled transparent a-IGZO based TFTs for flexible sensing applications *International Symposium on Photonics and Optoelectronics 2014*9233 ed Z Zhou (Bellingham, WA: SPIE), 92332A

[77] Butt M A, Kazanskiy N L and Khonina S N 2022 A revolution in flexible wearable electronics for temperature and pressure monitoring—a review *Electronics* **11** 716 66

[78] Honda W, Harada S and Ishida S *et al* 2015 High-performance, mechanically flexible, and vertically integrated 3D carbon nanotube and InGaZnO complementary circuits with a temperature sensor *Adv. Mater* **27** 4674

[79] Kim J and Jo S *et al* 2016 Ultrahigh detective heterogeneous photosensor arrays with in-pixel signal boosting capability for large-area and skin-compatible electronics *Adv. Mater* **28** 3078

[80] Li Z *et al* 2019 Flexible solar-blind Ga_2O_3 ultraviolet photodetectors with high responsivity and photo-to-dark current ratio *IEEE Photonics* J **11** 68

[81] Lee S H, Kim S B, Moon Y-J, Kim S M, Jung H J, Seo M S, Lee K M, Kim S-K and Lee S W 2017 *ACS Photonics* **4** 2937

[82] Zhao H *et al* 2022 Recent advances in flexible and wearable sensors for monitoring chemical molecules *Nanoscale* **14** 1653–69

[83] Knobelspies S, Daus A, Cantarella G, Petti L, Münzenrieder N, Tröster G and Salvatore G A 2016 Flexible a-IGZO phototransistor for instantaneous and cumulative UV-exposure monitoring for skin health *Adv. Electron. Mater* **2** 1600273

[84] Pierre A, Doris S E, Lujan R and Street R A 2018 Monolithic integration of ion-selective organic electrochemical transistors with thin film transistors on flexible substrates *Adv. Mater. Technol* 1800577

[85] Qian L-X, Wu Z-H, Zhang Y-Y, Lai P T, Liu X-Z and Li Y-R 2017 Ultrahigh-Responsivity, Rapid-Recovery, Solar-Blind Photodetector Based on Highly Nonstoichiometric Amorphous Gallium Oxide *ACS Photonics* **4** 2203

[86] Cui S, Mei Z, Zhang Y, Liang H and Du X 2017 Room-Temperature Fabricated Amorphous Ga2O3 High-Response-Speed Solar-Blind Photodetector on Rigid and Flexible Substrates *Adv. Opt. Mater* **5** 1700454

[87] Jiao S, Lu H, Wang X, Nie Y, Wang D, Gao S and Wang J 2019 The Structural and Photoelectrical Properties of Gallium Oxide Thin Film Grown by Radio Frequency Magnetron Sputtering *ECS J. Solid State Sci. Technol* **8** Q3086

[88] Li M-Z, Han S-T and Zhou Y 2020 Recent advances in flexible field-effect transistors toward wearable sensors , *Adv. Intell. Syst* **2** 2000113

[89] Pang C, Lee C and Suh K-Y 2013 Recent advances in flexible sensors for wearable and implantable devices *J. Appl. Polym. Sci.* **1429** 1429–41

[90] Heng W, Solomon S and Gao W 2022 Flexible electronics and devices as human–machine interfaces for medical robotics *Adv. Mater* **34** 2107902

[91] Shi R, Lei T, Wang S, Xia Z and Wong M 2021 High-Resolution Active-Matrix Organic Light-Emitting Diode Display Realized Using Elevated-Metal Metal-Oxide Transistor Technology *International Conference on Display Technology (ICDT 2021)* 52 (Campbell, CA: Society for Information Display) 2.2

[92] Tripathi A, Smits E P and Van Der Putten J *et al* 2011 Low-voltage gallium–indium–zinc–oxide thin film transistors based logic circuits on thin plastic foil: building blocks for radio frequency identification application *Appl. Phys. Lett.* **98** 162102

[93] Cantarella G, Ishida K and Petti L *et al* 2016 Flexible In–Ga–Zn–O based circuits with two and three metal layers: simulation and fabrication study *IEEE Electron Device Lett.* **37** 1582

[94] Myny K and Steudel S 2016 *Flexible thin-film NFC transponder chip exhibiting data rates compatible to ISO NFC standards using self-aligned metal-oxide TFTs 2016 IEEE International Solid-State Circuits Conference (ISSCC)* (Piscataway, NJ: IEEE), 298

[95] Chen Y, Geng D and Lin T *et al* 2016 Full-swing clock generating circuits on plastic using a-IGZO dual-gate TFTs with pseudo-CMOS and bootstrapping *IEEE Electron Device Lett.* **37** 882

[96] Perumal C, Ishida K and Shabanpour R *et al* 2013 A compact a-IGZO TFT model based on MOSFET SPICE Level=3 template for analog/RF circuit designs *IEEE Electron Device Lett.* **34** 1391

[97] Kim J, Fuentes-Hernandez C and Kim S J *et al* 2010 Flexible hybrid complementary inverters with high gain and balanced noise margins using pentacene and amorphous InGaZnO thin-film transistors *Org. Electron* **11** 1074

[98] Martins R, Ahnood A, Correia N, Pereira L, Barros R, Barquinha P, Costa R, Ferreira I, Nathan A and Fortunato E 2013 Recyclable, flexible, low-power oxide electronics *Adv. Funct. Mater* **23** 2153–61

[99] Dindar A, Kim J and Fuentes-Hernandez C *et al* 2011 Metal-oxide complementary inverters with a vertical geometry fabricated on flexible substrates *Appl. Phys. Lett.* **99** 172104

[100] Kim J, Fuentes-Hernandez C and Hwang D *et al* 2011 Vertically stacked hybrid organic–inorganic complementary inverters with low operating voltage on flexible substrates *Org. Electron* **12** 45

[101] Nomura K, Aoki T and Nakamura K *et al* 2010 Three-dimensionally stacked flexible integrated circuit: amorphous oxide/polymer hybrid complementary inverter using *n*-type a-In–Ga–Zn–O and *p*-type poly-(9,9-dioctylfluorene-co-bithiophene) thin-film transistors *Appl. Phys. Lett.* **96** 263509

[102] He Y *et al* 2018 Oxide-based thin film transistors for flexible electronics *J. Semicond.* **39** 011005

[103] Rockelé M, Pham D V and Steiger J *et al* 2012 Solution-processed and low-temperature metal oxide n-channel thin-film transistors and low-voltage complementary circuitry on large-area flexible polyimide foil *J. Soc. Inf. Disp.* **20** 499

[104] Oh M S, Choi W and Lee K *et al* 2008 Flexible high gain complementary inverter using n-ZnO and p-pentacene channels on polyethersulfone substrate *Appl. Phys. Lett.* **93** 033510

[105] Honda W, Arie T and Akita S *et al* 2015 Mechanically flexible and high-performance CMOS logic circuits *Sci. Rep.* **5** 15099

[106] Avis C and Jang J 2011 *J. Mater. Chem.* **21** 10649

[107] Kim J B, Fuentes-Hernandez C, Potscavage W J Jr, Zhang X-H and Kippelen B 2009 Low-voltage InGaZnO thin-film transistors with Al_2O_3 gate insulator grown by atomic layer deposition *Appl. Phys. Lett.* **94** 142107

[108] Kim J B, Fuentes-Hernandez C and Kippelen B 2008 High-performance InGaZnO thin-film transistors with high-ð amorphous $Ba_{0.5}Sr_{0.5}TiO_3$ gate insulator *Appl. Phys. Lett.* **93** 242111

[109] Zhou J, Wu G and Guo L *et al* 2013 Flexible transparent junctionless TFTs with oxygen-tuned indium–zinc–oxide channels *IEEE Electron Device Lett.* **34** 888

[110] Hong K, Kim S H and Lee K H *et al* 2013 Printed, sub-2 V ZnO electrolyte gated transistors and inverters on plastic *Adv. Mater* **25** 3413

[111] Liu Y H, Zhu L Q and Feng P *et al* 2015 Freestanding artificial synapses based on laterally proton-coupled transistors on chitosan membranes *Adv. Mater* **27** 5599

[112] Mitra S, Shukla A K and Sampath S 2001 Electrochemical capacitors with plasticized gel–polymer electrolytes *J. Power Sources* **101** 213

[113] Fujimoto T and Awaga K 2013 Electric-double-layer field-effect transistors with ionic liquids *Phys. Chem. Chem. Phys.* **15** 8983
[114] Du H, Lin X and Xu Z *et al* 2015 Electric double-layer transistors: a review of recent progress *J. Mater. Sci.* **50** 5641
[115] Ono S, Seki S and Hirahara R *et al* 2008 High-mobility, low-power, and fast-switching organic field-effect transistors with ionic liquids *Appl. Phys. Lett.* **92** 93
[116] Liu Y, Zhou H and Cheng R *et al* 2014 Highly flexible electronics from scalable vertical thin film transistors *Nano Lett.* **14** 1413
[117] Vidor F F, Meyers T and Wirth G I *et al* 2016 ZnO nanoparticle thinfilm transistors on flexible substrate using spray-coating technique *Microelectron. Eng* **159** 155
[118] Hsu H H, Chang C Y and Cheng C H *et al* 2014 High Mobility Field-Effect Thin Film Transistor Using Room-Temperature High-κ Gate Dielectrics *J. Disp. Technol* **10** 875–81
[119] Hsu H H, Chang C Y and Cheng C H 2013 A flexible IGZO thin-film transistor with stacked TiO_2-based dielectrics fabricated at room temperature *IEEE Electron Device Lett.* **34** 768
[120] Nathan A, Ahnood A and Cole M T *et al* 2012 Flexible electronics: the next ubiquitous platform *Proc. IEEE* **100** 1486
[121] Yang M, Chon M-W, Kim J-H, Lee S-H, Jo J, Yeo J, Ko S H and Choa S-H 2014 Mechanical and environmental durability of roll-to-roll printed silver nanoparticle film using a rapid laser annealing process for flexible electronics *Microelectron. Reliab* **54** 2871–80
[122] Frenzel H, Lajn A, Brandt M, Lorenz M, Schein F, Zhang Z and Grundmann M 2008 ZnO metal-semiconductor field-effect transistors with Ag-Schottky gates *Appl. Phys. Lett.* **92** 1921081–3
[123] Schlup P, Schein F-L, von Wenckstern H and Grundmann All Amorphous Oxide M 2015 Bipolar heterojunction diodes from abundant metals *Adv. Electron. Mater* **1** 1400023
[124] Reese S, Remo T, Green J and Zakutayev A 2019 How much will gallium oxide power electronics cost? *Joule* **3** 903–7
[125] Tolbert L M *et al* 2005 *Power Electronics for Distributed Energy Systems and Transmission and Distribution Applications: Assessing the Technical Needs for Utility Applications* ORNL/TM-2005/230 Oak Ridge National Laboratory https://info.ornl.gov/sites/publications/Files/Pub57485.pdf
[126] Pérez-Tomás A, Chikoizde E, Rogers D and Garrido J A 2021 A walk on the frontier of energy electronics with power ultra-wide bandgap oxides and ultra-thin neuromorphic 2D materials *Proc. SPIE* **116871** 116871Y
[127] Pérez-Tomás A, Chikoidze E, Jennings M R, Russell S A O, Teherani F H, Bove P, Sandana E V and Rogers D J 2018 Wide and ultra-wide bandgap oxides: where paradigm-shift photovoltaics meets transparent power electronics *Proc. SPIE* **10533** 105331Q-1
[128] Perez-Tomas A, Chikoidze E and Rogers D J 2021 A walk on the frontier of energy electronics with power ultra-wide bandgap oxides and ultra-thin neuromorphic 2D materials *Proc. SPIE* **11687** 116871Y
[129] Nouketcha F L L, Cui Y, Lelis A, Green R, Darmody C, Schuster J and Goldsman N 2020 Investigation of wide- and ultrawide-bandgap semiconductors from impact-ionization coefficients *IEEE Trans. Electron Devices* **67** 3999
[130] Green A J *et al* 2022 β-Gallium oxide power electronics *APL Mater* **10** 029201

[131] Chi Z, Asher J, Jennings M, Chikoidze E and Pérez-Tomás A 2022 Ga_2O_3 and related ultra-wide bandgap power semiconductor oxides: new energy electronics solutions for CO_2 emission mitigation *Materials* **15** 1164

[132] Víllora E G, Shimamura K, Yoshikawa Y, Aoki K and Ichinose N 2004 Large-size β-Ga_2O_3 single crystals and wafers *J. Cryst. Growth* **270** 420–6

[133] Kuramata A, Koshi K, Watanabe S, Yamaoka Y, Masui T and Yamakoshi S *et al* 2016 High-quality β-Ga_2O_3 single crystals grown by edge-defined film-fed growth *Jpn. J. Appl. Phys.* **55** 1202A2

[134] High-quality 100-mm β-type gallium oxide epitaxial wafers have been developed. Novel Crystal Technology, Inc https://novelcrystal.co.jp/eng/2021/887/

[135] Mastro M A, Kuramata A, Calkins J, Kim J, Ren F and Pearton S J 2017 Perspective—opportunities and future directions for Ga_2O_3 *ECS J. Solid State Sci. Technol.* **6** P356–9

[136] Kokubun Y, Miura K, Endo F and Nakagomi S 2007 Sol–gel prepared β-Ga_2O_3 thin films for ultraviolet photodetectors *Appl. Phys. Lett.* **90** 031912

[137] Rogers D J *et al* 2021 Sharp/tuneable UVC selectivity and extreme solar blindness in nominally undoped Ga_2O_3 MSM photodetectors grown by pulsed laser deposition *Proc. SPIE* **11687** 116872D-1

[138] Razeghi M *et al* 2018 A review of the growth, doping and applications of Ga_2O_3 thin films *Proc. SPIE* **10533** 105330R–1

[139] Tadjer M J, Lyons J L, Nepal N, Freitas J A, Koehler A D and Foster G M 2019 Theory and characterization of doping and defects in β-Ga_2O_3 *ECS J. Solid State Sci. Technol* **8** Q3187–94

[140] Leedy K D *et al* 2017 Films on (010) β-Ga_2O_3 by pulsed laser deposition *Appl. Phys. Lett.* **111** 012103

[141] Seryogin G, Alema F, Valente N, Fu H, Steinbrunner E, Neal A T, Mou S, Fine A and Osinsky A 2020 MOCVD growth of high purity Ga_2O_3 epitaxial films using trimethylgallium precursor *Appl. Phys. Lett.* **117** 262101

[142] Víllora E G, Shimamura K, Yoshikawa Y, Ujiie T and Aoki K 2008 Electrical conductivity and carrier concentration control in β-Ga_2O_3 by Si doping *Appl. Phys. Lett.* **92** 202120

[143] Suzuki N, Ohira S, Tanaka M, Sugawara T, Nakajima K and Shishido T 2007 Fabrication and characterization of transparent conductive Sn-doped β-Ga_2O_3 single crystal *Phys. Status Solidi* C **4** 2310

[144] Varley J B, Weber J R, Janotti A and Van de Walle C G 2010 Oxygen vacancies and donor impurities in β-Ga_2O_3 *Appl. Phys. Lett.* **97** 142106

[145] Perez-Tomas A *et al* 2019 Solar-blind conducting gallium oxide enables record bulk photovoltaic effect *Mater. Today Energy* **14** 100350

[146] Jung S, Jang S and Baik K H 2019 Ga_2O_3-based gas sensors *Gallium OxideMetal Oxides Series* (Amsterdam: Elsevier), 19 439–64

[147] Qian L-X, Zhang H-F, Lai P T, Wu Z-H and Liu X-Z 2017 High-sensitivity β-Ga_2O_3 solar-blind photodetector on high-temperature pretreated c-plane sapphire substrate *Opt. Mater. Express* **7** 3643–53

[148] Rogers D J, Teherani F H, Sandana V E, Arrateig X, Bove P, Razeghi M, McClintock R, Frisch E and Harel S 2018 The new oxide paradigm for solid state ultraviolet photodetectors *Proc. SPIE* **10533** 105331P-1

[149] Tsai T, Ou S, Hung M, Wuu D and Horng R 2011 MOCVD growth of GaN on sapphire using a Ga_2O_3 interlayer *J. Electrochem. Soc* **158** H1172–78

[150] Oshima T, Arai N, Suzuki N, Ohira S and Fujita S 2008 Surface morphology of homoepitaxial β-Ga_2O_3 thin films grown by molecular beam epitaxy *Thin Solid Films* 5768–71

[151] Liu G X, Shan F K, Lee W J, Shin B C, Kim S C, Kim H S and Cho C R 2007 Growth temperature dependence of Ga_2O_3 thin films deposited by plasma enhanced atomic layer deposition *Integr. Ferroelectr* **94** 11–20

[152] Nomura K, Goto K, Togashi R, Murakami H, Kumagai Y, Kuramata A, Yamakoshi S and Koukitu A 2014 Thermodynamic study of β-Ga_2O_3 growth by halide vapor phase epitaxy *J. Cryst. Growth* **405** 19–22

[153] Gogova D, Wagner G, Baldini M, Schmidbauer M, Irmscher K, Schewski R, Galazka Z, Albrecht M and Fornari R 2014 Structural properties of Si-doped β-Ga_2O_3 layers grown by MOVPE *J. Cryst. Growth* **401** 665–9

[154] Hu G C, Shan C X, Zhang N, Jiang M M, Wang S P and Shen D Z 2015 High gain Ga_2O_3 solar-blind photodetectors realized via a carrier multiplication process *Opt. Express* **23** 13554

[155] Sbrockey N M, Salagaj T, Coleman E, Tompa G S, Moon Y and Kim M S 2015 Large-area MOCVD growth of Ga_2O_3 in a rotating disc reactor *J. Electron. Mater* **44** 1357–60

[156] Teherani F H *et al* 2017 A study into the impact of sapphire substrate orientation on the properties of nominally-undoped Ga_2O_3 thin films grown by pulsed laser deposition *Proc. SPIE* **10105** 101051R-1

[157] Rafique S, Han L and Zhao H 2016 Synthesis of wide bandgap Ga_2O_3 ($E_g \sim 4.6$–4.7 eV) thin films on sapphire by low pressure chemical vapor deposition *Phys. Status Solidi* **213** 1002–9

[158] Hu D, Zhuang S, Ma Z, Dong X, Du G, Zhang B, Zhang Y and Yin J 2017 Study on the optical properties of β-Ga_2O_3 films grown by MOCVD *J. Mater. Sci. Mater. Electron* **28** 10997–1001

[159] Altuntas H, Donmez I, Ozgit-Akgun C and Biyikli N 2014 Effect of postdeposition annealing on the electrical properties of β-Ga_2O_3 thin films grown on p-Si by plasma enhanced atomic layer deposition *J. Vac. Sci. Technol.* A **32** 41504

[160] Higashiwaki M, Sasaki K, Kuramata A, Masui T and Yamakoshi S 2012 Gallium oxide (Ga_2O_3) metal–semiconductor field-effect transistors on single-crystal β-Ga_2O_3 (010) substrates *Appl. Phys. Lett.* **100** 013504

[161] Sasaki K, Higashiwaki M, Kuramata A, Masui T and Yamakoshi S 2013 Si-ion implantation doping in β-Ga_2O_3 and its application to fabrication of low-resistance ohmic contacts *Appl. Phys. Express* **6** 086502

[162] Wong M H, Nakata Y, Kuramata A, Yamakoshi S and Higashiwaki M 2017 Enhancementmode Ga_2O_3 MOSFETs with Si-ion-implanted source and drain *Appl. Phys. Express* **10** 41101

[163] Wong M H, Sasaki K, Kuramata A, Yamakoshi S and Higashiwaki M 2016 Field-plated Ga_2O_3 MOSFETs with a breakdown voltage of over 750 V *IEEE Electron Device Lett.* **37** 212

[164] Chabak K D *et al* 2016 Enhancement-mode Ga_2O_3 wrap-gate fin field-effect transistors on native (100) β-Ga_2O_3 substrate with high breakdown voltage *Appl. Phys. Lett.* **109** 1–6

[165] Hu Z, Nomoto K, Li W, Tanen N, Sasaki K, Kuramata A, Nakamura T, Jena D and Xing H G 2018 Enhancement-mode Ga_2O_3 vertical transistors with breakdown voltage >1 kV *IEEE Electron Device Lett.* **39** 869

[166] Hu Z *et al* 2018 Breakdown mechanism in 1 kA/cm^2 960 V E-mode β-Ga_2O_3 vertical transistors *Appl. Phys. Lett.* **113** 122103
[167] Gallium oxide vertical transistor with the world's highest breakdown voltage, Novel Crystal Technology, Inc https://novelcrystal.co.jp/eng/2021/911/
[168] Wong M H, Goto K, Morikawa Y, Kuramata A, Yamakoshi S, Murakami H, Kumagai Y and Higashiwaki M 2018 All-ion-implanted planar-gate current aperture vertical Ga_2O_3 MOSFETs with Mg-doped blocking layer *Appl. Phys. Express* **11** 064102
[169] Wong M H, Murakami H, Kumagai Y and Higashiwaki M 2020 Enhancement-mode β-Ga_2O_3 current aperture vertical MOSFETs with N-ion-implanted blocker *IEEE Electron Device Lett.* **41** 296
[170] Shahin D I, Tadjer M J, Wheeler V D, Koehler A D, Anderson T J, Eddy C R Jr and Christou A 2018 Electrical characterization of ALD HfO_2 high-k dielectrics on ($\bar{2}01$) β-Ga_2O_3 *Appl. Phys. Lett.* **112** 042107
[171] Sasaki K, Higashiwaki M, Kuramata A, Masui T and Yamakoshi S 2013 Ga_2O_3 Schottky barrier diodes fabricated by using single-crystal β-Ga_2O_3 (010) substrates *IEEE Electron Device Lett.* **34** 493–5
[172] Allen N, Xiao M, Yan X, Sasaki K, Tadjer M J, Ma J, Zhang R, Wang H and Zhang Y 2019 *IEEE Electron Device Lett.* **40** 1399
[173] Lu X, Zhang X, Jiang H, Zou X, Lau K M and Wang G 2020 Vertical β-Ga_2O_3 Schottky Barrier Diodes with Enhanced Breakdown Voltage and High Switching Performance *Phys. Status Solidi* A **217** 1900497
[174] Hu Z *et al* 2020 The Investigation of β-Ga_2O_3 Schottky Diode with Floating Field Ring Termination and the Interface States *ECS J. Solid State Sci. Technol* **9** 025001
[175] Joishi C, Rafique S, Xia Z, Han L, Krishnamoorthy S, Zhang Y, Lodha S, Zhao H and Rajan S 2018 Low-pressure CVD-grown beta-Ga_2O_3 bevel-field-plated Schottky barrier diodes *Appl. Phys. Express* **11** 031101
[176] Li W, Nomoto K, Hu Z, Jena D and Xing H G 2020 Field-plated Ga_2O_3 trench Schottky barrier diodes with a $BV^2/R_{on,sp}$ of up to 0.95 GW/cm^2 *IEEE Electron Device Lett.* **41** 107
[177] Sasaki K, Thieu Q T, Wakimoto D, Koishikawa Y, Kuramata A and Yamakoshi S 2017 First Demonstration of Ga2O3 Trench MOS-Type Schottky Barrier Diodes *Elect. Dev. Lett.* **38** 783–5
[178] Lv Y *et al* 2021 Demonstration of β-Ga_2O_3 junction barrier Schottky diodes with a Baliga's figure of merit of 0.85 GW/cm^2 or a 5 A/700 V handling capabilities *IEEE Trans. Power Electron* **36** 6179
[179] Pearton S J, Ren F, Tadjer M and Kim J 2018 Perspective: Ga_2O_3 for ultra-high power rectifiers and MOSFETS *J. Appl. Phys.* **124** 220901
[180] Zhang Y *et al* 2018 Demonstration of high mobility and quantum transport in modulation-doped β-$(Al_xGa_{1-x})_2O_3/Ga_2O_3$ heterostructures *Appl. Phys. Lett.* **112** 173502
[181] Cheng Z *et al* 2020 Thermal transport across ion-cut monocrystalline β-Ga_2O_3 thin films and bonded β-Ga_2O_3–SiC interfaces *ACS Appl. Mater. Interfaces* **12** 44943
[182] Matsumae T, Kurashima Y, Takagi H, Umezawa H and Higurashi E 2021 Low-temperature direct bonding of SiC and Ga_2O_3 substrates under atmospheric conditions *J. Appl. Phys.* **130** 085303
[183] Matsumae T, Kurashima Y, Umezawa H, Tanaka K, Ito T, Watanabe H and Takagi H 2020 Low-temperature direct bonding of β-Ga_2O_3 and diamond substrates under atmospheric conditions *Appl. Phys. Lett.* **116** 141602

[184] Cheng Z, Wheeler V D, Bai T, Shi J, Tadjer M J, Feygelson T, Hobart K D, Goorsky M S and Graham S 2020 Integration of polycrystalline Ga_2O_3 on diamond for thermal management *Appl. Phys. Lett.* **116** 062105

[185] Karim M R, Chen Z, Feng Z, Huang H-L, Johnson J M, Tadjer M J, Hwang J and Zhao H 2021 Two-step growth of β-Ga_2O_3 films on (100) diamond via low pressure chemical vapor deposition *J. Vac. Sci. Technol.* A **39** 023411

[186] Polyakov A Y *et al* 2021 Electrical properties and deep trap spectra in Ga_2O_3 films grown by halide vapor phase epitaxy on p-type diamond substrates *J. Appl. Phys.* **129** 185701

[187] Chaterjee *et al* 2019 *IEEE Trans. Compon. Packag. Manuf. Technol* **9** 2352

[188] Reese S and Zakutayev A 2020 Gallium oxide techno-economic analysis for the wide bandgap semiconductor market *Proc. SPIE* **11281** 112810H

[189] Shibata D, Kajitani R, Ogawa M, Tanaka K, Tamura S and Hatsuda T *et al* 2016 1.7 kV/1.0 mΩ cm^2 normally-off vertical GaN transistor on GaN substrate with regrown p-GaN/AlGaN/GaN semipolar gate structure *2016 IEEE Int. Electron Devices Meeting (IEDM), San Francisco, CA, US* A 10.1.1–4 pp

[190] Uemoto Y, Shibata D, Yanagihara M, Ishida H, Matsuo H and Nagai S *et al* 2007 8300 V blocking voltage AlGaN/GaN power HFET with thick poly-AlN passivation *2007 IEEE Int. Electron Devices Meeting, Washington, DC, US* A pp 861–4

[191] Ren S, Ma J, Zhao H and Fu X 2021 The effort of finding p-type β-Ga_2O_3 – a review of theoretical and experimental research *Proc. SPIE* **11763** 117631Y

[192] Chikoidze E *et al* 2017 p-Type β-gallium oxide: a new perspective for power and optoelectronic devices *Mater. Today Phys.* **3** 118–26

[193] Chikoidze E *et al* 2019 Enhancing the intrinsic p-type conductivity of the ultra-wide bandgap Ga_2O_3 semiconductor *J. Mater. Chem.* C **7** 10231–9

[194] Islam M M, Liedke M O and Winarski D *et al* 2020 Chemical manipulation of hydrogen induced high p-type and n-type conductivity in Ga_2O_3 *Sci. Rep.* **10** 6134

[195] Wu Z Y, Jiang Z X, Ma C C, Ruan W, Chen Y, Zhang H, Zhang G Q, Fang Z L, Kang J Y and Zhang T-Y 2021 Energy-driven multi-step structural phase transition mechanism to achieve high-quality p-type nitrogen-doped β-Ga_2O_3 films *Mater. Today Phys.* **17** 100356

[196] Watahiki T, Yuda Y, Furukawa A, Yamamuka M, Takiguchi Y and Miyajima S 2017 Heterojunction p-Cu_2O/n-Ga_2O_3 diode with high breakdown voltage *Appl. Phys. Lett.* **111** 222104

[197] Gong, H. H.,Chen, X. H., Xu, Y., Ren, F.-F., Gu, S. L., Ye, J. D., A 1.86-kV double-layered NiO/β-Ga_2O_3 vertical p–n heterojunction diode Appl. Phys. Lett. **117** 022104

[198] Lu X, Zhou X, Jiang H, Ng K W, Chen Z, Pei Y, Lau K M and Wang G 2020 Heterojunction diodes with an ultra-low leakage current below 1 μA/cm^2 *IEEE Electron Device Lett.* **41** 449–52

[199] Wang Y *et al* 2022 2.41 kV vertical p-NiO/n-Ga_2O_3 heterojunction diodes with a record Baliga's figure-of-merit of 5.18 GW/cm^2 *IEEE Trans. Power Electron.* **37** 4

[200] Zhang X, Liao F, Huang X, Yao D, Du Z, Ma X, Chen Y and Liu J 2022 Design of a 6 kV beta-Ga_2O_3 PN heterojunction diode with etched double-layered NiO with a figure of merit of 10 GW cm^{-2} *ECS J. Solid State Sci. Technol* **11** 4

[201] Razeghi M, Lee J, Gautam L, Leburton J-P, Teherani F H, Amiri P K, Grayson M, Dravid V P and Pavlidis D 2021 Microstrip array ring FETs with 2D p-Ga2O3 channels grown by MOCVD *Photonics* **8** 578

[202] Horng R-H, Huang C-Y, Ou S-L, Juang T-K and Liu P-L 2017 Epitaxial growth of $ZnGa_2O_4$: a new, deep ultraviolet semiconductor candidate *Crystal Growth Des.* **17** 6071–8
[203] Galazka Z *et al* 2019 Ultra-wide bandgap, conductive, high mobility, and high quality melt-grown bulk $ZnGa_2O_4$ single crystals *APL Mater* **7** 022512
[204] Peng H *et al* 2013 Doping rules and doping prototypes in A_2BO_4 spinel oxides *Adv. Funct. Mater* **23** 5267
[205] Chikoidze E *et al* 2020 p-Type ultrawide-band-gap spinel $ZnGa_2O_4$: new perspectives for energy electronics *Crystal Growth Des.* **20** 2535–46
[206] Wang C, Takei K, Takahashi T and Javey A 2013 Carbon nanotube electronics—moving forward *Chem. Soc. Rev.* **42** 2592–609 Liu, B.; Wang, C.; Liu, J.; Che, Y.; Zhou, C. 2013 Aligned carbon nanotubes: from controlled synthesis to electronic applications. *Nanoscale*, **5**, 9483–502
[207] Huang T C, Fukuda K and Lo C M *et al* 2011 Pseudo-CMOS: a design style for low-cost and robust flexible electronics *IEEE Trans. Electron Devices* **58** 141
[208] Chikoidze E *et al* 2020 Ultra-high critical electric field of 13.2 MV/cm for Zn-doped p-type β-Ga_2O_3 *Mater. Today Phys.* **15** 100263

Chapter 11

Conclusions

Ana Cremades and David Maestre

The achievement of clean and affordable energy, together with the development of more sustainable technologies and processes, are some of the driving forces for change in the 21st century, during which modern society must address crucial challenges. In fact, the attainment of efficient energy management is considered to be one of the main pillars of the current and foreseen technologies, which can be accomplished by the selection of appropriate nanomaterials.

Throughout this book, we have reviewed fundamental and applied issues related to the synthesis and properties of nanomaterials, together with their use in technological applications for the generation, storage, and transformation of energy, while also considering sustainability concerns.

In the first chapter, aspects of the synthesis of nanomaterials including tailored morphologies, dimensions, and properties along with their fundamental properties in sustainable energy-related applications were initially tackled. The synthesis of nanomaterials increasingly requires low-cost, effective, and environmentally friendly methods, among which, 'green chemistry' routes are gaining tremendous attention due to their underlying sustainable methodologies. Some examples of the green synthesis of metal nanoparticles (Au, Ag, Ni) and metal oxides (e.g. TiO_2, NiO) using plant extracts and other eco-friendly approaches were reviewed in this book. Moreover, eco-friendly physical and chemical routes with low energy consumption and faster processes that avoid toxic materials and subproducts are also gathering attention. As an example, resistive heating based on thermal evaporation methods can easily lead to a rapid and high yield of nanowires and plates of diverse oxide nanomaterials, such as ZnO, CuO, and TiO_2, in the range of tens of seconds or a few minutes. Moreover, in the search for improved synthesis methods, sustainability considerations should also entail demanding standards of efficiency and control of the required dimensions, morphology, and properties.

Some of the fundamental properties of the nanomaterials used in energy-related technologies were also reviewed in this book, with special emphasis on their

doi:10.1088/978-0-7503-3531-7ch11

© IOP Publishing Ltd 2023

crystalline structures, electronic band structures, and the role of defects. Optical and electronic properties at the nanoscale and the response of nanomaterials under diverse stimuli were also considered. Due to the large surface-to-volume ratio characteristic of nanomaterials, surface- and interface-related properties were also described in this book, mainly for semiconducting oxides, including both n-type (e.g. SnO_2, TiO_2, and ZnO) and p-type (e.g. NiO, SnO) oxides, as well as for graphene, 2D, and layered materials, which are attracting lot of attention due to their superb properties and technological interest.

The pursuit of alternative and efficient renewable energy sources is a paramount and urgent challenge at present. Hence, diverse approaches have been considered and explored to date in the effort to obtain efficient, sustainable, and economic alternatives which can lead to a real and compelling response to the global energy demand. In fact, in order to mitigate climate change and global warming, one of the main strategies is the development of renewable technologies for energy conversion and storage, which are also addressed in this book.

Among the energy-related applications, we first reviewed the use of nanomaterials in clean solar cells, providing an overview of the main challenges and problems that photovoltaic technology should address in the next years. Not only is the interest in perovskite solar cells increasing at a staggering pace, but this book also provides insights into the use of quantum dots and nanomaterials as charge transport layers for organic and dye-sensitized solar cells. In this regard, one of the main challenges that photovoltaic technology faces is the incorporation of nanomaterials and nanostructuring while retaining efficiency and stability but avoiding significant costs and environmental impact. The recovery and reuse of some of the nanomaterials used in solar cells are also considered in this book, together with some disruptive approaches to the use of nanomaterials in solar cells.

The urgent investment in sustainable and renewable resources must also include efficient ways to save energy. Energy storage systems are also essential in modern society; hence, different approaches are being considered to achieve energy storage through the use of highly efficient devices, such as Li-ion batteries, fuel cells, and supercapacitors, to name a few. The use of organic materials in batteries and supercapacitors is gaining attention in this field, as organic-based electrodes and self-healing materials can lead to devices that benefit from sustainability, easy processing, low cost, and scalability. A combination of high energy density, long lifetime, and sustainability is being pursued in the new generation of energy storage systems. These aspects are reviewed in this book, among which, the main focus is given to the promising self-healing mechanisms and alternative materials for future batteries that are playing a part in the search for more efficient, durable, and eco-friendly technology.

The efficient and economical generation of hydrogen by splitting water is also a promising technology, as hydrogen is envisioned as a clean energy vector for the future. In this regard, metal oxide semiconductor nanomaterials are receiving great attention based on their relevant photocatalytic properties, as reviewed in this book. Most of these materials exhibit advantages that include a strong oxidizing ability, hydrophilicity, stability, non-toxicity, low cost, and abundance. For this reason, an overview of some key physicochemical and photocatalytic features of metal oxide nanomaterials (e.g. TiO_2, WO_3, Fe_2O_3) and perovskite-type oxides (e.g. $SrTiO_3$, $CaTiO_3$), among others, was provided in this book.

A paradigm shift in the way in which energy resources are exploited is urgently required due to the growth in worldwide energy consumption. Alongside the use of clean and renewable energies, some other approaches leading to efficient energy processes have been pursued, such as the thermoelectric recovery of wasted energy in the form of heat. Concepts and breakthroughs in the use of nanomaterials in thermoelectric energy harvesting are reviewed in this book. Energy harvesting from heat leakages with diverse origins will support a reduction in global power consumption, promoting energy-saving processes. In this way, thermoelectric materials will surely play a major role. Concepts related to charge and energy transport in nanomaterials, such as nanowires, quantum dots, and 2D materials, were reviewed in this book. This field can also take advantage of phenomena that appear at the nanoscale, such as quantum interference effects and increased phonon scattering.

Furthermore, electro- and photochromic devices can lead to energy-efficient buildings using smart windows that can admit or reject solar radiation on demand, providing energy savings through chromogenic technologies. In this book, issues related to the fundamentals and recent advances in this technology, together with some relevant electrochromic and photochromic nanomaterials such as silver halides, transition-metal oxides, and rare-earth oxyhydrides, were considered.

An overview of high-quality light systems involving luminescent nanoparticles and high-efficiency systems was also included in this book. The current use of common inorganic phosphors in luminescent devices lead to high costs, a lack of efficient recycling, color-tuning difficulties, and thermal quenching. Hence, organic phosphors are becoming more attractive due to their low cost, easy tuning, and efficiency as compared to those of their inorganic counterparts. Insights into the basic concepts and architectures of nanosized phosphors for white light-emitting diodes are necessary to achieve improvements in these devices and to ensure their applicability in nanostructured emitters, fluorescent lamps, and plasma display panels, to name a few. Among the most relevant nanomaterials used in this technology, inorganic phosphors based on YAG:Ce nanoparticles, other metal-oxide-based nanophosphors, and hybrid nanomaterials stand out. Alongside the materials design and device applications of nano-oxides, recent challenges and progresses in this field were also reviewed in this book, including eco-friendly solutions.

Finally, electronic devices based on nano-oxides were also reviewed in this book. In particular, it provided insights into recent trends in flexible electronics and power electronic devices, as well as the materials adopted and state-of-the-art devices. In the former field, amorphous oxide semiconductors are playing a key role in the development of flexible electronics and have vast applications in solar cells, wearables, sensors, displays, and field-effect transistors, among others. This book reviewed current and emerging applications based on the underlying phenomena and selected materials, such as amorphous silicon, semiconducting oxides (e.g. ZnO, In_2O_3, ITO) and 2D materials (e.g. graphene, metal dichalcogenides, black phosphorous). On the other hand, power electronics are essential for advanced electrical energy applications that enable electrical transmission, generation, storage, and utilization. Loss reduction in power electronics can result in huge energy savings; hence, great efforts are currently being invested in replacing inefficient/lossy Si-based power electronics devices with other wide-bandgap materials, among which, Ga_2O_3 is generating increasing interest. In fact, future trends indicate that technology based on gallium oxide and other wide-bandgap and ultrawide-bandgap materials could become the mainstream high-power and high-frequency electronic technology in the coming decades, as reviewed in this book.

As our main conclusion, current trends to increase the sustainability of the planet entail the development of efficient energy-related applications, in which research into nanomaterials plays a key role. Great efforts are underway, but still there is room for improvement in the search for sustainable and efficient technologies, materials, and processes, such as those reviewed in this book.

www.ingramcontent.com/pod-product-compliance
Ingram Content Group UK Ltd.
Pitfield, Milton Keynes, MK11 3LW, UK
UKHW051907100926
8556IPUK00005B/794

9 780750 335294